T0155475

Mass Spectrometry

Third Edition

Mass Spectrometry

Principles and Applications

Third Edition

Edmond de Hoffmann
Université Catholique de Louvain, Belgium & Ludwig Institute for Cancer Research, Brussels, Belgium

Vincent Stroobant
Ludwig Institute for Cancer Research, Brussels, Belgium

John Wiley & Sons, Ltd

Other Wiley Editorial Offices

John Wiley & Sons Inc., 111 River Street, Hoboken, NJ 07030, USA

Jossey-Bass, 989 Market Street, San Francisco, CA 94103-1741, USA

Wiley-VCH Verlag GmbH, Boschstr. 12, D-69469 Weinheim, Germany

John Wiley & Sons Australia Ltd, 33 Park Road, Milton, Queensland 4064, Australia

John Wiley & Sons (Asia) Pte Ltd, 2 Clementi Loop #02-01, Jin Xing Distripark, Singapore 129809

John Wiley & Sons Canada Ltd, 6045 Freemont Blvd, Mississauga, Ontario, Canada L5R 4J3

Wiley also publishes its books in a variety of electronic formats. Some content that appears
in print may not be available in electronic books.

Anniversary Logo Design: Richard J. Pacifico

Library of Congress Cataloging-in-Publication Data

Hoffmann, Edmond de.
    [Spectrométrie de masse. English]
    Mass spectrometry : principles and applications. – 3rd ed. / Edmond de Hoffmann, Vincent Stroobant.
       p. cm.
    Includes bibliographical references and index.
    ISBN 978-0-470-03310-4
    1. Mass spectrometry.   I. Stroobant, Vincent.   II. Title.

    QD96.M3 H6413 2007
    573'.65 — dc22

                                                                            2007021691

British Library Cataloguing in Publication Data

A catalogue record for this book is available from the British Library

ISBN 978-0-470-03310-4 (H/B)
ISBN 978-0-470-03311-1 (P/B)

Typeset in 10/12pt Times by Aptara Inc., New Delhi, India

Contents

Preface to Third Edition

Following the first studies of J.J. Thomson (1912), mass spectrometry has undergone countless improvements. Since 1958, gas chromatography coupled with mass spectrometry has revolutionized the analysis of volatile compounds. Another revolution occurred in the 1980s when the technique became available for the study of non-volatile compounds such as peptides, oligosaccharides, phospholipids, bile salts, etc. From the discoveries of electrospray and matrix-assisted laser desorption in the late 1980s, compounds with molecular masses exceeding several hundred thousands of daltons, such as synthetic polymers, proteins, glycans and polynucleotides, have been analysed by mass spectrometry.

From the time of the second edition published in 2001 until now, much progress has been achieved. Several techniques have been improved, others have almost disappeared. New atmospheric pressure desorption ionization sources have been discovered and made available commercially. One completely new instrument, the orbitrap, based on a new mass analyser, has been developed and is now also available commercially. Improved accuracy in low-mass determination, even at low resolution, improvements in sensitivity, better detection limits and more efficient tandem mass spectrometry even on high-molecular-mass compounds are some of the main achievements. We have done our best to include them is this new edition.

As the techniques continue to advance, the use of mass spectrometry continues to grow. Many new applications have been developed. The most impressive ones arise in system biology analysis.

Starting from the very foundations of mass spectrometry, this book presents all the important techniques developed up to today. It describes many analytical methods based on these techniques and emphasizes their usefulness by numerous examples. The reader will also find the necessary information for the interpretation of data. A series of graduated exercises allows the reader to check his or her understanding of the subject. Numerous references are given for those who wish to go deeper into some subjects. Important Internet addresses are also provided. We hope that this new edition will prove useful to students, teachers and researchers.

We would like to thank Professor Jean-Louis Habib Jiwan and Alexander Spote for their friendly hospitality and competent help.

We would also like to acknowledge the financial support of the FNRS (Fonds National de la Recherche Scientifique, Brussels).

Many colleagues and friends have read the manuscript and their comments have been very helpful. Some of them carried out a thorough reading. They deserve special mention:

namely, Magda Claeys, Bruno Domon, Jean-Claude Tabet, and the late François Van Hoof. We also wish to acknowledge the remarkable work done by the scientific editors at John Wiley & Sons.

Many useful comments have been published on the first two editions, or sent to the editor or the authors. Those from Steen Ingemann were particularly detailed and constructive.

Finally, we would like to thank the Université Catholique de Louvain, the Ludwig Institute for Cancer Research and all our colleagues and friends whose help was invaluable to us.

Edmond de Hoffmann and Vincent Stroobant
Louvain-la-Neuve, March 2007

Introduction

Mass spectrometry's characteristics have raised it to an outstanding position among analytical methods: unequalled sensitivity, detection limits, speed and diversity of its applications. In analytical chemistry, the most recent applications are mostly oriented towards biochemical problems, such as proteome, metabolome, high throughput in drug discovery and metabolism, and so on. Other analytical applications are routinely applied in pollution control, food control, forensic science, natural products or process monitoring. Other applications include atomic physics, reaction physics, reaction kinetics, geochronology, inorganic chemical analysis, ion–molecule reactions, determination of thermodynamic parameters ($\Delta G^\circ{}_f$, K_a, etc.), and many others.

Mass spectrometry has progressed extremely rapidly during the last decade, between 1995 and 2005. This progress has led to the advent of entirely new instruments. New atmospheric pressure sources were developed [1–4], existing analysers were perfected and new hybrid instruments were realized by new combinations of analysers. An analyser based on a new concept was described: namely, the orbitrap [5] presented in Chapter 2. This has led to the development of new applications. To give some examples, the first spectra of an intact virus [6] and of very large non-covalent complexes were obtained. New high-throughput mass spectrometry was developed to meet the needs of the proteomic [7, 8], metabolomic [9] and other 'omics'.

Principles

The first step in the mass spectrometric analysis of compounds is the production of gas-phase ions of the compound, for example by electron ionization:

$$M + e^- \longrightarrow M^{\bullet+} + 2e^-$$

This molecular ion normally undergoes fragmentations. Because it is a radical cation with an odd number of electrons, it can fragment to give either a radical and an ion with an even number of electrons, or a molecule and a new radical cation. We stress the important difference between these two types of ions and the need to write them correctly:

$$
\begin{array}{c}
\qquad\quad \underset{\text{EVEN ION}}{\text{EE}^+} \;+\; \underset{\text{RADICAL}}{\text{R}^\bullet} \\[2pt]
M^{\bullet+} \\[2pt]
\qquad\quad \underset{\text{ODD ION}}{\text{OE}^{\bullet+}} \;+\; \underset{\text{MOLECULE}}{\text{N}}
\end{array}
$$

These two types of ions have different chemical properties. Each primary product ion derived from the molecular ion can, in turn, undergo fragmentation, and so on. All these ions are separated in the mass spectrometer according to their mass-to-charge ratio, and

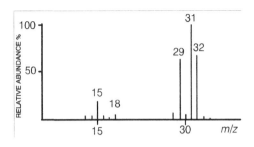

m/z	Relative abundance (%)	m/z	Relative abundance (%)
12	0.33	28	6.3
13	0.72	29	64
14	2.4	30	3.8
15	13	31	100
16	0.21	32	66
17	1.0	33	0.73
18	0.9	34	~ 0.1

Figure 1
Mass spectrum of methanol by electron ionization, presented as a graph and as a table.

are detected in proportion to their abundance. A mass spectrum of the molecule is thus
produced. It provides this result as a plot of ion abundance versus mass-to-charge ratio.
As illustrated in Figure 1, mass spectra can be presented as a bar graph or as a table. In
either presentation, the most intense peak is called the base peak and is arbitrarily assigned
the relative abundance of 100 %. The abundances of all the other peaks are given their
proportionate values, as percentages of the base peak. Many existing publications label the
y axis of the mass spectrum as number of ions, ion counts or relative intensity. But the term
relative abundance is better used to refer to the number of ions in the mass spectra.

Most of the positive ions have a charge corresponding to the loss of only one electron. For
large molecules, multiply charged ions also can be obtained. Ions are separated and detected
according to the mass-to-charge ratio. The total charge of the ions will be represented by
q, the electron charge by e and the number of charges of the ions by z:

$$q = ze \quad \text{and} \quad e = 1.6 \times 10^{-19}\,\text{C}$$

The x axis of the mass spectrum that represents the mass-to-charge ratio is commonly
labelled m/z. When m is given as the relative mass and z as the charge number, both of
which are unitless, m/z is used to denote a dimensionless quantity.

Generally in mass spectrometry, the charge is indicated in multiples of the elementary
charge or charge of one electron in absolute value ($1\,e = 1.602\,177 \times 10^{-19}$ C) and the mass
is indicated in atomic mass units ($1\,\text{u} = 1.660\,540 \times 10^{-27}$ kg). As already mentioned, the
physical property that is measured in mass spectrometry is the mass-to-charge ratio. When
the mass is expressed in atomic mass units (u) and the charge in elementary charge units

(*e*) then the mass-to-charge ratio has u/*e* as dimensions. For simplicity, a new unit, the Thomson, with symbol Th, has been proposed [10]. The fundamental definition for this unit is

$$1\,\mathrm{Th} = 1\,\mathrm{u}/e = 1.036\,426 \times 10^{-8}\,\mathrm{kg\,C^{-1}}$$

Ions provide information concerning the nature and the structure of their precursor molecule. In the spectrum of a pure compound, the molecular ion, if present, appears at the highest value of m/z (followed by ions containing heavier isotopes) and gives the molecular mass of the compound. The term molecular ion refers in chemistry to an ion corresponding to a complete molecule regarding occupied valences. This molecular ion appears at m/z 32 in the spectrum of methanol, where the peak at m/z 33 is due to the presence of the ^{13}C isotope, with an intensity that is 1.1 % of that of the m/z 32 peak. In the same spectrum, the peak at m/z 15 indicates the presence of a methyl group. The difference between 32 and 15, that is 17, is characteristic of the loss of a neutral mass of 17 Da by the molecular ion and is typical of a hydroxyl group. In the same spectrum, the peak at m/z 16 could formally correspond to ions $CH_4^{\bullet+}$, O^+ or even CH_3OH^{2+}, because they all have m/z values equal to 16 at low resolution. However, O^+ is unlikely to occur, and a doubly charged ion for such a small molecule is not stable enough to be observed.

The atomic mass units u or Da have the same fundamental definition:

$$1\,\mathrm{u} = 1\,\mathrm{Da} = 1.660\,540 \times 10^{-27}\,\mathrm{kg} \pm 0.59\,\mathrm{ppm}$$

However, they are traditionally used in different contexts: when dealing with mean isotopic masses, as generally used in stoichiometric calculations, Da will be preferred; in mass spectrometry, masses referring to the main isotope of each element are used and expressed in u.

There are different ways to define and thus to calculate the mass of an atom, molecule or ion. For stoichiometric calculations chemists use the average mass calculated using the atomic weight, which is the weighted average of the atomic masses of the different isotopes of each element in the molecule. In mass spectrometry, the nominal mass or the monoisotopic mass is generally used. The nominal mass is calculated using the mass of the predominant isotope of each element rounded to the nearest integer value that corresponds to the mass number, also called nucleon number. But the exact masses of isotopes are not exact whole numbers. They differ weakly from the summed mass values of their constituent particles that are protons, neutrons and electrons. These differences, which are called the mass defects, are equivalent to the binding energy that holds these particles together. Consequently, every isotope has a unique and characteristic mass defect. The monoisotopic mass, which takes into account these mass defects, is calculated by using the exact mass of the most abundant isotope for each constituent element.

The difference between the average mass, the nominal mass and the monoisotopic mass can amount to several Da, depending on the number of atoms and their isotopic composition. The type of mass determined by mass spectrometry depends largely on the resolution and accuracy of the analyser. Let us consider CH_3Cl as an example. Actually, chlorine atoms are mixtures of two isotopes, whose exact masses are respectively 34.968 852 u and 36.965 903 u. Their relative abundances are 75.77 % and 24.23 %. The atomic weight of chlorine atoms is the balanced average: $(34.968\,852 \times 0.7577 + 36.965\,903 \times 0.2423)$ $= 35.453$ Da. The average mass of CH_3Cl is $12.011 + (3 \times 1.007\,94) + 35.453 =$ 50.4878 Da, whereas its monoisotopic mass is $12.000\,000 + (3 \times 1.007\,825) + 34.968\,852 =$ 49.992 327 u. When the mass of CH_3Cl is measured with a mass spectrometer, two isotopic peaks will appear at their respective masses and relative abundances. Thus, two mass-to-charge ratios will be observed with a mass spectrometer. The first peak will be at m/z $(34.968\,852 + 12.000\,000 + 3 \times 1.007\,825) = 49.992\,327$ Th, rounded to m/z 50. The mass-to-charge value of the second peak will be $(36.965\,90 + 12.000\,000 + 3 \times$ $1.007\,825) = 51.989\,365$ Th, rounded to m/z 52. The abundance at this latter m/z value is $(24.23/75.77) = 0.3198$, or 31.98 % of that observed at m/z 50. Carbon and hydrogen also are composed of isotopes, but at much lower abundances. They are neglected for this example.

For molecules of very high molecular weights, the differences between the different masses can become notable. Let us consider two examples.

The first example is human insulin, a protein having the molecular formula $C_{257}H_{383}N_{65}O_{77}S_6$. The nominal mass of insulin is 5801 u using the integer mass of the most abundant isotope of each element, such as 12 u for carbon, 1 u for hydrogen, 14 u for nitrogen, 16 u for oxygen and 32 u for sulfur. Its monoisotopic mass of 5803.6375 u is calculated using the exact masses of the predominant isotope of each element such as $C = 12.0000$ u, $H = 1.0079$ u, $N = 14.0031$ u, $O = 15.9949$ u and $S = 31.9721$ u. These values can be found in the tables of isotopes in Appendices 4A and 4B. Finally, an average mass of 5807.6559 Da is calculated using the atomic weight for each element, such as $C = 12.011$ Da, $H = 1.0078$ Da, $N = 14.0067$ Da, $O = 15.9994$ Da and $S = 32.066$ Da.

The second example is illustrated in Figure 2. The masses of two alkanes having the molecular formulae $C_{20}H_{42}$ and $C_{100}H_{202}$ are calculated. For the smaller alkane, its nominal mass is $(20 \times 12) + (42 \times 1) = 282$ u, its monoisotopic mass is $(20 \times 12) + (42 \times 1.007\,825) = 282.3287$ u rounded to 282.33 and its average mass is $(20 \times 12.011) + (42 \times 1.007\,94) = 282.5535$ Da. The differences between these different types of masses are small but are more important for the heavier alkane. Indeed, its nominal mass is $(100 \times 12) + (202 \times 1) = 1402$ u, its monoisotopic mass is $(100 \times 12) + (202 \times 1.007\,825) = 1403.5807$ u rounded to 1403.58 and its average mass is $(100 \times 12.011) + (202 \times 1.007\,94) = 1404.7039$ Da.

In conclusion, the monoisotopic mass is used when it is possible experimentally to distinguish the isotopes, whereas the average mass is used when the isotopes are not distinguishable. The use of nominal mass is not recommended and should only be used for low-mass compounds containing only the elements C, H, N, O and S to avoid to making mistakes.

Diagram of a Mass Spectrometer

A mass spectrometer always contains the following elements, as illustrated in Figure 3: a sample inlet to introduce the compound that is analysed, for example a gas chromatograph or a direct insertion probe; an ionization source to produce ions from the sample; one or several mass analysers to separate the various ions; a detector to 'count' the ions emerging

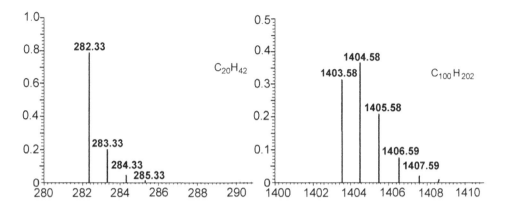

Figure 2
Mass spectra of isotopic patterns of two alkanes having the molecular formulae $C_{20}H_{42}$ and $C_{100}H_{202}$, respectively. The monoisotopic mass is the lighter mass of the isotopic pattern whereas the average mass, used by chemists in stoichiometric calculations, is the balanced mean value of all the observed masses.

from the last analyser; and finally a data processing system that produces the mass spectrum in a suitable form. However, some mass spectrometers combine the sample inlet and the ionization source and others combine the mass analyser and the detector.

A mass spectrometer should always perform the following processes:

1. Produce ions from the sample in the ionization source.

2. Separate these ions according to their mass-to-charge ratio in the mass analyser.

3. Eventually, fragment the selected ions and analyze the fragments in a second analyser.

4. Detect the ions emerging from the last analyser and measure their abundance with the detector that converts the ions into electrical signals.

5. Process the signals from the detector that are transmitted to the computer and control the instrument through feedback.

History

A large number of mass spectrometers have been developed according to this fundamental scheme since Thomson's experiments in 1897. Listed here are some highlights of this development [11, 12]:

1886: E. GOLDSTEIN discovers anode rays (positive gas-phase ions) in gas discharge [13].

1897: J.J. THOMSON discovers the electron and determines its mass-to-charge ratio. *Nobel Prize in 1906.*

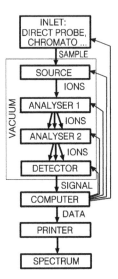

Figure 3
Basic diagram for
a mass spectrometer
with two analysers
and feedback control
carried out by a data
system.

1898: W. WIEN analyses anode rays by magnetic deflection and then establishes that these rays carried a positive charge [14]. *Nobel Prize in 1911.*

1901: W. KAUFMANN analyses cathodic rays using parallel electric and magnetic fields [15].

1909: R.A. MILLIKAN and H. FLETCHER determine the elementary unit of charge.

1912: J.J. THOMSON constructs the first mass spectrometer (then called a parabola spectrograph) [16]. He obtains mass spectra of O_2, N_2, CO, CO_2 and $COCl_2$. He observes negative and multiply charged ions. He discovers metastable ions. In 1913, he discovers isotopes 20 and 22 of neon.

1918: A.J. DEMPSTER develops the electron ionization source and the first spectrometer with a sector-shaped magnet (180°) with direction focusing [17].

1919: F.W. ASTON develops the first mass spectrometer with velocity focusing [18]. *Nobel Prize in 1922.* He measures mass defects in 1923 [19].

1932: K.T. BAINBRIDGE proves the mass–energy equivalence postulated by Einstein [20].

1934: R. CONRAD applies mass spectrometry to organic chemistry [21].

1934: W.R. SMYTHE, L.H. RUMBAUGH and S.S. WEST succeed in the first preparative isotope separation [22].

1940: A.O. NIER isolates uranium-235 [23].

1942: The Consolidated Engineering Corporation builds the first commercial instrument dedicated to organic analysis for the Atlantic Refinery Company.

1945: First recognition of the metastable peaks by J.A. HIPPLE and E.U. CONDON [24].

1948: A.E. CAMERON and D.F. EGGERS publish design and mass spectra for a linear time-of-flight (LTOF) mass spectrometer [25]. W. STEPHENS proposed the concept of this analyser in 1946 [26].

1949: H. SOMMER, H.A. THOMAS and J.A. HIPPLE describe the first application in mass spectrometry of ion cyclotron resonance (ICR) [27].

1952: Theories of quasi-equilibrium (QET) [28] and RRKM [29] explain the monomolecular fragmentation of ions. R.A. MARCUS receives the *Nobel Prize in 1992*.

1952: E.G. JOHNSON and A.O. NIER develop double-focusing instruments [30].

1953: W. PAUL and H.S. STEINWEDEL describe the quadrupole analyser and the ion trap or quistor in a patent [31]. W. PAUL, H.P. REINHARD and U. Von ZAHN, of Bonn University, describe the quadrupole spectrometer in *Zeitschrift für Physik* in 1958. PAUL and DEHMELT receive the *Nobel Prize in 1989* [32].

1955: W.L. WILEY and I.H. McLAREN of Bendix Corporation make key advances in LTOF design [33].

1956: J. BEYNON shows the analytical usefulness of high-resolution and exact mass determinations of the elementary composition of ions [34].

1956: First spectrometers coupled with a gas chromatograph by F.W. McLAFFERTY [35] and R.S. GOHLKE [36].

1957: Kratos introduces the first commercial mass spectrometer with double focusing.

1958: Bendix introduces the first commercial LTOF instrument.

1966: M.S.B. MUNSON and F.H. FIELD discover chemical ionization (CI) [37].

1967: F.W. McLAFFERTY [38] and K.R. JENNINGS [39] introduce the collision-induced dissociation (CID) procedure.

1968: Finnigan introduces the first commercial quadrupole mass spectrometer.

1968: First mass spectrometers coupled with data processing units.

1969: H.D. BECKEY demonstrates field desorption (FD) mass spectrometry of organic molecules [40].

1972: V.I. KARATEV, B.A. MAMYRIM and D.V. SMIKK introduce the reflectron that corrects the kinetic energy distribution of the ions in a TOF mass spectrometer [41].

1973: R.G. COOKS, J.H. BEYNON, R.M. CAPRIOLI and G.R. LESTER publish the book *Metastable Ions*, a landmark in tandem mass spectrometry [42].

1974: E.C. HORNING, D.I. CARROLL, I. DZIDIC, K.D. HAEGELE, M.D. HORNING and R.N. STILLWELL discover atmospheric pressure chemical ionization (APCI) [43].

1974: First spectrometers coupled with a high-performance liquid chromatograph by P.J. ARPINO, M.A. BALDWIN and F.W. McLAFFERTY [44].

1974: M.D. COMISAROV and A.G. MARSHALL develop Fourier transformed ICR (FTICR) mass spectrometry [45].

1975: First commercial gas chromatography/mass spectrometry (GC/MS) instruments with capillary columns.

1976: R.D. MACFARLANE and D.F. TORGESSON introduce the plasma desorption (PD) source [46].

1977: R.G. COOKS and T.L. KRUGER propose the kinetic method for thermochemical determination based on measurement of the rates of competitive fragmentations of cluster ions [47].

1978: R.A. YOST and C.G. ENKE build the first triple quadrupole mass spectrometer, one of the most popular types of tandem instrument [48].

1978: Introduction of lamellar and high-field magnets.

1980: R.S. HOUK, V.A. FASSEL, G.D. FLESCH, A.L. GRAY and E. TAYLOR demonstrate the potential of inductively coupled plasma (ICP) mass spectrometry [49].

1981: M. BARBER, R.S. BORDOLI, R.D. SEDGWICK and A.H. TYLER describe the fast atom bombardment (FAB) source [50].

1982: First complete spectrum of insulin (5750 Da) by FAB [51] and PD [52].

1982: Finnigan and Sciex introduce the first commercial triple quadrupole mass spectrometers.

1983: C.R. BLAKNEY and M.L. VESTAL describe the thermospray (TSP) [53].

1983: G.C. STAFFORD, P.E. KELLY, J.E. SYKA, W.E. REYNOLDS and J.F.J. TODD describe the development of a gas chromatography detector based on an ion trap and commercialized by Finnigan under the name Ion Trap [54].

1987: M. GUILHAUS [55] and A.F. DODONOV [56] describe the orthogonal acceleration time-of-flight (oa-TOF) mass spectrometer. The concept of this technique was initially proposed in 1964 by G.J. O'Halloran of Bendix Corporation [57].

1987: T. TANAKA [58] and M. KARAS, D. BACHMANN, U. BAHR and F. HILLENKAMP [59] discover matrix-assisted laser desorption/ionization (MALDI). TANAKA receives the *Nobel Prize in 2002*.

1987: R.D. SMITH describes the coupling of capillary electrophoresis (CE) with mass spectrometry [60].

1988: J. FENN develops the electrospray (ESI) [61]. First spectra of proteins above 20 000 Da. He demonstrated the electrospray's potential as a mass spectrometric technique for small molecules in 1984 [62]. The concept of this source was proposed in 1968 by M. DOLE [63]. FENN receives the *Nobel Prize in 2002*.

1991: V. KATTA and B.T. CHAIT [64] and B. GAMEN, Y.T. LI and J.D. HENION [65] demonstrate that specific non-covalent complexes could be detected by mass spectrometry.

1991: B. SPENGLER, D. KIRSCH and R. KAUFMANN obtain structural information with reflectron TOF mass spectrometry (MALDI post-source decay) [66].

1993: R.K. JULIAN and R.G. COOKS develop broadband excitation of ions using the stored-waveform inverse Fourier transform (SWIFT) [67].

1994: M. WILM and M. MANN describe the nanoelectrospray source (then called microelectrospray source) [68].

1999: A.A. MAKAROV describes a new type of mass analyser: the orbitrap. The orbitrap is a high-performance ion trap using an electrostatic quadro-logarithmic field [5,69].

The progress of experimental methods and the refinements in instruments led to spectacular improvements in resolution, sensitivity, mass range and accuracy. Resolution ($m/\delta m$) developed as follows:

	$m/\delta m$	
1913	13	Thomson [16]
1918	100	Dempster [17]
1919	130	Aston [18]
1937	2000	Aston [70]
1998	8 000 000	Marshall and co-workers [71]

A continuous improvement has allowed analysis to reach detection limits at the pico-, femto- and attomole levels [72,73]. Furthermore, the direct coupling of chromatographic techniques with mass spectrometry has improved these limits to the atto- and zeptomole levels [74,75]. A sensitivity record obtained by mass spectrometry has been demonstrated by using modified desorption/ionization on silicon DIOS method to measure concentration of a peptide in solution. This technique has achieved a lower detection limit of 800 yoctomoles, which corresponds to about 480 molecules [76].

Regarding the mass range, DNA ions of 10^8 Da were weighed by mass spectrometry [77]. In the same way, non-covalent complexes with molecular weights up to 2.2 MDa were measured by mass spectrometry [78]. Intact viral particles of tobacco mosaic virus with a theoretical molecular weight of 40.5 MDa were analysed with an electrospray ionization charge detection time-of-flight mass spectrometer [6].

The mass accuracy indicates the deviation of the instrument's response between the theoretical mass and the measured mass. It is usually expressed in parts per million (ppm) or in 10^{-3} u for a given mass. The limit of accuracy in mass spectrometry is about 1 ppm. The measurement of the atomic masses has reached an accuracy of better than 10^{-9} u [79].

In another field, Litherland et al. [80] succeeded in determining a $^{14}C/^{12}C$ ratio of $1:10^{15}$ and hence in dating a 40 000-year-old sample with a 1 % error. A quantity of ^{14}C corresponding to only 10^6 atoms was able to be detected in less than 1 mg of carbon [81].

Ion Free Path

All mass spectrometers must function under high vacuum (low pressure). This is necessary to allow ions to reach the detector without undergoing collisions with other gaseous molecules. Indeed, collisions would produce a deviation of the trajectory and the ion would lose its charge against the walls of the instrument. On the other hand, ion–molecule collisions could produce unwanted reactions and hence increase the complexity of the spectrum. Nevertheless, we will see later that useful techniques use controlled collisions in specific regions of a spectrometer.

According to the kinetic theory of gases, the mean free path L (in m) is given by

$$L = \frac{kT}{\sqrt{2}p\sigma} \tag{1}$$

where k is the Boltzmann constant, T is the temperature (in K), p is the pressure (in Pa) and σ is the collision cross-section (in m^2); $\sigma = \pi d^2$ where d is the sum of the radii of the stationary molecule and the colliding ion (in m). In fact, one can approximate the mean free path of an ion under normal conditions in a mass spectrometer ($k = 1.38 \times 10^{-21}$ J K^{-1}, $T \approx 300$ K, $\sigma \approx 45 \times 10^{-20}$ m^2) using either of the following equations where L is in centimetres and pressure p is, respectively, in pascals or milliTorrs:

$$L = \frac{0.66}{p} \tag{2}$$

$$L = \frac{4.95}{p} \tag{3}$$

Table 1 is a conversion table for pressure units. In a mass spectrometer, the mean free path should be at least 1 m and hence the maximum pressure should be 66 nbar. In instruments using a high-voltage source, the pressure must be further reduced to prevent the occurrence of discharges. In contrast, some trap-based instruments operate at higher pressure.

However, introducing a sample into a mass spectrometer requires the transfer of the sample at atmospheric pressure into a region of high vacuum without compromising the latter. In the same way, producing efficient ion–molecule collisions requires the mean free path to be reduced to around 0.1 mm, implying at least a 60 Pa pressure in a region of the

Table 1   Pressure units. The official SI unit is the pascal.

1 pascal (Pa) = 1 newton (N) per m^2
1 bar = 10^6 dyn cm^{-2} = 10^5 Pa
1 millibar (mbar) = 10^{-3} bar = 10^2 Pa
1 microbar (μbar) = 10^{-6} bar = 10^{-1} Pa
1 nanobar (nbar) = 10^{-9} bar = 10^{-4} Pa
1 atmosphere (atm) = 1.013 bar = 101 308 Pa
1 Torr = 1 mmHg = 1.333 mbar = 133.3 Pa
1 psi = 1 pound per square inch = 0.07 atm

spectrometer. These large differences in pressure are controlled with the help of an efficient pumping system using mechanical pumps in conjunction with turbomolecular, diffusion or cryogenic pumps. The mechanical pumps allow a vacuum of about 10^{-3} Torr to be obtained. Once this vacuum is achieved, the operation of the other pumping systems allows a vacuum as high as 10^{-10} Torr to be reached.

The sample must be introduced into the ionization source so that vacuum inside the instrument remains unchanged. Samples are often introduced without compromising the vacuum using direct infusion or direct insertion methods. For direct infusion, a capillary is employed to introduce the sample as a gas or a solution. For direct insertion, the sample is placed on a probe, a plate or a target that is then inserted into the source through a vacuum interlock. For the sources that work at atmospheric pressure and are known as atmospheric pressure ionization (API) sources, introduction of the sample is easy because the complicated procedure for sample introduction into the high vacuum of the mass spectrometer is removed.

References

1. Robb, D.B., Covey, T.R. and Bruins, A.P. (2000) *Anal. Chem.*, **72**, 3653.
2. Laiko, V.V., Baldwin, M.A. and Burlingame, A.L. (2000) *Anal. Chem.*, **72**, 652.
3. Takats, Z., Wiseman, J.M., Gologan, B. and Cooks, R.G. (2004) Mass spectrometry sampling under ambient conditions with desorption electrospray ionization. *Science*, **5695**, 471–3.
4. Cody, R.B., Laramée, J.A. and Durst, H.D. (2005) Versatile new ion source for analysis of materials in open air under ambient conditions. *Anal. Chem.*, **77**, 2297–302.
5. Makarov, A. (2000) Electrostatic axially harmonic orbital trapping: a high performance technique of mass analysis. *Anal. Chem.*, **72**, 1156–62.
6. Fuerstenauw, S., Benner, W., Thomas, J. *et al.* (2001) Whole virus mass spectrometry. *Angew. Chem. Int. Ed.*, **40**, 541–4.
7. Reinders, J., Lewandrowski, U., Moebius, J. *et al.* (2004) Challenges in mass spectrometry-based proteomics. *Proteomics*, **4**, 3686–703.
8. Naylor, S. and Kumar, R. (2003) Emerging role of mass spectrometry in structural and functional proteomics. *Adv. Protein Chem.*, **65**, 217–48.
9. Villas-Boas, S.G., Mas, S., Akesson, M. *et al.* (2005) Mass spectrometry in metabolome analysis. *Mass Spectrom. Rev.*, **24**, 613–46.
10. Cooks, R.G. and Rockwood, A.L. (1991) The 'Thomson'. A suggested unit for mass spectroscopists. *Rapid Commun. Mass Spectrom.*, **5**, 93.
11. Grayson, M.A. (2002) *Measuring Mass: from Positive Rays to Proteins*, Chemical Heritage Press, Philadelphia.
12. http://masspec.scripps.edu/MSHistory/mshisto.php (21 March 2007).
13. Goldstein, E. (1886) *Berl. Ber.*, **39**, 691.
14. Wien, W. (1898) *Verhanal. Phys. Ges.*, **17**.
15. Kaufmann, R.L., Heinen, H.J., Shurmann, L.W. and Wechsung, R.M. (1979) *Microbeam Analysis* (ed. D.E. Newburg), San Francisco Press, San Francisco, pp. 63–72.
16. Thomson, J.J. (1913) *Rays of Positive Electricity and Their Application to Chemical Analysis*, Longmans Green, London.
17. Dempster, A.J. (1918) *Phys. Rev.*, **11**, 316.
18. Aston, F.W. (1919) *Philos. Mag.*, **38**, 707.
19. Aston, F.W. (1942) *Mass Spectra and Isotopes*, 2nd edn, Edward Arnold, London.

20. Bainbridge, K.T. (1932) *Phys. Rev.*, **42**, 1; Bainbridge, K.T. and Jordan, E.B. (1936) *Phys. Rev.*, **50**, 282.
21. Conrad, R. (1934) *Trans. Faraday Soc.*, **30**, 215.
22. Smythe, W.R., Rumbaugh, L.H. and West, S.S. (1934) *Phys. Rev.*, **45**, 724.
23. Nier, A.O. (1940) *Rev. Sci. Instrum.*, **11**, 252.
24. Hipple, J.A. and Condon, E.U. (1946) *Phys. Rev.*, **69**, 347.
25. Cameron, A.E. and Eggers, D.F. (1948) *Rev. Sci. Instrum.*, **19**, 605.
26. Stephens, W. (1946) *Phys. Rev.*, **69**, 691.
27. Sommer, H., Thomas, H.A. and Hipple, J.A. (1949) *Phys. Rev.*, **76**, 1877.
28. Rosenstock, H.M., Wallenstein, M.B., Warhaftig, A.L. and Eyring, H. (1952) *Proc. Natl. Acad. Sci. USA.*, **38**, 667.
29. Marcus, R.A. (1952) *J. Chem. Phys.*, **20**, 359.
30. Johnson, E.G. and Nier, A.O. (1953) *Phys. Rev.*, **91**, 12.
31. Paul, W. and Steinwedel, H.S. (1953) *Z. Naturforsch.*, **8a**, 448.
32. Paul, W., Reinhard, H.P. and von Zahn, U. (1958) *Z. Phys.*, **152**, 143.
33. Wiley, W.L. and McLaren, I.H. (1955) *Rev. Sci. Instrum.*, **16**, 1150.
34. Beynon, J. (1956) *Mikrochim. Acta*, 437.
35. McLafferty, F.W. (1957) *Appl. Spectrosc.*, **11**, 148.
36. Gohlke, R.S. (1959) *Anal. Chem.*, **31**, 535.
37. Munson, M.S.B. and Field, F.H. (1966) *J. Am. Chem. Soc.*, **88**, 2681.
38. McLafferty, F.W. and Bryce, T.A. (1967) *Chem. Commun.*, 1215.
39. Jennings, K.R. (1968) *Int. J. Mass Spectrom. Ion Phys.*, **1**, 227.
40. Beckey, H.D. (1969) *Int. J. Mass Spectrom. Ion Phys.*, **2**, 500.
41. Karataev, V.I., Mamyrin, B.A. and Smikk, D.V. (1972) *Sov. Phys.–Tech. Phys.*, **16**, 1177.
42. Cooks, R.G., Beynon, J.H., Caprioli, R.M. and Lester, G.R. (1973) *Metastable Ions*, Elsevier, New York, p. 296.
43. Horning, E.C., Carroll, D.I., Dzidic, I. *et al.* (1974) *J. Chromatogr. Sci.*, **412**, 725.
44. Arpino, P.J., Baldwin, M.A. and McLafferty, F.W. (1974) *Biomed. Mass Spectrom.*, **1**, 80.
45. Comisarov, M.B. and Marshall, A.G. (1974) *Chem. Phys. Lett.*, **25**, 282.
46. Macfarlane, R.D. and Torgesson, D.F. (1976) *Science*, **191**, 920.
47. Cooks, R.G. and Kruger, T.L. (1977) *J. Am. Chem. Soc.*, **99**, 1279.
48. Yost, R.A. and Enke, C.G. (1978) *J. Am. Chem. Soc.*, **100**, 2274.
49. Houk, R.S., Fassel, V.A., Flesch, G.D. *et al.* (1980) *Anal. Chem.*, **52**, 2283.
50. Barber, M., Bardoli, R.S., Sedgwick, R.D. and Tyler, A.H. (1981) *J. Chem. Soc., Chem. Commun.*, 325.
51. McFarlane, R. (1982) *J. Am. Chem. Soc.*, **104**, 2948.
52. Barber, M. *et al.* (1982) *J. Chem. Soc., Chem. Commun.*, 936.
53. Blakney, C.R. and Vestal, M.L. (1983) *Anal. Chem.*, **55**, 750.
54. Stafford, G.C., Kelley, P.E., Syka, J.E. *et al.* (1984) *Int. J. Mass Spectrom. Ion Processes*, **60**, 85.
55. Dawson, J.H.J. and Guilhaus, M. (1989) *Rapid Commun. Mass Spectrom.*, **3**, 155–9.
56. Dovonof, A.F., Chernushevich, I.V. and Laiko, V.V. (1991) *Proceedings of the 12th International Mass Spectrometry Conference*, 26–30 August, Amsterdam, the Netherlands, p. 153.
57. O'Halloran, G.J., Fluegge, R.A., Betts, J.F. and Everett, W.L. (1964) Technical Documentary Report No. ASD-TDR-62-644, The Bendix Corporation, Research Laboratory Division, Southfield, MI.
58. Tanaka, T., Waki, H., Ido, Y. *et al.* (1988) *Rapid Commun. Mass Spectrom.*, **2**, 151.

59. Karas, M., Bachmann, D., Bahr, U. and Hillenkamp, F. (1987) *Int. J. Mass Spectrom. Ion Processes*, **78**, 53.
60. Olivares, J.A., Nguyen, N.T., Yonker, C.R. and Smith, R.D. (1987) *Anal. Chem.*, **59**, 1230.
61. Fenn, J.B., Mann, M., Meng, C.K. *et al.* (1989) *Science*, **246**, 64.
62. Yamashita, M. and Fenn, J.B. (1984) *J. Chem. Phys.*, **80**, 4451.
63. Dole, M., Mach, L.L., Hines, R.L. *et al.* (1968) *J. Chem. Phys.*, **49**, 2240.
64. Katta, V. and Chait, B.T. (1991) *J. Am. Chem. Soc.*, **113**, 8534.
65. Gamen, B., Li, Y.T. and Henion, J.D. (1991) *J. Am. Chem. Soc.*, **113**, 6294.
66. Spengler, B., Kirsch, D. and Kaufmann, R. (1991) *Rapid Commun. Mass Spectrom.*, **5**, 198.
67. Julian, R.K. and Cooks, R.G. (1993) *Anal. Chem.*, **65**, 1827.
68. Wilm, M. and Mann, M. (1994) *Proceedings of the 42nd ASMS Conference*, Chicago, IL, p. 770.
69. Makarov, A. (2000) Electrostatic axially harmonic orbital trapping: a high performance technique of mass analysis. *Anal. Chem.*, **72**, 1156–62.
70. Aston, F.W. (1937) *Proc. R. Soc., Ser. A*, **163**, 391.
71. Shi, S.D.H., Hendrickson, C.L. and Marshall, A.G. (1998) Counting individual sulfur atoms in a protein by ultrahigh-resolution Fourier transform ion cyclotron resonance mass spectrometry: experimental resolution of isotopic fine structure in proteins. *Proc. Natl. Acad. Sci. USA*, **95**, 11532–7.
72. Solouki, T., Marto, J.A., White, F.M. *et al.* (1995) Attomole biomolecule mass analysis by matrix-assisted-laser-desorption ionization Fourier-transform ion-cyclotron resonance. *Anal. Chem.*, **67**, 4139–44.
73. Morris, H.R., Paxton, T., Panico, M. *et al.* (1997) A novel geometry mass spectrometer, the Q-TOF, for low-femtomole/attomole-range biopolymer sequencing. *J. Protein Chem.*, **16**, 469–79.
74. Valaskovic, G.A., Kelleher, N.L. and McLafferty, F.W. (1996) Attomole protein characterization by capillary electrophoresis mass spectrometry. *Science*, **273**, 1199–1200.
75. Wang, P.G., Murugaiah, V., Yeung, B. Vouros, P. and Giese, R.W. (1996) 2-Phosphoglycolate and glycolate-electrophore detection, including detection of 87 zeptomoles of the latter by gas chromatography electron-capture mass spectrometry. *J. Chromatogr. A*, **721**, 289–96.
76. Trauger, S.A., Go, E.P., Shen, Z. *et al.* (2004) High sensitivity and analyte capture with desorption/ionization mass spectrometry on silylated porous silicon. *Anal. Chem.*, **76**, 4484–9.
77. Chen, R., Cheng, X, Mitchell, D. *et al.* (1995) *Anal. Chem.*, **67**, 1159.
78. Sanglier, S., Leize, E., Van Dorsselaer, A. and Zal, F. (2003) *J. Am. Soc. Mass Spectrom.*, **14**, 419.
79. Di Fillip, F., Natarajan, V., Bradley, M. *et al.* (1995) *Phys. Scr.*, **T59**, 144.
80. Litherland, A.E., Benkens, R.P., Lilius, L.R. *et al.* (1981) *Nucl. Instrum. Methods*, **186**, 463.
81. Vogel, J.S., Trteltaub, K.W., Finkel, R. and Nelson, D.E. (1995) *Anal. Chem.*, **67**, 353A.

1

Ion Sources

In the ion sources, the analysed samples are ionized prior to analysis in the mass spectrometer. A variety of ionization techniques are used for mass spectrometry. The most important considerations are the internal energy transferred during the ionization process and the physico-chemical properties of the analyte that can be ionized. Some ionization techniques are very energetic and cause extensive fragmentation. Other techniques are softer and only produce ions of the molecular species. Electron ionization, chemical ionization and field ionization are only suitable for gas-phase ionization and thus their use is limited to compounds sufficiently volatile and thermally stable. However, a large number of compounds are thermally labile or do not have sufficient vapour pressure. Molecules of these compounds must be directly extracted from the condensed to the gas phase.

These direct ion sources exist under two types: liquid-phase ion sources and solid-state ion sources. In liquid-phase ion sources the analyte is in solution. This solution is introduced, by nebulization, as droplets into the source where ions are produced at atmospheric pressure and focused into the mass spectrometer through some vacuum pumping stages. Electrospray, atmospheric pressure chemical ionization and atmospheric pressure photoionization sources correspond to this type. In solid-state ion sources, the analyte is in an involatile deposit. It is obtained by various preparation methods which frequently involve the introduction of a matrix that can be either a solid or a viscous fluid. This deposit is then irradiated by energetic particles or photons that desorb ions near the surface of the deposit. These ions can be extracted by an electric field and focused towards the analyser. Matrix-assisted laser desorption, secondary ion mass spectrometry, plasma desorption and field desorption sources all use this strategy to produce ions. Fast atom bombardment uses an involatile liquid matrix.

The ion sources produce ions mainly by ionizing a neutral molecule in the gas phase through electron ejection, electron capture, protonation, deprotonation, adduct formation or by the transfer of a charged species from a condensed phase to the gas phase. Ion production often implies gas-phase ion–molecule reactions. A brief description of such reactions is given at the end of the chapter.

1.1 Electron Ionization

The electron ionization (EI) source, formerly called electron impact, was devised by Dempster and improved by Bleakney [1] and Nier [2]. It is widely used in organic mass spectrometry. This ionization technique works well for many gas-phase molecules but induces extensive fragmentation so that the molecular ions are not always observed.

As shown in Figure 1.1, this source consists of a heated filament giving off electrons. The latter are accelerated towards an anode and collide with the gaseous molecules of

Mass Spectrometry: Principles and Applications, Third Edition Edmond de Hoffmann and Vincent Stroobant
© Copyright 2007, John Wiley & Sons Ltd

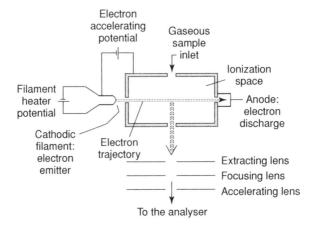

Figure 1.1
Diagram of an electron ionization source.

the analysed sample injected into the source. Gases and samples with high vapour pressure are introduced directly into the source. Liquids and solids are usually heated to increase the vapour pressure for analysis.

Each electron is associated to a wave whose wavelength λ is given by

$$\lambda = \frac{h}{mv}$$

where m is its mass, v its velocity and h Planck's constant. This wavelength is 2.7 Å for a kinetic energy of 20 eV and 1.4 Å for 70 eV. When this wavelength is close to the bond lengths, the wave is disturbed and becomes complex. If one of the frequencies has an energy hv corresponding to a transition in the molecule, an energy transfer that leads to various electronic excitations can occur [3]. When there is enough energy, an electron can be expelled. The electrons do not 'impact' molecules. For this reason, it is recommended that the term electron impact must be avoided.

Figure 1.2 displays a typical curve of the number of ions produced by a given electron current, at constant pressure of the sample, when the acceleration potential of the electrons (or their kinetic energy) is varied [4]. At low potentials the energy is lower than the molecule ionization energy. At high potentials, the wavelength becomes very small and molecules become 'transparent' to these electrons. In the case of organic molecules, a wide maximum appears around 70 eV. At this level, small changes in the electron energy do not significantly affect the pattern of the spectrum.

On average, one ion is produced for every 1000 molecules entering the source under the usual spectrometer conditions, at 70 eV. Furthermore, between 10 and 20 eV is transferred to the molecules during the ionization process. Since approximately 10 eV is enough to ionize most organic molecules, the excess energy leads to extensive fragmentation. This fragmentation can be useful because it provides structural information for the elucidation of unknown analytes.

At a given acceleration potential and at constant temperature, the number of ions I produced per unit time in a volume V is linked to the pressure p and to the electron current

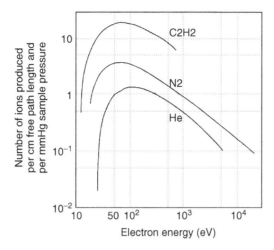

Figure 1.2
Number of ions produced as a function of the electron energy. A wide maximum appears around 70 eV.

i through the following equation, where N is a constant proportionality coefficient:

$$I = Npi\,V$$

This equation shows that the sample pressure is directly correlated with the resulting ionic current. This allows such a source to be used in quantitative measurements.

Figure 1.3 displays two EI spectra of the same β-lactam compound, obtained at 70 and 15 eV. Obviously, at lower energy there is less fragmentation. At first glance, the molecular ion is better detected at low energy. However, the absolute intensity, in arbitrary units, proportional to the number of detected ions, is actually lower: about 250 units at 70 eV and 150 units at 15 eV. Thus, the increase in relative intensity, due to the lower fragmentation, is illusory. Actually, there is a general loss of intensity due to the decrease in ionization efficiency at lower electron energy. This will generally be the rule, so that the method is not very useful for better detection of the molecular ion. However, the lowering of the ionization voltage may favour some fragmentation processes.

A modification implies desorbing the sample from a heated rhenium filament near the electronic beam. This method is called desorption electron ionization (DEI).

Under conventional electron ionization conditions, the formation of negative ions is inefficient compared with the formation of positive ions.

1.2 Chemical Ionization

Electron ionization leads to fragmentation of the molecular ion, which sometimes prevents its detection. Chemical ionization (CI) is a technique that produces ions with little excess energy. Thus this technique presents the advantage of yielding a spectrum with less fragmentation in which the molecular species is easily recognized. Consequently, chemical ionization is complementary to electron ionization.

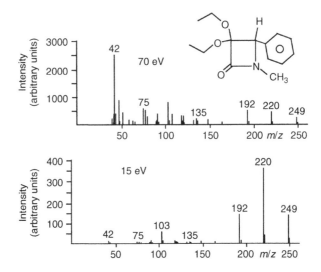

Figure 1.3
Two spectra of β-lactam. While the relative intensity of the
molecular ion peak is greater at lower ionization energy,
its absolute intensity, as read from the left-hand scale, is
actually somewhat reduced.

Chemical ionization [5] consists of producing ions through a collision of the molecule
to be analysed with primary ions present in the source. Ion–molecule collisions will thus
be induced in a definite part of the source. In order to do so, the local pressure has to
be sufficient to allow for frequent collisions. We saw that the mean free path could be
calculated from Equation 1 (see the Introduction). At a pressure of approximately 60 Pa,
the free path is about 0.1 mm. The source is then devised so as to maintain a local pressure
of that magnitude. A solution consists of introducing into the source a small box about 1 cm
along its side as is shown in Figure 1.4.

Two lateral holes allow for the crossing of electrons and another hole at the bottom
allows the product ions to pass through. Moreover, there is a reagent gas input tube and an
opening for the sample intake. The sample is introduced by means of a probe which will
close the opening.

This probe carries the sample within a hollow or contains the end part of a capillary
coming from a chromatograph or carries a filament on which the sample was deposited.
In the last case, we talk about desorption chemical ionization (DCI). The pumping speed
is sufficient to maintain a 60 Pa pressure within the box. Outside, the usual pressure in a
source, about 10^{-3} Pa, will be maintained.

Inside the box, the sample pressure will amount to a small fraction of the reagent
gas pressure. Thus, an electron entering the box will preferentially ionize the reagent gas
molecules through electron ionization. The resulting ion will then mostly collide with other
reagent gas molecules, thus creating an ionization plasma through a series of reactions. Both
positive and negative ions of the substance to be analysed will be formed by chemical reactions with ions in this plasma. This causes proton transfer reactions, hydride abstractions,
adduct formations, charge transfers, and so on.

This plasma will also contain low-energy electrons, called thermal electrons. These are
either electrons that were used for the first ionization and later slowed, or electrons produced

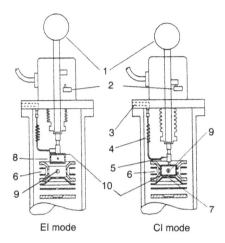

Figure 1.4
Combined EI and CI source. Lowering
the box 10 switches from the EI to CI
mode. (1) EI/CI switch; in EI mode, the
box serves as a pusher; (2) microswitch;
(3) entrance for the reagent gas; (4) flexible capillary carrying the reagent gas;
(5) diaphragm; (6) filament giving off
electrons; (7) path of the ions towards
the analyser inlet; (8) hole for the ionizing electrons in CI mode; (9) sample inlet; (10) box with holes, also named 'ion
volume'. From Finnigan MAT 44S documentation. Reprinted, with permission.

by ionization reactions. These slow electrons may be associated with molecules, thereby
yielding negative ions by electron capture.

Ions produced from a molecule by the abstraction of a proton or a hydride, or the
addition of a proton or of another ion, are termed ions of the molecular species or, less
often, pseudomolecular ions. They allow the determination of the molecular mass of the
molecules in the sample. The term molecular ions refers to $M^{\bullet+}$ or $M^{\bullet-}$ ions.

1.2.1 Proton Transfer

Among the wide variety of possible ionization reactions, the most common is proton transfer. Indeed, when analyte molecules M are introduced in the ionization plasma, the reagent
gas ions GH^+ can often transfer a proton to the molecules M and produce protonated
molecular ions MH^+. This chemical ionization reaction can be described as an acid–base
reaction, the reagent gas ions GH^+ and the analyte molecules being Brönsted acid (proton donor) and Brönsted base (proton acceptor) respectively. The tendency for a reagent
ion GH^+ to protonate a particular analyte molecule M may be assessed from its proton
affinity values. The proton affinity (PA) is the negative enthalpy change for the protonation
reaction (see Appendix 6). The observation of protonated molecular ions MH^+ implies
that the analyte molecule M has a proton affinity much higher than that of the reagent gas

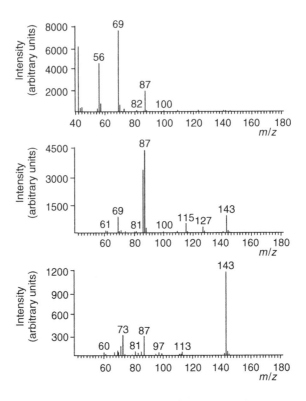

Figure 1.5
EI (top), methane CI (middle) and isobutane CI (bottom)
mass spectra of butyl methacrylate. The ionization techniques (EI vs CI) and the reagent gases (methane vs isobutane) influence the amount of fragmentation and the
prominence of the protonated molecular ions detected
at 143 Th.

(PA(M) > PA(G)). If the reagent gas has a proton affinity much higher than that of an analyte
(PA(G) > PA(M)), proton transfer from GH^+ to M will be energetically too unfavourable.

The selectivity in the types of compound that can be protonated and the internal energy
of the resulting protonated molecular ion depends on the relative proton affinities of
the reagent gas and the analyte. From the thermalizing collisions, this energy depends
also on the ion source temperature and pressure. The energetics of the proton transfer
can be controlled by using different reagent gases. The most common reagent gases are
methane (PA = 5.7 eV), isobutane (PA = 8.5 eV) and ammonia (PA = 9.0 eV). Not only are
isobutane and ammonia more selective, but protonation of a compound by these reagent
gases is considerably less exothermic than protonation by methane. Thus, fragmentation
may occur with methane while with isobutane or ammonia the spectrum often presents
solely a protonated molecular ion.

The differences between EI and CI spectra are clearly illustrated in Figure 1.5. Indeed,
the EI spectrum of butyl methacrylate displays a very low molecular ion at m/z 142. In
contrast, its CI spectra exhibit the protonated molecular ion at m/z 143, and very few

fragmentations. This example shows also the control of the fragmentation degree in CI by changing the reagent gas. Methane and isobutane CI mass spectra of butyl methacrylate give the protonated molecular ions, but the degree of fragmentation is different. With isobutane, the base peak is the protonated molecular ion at m/z 143, whereas with methane the base peak is a fragment ion at m/z 87.

1.2.2 Adduct Formation

In CI plasma, all the ions are liable to associate with polar molecules to form adducts, a kind of gas-phase solvation. The process is favoured by the possible formation of hydrogen bonds. For the adduct to be stable, the excess energy must be eliminated, a process which requires a collision with a third partner. The reaction rate equation observed in the formation of these adducts is indeed third order. Ions resulting from the association of a reagent gas molecule G with a protonated molecular ion MH^+ or with a fragment ion F^+, of a protonated molecular ion MH^+ with a neutral molecule, and so on, are often found in CI spectra. Every ion in the plasma may become associated with either a sample molecule or a reagent gas molecule. Some of these ions are useful in the confirmation of the molecular mass, such as

$$MH^+ + M \longrightarrow (2M + H)^+$$

$$F^+ + M \longrightarrow (F + M)^+$$

These associations are often useful to identify a mixture or to determine the molecular masses of the constituents of the mixture. In fact, a mixture of two species M and N can give rise to associations such as $(MH + N)^+$, $(F + N)^+$ with $(F + M)^+$, and so on. Adducts resulting from neutral species obtained by neutralization of fragments, or by a neutral loss during a fragmentation, are always at much too low concentrations to be observed.

It is always useful to examine the peaks appearing beyond the ions of the molecular species of a substance thought to be pure. If some peaks cannot be explained by reasonable associations, a mixture must be suspected.

Figure 1.6 shows an example of CI spectra for a pure sample and for a mixture. When interpreting the results, one must always keep in mind that a mixture that is observed may result from the presence of several constituents before the vaporization or from their formation after the vaporization.

The first spectrum contains the peaks of various adducts of the molecular ion of a pure compound. The second spectrum is that of a substance that is initially pure, as shown by other analysis, but appears as a mixture in the gas phase as it loses either hydrogen cyanide or water.

1.2.3 Charge-Transfer Chemical Ionization

Rare gases, nitrogen, carbon monoxide and other gases with high ionization potential react by charge exchange:

$$Xe + e^- \longrightarrow Xe^{\bullet+} + 2e^-$$

$$Xe^{\bullet+} + M \longrightarrow M^{\bullet+} + Xe$$

A radical cation is obtained, as in EI, but with a smaller energy content. Less fragmentation is thus observed. In practice, these gases are not used very often.

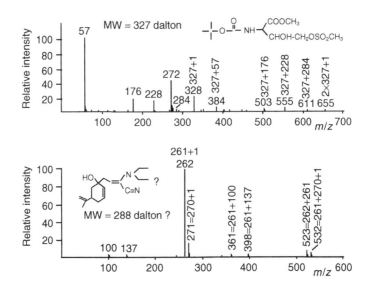

Figure 1.6
Two examples of chemical ionization (isobutane) spectra. The top
spectrum is that of a pure compound. The bottom spectrum is that
of a mixture of two compounds with masses 261 and 270. They
correspond respectively to the loss of hydrogen cyanide and water.

1.2.4 Reagent Gas

1.2.4.1 Methane as Reagent Gas

If methane is introduced into the ion volume through the tube, the primary reaction with
the electrons will be a classical EI reaction:

$$CH_4 + e^- \longrightarrow CH_4^{\bullet+} + 2e^-$$

This ion will fragment, mainly through the following reactions:

$$CH_4^{\bullet+} \longrightarrow CH_3^+ + H^\bullet$$
$$CH_4^{\bullet+} \longrightarrow CH_2^{\bullet+} + H_2$$

However, mostly, it will collide and react with other methane molecules yielding

$$CH_4^{\bullet+} + CH_4 \longrightarrow CH_5^+ + CH_3^\bullet$$

Other ion–molecule reactions with methane will occur in the plasma, such as

$$CH_3^+ + CH_4 \longrightarrow C_2H_5^+ + H_2$$

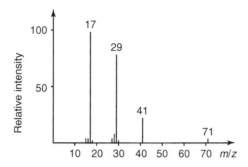

Figure 1.7
Spectrum of methane ionization plasma at
20 Pa. The relative intensities depend on the
pressure in the source.

A $C_3H_5^+$ ion is formed by the following successive reactions:

$$CH_2^{\bullet+} + CH_4 \longrightarrow C_2H_3^+ + H_2 + H^\bullet$$

$$C_2H_3^+ + CH_4 \longrightarrow C_3H_5^+ + H_2$$

The relative abundance of all these ions will depend on the pressure. Figure 1.7 shows the spectrum of the plasma obtained at 200 µbar (20 Pa). Taking CH_5^+, the most abundant ion, as a reference (100 %), $C_2H_5^+$ amounts to 83 % and $C_3H_5^+$ to 14 %.

Unless it is a saturated hydrocarbon, the sample will mostly react by acquiring a proton in an acid–base type of reaction with one of the plasma ions, for example

$$M + CH_5^+ \longrightarrow MH^+ + CH_4$$

A systematic study showed that the main ionizing reactions of molecules containing heteroatoms occurred through acid–base reactions with $C_2H_5^+$ and $C_3H_5^+$. If, however, the sample is a saturated hydrocarbon RH, the ionization reaction will be a hydride abstraction:

$$RH + CH_5^+ \longrightarrow R^+ + CH_4 + H_2$$

Moreover, ion–molecule adduct formation is observed in the case of polar molecules, a type of gas-phase solvation, for example

$$M + CH_3^+ \longrightarrow (M + CH_3)^+$$

The ions $(MH)^+$, R^+ and $(M + CH_3)^+$ and other adducts of ions with the molecule are termed molecular species or, less often, pseudomolecular ions. They allow the determination of the molecular mass of the molecules in the sample.

1.2.4.2 Isobutane as Reagent Gas

Isobutane loses an electron upon EI and yields the corresponding radical cation, which will fragment mainly through the loss of a hydrogen radical to yield a *t*-butyl cation, and to a

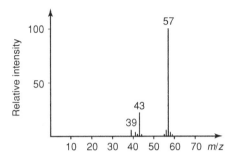

Figure 1.8
Spectrum of the isobutane plasma under
chemical ionization conditions at 200 μbar.

lesser extent through the loss of a methyl radical:

$$CH_3-\underset{\underset{CH_3}{|}}{\overset{\overset{CH_3}{|}}{C}}-H \ + \ e^- \ \longrightarrow \ \left[CH_3-\underset{\underset{CH_3}{|}}{\overset{\overset{CH_3}{|}}{C}}-H \right]^{\cdot +} \ + \ 2\,e^-$$

$$\left[CH_3-\underset{\underset{CH_3}{|}}{\overset{\overset{CH_3}{|}}{C}}-H \right]^{\cdot +} \ \Big\langle \ \begin{array}{l} CH_3-\underset{\underset{CH_3}{|}}{\overset{\overset{CH_3}{|}}{C}}{}^+ \ + \ H^{\cdot} \\[2em] CH_3-\underset{\underset{CH_3}{|}}{\overset{+}{C}}-H \ + \ CH_3^{\cdot} \end{array}$$

An ion with mass 39 Da is also observed in its spectrum (Figure 1.8) which corresponds to $C_3H_3^+$. Neither its formation mechanism nor its structure are known, but it is possible that it is the aromatic cyclopropenium ion.

Here again, the plasma ions will mainly react through proton transfer to the sample, but polar molecules will also form adducts with the t-butyl ions $(M+57)^+$ and with $C_3H_3^+$, yielding $(M+39)^+$ among others.

This isobutane plasma will be very inefficient in ionizing hydrocarbons because the t-butyl cation is relatively stable. This characteristic allows its use in order to detect specifically various substances in mixtures containing also hydrocarbons.

1.2.4.3 Ammonia as Reagent Gas

The radical cation generated by EI reacts with an ammonia molecule to yield the ammonium ion and the $NH_2^{\bullet}$ radical:

$$NH_3^{\bullet +} + NH_3 \longrightarrow NH_4^+ + NH_2^{\bullet}$$

An ion with mass 35 Da is observed in the plasma (Figure 1.9) which results from the association of an ammonium ion and an ammonia molecule:

$$NH_4^+ + NH_3 \longrightarrow (NH_4 + NH_3)^+$$

This adduct represents 15 % of the intensity of the ammonium ion at 200 μbar.

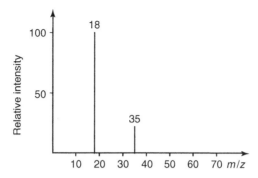

Figure 1.9
Spectrum of an ammonia ionization plasma at
200 μbar.

In this gas, the ionization mode will depend on the nature of the sample. The basic molecules, mostly amines, will ionize through a proton transfer:

$$RNH_2 + NH_4^+ \longrightarrow RNH_3^+ + NH_3$$

Polar molecules and those able to form hydrogen bonds while presenting no or little basic character will form adducts. In intermediate cases, two pseudomolecular ions $(M + 1)^+$ and $(M + 18)^+$ will be observed. Compounds that do not correspond to the criteria listed above, for example saturated hydrocarbons, will not be efficiently ionized. Alkanes, aromatics, ethers and nitrogen compounds other than amines will not be greatly ionized. Comparing spectra measured with various reagent gases will thus be very instructive. For example, the detection, in the presence of a wealth of saturated hydrocarbons, of a few compounds liable to be ionized is possible, as shown in Figure 1.10.

1.2.5 Negative Ion Formation

Almost all neutral substances are able to yield positive ions, whereas negative ions require the presence of acidic groups or electronegative elements to produce them. This allows some selectivity for their detection in mixtures. Negative ions can be produced by capture of thermal electrons by the analyte molecule or by ion–molecule reactions between analyte and ions present in the reagent plasma.

All CI plasmas contain electrons with low energies, issued either directly from the filament but deactivated through collisions, or mostly from primary ionization reactions, which produce two low-energy electrons through the ionization reaction. The interaction of electrons with molecules leads to negative ion production by three different mechanisms [5]:

$$AB + e^- \longrightarrow AB^{\bullet-} \quad \text{(associative resonance capture)}$$

$$AB + e^- \longrightarrow A^\bullet + B^- \quad \text{(dissociative resonance capture)}$$

$$AB + e^- \longrightarrow A^+ + B^- + e^- \quad \text{(ion pair production)}$$

These electrons can be captured by a molecule. The process can be associative or dissociative. The associative resonance capture that leads to the formation of negative molecular

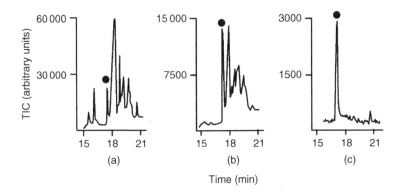

Figure 1.10
GC/MS TIC traces of butyl methacrylate dissolved in C11–C12 saturated hydrocarbon. (a) EI. The peak corresponding to butyl methacrylate is marked by a dot. The peaks following are C11 saturated hydrocarbons. (b) Same trace obtained by CI usig methane as reagent gas. Butyl methacrylate (dot) is still well detected, but the trace of the hydrocarbons is atenuated. (c) CI using isobutane as reagent gas. Butyl methacrylate is well detected, while the hydrocarbons are almost not detected.

ions needs electrons in the energy range 0–2 eV, whereas the dissociative resonance capture is observed with electrons of 0–15 eV and leads to the formation of negative fragment ions.

The associative resonance capture is favoured for molecules with several electronegative atoms or with possibilities to stabilize ions by resonance. The energy to remove an electron from the molecular anion by autodetachment is generally very low. Consequently, any excess of energy from the negative molecular ion as it is formed must be removed by collision. Thus, in CI conditions, the reagent gas serves not only for producing thermal electrons but also as a source of molecules for collisions to stabilize the formed ions.

Ion pair production is observed with a wide range of electron energies above 15 eV. It is principally this process that leads to negative ion production under conventional EI conditions. Ion pair production forms structurally insignificant very low-mass ions with a sensitivity that is 3–4 orders of magnitude lower than that for positive ion production.

The need to detect tetrachlorodioxins with high sensitivity has contributed to the development of negative ion CI. In some cases, the detection sensitivity with electron capture is better than with positive ions. It can be explained by the high mobility of the electron that ensures a greater rate of electron attachment than the rate of ion formation involving transfer of a much larger particle. However, electron capture is very dependent on the experimental conditions and thus can be irreproducible.

Note the different behaviours that the electron can adopt towards the molecules. Electrons at thermal equilibrium, that is those whose kinetic energy is less than about 1 eV ($1\,eV = 98\,kJ\,mol^{-1}$), can be captured by molecules and yield negative radical anions. Those whose energy lies between 1 and a few hundred electronvolts behave as a wave and transfer energy to molecules, without any 'collisions'. Finally, molecules will be 'transparent' to the electrons with higher energies: here we enter the field of electron microscopy.

As already discussed, negative ions can also be formed through ion–molecule reactions with one of the plasma ions. These reactions can be an acid–base reaction or an addition reaction through adduct formation.

The mixture, CH_4–N_2O 75:25, is very useful because of the following reactions involving low-energy electrons:

$$N_2O + e^- \longrightarrow N_2O^{\bullet-}$$

$$N_2O^{\bullet-} \longrightarrow N_2 + O^{\bullet-}$$

$$O^{\bullet-} + CH_4 \longrightarrow CH_3^{\bullet} + OH^-$$

This plasma also contains other ions. The advantage derives from the simultaneous presence of thermal electrons, allowing the capture of electrons, and of a basic ion, OH^-, which reacts with acidic compounds in the most classical acid–base reaction. Because of the presence of the methane, the same mixture is also suitable for positive ion production.

The energy balance for the formation of negative ions appears as in the following example (PhOH = phenol) [6]:

$$H^+ + OH^- \longrightarrow H_2O \qquad\qquad \Delta H^\circ = -1634.7 \text{ kJ mol}^{-1}$$
$$PhOh \longrightarrow PhO^- + H^+ \qquad\qquad \Delta H^\circ = +145 \text{ kJ mol}^{-1}$$

$$PhOH + OH^- \longrightarrow PhO^- + H_2O \qquad \Delta H^\circ = -178.7 \text{ kJ mol}^{-1}$$

The exothermicity of the reaction is the result of the formation of H_2O, which is neutral and carries off the excess energy. The product anion will be 'cold'.

However, during a positive ionization, the following will occur:

$$CH_5^+ \longrightarrow CH_4 + H^+ \qquad\qquad \Delta H^\circ = +543.5 \text{ kJ mol}^{-1}$$
$$(C_2H_5)_2S + H^+ \longrightarrow (C_2H_5)_2SH^+ \qquad\qquad \Delta H^\circ = -856.7 \text{ kJ mol}^{-1}$$

$$CH_5^+ + (C_2H_5)_2S \longrightarrow (C_2H_5)_2SH^+ + CH_4 \qquad \Delta H^\circ = -313.2 \text{ kJ mol}^{-1}$$

In this case, the exothermicity comes mainly from the association of the proton with the molecule to be ionized. The resulting cation will contain an appreciable level of excess energy.

1.2.6 Desorption Chemical Ionization

Baldwin and McLafferty [7] noticed that introducing a sample directly into the CI plasma on a glass or a metal support allowed the temperature for the observation of the mass spectrum to be reduced, sometimes by as much as 150 °C. This prevents the pyrolysis of non-volatile samples. A drop of the sample in solution is applied on a rhenium or tungsten wire. The solvent is then evaporated and the probe introduced into the mass spectrometer source. The sample is desorbed by rapidly heating the filament by passing a controllable electric current through the wire. The ion formation from these compounds sometimes last for only a very short time, and the ions of the molecular species are observed for only a few seconds. The spectrum appearance generally varies with the temperature. Figure 1.11 displays the spectrum of mannitol obtained by the desorption chemical ionization (DCI) technique.

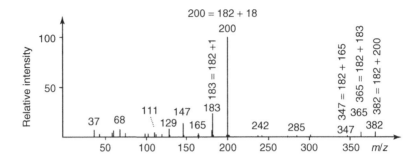

Figure 1.11
DCI spectrum of mannitol, a non-volatile compound, with H_2O as an reagent
gas. Note that water yields radical cation adducts $(M + H_2O)^{\bullet+}$.

The observed spectrum probably results from the superposition of several phenomena:
evaporation of the sample with rapid ionization, direct ionization on the surface of the
filament, direct ion desorption and, at higher temperature, pyrolysis followed by ionization.

Generally, the molecular species ion is clearly detected in the case of non-volatile
compounds. The method can be useful, for example, for tetrasaccharides, small peptides,
nucleic acids and other organic salts, which can be detected in either the positive or negative
ion mode.

1.3 Field Ionization

Field ionization (FI) is a method that uses very strong electric fields to produce ions from
gas-phase molecules. Its use as a soft ionization method in organic mass spectrometry is
principally due to Beckey [8]. Like EI or CI, FI is only suitable for gas-phase ionization.
Therefore, the sample is introduced into the FI source by the same techniques that are
commonly used in EI and CI sources, for example using a direct probe that can be heated
or the eluent from a gas chromatograph.

The intense electric fields used in this ionization method are generally produced by a
potential difference of 8–12 kV that is applied between a filament called the emitter and
a counter-electrode that is a few millimetres distant. Sample molecules in the gas phase
approach the surface of the emitter that is held at high positive potential. If the electric field
at the surface is sufficiently intense, that is if its strength reaches about 10^7–10^8 V cm^{-1},
one of the electrons from the sample molecule is transferred to the emitter by quantum
tunnelling, resulting in the formation of a radical cation $M^{\bullet+}$. This ion is repelled by the
emitter and flies towards the negative counter-electrode. A hole in the counter-electrode
allows the ion to pass into the mass analyser compartment. In order to achieve the high
electric field necessary for ionization, the emitter constituted of tungsten or rhenium filament
is covered with thousands of carbon microneedles on its surface. It is at the tips of these
microneedles that the electric field strength reaches its maximum. FI leads to the formation
of $M^{\bullet+}$ and/or MH^+ ions depending on the analyte. The formation of protonated molecular
species results from ion–molecule reactions that can occur between the initial ion and the
sample molecules close to the surface of the emitter. It is not unusual to observe both $M^{\bullet+}$
and MH^+ in the FI spectrum.

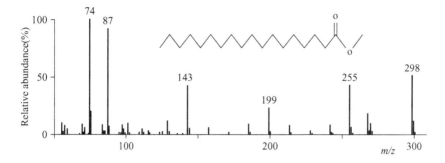

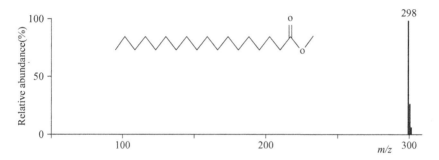

Figure 1.12
Comparison of EI (top) and FI (bottom) spectra of methyl stearate. Field ionization
yields simple spectrum that shows intense molecular ion detected at m/z 298 without fragmentation. Reproduced, with permission from Micromass documentation.

The energy transferred in the FI process corresponds to a fraction of 1 electronvolt. So,
this ionization source generates ions with an extremely low excess of internal energy thus
exhibiting no fragmentation, as shown in Figure 1.12. As the internal energy of the ions
is much lower than that resulting in EI and CI processes, FI is one of the softest methods
to produce ions from organic molecules. However, thermal decomposition of the analyte
can occur during its evaporation prior to ionization. Therefore, FI allows molecular species
to be easily recognized only for compounds sufficiently volatile and thermally stable. For
instance, FI has proven to be a method of choice for the analysis of highly complex mixtures
such as fossil fuels. Generally, FI is complementary to EI and CI. It is used when EI and CI
fail to give ions of the molecular species, although it is not as sensitive. Indeed, this process
has very low ionization efficiency.

1.4 Fast Atom Bombardment and Liquid Secondary Ion Mass Spectrometry

Secondary ion mass spectrometry (SIMS) analyses the secondary ions emitted when a
surface is irradiated with an energetic primary ion beam [9, 10]. Ion sources with very
low-current primary ion beams are called static sources because they do not damage the

surface of the sample, as opposed to dynamic sources that produce surface erosion. Static SIMS causes less damage to any molecules on the surface than dynamic SIMS and gives spectra that can be similar to those obtained by plasma desorption [11]. This technique is mostly used with solids and is especially useful to study conducting surfaces. High-resolution chemical maps are produced by scanning a tightly focused ionizing beam across the surface.

Fast atom bombardment (FAB) [12] and liquid secondary ion mass spectrometry (LSIMS) [13] are techniques that consist of focusing on the sample a high primary current beam of neutral atoms/molecules or ions, respectively. Essential features of these two ionization techniques are that the sample must be dissolved in a non-volatile liquid matrix. In practice, glycerol is most often used, while m-nitrobenzylic alcohol (MNBA) is a good liquid matrix for non-polar compounds, and di- and triethanolamine are efficient, owing to their basicity, in producing negative ions. Thioglycerol and a eutectic mixture of dithiothreitol and dithioerythritol (5:1 w/w), referred to as magic bullet, are alternatives to glycerol.

These techniques use current beams that are in dynamic SIMS high enough to damage the surface. But they produce ions from the surface, as in static SIMS, because convection and diffusion inside the matrix continuously create a fresh layer from the surface for producing new ions. The energetic particles hit the sample solution, inducing a shock wave which ejects ions and molecules from the solution. Ions are accelerated by a potential difference towards the analyser. These techniques induce little or no ionization. They generally eject into the gas phase ions that were already present in the solution.

Under these conditions, both ion and neutral bombardment are practical techniques. The neutral atom beam at about 5 keV is obtained by ionizing a compound, most often argon, sometimes xenon. Ions are accelerated and focused towards the compound to be analysed under several kilovolts. They then go through a collision cell where they are neutralized by charge exchange between atoms and ions. Their momentum is sufficient to maintain the focusing. The remaining ions are then eliminated from the beam as it passes between electrodes. A diagram of such a source is shown in Figure 1.13. The reaction may be written as follows:

$$Ar^{\bullet+}{}_{(rapid)} + Ar_{(slow)} \longrightarrow Ar^{\bullet+}{}_{(slow)} + Ar_{(rapid)}$$

Using a 'caesium gun', one produces a beam of Cs^+ ions at about 30 keV. It is claimed to give better sensitivity than a neutral atom beam for high molecular weights. However, the advantage of using neutral molecules instead of ions lies in the avoidance of an accumulation of charges in the non-conducting samples.

This method is very efficient for producing ions from polar compounds with high molecular weights. Ions up to 10 000 Da and above can be observed, such as peptides and nucleotides. Moreover, it often produces ion beams that can be maintained during long periods of time, sometimes several tens of minutes, which allows several types of analysis to be carried out. This advantage is especially appreciated in measurements using multiple analysers (MS/MS). Other desorption techniques, such as DCI or field desorption (FD) (see later), generally give rise to transient signals, lasting only a few seconds at most. However, FAB or LSIMS requires a matrix such as glycerol, whose ions make the spectrum more complex. DCI and FD do not impose this inconvenience.

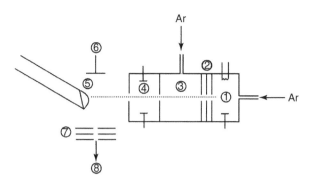

Figure 1.13
Diagram of an FAB gun. 1, Ionization of argon; the resulting ions are accelerated and focused by the lenses 2.
In 3, the argon ions exchange their charge with neutral
atoms, thus becoming rapid neutral atoms. As the beam
path passes between the electrodes 4, all ionic species are
deflected. Only rapid neutral atoms reach the sample dissolved in a drop of glycerol, 5. The ions ejected from the
drop are accelerated by the pusher, 6, and focused by the
electrodes, 7, towards the analyser, 8.

Figure 1.14 displays an example of FABMS and MS/MS applied to the detection of
peptides in a mixture and the sequence determination of one of them.

1.5 Field Desorption

The introduction of field desorption (FD) as a method for the analysis of non-volatile
molecules is principally due to Beckey [14].

Based on FI, already described, FD has been developed as the first method that combines
desorption and ionization of the analyte. There is no need for evaporation of the analyte prior
to ionization. Consequently, FD is particularly suitable for analysing high-molecular-mass
and/or thermally labile compounds.

In FD, the sample is deposited, through evaporation of a solution also containing a
salt, on a tungsten or rhenium filament covered with carbon microneedles. A potential
difference is set between this filament and an electrode so as to obtain a field that can
go up to 10^8 V cm^{-1}. The filament is heated until the sample melts. The ions migrate
and accumulate at the tip of the needles, where they end up being desorbed, carrying along
molecules of the sample. Ionization occurs in the condensed phase or near the surface of the
filament by interaction with the high electric field according to the same mechanism as FI.
FD produces ions of extremely low internal energy thus exhibiting almost no fragmentation
and an abundant molecular species.

The technique is demanding and requires an experienced operator. It has now been
largely replaced by other desorption techniques. However, it remains an excellent method
to ionize high-molecular-mass non-polar compounds such as polymers.

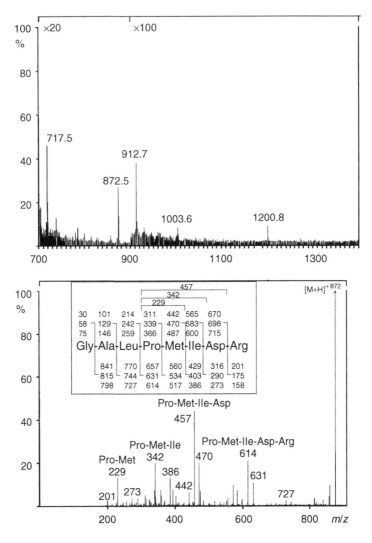

Figure 1.14
Top: FAB mass spectrum of a mixture of five peptides. The *m/z* of
the protonated molecular ion $(M + H)^+$ of each of them is observed.
Bottom: product ion tandem mass spectrum of the $(M + H)^+$ ion with
m/z 872, giving the sequence of this peptide alone. The values within
the frame are the masses of the various possible fragments for the
indicated sequence.

1.6 Plasma Desorption

Plasma desorption (PD) was introduced by Mcfarlane and Torgesson [15]. In this ionization
technique, the sample deposited on a small aluminized nylon foil is exposed in the source
to the fission fragments of ^{252}Cf, having an energy of several mega-electronvolts.

The shock waves resulting from a bombardment of a few thousand fragments per second induce the desorption of neutrals and ions. This technique has allowed the observation of ions above 10 000 Da [16]. However, nowadays it is of limited use and has been replaced mainly by matrix-assisted laser desorption ionization.

1.7 Laser Desorption

Laser desorption (LD) is an efficient method for producing gaseous ions. Generally, laser pulses yielding from 10^6 to 10^{10} W cm^{-2} are focused on a sample surface of about 10^{-3}–10^{-4} cm^2, most often a solid. These laser pulses ablate material from the surface, and create a microplasma of ions and neutral molecules which may react among themselves in the dense vapour phase near the sample surface. The laser pulse realizes both the vaporization and the ionization of the sample.

This technique is used in the study of surfaces and in the analysis of the local composition of samples, such as inclusions in minerals or in cell organelles. It normally allows selective ionization by adjusting the laser wavelength. However, in most conventional infrared LD modes, the laser creates a thermal spike, and thus it is not necessary to match the laser wavelength with the sample.

Since the signals are very short, simultaneous detection analysers or time-of-flight analysers are required. The probability of obtaining a useful mass spectrum depends critically on the specific physical proprieties of the analyte (e.g. photoabsorption, volatility, etc.). Furthermore, the produced ions are almost always fragmentation products of the original molecule if its mass is above approximately 500 Da. This situation changed dramatically with the development of matrix-assisted laser desorption ionization (MALDI) [17, 18].

1.8 Matrix-Assisted Laser Desorption Ionization

This was introduced in 1988 principally by Karas and Hillenkamp [19–21]. It has since become a widespread and powerful source for the production of intact gas-phase ions from a broad range of large, non-volatile and thermally labile compounds such as proteins, oligonucleotides, synthetic polymers and large inorganic compounds. The use of a MALDI matrix, which provides for both desorption and ionization, is the crucial factor for the success of this ionization method. The method is characterized by easy sample preparation and has a large tolerance to contaminantion by salts, buffers, detergents, and so on [22, 23].

1.8.1 Principles of MALDI

MALDI is achieved in two steps. In the first step, the compound to be analysed is dissolved in a solvent containing in solution small organic molecules, called the matrix. These molecules must have a strong absorption at the laser wavelength. This mixture is dried before analysis and any liquid solvent used in the preparation of the solution is removed. The result is a 'solid solution' deposit of analyte-doped matrix crystals. The analyte molecules are embedded throughout the matrix so that they are completely isolated from one another.

The second step occurs under vacuum conditions inside the source of the mass spectrometer. This step involves ablation of bulk portions of this solid solution by intense laser

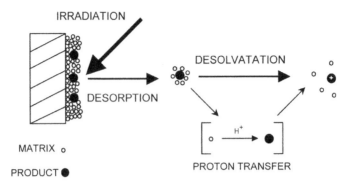

Figure 1.15
Diagram of the principle of MALDI.

pulses over a short duration. The exact mechanism of the MALDI process is not completely elucidated [24,25]. However, irradiation by the laser induces rapid heating of the crystals by the accumulation of a large amount of energy in the condensed phase through excitation of the matrix molecules. The rapid heating causes localized sublimation of the matrix crystals, ablation of a portion of the crystal surface and expansion of the matrix into the gas phase, entraining intact analyte in the expanding matrix plume [26].

Ionization reactions can occur under vacuum conditions at any time during this process but the origin of ions produced in MALDI is still not fully understood [27,28]. Among the chemical and physical ionization pathways suggested for MALDI are gas-phase photoionization, excited state proton transfer, ion–molecule reactions, desorption of preformed ions, and so on. The most widely accepted ion formation mechanism involves proton transfer in the solid phase before desorption or gas-phase proton transfer in the expanding plume from photoionized matrix molecules. The ions in the gas phase are then accelerated by an electrostatic field towards the analyser. Figure 1.15 shows a diagram of the MALDI desorption ionization process.

MALDI is more sensitive than other laser ionization techniques. Indeed, the number of matrix molecules exceeds widely those of the analyte, thus separating the analyte molecules and thereby preventing the formation of sample clusters that inhibit the appearance of molecular ions. The matrix also minimizes sample damage from the laser pulse by absorbing most of the incident energy and increases the efficiency of energy transfer from the laser to the analyte. So the sensitivity is also highly increased. MALDI is also more universal than the other laser ionization techniques. Indeed, it is not necessary to adjust the wavelength to match the absorption frequency of each analyte because it is the matrix that absorbs the laser pulse. Furthermore, because the process is independent of the absorption properties and size of the compound to be analysed, MALDI allows the desorption and ionization of analytes with very high molecular mass in excess of 100 000 Da. For example, MALDI allows the detection of femtomoles of proteins with molecular mass up to 300 000 Da [29,30].

MALDI mass spectrometry has become a powerful analytical tool for both synthetic polymers and biopolymers [31]. Typical MALDI spectra include mainly the monocharged molecular species by protonation in positive ion mode. More easily deprotonated compounds are usually detected in negative ion mode. Some multiply charged ions, some

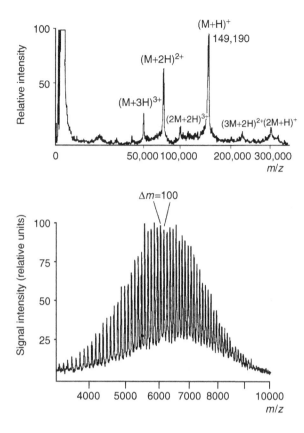

Figure 1.16
MALDI spectra of a monoclonal antibody (above) and of
a polymer PMMA 7100 (below). Reproduced (modified)
from Hillenkamp F.H. and Kras M., Meth. Enzymol.,
193, 280–295, 1990 and from Finnigan MAT documentation, with permission.

multimers and very few fragments can also be observed. Compounds that are not easily
protonated can be cationized instead, often by adding a small quantity of alkali, copper or
silver cations to the sample. As MALDI spectra are simple, complex mixtures can be easily
analysed. Figure 1.16 shows the MALDI spectrum of a monoclonal antibody [32] of about
150 kDa. This figure also presents the MALDI spectrum of a synthetic polymer corresponding to polymethyl methacrylate (PMMA 7100) with an average mass of about 7100 Da.

The use of MALDI to image biological materials is another interesting application
[33, 34]. Indeed, as with LD and SIMS, MALDI has been used to map the distribution of
targeted biomolecules in tissue. It allows for example the study of peptides, proteins and
other biomolecules directly on tissue sections.

Contrary to most other ionization sources that yield a continuous ion beam, MALDI is
a pulsed ionization technique that produces ions in bundles by an intermittent process. The
pulsed nature of the MALDI source is well suited for the time-of-flight (TOF) analyser. In
addition, the TOF analyser has the ability to analyse ions over a wide mass range and thus

Table 1.1   Some common lasers used for MALDI.

Laser	Wavelength	Energy (eV)	Pulse width
Nitrogen	337 nm	3.68	<1 ns to a few ns
Nd:YAG μ3	355 nm	3.49	5 ns
Nd:YAG μ4	266 nm	4.66	5 ns
Er:YAG	2.94 μm	0.42	85 ns
CO_2	10.6 μm	0.12	100 ns + 1 μs tail

can analyse the high-mass ions generated by MALDI. Altogether, this explains why most MALDI spectra have been obtained with MALDI-TOF spectrometers. However, there is no fundamental reason to limit the use of MALDI sources with TOF analysers. MALDI sources have also been coupled to other mass analysers, such as ion trap or Fourier transform mass spectrometers. These instruments allow MS/MS analysis to be performed much more powerfuly and easier realized than using TOF instruments. Furthermore, Fourier transform mass spectrometers reach high resolutions.

1.8.2 Practical Considerations

Among the different lasers used, UV lasers are the most common because of their ease of operation and their low price. N_2 lasers ($\lambda = 337$ nm) are considered as the standard, though Nd:YAG lasers ($\lambda = 266$ or 355 nm) are also used. MALDI can also use IR lasers like Er:YAG lasers ($\lambda = 2.94$ μm) or CO_2 lasers ($\lambda = 10.6$ μm). A summary of laser wavelengths and pulse widths usually used for MALDI is listed in Table 1.1. It is not the power density that is the most important parameter to produce significant ion current but the total energy in the laser pulse at a given wavelength [35]. Generally, the power density required corresponds to an energy flux of 20 mJ cm^{-2}. The pulse widths of lasers vary from a few tens of nanoseconds to a few hundred microseconds. The laser spot diameter at the surface of the sample varies from 5 to 200 μm. It is important to determine the threshold irradiance, the laser pulse power that results in the onset of desorption of the matrix. Molecular species of the analyte are generally observed at slightly higher irradiances but higher laser power leads to more extensive fragmentation and induces a loss of mass resolution.

MALDI spectra obtained with UV or IR lasers are essentially identical for most analysed samples. There are only very small differences. Indeed, when an IR laser is used, only less fragmentation is observed, indicating that the IR-MALDI is somewhat cooler. On the other hand, IR-MALDI induces a larger depth of vaporization per shot that leads to shorter lifetime of the sample. Compared with UV-MALDI, a somewhat lower sensitivity is observed.

Matrix selection and optimization of the sample preparation protocol are the most important steps in the analysis because the quality of the results depends on good sample preparation. However, the preparation procedures are still empirical. The MALDI matrix selection is based on the laser wavelength used. In addition, the most effective matrix is strongly related to the class of analyte and may differ for analytes that have apparently similar structures. The MALDI matrices must meet a number of requirements simultaneously. These are strong absorbance at the laser wavelength, low enough mass to be sublimable,

Table 1.2   Some common UV-MALDI matrices.

Analyte	Matrix	Abbreviation
Peptides/proteins	α-Cyano-4-hydroxycinnamic acid	CHCA
	2,5-Dihydroxybenzoic acid (gentisic)	DHB
	3,5-Dimethoxy-4-hydroxycinnamic acid (sinapic)	SA
Oligonucleotides	Trihydroxyacetophenone	THAP
	3-Hydroxypicolinic acid	HPA
Carbohydrates	2,5-Dihydroxybenzoic acid	DHB
	α-Cyano-4-hydroxycinnamic acid	CHCA
	Trihydroxyacetophenone	THAP
Synthetic	Trans-3-indoleacrylic acid	IAA
polymers	Dithranol	DIT
	2,5-Dihydroxybenzoic acid	DHB
Organic	2,5-Dihydroxybenzoic acid	DHB
molecules		
Inorganic	Trans-2-(3-(4-tert-Butylphenyl)-2methyl-2-	DCTB
molecules	propenyliedene)malononitrile	
Lipids	Dithranol	DIT

vacuum stability, ability to promote analyte ionization, solubility in solvents compatible with analyte and lack of chemical reactivity. However, these general guidelines for matrix selection are not sufficient to predict a good matrix. Indeed, numerous matrix candidates have been inspected and their ability to function as a MALDI matrix has been exemplified, but only very few are good matrices.

Common UV-MALDI matrices are listed in Table 1.2 with the class of compounds with which they are used. The matrices used with IR lasers, such as urea, caboxylic acids, alcohols and even water, are often closer to the natural solutions than the highly aromatic UV-MALDI matrices. In addition, there are many more potential matrices for IR-MALDI owing to the strong absorption of molecular compounds at IR wavelengths, even if the correlation between ion formation and matrix absorption in IR-MALDI is not clear [36].

A number of different sample preparation methods have been described in the literature [37, 38]. A collection of these protocols is accessible on the Internet [39, 40]. The original method that is always the most widely used has been called dried-droplet. This method consists of mixing some saturated matrix solution (5–10 µl) with a smaller volume (1–2 µl) of an analyte solution. Then, a droplet (0.5–2 µl) of the resulting mixture is placed on the MALDI probe, which usually consists of a metal plate with a regular array of sites for sample application. The droplet is dried at room temperature and when the liquid has completely evaporated to form crystals, the sample may be loaded into the mass spectrometer.

MALDI suffers from some disadvantages such as low shot-to-shot reproducibility and strong dependence on the sample preparation method. Each laser shot ablates a few layers of the deposit at the spot where the laser irradiates. This can produce variation in the shot-by-shot spectrum. Also, the impact position on the surface of the deposit can lead to spectral variations. Improvement of the deposit homogeneity gives a better reproducibility of the signal. This is very important if precise quantitative results must be obtained. A given position may become depleted after approximately 50 shots but a few laser shots are usually sufficient to acquire a reasonable spectrum. When a long and stable signal is needed, the target plate moves during spectra acquisition to expose fresh sample continuously to the laser irradiation spot.

MALDI is relatively less sensitive to contamination by salts, buffers, detergents, and so on in comparison with other ionization techniques [41]. The analyte must be incorporated into the matrix crystals. This process may generally serve to separate in solid phase the analyte from contaminants. However, high concentrations of buffers and other contaminants commonly found in analyte solutions can interfere with the desorption and ionization process of samples. Prior purification to remove the contaminants leads to improvements in the quality of mass spectra. For instance, the removal of alkali ions has proven to be very important for achieving high desorption efficiency and mass resolution.

On-probe purification using derivatized MALDI probe surfaces has been described to simplify the sample preparation process. Various developments in this field have allowed the introduction of new techniques such as the surface-enhanced laser desorption ionization (SELDI) [42]. The surface of the probe plays an active role in binding the analyte by hydrophobic or electrostatic interactions, while contaminants are rinsed away. In the same way, this technique uses targets with covalently coupled antibodies directed against a protein, allowing its purification from biological samples as urine or plasma. Subsequent addition of a droplet of matrix solution allows MALDI analysis.

In MALDI, the laser typically irradiates the analyte on the front side of an opaque surface (reflection geometry). Another configuration consisting of laser irradiation through the back of the sample (transmission geometry) has been used. However, the use of this configuration is limited. The sample consumption for MALDI is much lower than the amount required for analysis because only a small fraction of the surface of the sample on the MALDI probe is irradiated by the laser during the acquisition. After analysis, the remaining sample can be recovered for other experiments.

Matrix-free direct laser desorption ionization of analyte has been studied on different kinds of surfaces without real success because degradation of the sample is usually observed. However, good results were obtained with the method called surface-activated laser desorption ionization (SALDI) [43] which uses graphite as the surface. But the use of porous silicon as a new surface is more promising and has led to the development of a new method called desorption ionization on silicon (DIOS) [44]. Unlike the other matrix-free laser desorption ionization methods, DIOS allows ion formation from analyte with little or no degradation.

As already mentioned, DIOS is a matrix-free method that uses pulsed laser desorption ionization on porous silicon. Indeed, this method simply consists of depositing the sample in solution on porous silicon, without any added organic matrix. The structure of porous silicon allows the analyte molecules to be retained while its strong UV absorption allows the desorption ionization of the sample under UV laser irradiation. DIOS has most of its characteristics in common with MALDI but has several advantages to MALDI because it does not use a matrix. DIOS mass spectra do not present interference in the low-mass range, while signals due to the matrix are observed in MALDI. It allows small molecules to be easily analysed. For the same reason, the deposition of the sample in aqueous solution is uniform and the preparation of the sample is simplified. Furthermore, DIOS is equivalent to MALDI in sensitivity, but is more tolerant of the presence of salts or buffers. This method is useful for a large range of small-size compounds (100–3000 Da) such as organic compounds or biomolecules including peptides and oligosaccharides. With all of these characteristics, DIOS is a complementary ionization method to MALDI.

1.8.3 Fragmentations

The MALDI process can lead to fragmentations that occur as a result of the excess energy that is imparted to the analyte during the desorption ionization process. There are essentially three different types of fragmentations that generate fragment ions in MALDI spectra. These fragmentations are discriminated according to the place where they occur.

Fragmentations taking place in the source are called in-source decay (ISD) fragmentations. To be precise, fragmentation at the sample surface that occurs before or during the desorption event (on a time scale of a few picoseconds to nanoseconds) is called prompt fragmentation. Fragmentation occurring in the source after the desorption event but before the acceleration event (on a time scale of a few nanoseconds to microseconds) is called fast fragmentation. Fragmentation that occurs after the acceleration region of the mass spectrometer is called post-source decay (PSD) fragmentation. It corresponds to the fragmentation of metastable ions, which are stable enough to leave the source but contain enough excess energy to allow their fragmentation before they reach the detector.

There are many mechanisms involved in the activation of ions produced in MALDI. Acquisition of an excess of internal energy can be due to the direct interaction photon/molecule, to ionization energy and to activation of molecules in solid state. Another important mechanism consists of the multiple collisions that ions undergo in the source. Indeed, the laser pulse provokes expansion of the matrix into the gas phase, carrying intact analyte in the expanding matrix plume. Desorption of these small molecules induces the formation of a very dense cloud of neutral molecules located just over the surface. The acceleration of the ions through this cloud during their extraction from the source causes many collisions that increase their internal energy. These collisions can be controlled by the strength of the electric field used to extract the ions from the source. By increasing the electric field, the ion-neutral collision energy is increased and thus the internal energy of the ions increases too. On the other hand, the internal energy of the ions increases less in a weaker electric field because the collision energy decreases in the expanding cloud.

ISD fragmentations lead to product ions that are always apparent in the MALDI spectra, whereas the observation of product ions from PSD fragmentation needs certain instrumental conditions. For example, a MALDI source coupled to a linear TOF analyser allows detection of fragment ions produced in the source at their appropriate m/z ratio. On the contrary, fragment ions produced after the source cannot be resolved from their precursor ions and are detected at the same apparent m/z ratio. This induces a broadening of the peaks with a concomitant loss of mass resolution and sensitivity.

1.8.4 Atmospheric Pressure MALDI

In 2000, various developments in the field of MALDI led to the advent of new methods such as the atmospheric pressure MALDI (AP-MALDI) source. This method combines the atmospheric pressure (AP) source and MALDI [45–47]. Indeed, this source produces ions of analytes under normal atmospheric pressure conditions from analyte-doped matrix microcrystals by irradiating these crystals with laser pulses.

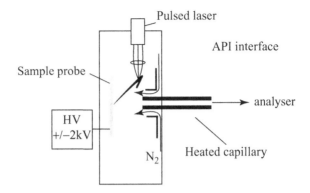

Figure 1.17
Diagram of an AP-MALDI source. Ions are transferred
into the mass analyser using the atmospheric pressure
interface.

The AP-MALDI source is illustrated in Figure 1.17. It works in a similar manner to
the conventional MALDI source. The same sample preparation techniques and the same
matrices used for conventional vacuum MALDI can be used successfully for AP-MALDI.
The main difference is the pressure conditions where ions are produced. Conventional
MALDI is a vacuum ionization source where analyte ionization takes place inside the
vacuum of the mass spectrometer whereas AP-MALDI is an atmospheric ionization source
where ionization occurs under atmospheric pressure conditions outside of the instrument
vacuum.

The ions are transferred into the vacuum of the mass analyser using an atmospheric
pressure interface (API). To assist the transport of ions produced from the atmospheric
pressure ionization region towards the high vacuum, a high voltage (typically, 2–3 kV) is
applied on the surface of the target plate (MALDI probe) and a stream of dry nitrogen
is applied to the area surrounding the target plate. As the transfer of ions into the mass
spectrometer is relatively inefficient, the total sample consumption is higher for AP-MALDI
than for vacuum MALDI. However, sensitivity of this ion source is not decreased because
the sample consumption for MALDI is much lower than the amount required for analysis.
Indeed, a large fraction of the sample is not used during data acquisition.

The mechanism of AP-MALDI ion production is similar to that of conventional MALDI.
Thus, the AP-MALDI source, like the conventional MALDI source, produces mainly
monocharged molecular species but with a narrower mass range. But, because of the
fast and efficient thermalization of the ion internal energy at atmospheric conditions, AP-
MALDI is a softer ionization technique compared with conventional vacuum MALDI and
even softer than vacuum IR-MALDI. Ions produced by this method generally exhibit no
fragmentation but tend to form clusters with the matrix. These unwanted adducts between
matrix and analyte can be eliminated by increasing the energy transferred to the ions in the
source. For instance, increasing the laser energy or some API parameters, such as capillary
temperature, increases the analyte-matrix dissociation process.

The advantages of AP-MALDI include those advantages typically associated with a
MALDI source but without some of the drawbacks. Indeed, AP-MALDI does not require
a vacuum region and is decoupled from the mass analyser, allowing it to be coupled with
any mass spectrometer equipped with API. It is also easily interchangeable with other

atmospheric pressure sources, such as electrospray ionization (see later). As a result of the almost complete decoupling of the ion desorption from the mass analyser, the performance of the instrument (calibration, resolution and mass accuracy) is not affected by source conditions (type of sample matrix, sample preparation method and location of the laser spot on the sample). This allows much greater experimental flexibility. It is possible, for instance, to use long-pulse lasers to increase the overall sensitivity without observing deterioration in resolution. AP-MALDI sources, like the conventional MALDI source, can be used to image biological materials. But sampling at atmospheric pressure allows the examination of samples that are vacuum sensitive, such as samples containing volatile solvents. This method can be easily applied to spatial analysis of native surfaces of biological tissues.

1.9 Thermospray

The principle of the thermospray (TSP), proposed by Blakney and Vestal [48,49] in 1983, is shown in Figure 1.18.

A solution containing a salt and the sample to be analysed is pumped into a steel capillary, which is heated to high temperature allowing the liquid to heat quickly. The solution passes through a vacuum chamber as a supersonic beam. A fine-droplet spray occurs, containing ions and solvent and sample molecules. The ions in the solution are extracted and accelerated towards the analyser by a repeller and by a lens focusing system. They are desorbed from the droplets carrying one or several solvent molecules or dissolved compounds. It is thus not necessary to vaporize before ionization: ions go directly from the liquid phase to the vapour phase. To improve the ion extraction, the droplets at the outlet of the capillary may be charged by a corona discharge. The droplets remain on their supersonic trip to the outlet where they are pumped out continuously through an opening located in front of the supersonic beam. Large vapour volumes from the solvent are thus avoided.

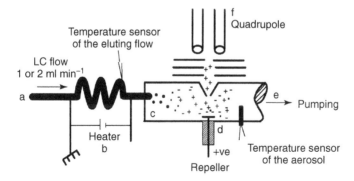

Figure 1.18
Diagram of a thermospray source. The chromatographic effluent comes in at (a) the transfer line is suddenly heated at (b) and the spray is formed under vacuum at (c). At (d) the spray goes between a pusher with a positive potential and a negative cone for positive ions. The ions are thus extracted from the spray droplets and accelerated towards the spectrometer (f). At (e), a high-capacity pump maintains the vacuum.

In order to avoid the freezing of the droplets under vacuum, the liquid must be heated during the injection. This heating is programmed by feedback from a thermocouple which measures the beam temperature under vacuum. The heating is achieved by having a current pass through the capillary that carries the liquid and thus also acts as a heating resistance.

1.10 Atmospheric Pressure Ionization

Besides AP-MALDI, already described earlier, electrospray (ESI), atmospheric pressure chemical ionization (APCI), atmospheric pressure photoionization ionization (APPI), DESI and DART are other examples of atmospheric pressure ionization (API) sources.

Such sources ionize the sample at atmospheric pressure and then transfer the ions into the mass spectrometer. An atmospheric pressure interface is then used to transfer ions into the high vacuum of the mass analyser. The problem lies in coupling an atmospheric pressure source compartment with an analyser compartment that must be kept at a very low pressure or at a very high vacuum (10^{-5} Torr).

This problem is solved by adopting a differential pumping system. Usually two intermediate vacuum compartments are used between the source compartment and the analyser compartment because the pressure difference is quite large. The compartments are connected between them by lenses with very small orifices (called skimmers or cones). The pressures of the intermediate vacuum compartments are gradually reduced by using several differential stages of high-capacity pumps. Ions go across the compartments in the order of higher to lower pressure through these small orifices to reach the analyser compartment. This orifice must be wide enough to allow the introduction of as many ions as possible in order to enhance the sensitivity. But, on the other hand, the orifice must not be too wide to maintain a correct vacuum in the analyser compartment. A transfer optics system including focusing lenses or focusing multipole lenses is provided in the intermediate-vacuum compartments to inject ions effectively into the orifices. The scheme of an atmospheric pressure interface is illustrated in Figure 1.19.

Another problem lies in the cooling caused by the sample and the solvent adiabatic expansion that favours the appearance of ion clusters. Consequently, ion desolvation is also an important aspect of atmospheric pressure interface design. Efficient desolvation is provided by the introduction of a heated metallized transfer tube (about 200°C) or by applying a counter-current flow of heated dry gas also called curtain gas. The desolvation is also improved by accelerating the ions in a region of the interface where the pressure is in the millibar range. The acceleration is obtained by applying a voltage between the different

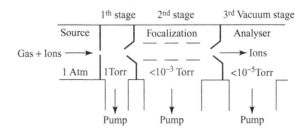

Figure 1.19
Scheme of an atmospheric interface with differential pumping system using three stages.

extraction lenses. The ions collide with residual gas molecules, increase their internal energy, which induces their final desolvation, and the ion clusters disappear. However, these collisions can also give enough energy to induce ion fragmentation. This kind of fragmentation is called in-source fragmentation. The desolvation should be maximized to get a gain in sensitivity because the distribution of an analyte in different species, which are clusters, leads to a decrease in detection sensitivity. Furthermore, the formation of clusters should be controlled when doing quantitative analysis because the analyte of interest should be in a well-defined and stable form.

The transfer of ions from atmospheric pressure to the vacuum of a spectrometer necessarily induces ion losses. But these losses are compensated by the higher total ion yield in the API source due to fast thermal stabilization at atmospheric conditions. Indeed, when the sample ionization is performed under atmospheric pressure [50,51], an ionization efficiency 10^3 to 10^4 times as great as in a reduced-pressure CI source is obtained.

The early API source designs used an axial configuration. The ions were produced in the axis of the orifice. Designs have changed, however. Now the orthogonal configuration to introduce the ions into the interface is used in many API sources. The main advantage is that the orifice is no longer saturated by solvent. Instead, only ions are directed towards the inlet. Consequently, orifices can be larger than in the axial configuration. The combination of larger orifices and noise reduction largely compensates for transmission losses due to the orthogonal geometry, giving a large gain in sensitivity. Another advantage of this configuration is that the flow rates can be increased. Furthermore, this configuration gives better protection of the orifice against contamination or clogging, giving a gain in robustness. However, the orthogonal configuration with indirect trajectory of analyte ions also introduces unwanted discrimination based on mass or charge.

The most important advantage of API sources is the simplicity for the direct on-line coupling of separation techniques (HPLC, CE, etc.) to the mass spectrometer. Another attractive aspect of these sources is the easy introduction of the sample into a mass spectrometer because the operation at atmospheric pressure outside of the mass spectrometer eliminates the complicated procedure of introducing the sample into its high vacuum.

1.11 Electrospray

In the literature, electrospray is abbreviated to either ESI or ES. Because ES is ambiguous, we prefer to use ESI. The success of ESI started when Fenn *et al.* [52, 53] showed that multiply charged ions were obtained from proteins, allowing their molecular weight to be determined with instruments whose mass range is limited to as low as 2000 Th. At the beginning, ESI was considered as an ionization source dedicated to protein analysis. Later on, its use was extended not only to other polymers and biopolymers, but also to the analysis of small polar molecules. It appeared, indeed, that ESI allows very high sensitivity to be reached and is easy to couple to high-performance liquid chromatography HPLC, μHPLC or capillary electrophoresis. ESI principles and biological applications have been extensively reviewed [54–56]. Several edited books on this subject also appeared in 1996 and 1997 [57,58].

ESI [59–64] is produced by applying a strong electric field, under atmospheric pressure, to a liquid passing through a capillary tube with a weak flux (normally $1-10\,\mu l\,min^{-1}$). The electric field is obtained by applying a potential difference of $3-6\,kV$ between this capillary and the counter-electrode, separated by $0.3-2\,cm$, producing electric fields of the

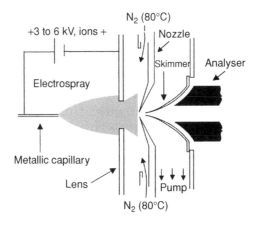

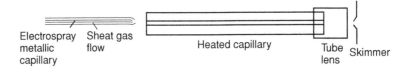

Figure 1.20
Diagram of electrospray sources, using skimmers for ion focalization and a curtain of heated nitrogen gas for desolvation (top), or with a heated capillary for desolvation (bottom).

order of $10^6\,V\,m^{-1}$ (Figure 1.20). This field induces a charge accumulation at the liquid surface located at the end of the capillary, which will break to form highly charged droplets. A gas injected coaxially at a low flow rate allows the dispersion of the spray to be limited in space. These droplets then pass either through a curtain of heated inert gas, most often nitrogen, or through a heated capillary to remove the last solvent molecules.

The spray starts at an 'onset voltage' that, for a given source, depends on the surface tension of the solvent. In a source which has an onset voltage of 4 kV for water (surface tension $0.073\,N\,m^{-2}$), 2.2 kV is estimated for methanol ($0.023\,N\,m^{-2}$), 2.5 kV for acetonitrile ($0.030\,N\,m^{-2}$) and 3 kV for dimethylsulfoxide ($0.043\,N\,m^{-2}$) [65]. If one examines with a microscope the nascent drop forming at the tip of the capillary while increasing the voltage, as schematically displayed in Figure 1.21, at low voltages the drop appears spherical, then elongates under the pressure of the accumulated charges at the tip in the stronger electric field; when the surface tension is broken, the shape of the drop changes to a 'Taylor cone' and the spray appears.

Gomez and Tang [66] were able to obtain photographs of droplets formed and dividing in an ESI source. A drawing of a decomposing droplet is displayed in Figure 1.22. From their observations, they concluded that breakdown of the droplets can occur before the limit given by the Rayleigh equation is reached because the droplets are mechanically deformed, thus reducing the repulsion necessary to break down the droplets.

The solvent contained in the droplets evaporates, which causes them to shrink and their charge per unit volume to increase. Under the influence of the strong electric field,

Figure 1.21
Effect of electrospray potential on the drop at the tip of the capillary, as observed with binoculars while increasing the voltage. Left: at low voltage, the drop is almost spherical. Centre: at about 1 or 2 kilovolts, but below the onset potential, the drop elongates under the pressure of the charges accumulating at the tip. Right: at onset voltage, the pressure is higher than the surface tension, the shape of the drop changes at once to a Taylor cone and small droplets are released. The droplets divide and explode, producing the spray.

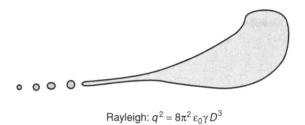

Rayleigh: $q^2 = 8\pi^2 \varepsilon_0 \gamma D^3$

Figure 1.22
A decomposing droplet in an electrospray source, according to [66]; q, charge; ε_0, permittivity of the environment; γ, surface tension and D, diameter of a supposed spherical droplet.

deformation of the droplet occurs. The droplet elongates under the force resulting from the accumulation of charge, similarly to what occurred at the probe tip, and finally produces a new Taylor cone. From this Taylor cone, about 20 smaller droplets are released. Typically a first-generation droplet from the capillary will have a diameter of about 1.5 µm and will carry around 50 000 elementary charges, or about 10^{-14} C. The offspring droplets will have a diameter of 0.1 µm and will carry 300 to 400 elementary charges. The total volume of the offspring droplets is about 2 % of the precursor droplet but contain 15 % of the charge. The charge per unit volume is thus multiplied by a factor of seven. The precursor droplet will shrink further by solvent evaporation and will produce other generations of offspring.

These small, highly charged droplets will continue to lose solvent, and when the electric field on their surface becomes large enough, desorption of ions from the surface occurs [65]. Charges in excess accumulate at the surface of the droplet. In the bulk, analytes as well as electrolytes whose positive and negative charges are equal in number are present at a somewhat higher concentration than in the precursor droplet. The desorption of charged molecules occurs from the surface. This means that sensitivity is higher for compounds whose concentration at the surface is higher, thus more lipophilic ones. When mixtures of compounds are analysed, those present at the surface of droplets can mask, even completely, the presence of compounds which are more soluble in the bulk. When the droplet contains very large molecules, like proteins for example, the molecules will not desorb, but are freed by evaporation of the solvent. This seems to occur when the molecular weight of the compounds exceeds 5000 to 10 000 Da.

The ions obtained from large molecules carry a greater number of charges if several ionizable sites are present. Typically, a protein will carry one charge per thousand daltons

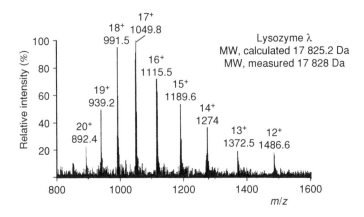

Figure 1.23
ESI spectrum of phage λ lysozyme; m/z in Th and the number of
charges are indicated on each peak. The molecular mass is measured
as being $17\,828 \pm 2.0\,\mathrm{Da}$.

approximately, less if there are very few basic amino acids. As an example, the ESI
spectrum of phage lambda lysozyme is shown in Figure 1.23. Small molecules, say less
than a thousand daltons, will produce mainly monocharged ions. ESI can also be used in the
case of molecules without any ionizable site through the formation of sodium, potassium,
ammonium, chloride, acetate or other adducts.

ESI has important characteristics: for instance, it is able to produce multiply charged ions
from large molecules. The formation of ions is a result of the electrochemical process and of
the accumulation of charge in the droplets. The ESI current is limited by the electrochemical
process that occurs at the probe tip and is sensitive to concentration rather than to total
amount of sample.

1.11.1 Multiply Charged Ions

Large molecules with several ionizable sites produce by ESI multiply charged ions, as
shown for lysozyme positive ions in Figure 1.23.

Obtaining multiply charged ions is advantageous as it improves the sensitivity at the
detector and it allows the analysis of high-molecular-weight molecules using analysers with
a weak nominal mass limit. Indeed, the technical characteristics of mass spectrometers are
such that the value being measured is not the mass, but the mass-to-charge ratio m/z.

The ESI mass spectra of biological macromolecules normally correspond to a statistical
distribution of consecutive peaks characteristic of multiply charged molecular ions obtained
through protonation $(M + zH)^{z+}$, or deprotonation $(M - zH)^{z-}$, with minor if any contri-
butions of ions produced by dissociations or fragmentations. However, as the measured
apparent mass is actually m/z, to know m one needs to determine the number of charges z.

Consider a positive ion with charge z_1 whose mass-to-charge ratio is measured as being
m_1 Th, issued from a molecular ion with mass M Da to which z_1 protons have been added.
We then have

$$z_1 m_1 = M + z_1 m_p$$

where m_p is the mass of the proton.

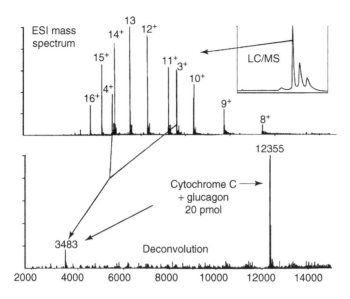

Figure 1.24
Deconvolution of an ESI spectrum of a protein mixture. From
Finnigan documentation. Reprinted, with permission.

An ion separated from the first one by $(j - 1)$ peaks, in increasing order of mass-to-charge
ratio, has a measured ratio of m_2 Th and a number of charges $z_1 - j$, so that

$$m_2(z_1 - j) = M + (z_1 - j)m_p$$

These two equations lead to

$$z_1 = \frac{j(m_2 - m_p)}{(m_2 - m_1)} \quad \text{and} \quad M = z_1(m_1 - m_p)$$

In the case of negative multiply charged ions, analogous equations lead to

$$z_1 = \frac{j(m_2 + m_p)}{(m_2 - m_1)} \quad \text{and} \quad M = z_1(m_1 + m_p)$$

In the example shown in Figure 1.23 using the peaks at m/z 939.2 and 1372.5 $(j = 6)$,
we obtain $z_1 = 6(1372.5 - 1.0073)/(1372.5 - 939.2) = 19$ and we can number all the peaks
measured according to the number of charges. M can be calculated from their mass. The
average value obtained from all of the measured peaks is 17 827.9 Da with a mean error
of 2.0 Da. This technique allowed the determination of the molecular masses of proteins
above 130 kDa with a detection limit of about 1 pmol using a quadrupole analyser.

A variety of algorithms have been developed to allow the determination of the molecular
mass through the transformation of the multiply charged peaks present in the ESI spectrum
into singly charged peaks. Some of them also allow the deconvolution ESI spectra of
mixtures, as is shown in Figure 1.24. However, the complexity of the spectra obtained for
a single compound is such that only simple mixtures can be analysed.

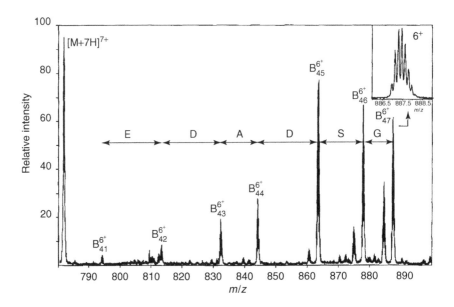

Figure 1.25
Product ion spectrum of the $[M + 7H]^{7+}$ ion from the following peptide: ALVRQG-
LAKVAYVYKPNNTHEQHLRKSEAQAKKEKLLNIWSEDNADSGQ. Notice that
fragment ions having lower charge number z may appear at higher m/z values than
the precursor, which indeed occurs in the spectrum shown. The inset shows that,
owing to the high resolution, the isotopic peaks are observed separated by 1/6 Th,
and thus $1/z = 1/6$ or $z = 6$. From Andersen J., Molina H., Moertz E., Krogh T.N.,
Chernuchevich I., Taylor L., Vorm O. and Mann M., 'Quadrupole-TOF Hybrid
Mass Spectrometers Bring Improvements to Protein Indentification and MS/MS
Analysis of Intact Proteins' The 46th Conference on Mass Spectrometry and Allied Topics, Orlando, Florida, 1998, p. 978. Reprinted, with permission.

At high resolution, the individual peaks with different charge states observed at low
resolution are each split into several peaks corresponding to the isotope distribution. As
neighbour peaks differ by 1 Da, the observed distance between them will be $1/z$, allowing
the direct determination of the charge state of the corresponding ion. This is important for
MS/MS spectra of multiply charged ions, as the preceding rules to assign the z value can
no longer be applied. An example is displayed in Figure 1.25 [67].

The ability of this ionization method for the determination of very high molecular weights
is illustrated in Figure 1.26 [68]. The spectrum displayed is obtained from assemblies of
vanillyl alcohol oxidase containing respectively 16 and 24 proteins. The spectrum was
obtained with a hybrid quadrupole TOF instrument, Q-TOF Micromass, equiped with a
micro-ESI source. To obtain such a spectrum one needs not only a mass spectrometer with
sufficient mass range and resolution, but also high skill in protein purification.

1.11.2 Electrochemistry and Electric Field as Origins of Multiply Charged Ions

Charges of ions generated by ESI do not reflect the charge state of compounds in the
analysed solution, but are the result of both charge accumulation in the droplets and charge

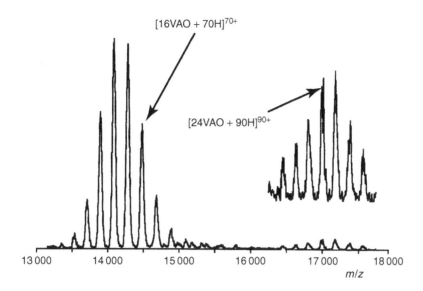

Figure 1.26
MicroESI Q-TOF spectrum of assemblies of vanillyl alcohol oxidase obtained
by W.J.H. van Berkel and co-workers [68]. At the left, an octamer–dimer
of 1.02 MDa molecular weight, and at the right, the octamer–trimer of
1.53 MDa. Reproduced courtesy of Dr. A.J.R. Heck.

modification by electrochemical process at the probe tip. This is clearly demonstrated by
a convincing experiment reported by Fenselau and co-workers, and illustrated in Figure
1.27 [69]. They showed that negative ions of myoglobin can be observed at pH 3, while a
calculation based on known pK values predicts that only 1 molecule per 3500 approximately
would have one negative charge in the original solution. The results point out the role of the
charge accumulation in droplets under the influence of the electric field on the formation of
multiply charged ions. Furthermore, the 'pumping out' of the negative charges can only be
performed if, at the same time, the same number of positive charges is electrochemically
neutralized at the probe tip.

Moreover, it is worth noting that the negative ion spectrum of myoglobin at pH 3 shows
a better signal-to-noise ratio than the same spectrum at pH 10 (Figure 1.27). This results
from the fact that protons have a high electrochemical mobility, and is the first indication
of the importance of the reduction process, when negative ions are analysed, that occurs
at the probe tip.

At pH 10, positive ions can be observed too. Additional peaks in the spectra result from
a modification of the protein at basic pH (loss of heme group).

Thus, it is worthy of consideration to try to acidify a solution with a view to a better
detection of negative ions, and vice versa. Indeed, both H_3O^+ and OH^- have high limit
equivalent conductivities, as shown in Table 1.3.

1.11.3 Sensitivity to Concentration

Another feature of ESI is its sensitivity to concentration, and not to the total quantity of
sample injected in the source, as is the case for most other sources. This is shown in

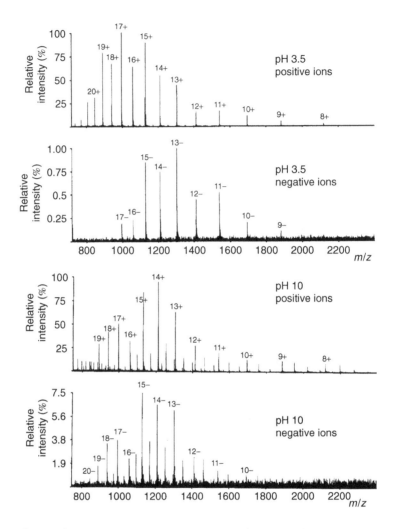

Figure 1.27
ESI spectra of myoglobin have been acquired in the positive and negative
ion mode, at pH 3 and at pH 10. At pH 3, negative ions are observed,
the most intense ions bearing from 13 to 15 charges. A calculation based
on known pK values shows that in the original solution, only 1 molecule
in approximately 3500 bears one negative charge. From Kelly M.A.,
Vestling M.M., Fenselau C. and Smith P.B., Org. Mass Spectrom., 27,
1143, 1992. Reproduced, with permission.

Figure 1.28 [70]. The intensity of the signals from the two monitored compounds is
measured while injecting the total flow from an HPLC column, 400 µl min^{-1}, or a part
of this flow, after splitting. As can be seen, the sensitivity increases somewhat when the
flow entering the source is reduced. This remains true up to flows as low as some tens
of nanolitres per minute. When flow rates higher than about 500 µl min^{-1} are used, the
sensitivity is reduced. Lower flow rates also allow less analyte and buffer to be injected
in the source, reducing contamination. Furthermore, for the same amount of sample, an

Table 1.3 Some values of equivalent conductivity.

Cation	λ_0^+	Anion	λ_0^-
H_3O^+	350	OH^-	200
NH_4^+	74	Br^-	78
K^+	74	I^-	77
Na^+	50	Cl^-	76

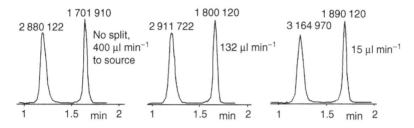

Figure 1.28
HPLC on a 2.1 mm column at $400\,\mu l\,min^{-1}$ flow. Two drugs are monitored by selected ion monitoring. Left: $400\,\mu l\,min^{-1}$ is injected in the source. Centre: $132\,\mu l\,min^{-1}$ is split to the source. Right: $15\,\mu l\,min^{-1}$ is split to the source. The integration values are displayed on top of the peaks and show that at reduced flow rates the sensitivity is slightly increased. Reproduced with data from Covey T., 'Analytical Characteristics of the Electrspray Ionization Process' pp. 21–59 in 'Biochemical and Biotechnological Applications of Eletrospray Ionization Mass Spectrometry' Snyder A.P., ed., ACS Symposium Series 619, American Chemical Society, 1996.

HPLC column with a lower diameter, and using smaller flow rates, will give an increased sensitivity because the concentration of the sample in the elution solvent is increased.

Based on this concentration dependence, modifications of the technique, called microelectrospray (μESI), or nanospray (nESI), which use much lower flow rates down to some tens of nanolitres per minute, have been developed using adapted probe tips [71–73]. Detection limits in the range of attomoles (10^{-15} moles) injected have been demonstrated.

1.11.4 Limitation of Ion Current from the Source by the Electrochemical Process

As may be seen from the electric circuit in Figure 1.29, when positive ions are extracted for analysis, electrons have to be provided in the circuit from the capillary. This will occur through oxidation of species in the solution at the capillary tip, mainly ions having a sufficient mobility. In other words, the same number of negative charges must be 'pumped' out of the solution as positive charges are extracted to the analyser. Thus, ESI is truly an electrochemical process, with its dependence on ion concentration and mobility as well as on polarization effects at the probe tip. For the detection of positive ions, electrons have to be provided by the solution, and thus an oxidation occurs. For negative ions, electrons

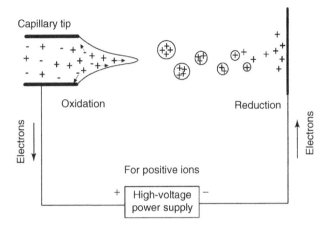

Figure 1.29
Schematic representation of electrochemical process in ESI.

have to be consumed, and thus a reduction occurs. These electrochemical reactions occur in the last micrometres of the metallic capillary. A major consequence is that the total number of ions per unit time that can be extracted to the spectrometer is actually limited by the electric current produced by the oxidation or reduction process at the probe tip. This limiting current is not dependent on the flow rate, up to very low flow, and this explains why ESI is only concentration dependent. In practice, the total ion current is limited to a maximum of about 1 μA.

In ESI, the number of variables is large, including nature of the solvent, flow, nature and size of the capillary, distance to the counter-electrode, applied potentiel, and so on. Furthermore, the ionization process includes many parameters, such as surface tension, nature of analyte and electrolytes, presence or not of other analytes, electrochemical processes at the probe tip, and so on.

Furthermore, the ESI source is a constant-current electrochemical cell [74]. Figure 1.30 shows the analogy between a classical constant-current electrochemical cell and an ESI source. The important consequence is that there will be a constant current I_M carried by the ions. If there are too many ions from salts in the flow, they will suffice to produce I_M and the ions of the sample will be either at low abundances or not observed. On the other hand, if the solution is very dilute and at very low flow (below 1 μl), the ion flow from the capillary can be insufficient to provide I_M. The electrochemical process at the probe tip will then produce additional ions by oxidation (or reduction in negative ion mode) of either the solvent or the sample depending on their respective oxidation (reduction) potentials. This will lead to the observation of radical cations or radical anions in the spectrum.

A paper by Fenn *et al.* [75] makes a critical comparison of the various theories about ESI. A simplified theory will be presented here for the relation between analyte concentration and abundances of the ions.

Ions, either positive or negative, of an analyte A will be desorbed from the droplets, producing a theoretical ion current $I_A = k_A[A]$, where k_A is a rate constant depending on the nature of A. Let us suppose that another ion B is produced from the buffer, at a rate $I_B = k_B[B]$, and that no other ions are sprayed. The total ion current for these two ions

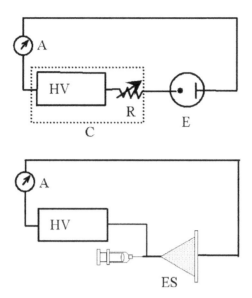

Figure 1.30
Top: scheme of a classical constant-current electrochemical cell, combining a high-voltage power supply with a large resistance. Bottom, an ESI source. The large resistance results here from the ion flow in the air. Reproduced with data from Van Berkel G.J. and Zhou F., Anal. Chem., 67, 2916, 1995.

is $I_T = (I_A + I_B)$, but this total ion current is limited by the oxidation, if positive ions are desorbed, or reduction process that occurs at the probe tip. This limiting current is symbolized I_M, and $I_T = I_M$ if no other ionic species are present.

The current for each ion will be proportional to its relative desorption rate, and the pertinent equations are

$$I_A = I_M \frac{k_A [A]}{k_A [A] + k_B [B]} \qquad I_B = I_M \frac{k_B [B]}{k_A [A] + k_B [B]} \tag{1.1}$$

Let us consider that $[B]$ remains constant, but the analyte concentration $[A]$ varies; then two limiting cases are to be considered. First, for $k_A[A] \gg k_B[B]$,

$$I_A \approx I_M \frac{k_A [A]}{k_B [B]} \qquad I_B \approx I_M \frac{k_B [B]}{k_B [B]} \approx I_M \tag{1.2}$$

This means that the intensity detected for A will be proportional to its concentration, but the sensitivity will be inversely proportional to $[B]$.

The other extreme case leads to

$$I_A \approx I_M \frac{k_A [A]}{k_A [A]} \approx I_M \qquad I_B \approx I_M \frac{k_B [B]}{k_A [A]} \tag{1.3}$$

I_A remains constant, and quantitation of $[A]$ is no longer possible. The intensity of the signal for B will become weaker as $[A]$ increases.

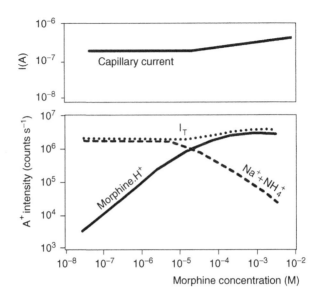

Figure 1.31
Influence of concentration on observed ion abundances, when increasing concentrations of morphine.HCl is injected in a solvent containing a constant concentration of sodium and ammonium salts. Linearity is observed at low concentrations, but from about 5×10^{-6} significant curvature is observed (note that the scales are logarithmic). At still higher concentrations, the intensity levels out. Reproduced (modified) from Kebarle P. and Tang L., Anal. Chem., 65, 972A, 1993.

This is shown by an experimental example from [65] in Figure 1.31. In a solvent containing NH_4^+ and Na^+ ions at constant concentrations, an increasing amount of morphine chlorhydrate is added. The graph shows on top the number of amperes at the capillary tip, and below the intensity monitored at the mass of protonated morphine and the sum of the intensities for the NH_4^+ and Na^+ ions. At low concentrations of morphine chlorhydrate, Equation (1.2) pertains, and linearity towards morphine concentration is observed. At high concentrations, the intensity for morphine is constant, and the signal for the other ions diminishes, in agreement with Equation (1.3). At intermediate values, the general Equation (1.1) applies.

1.11.5 Practical Considerations

To observe a stable spray, a minimum amount of electrolyte in the solvent is required, but this is so low that normal solvents contain enough electrolytes for this purpose. On the other hand, the maximum tolerable total concentration of electrolytes still to have a good sensitivity is about 10^{-3} M. Furthermore, volatile electrolytes are preferred to avoid contamination of the source. With most samples, the problem is more to remove the salts rather than to add some. Also, sample dilution is often performed. Often HPLC methods have to be modified for ESI when they use high concentrations of buffers.

When it is believed that it could be better to add an electrolyte to improve sample detection, one should think about the electrochemical process when selecting it. We have seen, for instance, that adding an acid can improve the detection of negative ions. But the ESI process is not simple, and many trials are often needed.

1.12 Atmospheric Pressure Chemical Ionization

APCI is an ionization technique which uses gas-phase ion–molecule reactions at atmospheric pressure [76, 77]. It is a method analogous to CI (commonly used in GC–MS) where primary ions are produced by corona discharges on a solvent spray. APCI is mainly applied to polar and relatively non-polar compounds with moderate molecular weight up to about 1500 Da and gives generally monocharged ions. The principle governing an APCI source is shown in Figure 1.32.

The analyte in solution from a direct inlet probe or a liquid chromatography eluate at a flow rate between 0.2 and 2 ml min^{-1}, is directly introduced into a pneumatic nebulizer where it is converted into a thin fog by a high-speed nitrogen beam. Droplets are then displaced by the gas flow through a heated quartz tube called a desolvation/vaporization chamber. The heat transferred to the spray droplets allows the vaporization of the mobile phase and of the sample in the gas flow. The temperature of this chamber is controlled, which makes the vaporization conditions independent of the flow and from the nature of the mobile phase. The hot gas (120°C) and the compounds leave this tube. After desolvation, they are carried along a corona discharge electrode where ionization occurs. The ionization processes in APCI are equivalent to the processes that take place in CI but all of these occur under atmospheric pressure. In the positive ion mode, either proton transfer or adduction of reactant gas ions can occur to produce the ions of molecular species, depending on the relative proton affinities of the reactant ions and the gaseous analyte molecules. In the negative mode, the ions of the molecular species are produced by either proton abstraction or adduct formation.

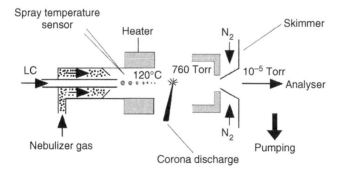

Figure 1.32
Diagram of an APCI source.

Generally, the evaporated mobile phase acts as the ionizing gas and reactant ions are produced from the effect of a corona discharge on the nebulized solvent. Typically, the corona discharge forms by electron ionization primary ions such as $N_2^{\bullet+}$ or $O_2^{\bullet+}$. Then, these ions collide with vaporized solvent molecules to form secondary reactant gas ions.

The electrons needed for the primary ionization are not produced by a heated filament, as the pressure in that part of the interface is atmospheric pressure and the filament would burn, but rather using corona discharges or β^- particle emitters. These two electron sources are fairly insensitive to the presence of corrosive or oxidizing gases.

As the ionization of the substrate occurs at atmospheric pressure and thus with a high collision frequency, it is very efficient. Furthermore the high frequency of collisions serves to thermalize the reactant species. In the same way, the rapid desolvation and vaporization of the droplets reduce considerably the thermal decomposition of the analyte. The result is production predominantly of ions of the molecular species with few fragmentations.

The ions produced at atmospheric pressure enter the mass spectrometer through a tiny inlet, or through a heated capillary, and are then focused towards the analyser. This inlet must be sufficiently wide to allow the entry of as many ions as possible while keeping a correct vacuum within the instrument so as to allow the analysis. The most common solution to all these constraints consists of using the differential pumping technique on one or several stages, each one separated from the others by skimmers [78]. In the intermediate-pressure region, an effective declustering of the formed ions occurs.

APCI has become a popular ionization source for applications of coupled HPLC–MS. Figure 1.33 shows an example of an application of HPLC–APCI coupling [79]. It shows the analysis obtained from extracts of maize plants. Six compounds are identified by mass spectrometry. These compounds have been identified as glucoconjugated DIMBOA (2,4-dihydroxy-7-methoxy-1,4-benzoxazin-3-one) and similar molecules that differ by the number of methoxy groups in the benzene ring and/or by the N-O methylation of the hydroxamate function. This example clearly shows the influence of the analyte on the type of observed molecular species. Indeed, the presence of an acidic group in the compound from peak 1 allows mainly the detection of deprotonated molecular ions, whereas the compound from peak 4 does not contain an acid group and thus leads only to the formation of adduct ions.

1.13 Atmospheric Pressure Photoionization

The APPI source is one of the last arrivals of atmospheric pressure sources [80, 81]. The principle is to use photons to ionize gas-phase molecules. The scheme of an APPI source is shown in Figure 1.34. The sample in solution is vaporized by a heated nebulizer similar to the one used in APCI. After vaporization, the analyte interacts with photons emitted by a discharge lamp. These photons induce a series of gas-phase reactions that lead to the ionization of the sample molecules. The APPI source is thus a modified APCI source. The main difference is the use of a discharge lamp emitting photons rather than the corona discharge needle emitting electrons. Several APPI sources have been developed since 2005 and are commercially available. The interest in the photoionization is that it has the potential to ionize compounds that are not ionizable by APCI and ESI, and in particular, compounds that are non-polar.

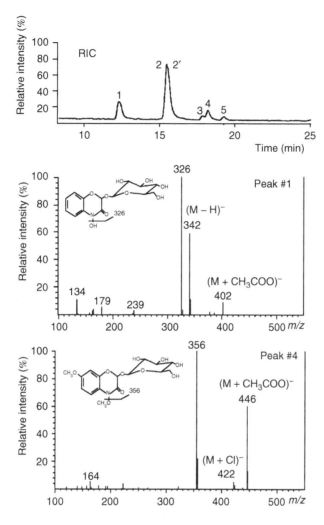

Figure 1.33
HPLC-APCI analysis of a mixture of glucoconjugated compounds related to DIMBOA. Spectrum from peak 4 does not display the deprotonated molecular species. The molecular mass (387 Da) is deduced from the adducts. The sample is obtained from extracts of maize plants. Reproduced (modified) from Cambier V., Hance T., and de Hoffmann E., Phytochem. Anal., 10, 119–126, 1999.

UV lamps generally employed provide photons at higher energy than the ionization potentials of the analytes but lower than those of the atmospheric gas and of the used solvents. This allows ions from the analytes to be selectively produced without ionizing the solvent, thus considerably reducing the background noise. As suggested by the data in Figure 1.35, the best lamp is the krypton discharge lamp emitting photons

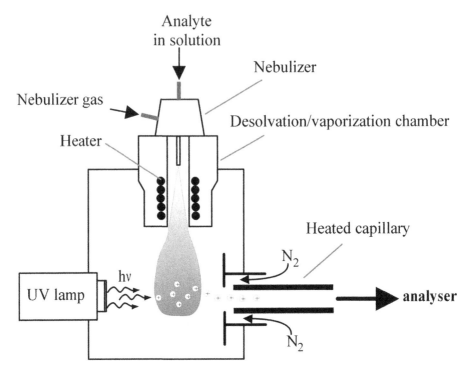

Figure 1.34
Scheme of an APPI source. The sample in solution is introduced perpendicular to the axis
of the analyser. The lamp is located in front of the entrance hole of the analyser. This
source has been designed initially for direct APPI (see text).

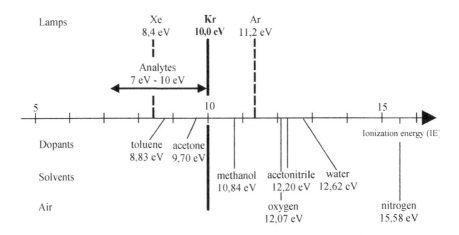

Figure 1.35
Ionization energy of molecules frequently present in APPI sources (solvents, doping
compounds, air components).

at 10.0 and 10.6 eV. Indeed, most analytes have ionization energies between 7 and 10 eV. On the other hand, air components (nitrogen and oxygen) and most of the common solvents (methanol, water, acetonitrile, etc.) have higher ionization potentials.

However, the direct ionization of the analyte is generally characterized by a weak efficiency. This can be partially explained by the solvent property to absorb photons producing photoexcitation without ionization. This reduces the number of photons available for the direct ionization of the sample, thus reducing the ionization efficiency. Consequently, ionization using doping molecules has also been described. It has indeed been shown that dopant at relatively high concentrations in comparison with the sample allows generally an increase in the efficiency of ionization from 10 to 100 times. This indicates that the process is initiated by the photoionization of the dopant. The dopant must be photoionizable and able to act as intermediates to ionize the sample molecules. The most commonly used dopants are toluene and acetone. Thus, two distinct APPI sources have been described: direct APPI and dopant APPI.

The mass spectra obtained by the APPI source in the positive ion mode are characterized by the presence of two main types of ions of the molecular species that may coexist [82]: the radical cation $M^{\bullet+}$ and the protonated molecule $[M+H]^+$. In direct APPI, the reaction is the classical photoionization leading to the radical cation of the molecular species:

$$M + h\nu \longrightarrow M^{\bullet+} + e^-$$

However, the often dominant presence of the protonated molecule suggests gas-phase ion–molecule reactions after the photoionization. The most likely reaction is the abstraction by the molecular ion of an hydrogen atom from a solvent molecule [83]:

$$M^{\bullet+} + S \longrightarrow [M+H]^+ + (S-H)^\bullet$$

In the presence of a dopant in the positive ion mode, the first step of the ionization process is the production of a radical ion from the dopant molecule by direct photoionization:

$$D + h\nu \longrightarrow D^{\bullet+} + e^-$$

This radical cation may then ionize a solvent molecule by proton transfer, if the proton affinity of the solvent molecule is higher than that of the deprotonated radical cation. It seems that the solvent acts as aggregates, having then a higher proton affinity. These protonated solvent molecules may then ionize analyte molecules by proton transfer if these last have a higher proton affinity than the solvent molecules:

$$D^{\bullet+} + S \longrightarrow [S+H]^+ + (D-H)^\bullet$$
$$M + [S+H]^+ \longrightarrow [M+H]^+ + S$$

Alternatively, if the ionization energy of the analyte is lower than that of the dopant, the radical cation of the dopant can directly interact with an analyte molecule and ionization by charge exchange of the analyte can occur to produce the molecular radical cation of the analyte:

$$D^{\bullet+} + M \longrightarrow M^{\bullet+} + D$$

Thus, besides the direct photoionization, the analytes in positive APPI mode are ionized either by charge exchange or by proton transfer. The direct ionization and the charge exchange processes allow the ionization of non-polar compounds. This is not possible either with APCI or ESI.

The formation of either the radical cation $M^{\bullet+}$ or the protonated $[M+H]^+$ molecule, or both together, will depend on the relative ionization energies or proton affinities of the sample molecules and the solvent components. Concerning the solvent, the charge exchange is favoured for solvents with low proton affinity (water, chloroform, cyclohexane, etc.), while solvents with higher proton affinities (methanol, acetonitrile, etc.) will favour proton transfer.

APPI is also efficient in negative ion mode, in so far as a dopant is used [84]. Indeed, in this ionization mode, analysis made without dopant dramatically decreases the sensitivity. All the reactions leading to the ionization of the analytes are initiated by thermal electrons produced with the photoionization of the dopant. Hence, solvents of high positive electron affinity, for example halogenated solvents, inhibit the ionization of all the analytes because these solvents capture all the available thermal electrons in the source.

Production of the negative molecular ion $M^{\bullet-}$ by electron capture is possible for any analyte presenting a positive electron affinity:

$$M + e^- \longrightarrow M^{\bullet-}$$

This molecular radical anion can also be produced by charge exchange reaction with $O_2^{\bullet-}$, itself produced in the source by electron capture of an oxygen molecule from the atmosphere. In this case, the electron affinity of the analyte must be higher than 0.45 eV, this value corresponding to the electron affinity of the O_2 molecule:

$$O_2 + e^- \longrightarrow O_2^{\bullet-}$$

$$M + O_2^{\bullet-} \longrightarrow M^{\bullet-} + O_2$$

Formation of the deprotonated molecular ion is also possible for analytes with high gas-phase acidity. This is due to the presence in the source of basic species often produced from the solvent. Solvent evaporated in the source acts as the ionization gas in chemical ionization, yielding species that can react with the analyte by proton transfer. This reaction is, however, only possible if the acidity of the analyte is higher than that of the solvent:

$$M + [S - H]^- \longrightarrow [M - H]^- + S$$

Besides the solvent, other species may participate in the proton abstraction from the analyte molecules. For instance, $O_2^{\bullet-}$ has a strong basicity in the gas phase and may react with other molecules from the solvent or the analyte by proton transfer. These molecules must have acidity lower than 1451 kJ mol^{-1}, this value corresponding to the gas-phase acidity of the $HO_2^{\bullet}$ species:

$$S + O_2^{\bullet-} \longrightarrow [S - H]^- + HO_2^{\bullet}$$

$$M + O_2^{\bullet-} \longrightarrow [M - H]^- + HO_2^{\bullet}$$

Ionization of analytes by proton abstraction is suppressed if the solvent molecules have higher gas-phase acidities than the analytes, because then transfer of the proton occurs in the opposite way, from the solvent to the analyte. In the same way, the presence of compounds of high gas-phase acidity from the solvent or other additives will suppress the ionization of the analyte, even if they have a positive electronic affinity. This is probably due to the neutralization by proton transfer of the $O_2^{\bullet-}$ ions, leading to the inhibition of all charge exchange reactions.

Thus, the negative APPI mode will produce molecular ions from the analytes by charge exchange or by electron capture if they have a sufficient electron affinity. Analytes that have high gas-phase acidity will be mainly ionized by proton transfer to yield deprotonated molecular ions.

Compared with APCI, APPI is more sensitive to the experimental conditions. Properties of solvents, additives, dopants or buffer components can strongly influence the selectivity or sensitivity of the detection of analytes. Nevertheless, this technique allows the ionization of compounds not detectable in APCI or ESI, mainly non-polar compounds. For these last compounds, APPI is a valuable alternative. Thus, APPI is a complementary technique to APCI and ESI. However, for a given substance it remains difficult to predict which ionization source (APPI, APCI or ESI) will give the best results. Only preliminary tests will allow the choice of the best ionization source. APPI appears to be efficient for some compound classes such as flavonoids, steroids, drugs and their metabolites, pesticides, polyaromatic hydrocarbons, etc. [85].

1.14 Atmospheric Pressure Secondary Ion Mass Spectrometry

Some ionization methods have been developed in the last few years that allow a sample at atmospheric pressure and at ground potential to be directly exposed to an ionizing beam. Some of them allow the exposure of living material, for example a finger.

1.14.1 Desorption Electrospray Ionization

A new ionization method called desorption electrospray ionization (DESI) was described by Cooks and his co-workers in 2004 [86]. This direct probe exposure method based on ESI can be used on samples under ambient conditions with no preparation. The principle is illustrated in Figure 1.36. An ionized stream of solvent that is produced by an ESI source is sprayed on the surface of the analysed sample. The exact mechanism is not yet established, but it seems that the charged droplets and ions of solvent desorb and extract some sample material and bounce to the inlet capillary of an atmospheric pressure interface of a mass spectrometer. The fact is that samples of peptides or proteins produce multiply charged ions, strongly suggesting dissolution of the analyte in the charged droplet. Furthermore, the solution that is sprayed can be selected to optimize the signal or selectively to ionize particular compounds.

The sample may be exposed as such or deposited on a sample holder that may be made of any material, conducting or non-conducting, provided it does not produce background noise. Furthermore, the sample can be freely moved or manipulated during the experiment. An interesting feature is that reactants can be introduced in the spray solution to react with the sample. Another interesting feature of DESI is the ability to map the position of analytes of the native surfaces, such as plant or animal tissues. It has also been demonstrated that DESI can be used with thin-layer chromatography [87].

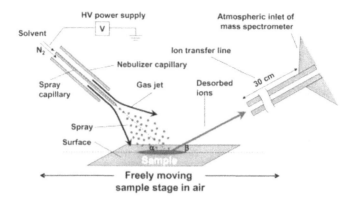

Figure 1.36
A pneumatically assisted ion spray source is oriented at 45° towards a sample. The nitrogen spray is adjusted so as to have a linear speed of the droplets of about 35 m s^{-1}. From Takats Z., Wiseman J.M., Gologan B. and Cooks R.G., Science, 306, 471–473, 2004. Reprinted, with permission.

A broad range of compounds, including small non-polar molecules or large polar molecules like peptides and proteins, have been analysed successfully by DESI. This method can also detect drug molecules on the surface of the skin. As an example, Figure 1.37 shows the spectrum obtained by exposing the finger of a person after intake of 10 mg of the antihistaminic Loratadine. This drug is characterized by a protonated molecular ion at m/z 383 with the isotopic distribution at m/z 383/385 due to one chlorine atom. Other applications include explosives on tanned porcine leather, sections of stems or seeds from vegetals, and so on. The resulting mass spectra are similar to normal ESI mass spectra. They show mainly singly or multiply charged molecular ions from small or large analytes, respectively.

1.14.2 Direct Analysis in Real Time

The direct analysis in real time (DART) method has been described by Cody *et al.* [88] and commercialized by JEOL. This method allows direct detection of chemicals on surfaces, in liquids and in gases without the need for sample preparation. All of these analyses take place under ambient conditions in a space just in front of the inlet of the mass spectrometer. The sample is not altered because no exposure to high voltage or to vacuum is required.

In the source, a gas such as helium or nitrogen is introduced and submitted to a beam of electrons by applying a high-voltage potential between two electrodes as shown in Figure 1.38. Ions, electrons and neutral species with electronic or vibronic excitation are produced. This resulting plasma passes through a series of electrodes designed to remove any charged species, leaving only neutral species that then interact with the sample and the atmosphere. It seems that mainly these excited neutral species produce the ionization of the sample molecules.

Several mechanisms involved in ion formation are possible, depending on the analysed molecule and the operating conditions like the polarity and the gas used. In positive ion

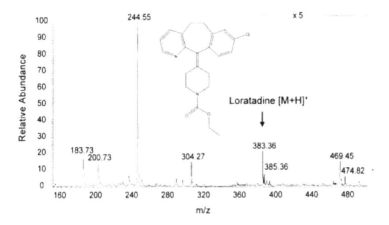

Figure 1.37
Spectrum obtained by DESI exposure of the finger of a person after intake of 10 mg Loratadine. From Takats Z., Wiseman J.M., Gologan B. and Cooks R.G., Science, 306, 471–473, 2004. Reprinted, with permission.

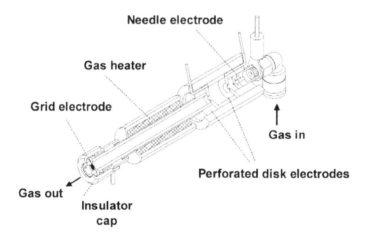

Figure 1.38
Cutaway view of a DART source. Discharge is produced in the first chamber and the gas then flows into a second chamber where the ions can be discharged. The gas flows through a tube that can optionally be heated and then flows out to the sample through a grid that allows removal of ions of opposite polarity. From Cody R.B., Laramee J.A. and Dupont Durst H., Anal. Chem., 77, 2297–2302, 2005. Reprinted, with permission.

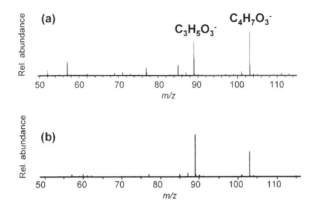

Figure 1.39
Detection of γ-hydroxybutyrate (a) in gin at 10 ppm
and (b) as sodium salt on the rim of a drinking glass.
From Cody R.B., Laramee J.A. and Dupont Durst H.,
Anal. Chem., 77, 2297–2302, 2005. Reprinted, with
permission.

mode, the simplest process is Penning ionization involving transfer of energy from the
excited gas to an analyte, leading to the radical cation of the molecular species:

$$G^* + M \longrightarrow M^{\bullet +} + G + e^-$$

Another ionization process that is the main process when helium is used as the gas is proton
transfer. This type of ionization occurs when metastable helium atoms react with water in
the atmosphere to produce ionized water clusters that can protonate the sample molecule,
leading to the protonated molecule.

Negative ions can be formed by electron capture due to the presence of thermalized
electrons produced by Penning ionization or by surface Penning ionization. Negative ions
can also be obtained by reactions of analyte molecules with negative ions formed from
atmospheric water and oxygen to produce the deprotonated molecule.

The mechanism involved in desorption of materials from surfaces by this method is
not well understood. One of these mechanisms is the thermal desorption because heating
the gas helps desorption of some analytes. However, the successful analysis by DART
of analytes having little or no vapour pressure indicates that other processes occur. The
transfer of energy to the surface by metastable atoms and molecules has been proposed as
the mechanism to facilitate desorption and ionization of these analytes.

Hundreds of compounds on different surfaces have been tested, many examples concerning mainly explosives on different surfaces, as well as the analysis of a urine sample
containing the drug ranitidine and the analysis of capsaicin in different parts of a pepper
pod. Detection of γ-hydroxybutirate, an illegal drug classified as a sedative–hypnotic, is
shown in Figure 1.39.

DART produces relatively simple mass spectra characterized by the presence of two main
types of ions of the molecular species: $M^{\bullet +}$ or $[M + H]^+$ in positive ion mode and $M^{\bullet -}$
or $[M - H]^-$ in negative ion mode. Fragmentation is observed for most of the compounds.
These results appear similar to those obtained with DESI, but no multiply charged ions are

produced. Consequently, the range of analytes that can be analysed by DART is less broad than by DESI. Furthermore, DART cannot be used for the spatial analysis of surfaces.

1.15 Inorganic Ionization Sources

Mass spectrometry is not only an indispensable tool in organic and biochemical analysis, but also a powerful technique for inorganic analysis [89–91]. Indeed, over the last 20 years the application of mass spectrometry to inorganic and organometallic compounds has revolutionized the analysis of these compounds. Important advances have been made in the diversification of ionization sources, in the commercial availability of the instruments and in the fields of applications.

Mass spectrometry is now widely used for inorganic characterization and microsurface analysis. EI is the preferred ionization source for volatile inorganic compounds, whereas the others which are non-volatile may be analysed using ionization sources already described such as SIMS, FD, FAB, LD or ESI [92–94]. The example in Figure 1.40 shows the analysis of orthorhombic sulfur (S_8 ring) and $Cr(CO)_2(dpe)_2$, respectively obtained with an EI source and an ESI source.

In addition, quantitative and qualitative elemental analysis of inorganic compounds with high accuracy and high sensitivity can be effected by mass spectrometry. For elemental analysis, atomization of the analysed sample that corresponds to the transformation of solid matter in atomic vapour and ionization of these atoms occur in the source. These atoms are then sorted and counted with the help of mass spectrometry. The complete decomposition of the sample in the ionization source into its constituent atoms is necessary because incomplete decomposition results in complex mass spectra in which isobaric overlap might cause unsuspected spectral interferences. Furthermore, the distribution of any element in different species leads to a decrease in sensitivity for this element.

Four techniques based on mass spectrometry are widely used for multi-elemental trace analysis of inorganic compounds in a wide range of sample types. These techniques are thermal ionization (TI), spark source (SS), glow discharge (GD) and inductively coupled plasma (ICP) mass spectrometry. In these techniques, atomization and ionization of the analysed sample are accomplished by volatilization from a heated surface, attack by electrical discharge, rare-gas ion sputtering and vaporization in a hot flame produced by inductive coupling.

All of these ionization sources are classical sources used also in optical spectroscopy. The only fundamental difference is that these sources are not used for atomization/excitation processes to generate photons but to generate ions.

1.15.1 Thermal Ionization Source

Thermal ionization is based on the production of atomic or molecular ions at the hot surface of a metal filament [95, 96]. In this ionization source, the sample is deposited on a metal filament (W, Pt or Re) and an electric current is used to heat the metal to a high temperature. The ions are formed by electron transfer from the atom to the filament for positive species or from the filament to the atom for negative species. The analysed sample can be fixed to the filament by depositing drops of the sample solution on the filament surface followed by evaporation of the solvent to complete dryness, or by using electrodeposition methods.

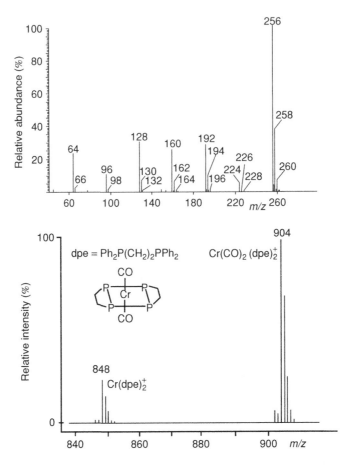

Figure 1.40
Analysis of inorganic compounds by mass spectrometry. Top: EI
spectrum of orthorhombic sulfur (S_8). Bottom: ESI spectrum of
$Cr(CO)_2(dpe)_2$. The last spectrum is redrawn from data taken
from Traeger J.C. and Colton R., Adv. Mass Spectrom., 14, 637–
659, 1998.

Single, double and triple filaments have been broadly used in thermal ionization sources.
In a single filament source, the evaporation and ionization process of the sample are
carried out on the same filament surface. Using a double filament source, the sample is
placed on one filament used for the evaporation while the second filament is left free
for ionization. In this way, it is possible to set the sample evaporation rate and ionization
temperature independently, thus separating the evaporation from the ionization process. This
is interesting when the vapour pressure of the studied elements reaches high values before
a suitable ionization temperature can be achieved. A triple filament source can be useful to
obtain a direct comparison of two different samples under the same source conditions.

Positive and negative ions can be obtained by the thermal ionization source. High yields
of positive and negative ions are obtained for atoms or molecules with low ionization
potential and with high electron affinity, respectively. Owing to the ionization process and

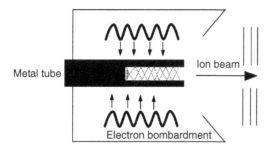

Figure 1.41
Schematic representation of a thermal ionization
cavity source.

their physical parameters, metals can be analysed in positive ion mode, whereas many non-metals, semimetals and transition metals or their oxides are able to form negative thermal ions.

In thermal ionization sources, the most abundant ions are usually the singly charged atomic ions. No multiply charged ions can be observed under normal ionization conditions. Cluster ions occur very seldom. However, some metal compounds lead to abundant metal oxide ions.

For thermal ionization filament sources, the ionization efficiency varies from less than 1 % to more than 10 %, depending on the analysed elements. A significant improvement of ionization efficiency can be observed with the use of a thermal ionization cavity (TIC) source [97]. In this type of thermal source, as schematically displayed in Figure 1.41, a refractory metal tube is used instead of a filament to evaporate and ionize the sample. High ionization temperatures are achieved by using high-energy electron bombardment to heat the tube. As the sample evaporates, the gaseous atoms interact with each other and with the inner wall of the cavity to produce ions. Compared with filament sources, the cavity sources can provide orders-of-magnitude enhancement of the ionization efficiency.

1.15.2 Spark Source

In spark sources, electrical discharges are used to desorb and ionize the analytes from solid samples [98]. As shown in Figure 1.42, this source consists of a vacuum chamber in which two electrodes are mounted. A pulsed 1 MHz radio-frequency (RF) voltage of several kilovolts is applied in short pulses across a small gap between these two electrodes and produces electrical discharges. If the sample is a metal it can serve as one of the two electrodes, otherwise it can be mixed with graphite and placed in a cup-shaped electrode.

Atomization is accomplished by direct heating of the electrode by the electron component of the discharge current and by the discharge plasma. Then, ionization of these atoms which occurs in the plasma is due mainly to plasma heating by electrons accelerated by the electric field. Chemical reactions can also take place in the plasma, leading to the formation of clusters.

Various types of positive ions are produced in a spark discharge such as singly and multiply charged atomic ions, polymer ions and heterogeneous compound ions. A spark source mass spectrum is always characterized by singly and multiply charged ions of the

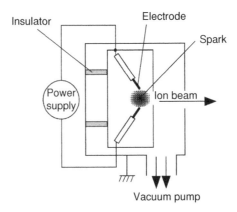

Figure 1.42
Typical spark ion source.

major constituents, with a decrease of their intensities when their charges increase. Also, the singly charged species are always the most intense for minor constituents and are usually the only ones used for analytical purposes. Another abundant class of ions detected in a spark source is heterogeneous compounds ions formed by the association of the matrix with hydrogen, carbon, nitrogen and oxygen.

It must be noted that this technique does not provide accurate quantitative analysis.

1.15.3 Glow Discharge Source

A glow discharge (GD) source is particularly effective at sputtering and ionizing compounds from solid surfaces [99–101]. This source is indicated schematically in Figure 1.43. A glow discharge source consists of one cathode and one anode in a low-pressure gas (0.1–10 Torr), usually one of the noble gases. Argon is the most commonly used gas because of its low cost and its high sputtering efficiency. The sample is introduced in this source as the cathode. Application of an electric current across the electrodes causes breakdown of the gas and the acceleration of electrons and positive ions towards the oppositely charged electrodes. Argon ions from the resulting plasma attack by bombardment the analyte at the surface of the cathode. Collisions of these energetic particles on the surface transfer their kinetic energy. The species near the surface can receive sufficient energy to overcome the lattice binding and be ejected mainly as neutral atoms. The sample atoms liberated are then carried into the negative glow region of the discharge where they are ionized mainly by electron impact and Penning ionization (Figure 1.43).

A simple DC power supply at voltages of 500–1000 V and currents of 1–5 mA suffices to obtain glow discharge. However, pulsed DC discharges allow use of higher peak voltages and currents, whereas an RF discharge allows the analysis of poorly conducting samples directly.

The principal application of glow discharge is in bulk metal analysis. The conducting solid samples can be made into an electrode whereas non-conducting materials are compacted with graphite into the electrode for analysis. Solution samples can also be analysed by drying the sample on a graphite electrode. Other applications such as the examination of thin films or solution residues are also possible. Glow discharge ion sources can also be

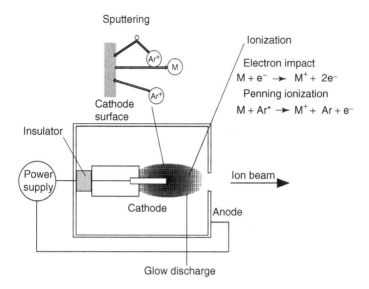

Figure 1.43
Schematic diagram of a glow discharge source with sputtering at the cathode surface and ionization in the discharge.

used in conjunction with laser ablation. The pulse of ablated material enters directly into the plasma of the glow discharge where it undergoes ionization.

Owing to the ionization process, only positive ions can be obtained by the glow discharge ionization source. Consequently, the spectra obtained with the glow discharge source are characterized by singly charged positive ions of the sputtered cathode atoms. Almost no doubly charged ions from the sample are observed. Some diatomic cluster ions are formed but are normally observed only for major constituents. The noble gas used (usually Ar) and residual gases such as nitrogen, oxygen and water vapor are always observed.

Only a low net ionization in the discharge is produced. The ionization efficiency is estimated to be 1 % or less. However, glow discharge produces an atomic vapour representative of the cathode constituents and the discharge ionization processes are also relatively non-selective. Also, because most elements are sputtered and ionized with almost the same efficiency in the source, quantitative analysis without a standard is possible.

It must be noted that glow discharge requires several minutes for the extracted ions to reach equilibrium with elemental concentrations in the analysed sample. Thus, the sample throughput with this technique is relatively low.

1.15.4 Inductively Coupled Plasma Source

An inductively coupled plasma source is made up of a hot flame produced by inductive coupling in which a solution of the sample is introduced as a spray [102–104]. This source consists of three concentric quartz tubes through which streams of argon flow. As shown in Figure 1.44, a cooled induction coil surrounds the top of the largest tube. This coil is powered by an RF generator that produces between 1.5 and 2.5 kW at 27 or 40 MHz typically. The gas at atmospheric pressure that sustains the plasma is initially made

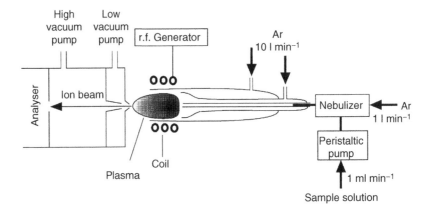

Figure 1.44
Schematic diagram of an inductively coupled plasma source.

electrically conductive by Tesla sparks which lead to the ionization of the flowing argon. Then, the resulting ions and the produced electrons which are present in the discharge interact with the high-frequency oscillating inductive field created by the RF current in the coil. They are consequently accelerated, collide with argon atoms and ionize them. The released products by this ionization then undergo the same events until the argon ionization process is balanced by the opposing process corresponding to ion–electron recombination. These colliding species cause heating of the plasma to a temperature of about 10 000 K. This temperature requires thermal isolation from the outer quartz tubes by introducing a high-velocity flow of argon of about $10 \, \text{l min}^{-1}$ tangentially along the walls of these tubes.

The sample is carried into the hot plasma as a thermally generated vapour or a finely divided aerosol of droplets or microparticulates by argon flowing at about $1 \, \text{l min}^{-1}$ through the central tube. The high temperatures rapidly desolvate, vaporize and largely atomize the sample. Furthermore, this plasma at high temperature and at atmospheric pressure is a very efficient excitation source. The resulting atoms may spend several milliseconds in a region at a temperature between 5000 and 10 000 K. Under these conditions, most elements are ionized to singly charged positive ions with an ionization efficiency close to 100 %. Some elements can be ionized to higher charged states but with very low abundance. Ions are extracted from the plasma and introduced in the mass analyser through a two-stage vacuum-pumped interface containing two cooled metal skimmers.

The most common introduction of the samples in this source consists of a pneumatic nebulizer which is driven by the same flow of argon which carries the resulting droplets in the plasma. An ultrasonic nebulizer and heated desolvation tube are also used because they allow a better droplet size distribution which increases the load of sample into the plasma. Generally, the sample solutions are continuously introduced in the nebulizer at the rate of about $1 \, \text{ml min}^{-1}$ with the help of a peristaltic pump. However, this is not acceptable with small-sample solutions. Therefore an alternative method using the flow injection technique is employed to introduce a small sample of about $100 \, \mu\text{l}$. The sample solution is injected into a reference blank flow so that the sample is transported in the nebulizer and a transitory signal is observed.

Other alternative methods of sample introduction have been applied. A very small volume of liquid ($<10 \, \mu\text{l}$) or solid sample may be introduced in the plasma as a vapour produced

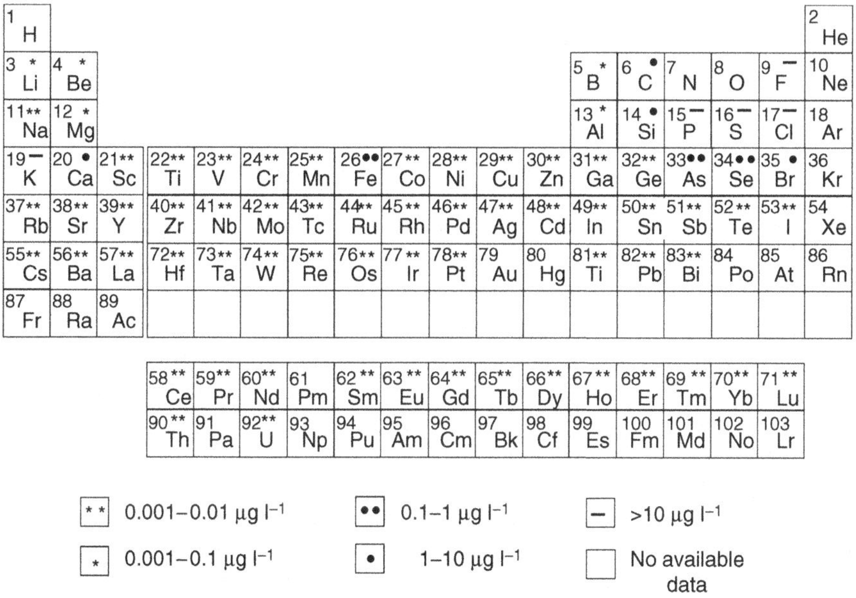

1 H																	2 He
3 * Li	4 * Be											5 * B	6 • C	7 N	8 O	9 — F	10 Ne
11** Na	12 * Mg											13 * Al	14 • Si	15 — P	16 — S	17 — Cl	18 Ar
19 — K	20 • Ca	21** Sc	22** Ti	23** V	24** Cr	25** Mn	26•• Fe	27** Co	28** Ni	29** Cu	30** Zn	31** Ga	32** Ge	33•• As	34•• Se	35 • Br	36 Kr
37** Rb	38** Sr	39** Y	40** Zr	41** Nb	42** Mo	43** Tc	44** Ru	45** Rh	46** Pd	47** Ag	48** Cd	49** In	50** Sn	51** Sb	52** Te	53** I	54 Xe
55** Cs	56** Ba	57** La	72** Hf	73** Ta	74** W	75** Re	76** Os	77** Ir	78** Pt	79 Au	80 Hg	81** Ti	82** Pb	83** Bi	84 Po	85 At	86 Rn
87 Fr	88 Ra	89 Ac															

58** Ce	59** Pr	60** Nd	61 Pm	62 ** Sm	63 ** Eu	64** Gd	65** Tb	66** Dy	67 ** Ho	68** Er	69 ** Tm	70** Yb	71 ** Lu
90 ** Th	91 Pa	92** U	93 Np	94 Pu	95 Am	96 Cm	97 Bk	98 Cf	99 Es	100 Fm	101 Md	102 No	103 Lr

** 0.001–0.01 µg l⁻¹ •• 0.1–1 µg l⁻¹ — >10 µg l⁻¹

* 0.001–0.1 µg l⁻¹ • 1–10 µg l⁻¹ (blank) No available data

Figure 1.45
Analytical sensitivity of the elements by ICP-MS.

by electrothermal vaporization (ETV), which vaporizes the sample by flash evaporation from a filament heated by a current pulse. In this case, the sample vapour is produced in a short time of a few seconds and is transferred to the plasma as a gas. Solid samples at atmospheric pressure can also be introduced by laser ablation. This technique also allows spatial analysis of the surface of a sample.

It must be noted that handling samples in solution allows automation and high sample throughput. Indeed, the sample throughput for an ICP-MS instrument is typically 20–30 elemental determinations in a few minutes, depending on such factors as the concentration levels and precision required. Handling solution samples includes other advantages such as the record and the subtraction of a true blank spectrum, the adjustment of the dynamic range by dilution, the simplicity of adding internal standards, etc.

1.15.5 Practical Considerations

When compared with optical spectrometric techniques of elemental analysis, the techniques based on mass spectrometry provide an increase in sensitivity and in analytical working range of some orders of magnitude. For instance, the detection limits with ICP-MS are three orders of magnitude better than ICP-optical emission spectrometry (ICP-OES). Figure 1.45 shows the maximum sensitivity obtained for the different elements, using an ICP-MS coupling with a quadrupole.

These techniques are relatively interference free but there are, nevertheless, two major types of interferences. A first type is the matrix interferences which induce suppression or enhancement of the analyte signal. Such interferences are due to identity and composition of the sample itself. As they cause differences between samples and standards for a particular element concentration, they lead to quantitative errors. Matrix effects are generally more serious with these techniques than with optical spectrometric techniques.

However, the most common interferences are the spectral interferences, also called isobaric interferences. They are due to overlapping peaks which can mask the analyte of interest and can give erroneous results. Such interferences may occur from ions of other elements within the sample matrix, elemental combination, oxide formation, doubly charged ions, and so on.

A solution to the overlapping peaks problem consists of identifying the interfering species and applying corrections for their contribution to the signal from the analyte. Corrections are based on identifying an isotope of the suspected interfering element that does not itself suffer from spectral interference and that can be measured with a sufficiently accurate signal. Then, knowing the relative natural abundance of the isotopes, the contribution of this interfering element to the signal at the mass of interest can be calculated. This correction is not appropriate when the interference is many times more abundant than the analyte because the error on the measurement can be too large.

Another very straightforward solution to the overlapping peaks problem is to increase the resolution power of the mass spectrometer. High-resolution mass spectrometers have the ability to resolve many isobaric interferences from the analyte and thus allow unambiguous quantitative analysis to be carried out. However, it should be noted that the sensitivity decreases when the resolution increases.

1.16 Gas-Phase Ion–Molecule Reactions

While EI does not imply any ion–molecule reactions, the latter provide the whole basis of CI and of all the API methods. Other ionization methods give rise to ion–molecule reactions as secondary processes. We will now emphasize some characteristics of these gas-phase reactions and compare them with condensed-phase reactions.

Figure 1.46 displays the energy characteristics of both the gas-phase and aqueous solution phase for the substitution reaction [105]:

$$Cl^- + CH_3Br \longrightarrow CH_3Cl + Br^-$$

In the condensed phase, molecules and *a fortiori* ions are strongly solvated. This solvation cage must be at least partially destroyed in order to form the activated complex. This requires a lot of energy and high activation energies are thus observed. Furthermore, each species mingles with the surrounding molecules and continuously exchanges energy.

In the gas phase the opposite occurs: the naked ion interaction with the molecule is exothermic, which leads to the formation of an ion–molecule complex. Some activation energy thus becomes necessary in order to transform the $Cl^- \cdot CH_3Br$ complex into the activated complex and finally into $Br^- \cdot CH_3Cl$. As can be seen in Figure 1.46, this activation energy is generally lower than the energy produced by the first step. Thus, if compared with the starting reactants, the actual activation energy is negative. Another very important difference must be noted: the mass spectrometry literature refers to the gas phase but this association with conventional gases is incorrect. Indeed, the reacting species generally considered in mass spectrometry do not interact with each other, owing to the low pressure, and thus do not continuously exchange energy with surrounding molecules. They are never present as an equilibrium phase. Hence the energy produced by the association of the ion

Table 1.4 Rate constants for the reaction $CH_3Br + OH^- \rightleftarrows CH_3OH + Br^-$ for naked and solvated ions in the gas phase and in aqueous solution.

Hydroxide ion	Rate constant ($cm^3 \, mol^{-1} \, s^{-1}$)
OH^-	$(1.0 \pm 0.2) \times 10^{-9}$
$OH^- \cdot H_2O$	$(6.3 \pm 2.5) \times 10^{-10}$
$OH^- \cdot (H_2O)_2$	$(2 \pm 1) \times 10^{-12}$
$OH^- \cdot (H_2O)_3$	$<2 \times 10^{-13}$
Aq. sol. $OH^- \cdot (H_2O)_x$	2.3×10^{-25}

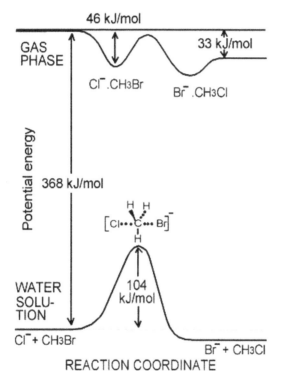

Figure 1.46
Potential energy diagram for a substitution reaction in the gas phase and in solution in water. Reproduced (modified), with permission from McIver R.T., Sci. Am., 243, 148, 1980.

with the molecule remains in the complex, thereby allowing it to overcome the activation barrier. Deactivation through radiation is a relatively slow process. This energy further allows the separation of the products into free the Br^- ion and CH_3Cl molecules.

Figure 1.47 displays the results obtained by measuring the kinetic and thermodynamic parameters for the following reaction [106]:

$$OH^- + CH_3Br \rightleftarrows Br^- + CH_3OH$$

Table 1.5   Definition of most important gas-phase thermodynamic data.

Gas-phase basicity	Gas-phase acidity
$M + H^+ \rightleftarrows MH^+$	$AH \rightleftarrows A^- + H^+$
$PA = -\Delta H^\circ \, GB = -\Delta G^\circ$	ΔH°_{ACID} and ΔG°_{ACID}
Exothermic	Endothermic
Ionization energy (IE)	**Electron affinity (EA)**
$M \rightleftarrows M^{\bullet+} + e^-$	$M + e^- \rightleftarrows M^{\bullet-}$
$IE = \Delta H^\circ$	$EA = -\Delta H^\circ$
Endothermic	Exothermic or weakly endothermic

Note. The same values are available for radicals instead of molecules.

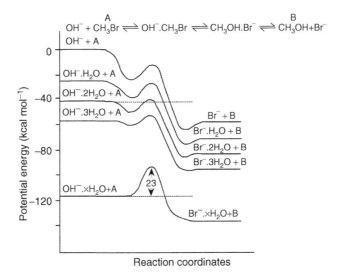

Figure 1.47
(A) CH_3Br; (B) CH_3OH. Potential energy profile for differently solvated ions reacting with molecules. Reproduced (modified), with permission from Bohme D.K. and Mackay G.I., J. Am. Chem. Soc., 103, 978, 1981.

In this case, the authors succeeded in measuring separately the reaction characteristics both for the naked OH^- ion and for the ion solvated by one to three water molecules. As the number of water molecules increases, the shape of the curves approaches that for the reaction in aqueous solution. The observed rate constants are given in Table 1.4. The rate constant for the gas-phase ion–molecule reaction is 10^{16} times greater than the rate observed in solution. This experimentally observed factor goes up to 10^{20} in the case of other reactions.

Let us now address another problem: are all the ion–molecule reactions possible?

The aim of this section is to give an overview of the factors determining the formation of the various types of ions encountered in the different ionization modes. Exceptions and examples are given in the sections dedicated to the ionization methods and to the fragmentations.

Table 1.6 Gas-phase basicities (kJ mol^{-1}) showing the influence of the protonated atoms and of the substituents. Note that the phenyl group is an electron donor in the gas phase.

kJ mol^{-1}	GB		GB		GB
NH_3	819	H_2O	660	H_2S	674
CH_3NH_2	864	CH_3OH	724	CH_3SH	742
$(C_2H_5)_2NH$	878	$(C_2H_5)_2O$	801	$(CH_3)_2S$	801
$PhNH_2$	851	$PhOH$	786	CH_3CHO	736

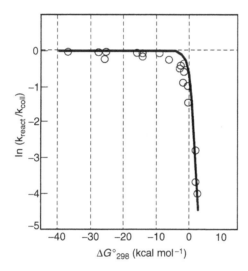

Figure 1.48
The natural logarithm of the number of reactions per collision ratio indicates that the proton transfer is almost 100 % efficient when the process is exergonic. When it becomes endergonic, the efficiency drops sharply. Reproduced (modified), with permission from Bohme D., Mackay G.I. and Schiff H.I., J. Chem. Phys., 73, 4976, 1980.

This section will use gas-phase thermochemical data from Appendices 6 for molecules and 7 for radicals. These data include ionization energy (IE), electron affinity (EA), proton affinity (PA), gas-phase basicity (GB) and gas-phase acidity. Definitions of these parameters are given in Table 1.5. Some values of gas-phase basicities are given in Table 1.6.

Figure 1.48 displays experimental results of proton transfer under chemical ionization conditions. Exergonic reactions, that is $G° < 0$, are highly efficient, almost every collision giving rise to a proton transfer. However, the efficiency decreases sharply when the process becomes endergonic [107].

$G°$ values for some acids and bases can be found in the appendices to this book. Extensive values can be found in reference [108] or by internet at www.webbook.nist.gov.

As an example, let us examine whether or not the proton transfer from protonated ammonia to neutral aniline is efficient:

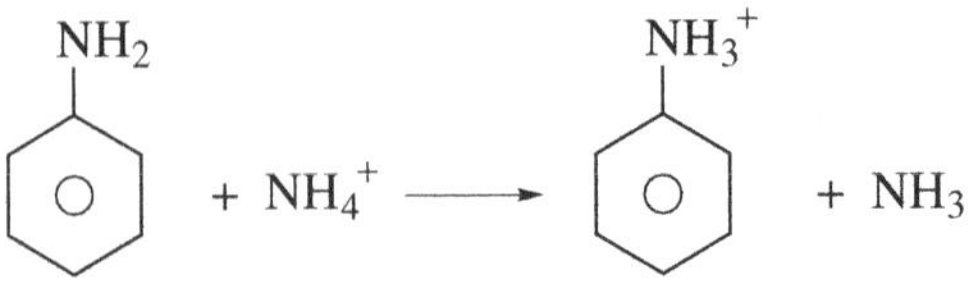

$$C_6H_5\text{-}NH_2 + NH_4^+ \longrightarrow C_6H_5\text{-}NH_3^+ + NH_3$$

From the individual $\Delta G°$ values, equal to $-GB$, the $\Delta G°$ value for this reaction is calculated:

$$NH_4^+ \longrightarrow NH_3 + H^+ \qquad \Delta G° = -(-819)\ \text{kJ mol}^{-1}$$

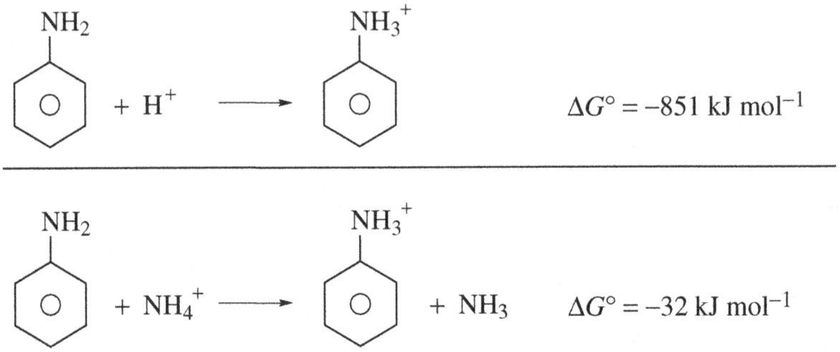

$$C_6H_5\text{-}NH_2 + H^+ \longrightarrow C_6H_5\text{-}NH_3^+ \qquad \Delta G° = -851\ \text{kJ mol}^{-1}$$

$$C_6H_5\text{-}NH_2 + NH_4^+ \longrightarrow C_6H_5\text{-}NH_3^+ + NH_3 \qquad \Delta G° = -32\ \text{kJ mol}^{-1}$$

In a standard source, the reaction being exergonic, the proton transfer from ammonium to aniline will be very efficient. Note that in the source we are dealing here with efficiency at each collision, not with equilibrium. Under the high-vacuum conditions, equilibrium is not established. This example was selected because it shows that, in the gas phase, aniline is actually a stronger base than ammonia. The importance of solvation is thus emphasized once again. On the other hand, the methylamine is more basic than aniline:

$$CH_3\text{-}NH_2 + H^+ \longrightarrow CH_3\text{-}NH_3^+ \qquad \Delta G° = -864\ \text{kJ mol}^{-1}$$

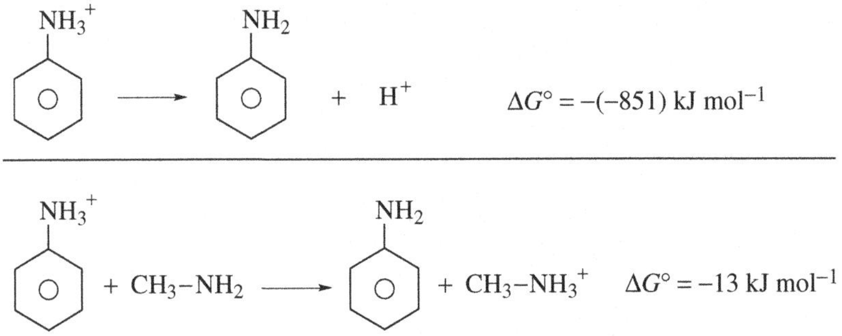

$$C_6H_5\text{-}NH_3^+ \longrightarrow C_6H_5\text{-}NH_2 + H^+ \qquad \Delta G° = -(-851)\ \text{kJ mol}^{-1}$$

$$C_6H_5\text{-}NH_3^+ + CH_3\text{-}NH_2 \longrightarrow C_6H_5\text{-}NH_2 + CH_3\text{-}NH_3^+ \qquad \Delta G° = -13\ \text{kJ mol}^{-1}$$

1.17 Formation and Fragmentation of Ions: Basic Rules

The aim of this section is to give an overview of the factors determining the formation of the various types of ions encountered in the different ionization modes. Exceptions and examples are given in the sections dedicated to the ionization methods and to the fragmentations.

1.17.1 Electron Ionization and Photoionization Under Vacuum

These reactions occur under high vacuum. Thus, no ion–molecule reaction occurs. The species formed during the ionization process is a radical cation. Ionization efficiency depends on the ionization energy of the molecule. The presence or not of the molecular ions also depends on how easy it fragments.

Fragmentation often produces both a radical and a cation. This can be represented by the following equation:

$$R-R'^{\cdot+} \begin{cases} R^{\cdot} + R'^{+} \\ R^{+} + R'^{\cdot} \end{cases}$$

The factor that determines which of the fragments is a radical or a cation can be emphasized as a competition between two cations to capture the electron:

$$R^{+} \cdots e^{-} \cdots R'^{+}$$

As the fragment with the higher propensity to retain the electron should have the higher ionization energy, the fragment observed in the spectrum as a cation is the one having the lowest ionization energy. The other one, having the highest ionization energy, takes the electron to be a radical. This is the origin of the Stevenson rule that will be explained in Chapter 7 on fragmentation.

1.17.2 Ionization at Low Pressure or at Atmospheric Pressure

The CI source operates at low pressure. Ion–molecule reactions occur and are needed for sample ionization. The MALDI source is under vacuum, but during the ionization process the pressure increases in the plume close to the target and ion–molecule reactions occur. The various sources operating at atmospheric pressure include ESI, APCI, APPI and AP-MALDI. All these sources operate at sufficient pressure to have numerous collisions between ions and molecules, and reactions between these species are observed.

It is worth noting that reactions between neutrals produced by fragmentations and ions are not observed. This is due to the fact that, whatever the ionization method, only a small fraction of the analyte molecules are ionized, and their fragments are at even lower concentrations. The probability of a collision is thus too low. Similarly, under normal conditions, no collision between ions is observed.

However, reactions may be observed between an ion and a neutral both resulting from the fragmentation of one precursor ion, immediately after cleavage, provided they remain associated for some time. This time is rarely more than a few microseconds. This can occur as well under vacuum as at higher pressure.

1.17.3 Proton Transfer

Proton transfer to produce a cation or an anion is the most often observed ion–molecule reaction in sources that allow collisions. The general rule is that the proton affinity of the proton acceptor (neutral or anion) has to be higher than the proton affinity of the donor (cation or neutral). If there is a difference in proton affinity such that the reaction is exergonic, the transfer occurs at each collision (see Figure 1.48).

The protonated molecule fragments, if necessary after activation. The fragments do not always result from the cleavage of only one bond, as this can lead to the formation of a radical fragment and a radical cation, a very unfavorable process. The pathway is thus often more complicated than for radical cations. It can be represented as follows:

$$M + H^+ \longrightarrow MH^+$$

$$MH^+ \left\langle \begin{array}{l} Fl + F2H^+ \\ FlH^+ + F2 \end{array} \right.$$

$$Fl \cdots H^+ \cdots F2$$

Here the competition is between two fragments for a proton. The fragment with the highest gas-phase basicity gets the proton. For negative ions a similar rule applies, but now it is the most acidic species that carries the negative charge. This is analogous to the competition seen before, about EI, between two ions for an electron.

A similar competition already exists at the ion formation stage in the source. For this reason, in the presence of a solvent having a certain proton affinity, it is not possible to see the protonated cation of an analyte having a lower proton affinity. It is, however, possible to observe an adduct with another cation, such as sodium, ammonium, and so on. The reverse is true for negative ions. Here too, adducts with anions as chloride, acetate, and so on may be observed.

Similarly, if two analytes in a mixture have a marked difference of acidity or basicity, only one is observed in the spectrum: the best proton acceptor in positive ion mode, or the best proton donor in negative ion mode. However, at low concentrations the competition is less obvious, and both ions can sometimes be observed together.

1.17.4 Adduct Formation

An adduct is an ion formed by direct combination of a neutral molecule and an 'ionizing' ion other than the proton. In positive ion mode the most often observed is the sodium adduct, producing an ion with 22 mass units higher than the protonated molecule, that is $(M + 23)^+$ instead of $(M + 1)^+$. It is often accompanied by a potassium adduct, another 16 u higher:

$$M + Na^+ \longrightarrow (M + Na)^+$$

Extended tables of gas-phase proton affinity exist. This is not true for the affinity towards metal ions. To attach a proton a basic site is needed. Binding a sodium ion requires the availability of several electron pairs in its surrounding. Sugars for example are not basic, but a sodium ion may find many electron pairs. This is why the protonated molecule is difficult to observe in the mass spectrum if the solution is not carefully desalted. Otherwise, the sodium adduct is dominant. If ammonium salt is present it can also form adducts $(M + NH_4)^+$ because of its ability to form hydrogen bonds.

In the negative ion mode, the chloride adduct is often observed yielding $(M + 35)^-$ and $(M + 37)^-$. As for the sodium, the chloride ion is always present if the solution is not desalted. However, it produces fewer adducts than the sodium. The acetate ions, if present, produce $(M + 59)^-$ owing to their ability to form hydrogen bonds.

The addition of ammonium acetate, at low concentration, in API methods can be interesting to produce protonated or deprotonated species. Indeed, in the heated gas or heated capillary interface, ammonia or acetic acid evaporates, leaving the corresponding protonated or deprotonated species. The interest of nitrate adducts in the analysis of sugars has been recently demonstrated [109, 110].

1.17.5 Formation of Aggregates or Clusters

'Dimer' ions such as $(M + M + H)^+$ or of higher order $(nM + H)^+$ are often observed. The proton can be replaced by another cation. 'Heterodimers' of the general formula $(M + M' + H)^+$, or with a metal cation or of higher order, are also observed. The corresponding ions are also observed in negative ion mode.

It should be noted that the formation of such aggregates in the gas phase causes a diminution of entropy. To be possible, the formation of such aggregates must be exothermic. Furthermore, if the partners have similar basicities or affinities for the cation, the cluster is more stable. Otherwise it dissociates, one of the partners taking the proton, or the cation, according to the relative stability. This is why associations of two or more identical molecules are observed more often or at higher abundances. Indeed, they have the same affinities of course.

The formation of oligomers has as a consequence the diminution of the number of molecules in the gas phase, and thus occurs with a diminution of the entropy. As the reaction must be exergonic to occur, it must be sufficiently exothermic, at least to compensate for the entropy loss. Once formed, the internal energy of the oligomer should be reduced, since it contains a sufficient amount of energy to dissociate. This needs a collision with a third partner, and this requires a sufficient pressure.

As a general rule, the abundance is reduced when the number of associated molecules increases. However, some specific aggregates, resulting from particularly important interactions, are present at particularly high abundance. This occurs often with organometallic compounds, as the metal tries to complete its electronic shell.

Aggregates are rarely observed in the negative ion mode, because the presence of the negative charge causes an expansion of the electronic shell, thus reducing the electric field around the negatively charged ion. This reduces the interactions between the partners.

1.17.6 Reactions at the Interface Between Source and Analyser

Atmospheric pressure sources need to have a device for the desolvation of the ions, such as a heated capillary, a heated gas curtain, collisions at intermediate pressure in a focusing multipole. The increase in internal energy induces the dissociation of the associated ion–molecules. Furthermore, the ions are at low pressure and thus the probability of the reverse reaction is strongly decreased. For example, if a molecule forms a complex with an ammonium ion, the ammonia can be evacuated by pumping:

$$(M + NH_4)^+ \longrightarrow (M + H)^+ + NH_3$$

This reaction occurs even if M is less basic than ammonia. So, the protonated molecular ion of a sugar may be formed by desolvatation of its ammoniacal complex.

In negative ion mode, the acetate ion is used and acetic acid can be evacuated by pumping to obtain the deprotonated molecular ion of the analyte:

$$(M + AcO)^- \longrightarrow (M - H)^- + AcOH$$

References

1. Bleakney, W. (1929) *Phys. Rev.*, **34**, 157.
2. Nier, A.O. (1947) *Rev. Sci. Instrum.*, **18**, 415.
3. Bentley, T.W. and Johnstone, R.A.W. (1970) Mechanism and structure in mass spectrometry: a comparison with other chemical processes, in *Advances in Physical Organic Chemistry*, vol. **8** (ed. V. Gold), Academic Press, London.
4. Kienitz, H. (1968) *Massenspektrometrie*, Verlag Chemie, Weinheim, p. 45.
5. Harrison, A.G. (1983) *Chemical Ionisation Mass Spectrometry*, CRC Press, Boca Raton, FL.
6. Thermodynamic data from: Lias, S.G., Bartmess, J.E., Liebman, J.F. *et al.* (1988) Gas-phase ion and neutral thermochemistry. *J. Phys. Chem. Ref. Data*, **17** (Suppl. 1). Also available on laser disk from NIST (National Institute of Standards and Technology), Washington, DC, USA).
7. Baldwin, M.A. and McLafferty, F.W. (1973) *Org. Mass Spectrom.*, **7**, 1353.
8. Beckey, H.D. (1969) Field ionization mass spectrometry. *Research/Development*, **20**, 26–9.
9. Van Vaeck, L., Adriaens, A. and Gijbels, R. (1999) Static secondary ion mass spectrometry: (S-SIMIS) Part 1. Methodology and structural information. *Mass Spectrom. Rev.*, **18** (1), 1–47.
10. Adriaens, A., Van Vaeck, L. and Adams, F. (1999) Static secondary ion mass spectrometry: (S-SIMIS) Part 2. Material sciences applications. *Mass Spectrom. Rev.*, **18** (1), 48–81.
11. Benninghoven, A. and Sichtermann, W.K. (1978) *Anal. Chem.*, **50**, 1180.
12. Barber, M., Bardoli, R.S., Sedgwick, R.D. and Tyler, A.H. (1981) *J. Chem. Soc., Chem. Commun.*, 325.
13. Aberth, W., Straub, K.M. and Burlingame, A.L. (1982) *Anal. Chem.*, **54**, 2029.
14. Beckey, H.D. (1977) *Principles of Field Ionization and Field Desorption in Mass Spectrometry*, Pergamon Press, Oxford.
15. Mcfarlane, R.D. and Torgesson, T.F. (1976) *Science*, **191**, 970.
16. McNeal, C.J. and McFarlane, R.D. (1981) *J. Am. Chem. Soc.*, **103**, 1609.
17. Karas, M. and Hillenkamp, F.H. (1988) *Anal. Chem.*, **60**, 2229.
18. Hillenkamp, F., Karas, M., Ingeldoh, A. and Stahl, B. (1990) Matrix assisted UV-laser desorption ionization: a new approach to mass spectrometry of large molecules, in *Biological Mass Spectrometry*, p. 49 (eds A.L. Burlingame and J.A. McCloskey), Elsevier, Amsterdam.
19. Karas, M., Bachmann, D., Bahr, U. and Hillenkamp, F. (1987) *Int. J. Mass Spectrom. Ion Processes*, **78**, 53.
20. Karas, M. and Hillenkamp, F.H. (1988) *Anal. Chem.*, **60**, 2229.

21. Hillenkamp, F., Karas, M., Ingeldoh, A. and Stahl, B. (1990) Matrix assisted UV-laser desorption ionization: a new approach to mass spectrometry of large molecules, in *Biological Mass Spectrometry* (eds A.L. Burlingame and J.A. McCloskey), Elsevier, Amsterdam, p. 49.

22. Stump, M.J., Fleming, R.C., Gong, W.-H. *et al.* (2002) *Appl. Spectrosc. Rev.*, **37**, 275.

23. Chen, X.J., Carroll, J.A. and Beavis, R.C. (1998) Near-UV-induced matrix-assisted laser desorption/ionization as a function of wavelength. *J. Am. Soc. Mass Spectrom.*, **9** (9), 885–91.

24. Knochenmuss, R. (2002) A quantitative model of UV matrix-assisted laser desorption/ionization. *J. Mass Spectrom.*, **37**, 867.

25. Karas, M. and Kruger, R. (2003) *Chem. Rev.*, **103**, 427.

26. Dreisewerd, K. (2003) *Chem. Rev.*, **103**, 395.

27. Zenobi, R. and Knochenmuss, R. (1998) Ion formation in MALDI mass spectrometry. *Mass Spectrom. Rev.*, **17** (5), 337–66.

28. Knochenmuss, R. and Zenobi, R. (2003) *Chem. Rev.*, **103**, 441.

29. Karas, M. and Hillenkamp, F.H. (1988) *Ion Formation from Organic Solids IV* (ed. A. Benninghoven), John Wiley & Sons, Inc., New York, p. 103.

30. Spengler, B. and Cotter, R.J. (1990) *Anal. Chem.*, **62**, 793.

31. Pasch, H. and Schrepp, W. (2003) *MALDI-TOF Mass Spectrometry of Synthetic Polymers*, Series: Springer Laboratory XVIII.

32. Hillenkamp, F.H. and Karas, M. (1990) *Methods in Enzymology*, vol. **193** (ed. J.A. McCloskey), Academic Press, New York, pp. 280–95.

33. Stoeckli, M., Chaurand, P., Hallahan, D.E. and Caprioli, R.M. (2001) *Nat. Med.*, **7**, 493.

34. Pierson, J., Norris, J.L., Aerni, H.R. *et al.* (2004) *J. Proteome Res.*, **3**, 289.

35. Demirev, P., Westman, A., Reimann, C.T. *et al.* (1992) *Rapid Commun. Mass Spectrom.*, **6**, 187.

36. Talrose, V. L., Person, M.D., Whittal, R.M. *et al.* (1999) *Rapid Commun. Mass Spectrom.*, **13**, 2191.

37. Vorm, O., Roepstorff, P. and Mann, M. (1994) *Anal. Chem.*, **66**, 3281.

38. Xiang, F. and Beavis, R.C. (1994) *Rapid Commun. Mass Spectrom.*, **8**, 199.

39. http://www.chemistry.wustl.edu/~msf/damon/sample_prep_toc.html" (21 March 2007).

40. http://www.nist.gov/maldi (21 March 2007).

41. Vorm, O., Chait, B.T. and Roepstorff, P. (1994) *Mass Spectrometry of Protein Samples Containing Detergents*, 41th ASMS Conference Proceedings, pp. 621a–b.

42. Merchant, M. and Weinberger, S.R. (2000) *Electrophoresis*, **21**, 1164.

43. Chen, Y.C., Shiea, J. and Sunner, J. (1998) *J. Chromatogr. A*, **826**, 77.

44. Shen, Z.X., Thomas, J.J., Averbuj, C. *et al.* (2001) *Anal. Chem.*, **73**, 612.

45. Laiko, V.V., Baldwin, M.A. and Burlingame, A.L. (2000) *Anal. Chem.*, **72**, 652–7.

46. Laiko, V.V., Moyer, S.C. and Cotter, R.J. (2000) *Anal. Chem.*, **72**, 5239–43.

47. Moyer, S.C. and Cotter, R.J. (2002) *Anal. Chem.*, **74**, 469A.

48. Vestal, M.L. (1983) *Mass Spectrom. Rev.*, **2**, 447.

49. Blakney, C.R. and Vestal, M.L. (1983) *Anal. Chem.*, **55**, 750.

50. Huang, E.C., Wachs, T., Conboy, J.J. and Henion, J.D. (1990) *Anal. Chem.*, **62**, 713A.

51. Bruins, A.P. (1991) *Mass Spectrom. Rev.*, **10**, 53.

52. Mann, M., Meng, C.K. and Fenn, J.B. (1989) Interpreting mass spectra of multiply charged ions. *Anal. Chem.*, **61**, 1702–8.

53. Fenn, J.B., Mann, M., Meng, C.K. *et al.* (1989) *Science*, **246**, 64.

54. de la Mora, J.F., Van Berkel, G.J., Enke, C.G. *et al.* (2000) Electrochemical processes in electrospray ionization mass spectrometry. *J. Mass. Spectrom.*, **35**, 939–52.

55. Cech, N.B. and Enke, C.G. (2001) Practical implications of some recent studies in electrospray ionization fundamentals. *Mass Spectrom. Rev.*, **20**, 362–87.

56. Rohner, T.C., Lion, N. and Girault, H.H. (2004) Electrochemical and theoretical aspects of electrospray ionisation. *Phys. Chem. Chem. Phys.*, **6**, 3056–68.

57. Snyder, A.P. (ed.) (1996) *Biochemical and Biotechnological Applications of Electrospray Ionization Mass Spectrometry*, ACS Symposium Series 619, American Chemical Society, Washington, DC.

58. Cole, R.B. (ed.) (1997) *Electrospray Ionisation Mass Spectrometry*, John Wiley & Sons, Ltd., Chichester.

59. Yamashita, M. and Fenn, J.B. (1988) *Phys. Chem.*, **88**, 4451.

60. Yamashita, M. and Fenn, J.B. (1988) *Phys. Chem.*, **88**, 4671.

61. Loo, J.A., Udseth, H.R. and Smith, R.D. (1989) *Anal. Biochem.*, **179**, 404.

62. Ikonomou, M.G., Blades, A.T. and Kebarle, P. (1990) *Anal. Chem.*, **62**, 957.

63. Smith, R.D., Loo, J.A., Edmons, C.G. *et al.* (1990) *Anal. Chem.*, **62**, 882.

64. Mann, M. (1990) *Org. Mass Spectrom.*, **25**, 575.

65. Kebarle, P. and Tang, L. (1993) From ions in solution to ions in the gas phase. The mechanism of electrospray mass spectrometry. *Anal. Chem.*, **65**, 972A.

66. Gomez, A. and Tang, K. (1994) *Phys. Fluids*, **6**, 404.

67. Andersen, J., Molina, H., Moertz, E. *et al.* (1998) Quadrupole-TOF hybrid mass spectrometers bring improvements to protein identification and MS/MS analysis of intact proteins. *The 46th Conference on Mass Spectrometry and Allied Topics*, Orlando, Florida, p. 978.

68. van Berkel, W.J.H., van den Heuvel, R.H.H., Versluis, C. and Heck, A.J.R. (2000) Detection of intact megaDalton protein assemblies of vanillyl-alcohol oxidase by mass spectrometry. *Protein Sci.*, **9**, 435–9.

69. Kelly, M.A., Vestling, M.M., Fenselau, C. and Smith, P.B. (1992) *Org. Mass Spectrom.*, **27**, 1143.

70. Covey, T. (1996) Analytical characteristics of the electrospray ionization process, in *Biochemical and Biotechnological Applications of Electrospray Ionization Mass Spectrometry*, pp. 21–59 (ed. A.P. Snyder), ACS Symposium Series 619, American Chemical Society, Washington, DC.

71. Emmett, M.R. and Caprioli, R.M. (1994) *J. Am. Soc. Mass Spectrom.*, **5**, 605.

72. Wilm, M. and Mann, M. (1994) *Proceedings of the 42nd ASMS Conference*, Chicago, p. 770.

73. Wilm, M., Shevchenko, A., Houthaeve, T. *et al.* (1996) Femtomole sequencing of proteins from polyacrylamide gels by nano-electrospray mass spectrometry. *Nature*, **379** (6564), 466–9.

74. Van Berkel, G.J. and Zhou, F. (1995) *Anal. Chem.*, **67**, 2916.

75. Fenn, J.B., Rosell, J., Nohmi, T. *et al.* (1996) Electrospray ion formation: desorption versus desertion, in *Biochemical and Biotechnological Applications of Electrospray Ionization Mass Spectrometry*, pp. 60–80 (ed. A.P. Snyder), ACS Symposium Series 619, American Chemical Society, Washington, DC.

76. Carroll, D.I., Dzidic, I., Stillwell, R.N. *et al.* (1975) *Anal. Chem.*, **47**, 2369.

77. French, J.B. and Reid, N.M. (1980) *Dynamic Mass Spectrometry*, vol. **6**, Heyden, London, pp. 220–33.

78. Bruins, A.P. (1991) *Mass Spectrom. Rev.*, **10**, 53.

79. Cambier, V., Hance, T. and de Hoffmann, E. (1999) Non-injured maize contains several 1,4-benzoxazin-3-one related compounds but only as glucoconjugates. *Phytochem. Anal.*, **10**, 119–26.

80. Robb, D.B., Covey, T.R. and Bruins, A.P. (2000) *Anal. Chem.*, **72**, 3653.

81. Syage, J.A. and Evans, M.D. (2001) *Spectroscopy*, **16**, 14.
82. Kauppila, T.J., Kuuranne, T., Meurer, E.C. *et al.* (2002) *Anal. Chem.*, **74**, 5470.
83. Syage, J.A. (2004) *J. Am. Soc. Mass Spectrom.*, **15**, 1521.
84. Kauppila, T.J., Kotiaho, T., Kostiainen, R. and Bruins, A.P. (2004) *J. Am. Soc. Mass Spectrom.*, **15**, 203.
85. Raffaelli, A. and Saba, A. (2003) *Mass Spectrom. Rev.*, **22**, 318.
86. Takats, Z., Wiseman, J.M., Gologan, B. and Cooks, R.G. (2004) Mass spectrometry sampling under ambient conditions with electrospray ionization. *Science*, **306**, 471–3.
87. Kertezs, V., Ford, M.J. and Van Berkel, G.J. (2005) Automation of a surface sampling probe/electrospray mass spectrometry system. *Anal. Chem.*, **77**, 7183–9.
88. Cody, R.B., Laramee, J.A. and Dupont Durst, H. (2005) Versatile new ion source for the analysis of materials in open air under ambient conditions. *Anal. Chem.*, **77**, 2297–302.
89. Adams, F., Gijbels, R. and Van Grieken, R. (1988) *Inorganic Mass Spectrometry (Chemical Analysis: A Series of Monographs on Analytical Chemistry and Its Applications)*, vol. 95, John Wiley & Sons, Inc., New York.
90. Gijbels, R., Van Straaten, M. and Bogaerts, A. (1995) *Adv. Mass Spectrom.*, **13**, 241–56.
91. Becker, J.S. and Dietze, H.J. (1998) Inorganic trace analysis by mass spectrometry. *Spectrochim. Acta Part B – At. Spectrosc.*, **53** (11), 1475–1506.
92. Steward, I.I. and Horlick, G. (1996) Developments in the electrospray mass spectrometry of inorganic species. *Trends Anal. Chem.*, **15** (2), 80–90.
93. Traeger, J.C. and Colton, R. (1998) *Adv. Mass Spectrom.*, **14**, 637–659.
94. Colton, R., D'Agostino, A. and Traeger, J.C. (1999) Electrospray mass spectrometry applied to inorganic and organometallic chemistry. *Mass Spectrom. Rev.*, **14** (2), 79–106.
95. Inghram, M.G. and Chupka, W.A. (1953) *Rev. Sci. Instrum.* **24**, 518.
96. Heumann, K.G., Eisenhut, S., Gallus, S. *et al.* (1995) *Analyst*, **120** (5), 1291–9.
97. Johnson, P.G., Bolson, A. and Henderson, C.M. (1973) *Nucl. Instrum. Methods*, **106**, 83.
98. Dempster, A.J. (1936) *Rev. Sci. Instrum.*, **7**, 46.
99. Harrison, W.W. and Hang, W. (1996) Powering the analytical glow discharge. *Fresenius J. Anal. Chem.*, **355** (7–8), 803–7.
100. Bogaerts, A. and Gijbels, R. (1998) Fundamental aspects and applications of glow discharge spectrometric techniques. *Spectrochim. Acta Part B – At. Spectrosc.*, **53** (1), 1–42.
101. Hang, W., Yan, X.M., Wayne, D.M. *et al.* (1999) Glow discharge source interfacing to mass analyzers: Theoretical and practical considerations. *Anal. Chem.*, **71** (15), 3231–7.
102. Gray, A.L. (1985) *Spectrochim. Acta Part B – At. Spectrosc.*, **40**, 1525.
103. Zoorob, G.K., McKiernan, J.W. and Caruso, J.A. (1998) ICP-MS for elemental speciation studies. *Mikrochim. Acta*, **128** (3–4), 145–68.
104. Hill, S.J. (1999) *ICP Spectrometry and its Applications*, Sheffield Academic Press, Sheffield, p. 320.
105. McIver, R.T. (1980) *Sci. Am.*, **243**, 148.
106. Bohme, D.K. and Mackay, G.I. (1981) *J. Am. Chem. Soc.*, **103**, 978.
107. Bohme, D., Mackay, G.I. and Schiff, H.I. (1980) *J. Chem. Phys.*, **73**, 4976.
108. Lias, S.G., Bartmess, J.E., Liebman, J.F. *et al.* (1988) Gas-phase ion and neutral thermochemistry. *J. Phys. Chem. Ref. Data*, **17** (Suppl. 1).
109. Jiang, Y. and Cole, R.B. (2005) Oligosaccharide analysis using anion attachment in negative mode electrospray mass spectrometry. *J. Am. Soc. Mass Spectrom.*, **16** (1), 60–70.
110. Harvey, D.J. (2005) Fragmentation of negative ions from carbohydrates: Part 1. Use of nitrate and other anionic adducts for the production of negative ion electrospray spectra from N-linked carbohydrates. *J. Am. Soc. Mass Spectrom.*, **16** (5), 622–30.

2

Mass Analysers

Once the gas-phase ions have been produced, they need to be separated according to their masses, which must be determined. The physical property of ions that is measured by a mass analyser is their mass-to-charge ratio (m/z) rather than their mass alone. Therefore, it should be mentioned that for multiply charged ions the apparent m/z values are fractional parts of their actual masses.

As there are a great variety of sources, several types of mass analysers have been developed. Indeed, the separation of ions according to their mass-to-charge ratio can be based on different principles (Table 2.1). All mass analysers use static or dynamic electric and magnetic fields that can be alone or combined. Most of the basic differences between the various common types of mass analyser lie in the manner in which such fields are used to achieve separation.

Each mass analyser has its advantages and limitations. Analysers can be divided into two broad classes on the basis of many properties. Scanning analysers transmit the ions of different masses successively along a time scale. They are either magnetic sector instruments with a flight tube in the magnetic field, allowing only the ions of a given mass-to-charge ratio to go through at a given time, or quadrupole instruments. However, other analysers allow the simultaneous transmission of all ions, such as the dispersive magnetic analyser, the TOF mass analyser and the trapped-ion mass analysers that correspond to the ion traps, the ion cyclotron resonance or the orbitrap instruments. Analysers can be grouped on the basis of other properties, for example ion beam versus ion trapping types, continuous versus pulsed analysis, low versus high kinetic energies.

A new type of mass analyser, the orbitrap, was introduced on the market in 2005. However, the existing mass analysers have continued to be improved. Indeed, to overcome the limitations of conventional ion traps, linear ion traps (LITs), which have geometries more closely related to those of quadrupole mass analysers, have been developed. In the same way, the TOF mass analysers have been improved significantly with the development of numerous techniques, such as delayed extraction and the reflectron, which are based on concepts known since the 1960s. The TOF mass analyser requires the ions to be produced in bundles and is thus especially well suited for pulsed laser sources. However, the advent of ion introduction techniques such as orthogonal acceleration allows the analysis of ions from a continuous source with a TOF analyser. Fourier transform (FT) ion cyclotron resonance spectrometers have shown impressive performances at high masses and high resolution. The new orbitrap mass analyser is capable of similar high performance but has more modest size, complexity and cost.

Another trend in mass analyser development is to combine different analysers in sequence in order to increase the versatility and allow multiple experiments to be performed.

Mass Spectrometry: Principles and Applications, Third Edition Edmond de Hoffmann and Vincent Stroobant
© Copyright 2007, John Wiley & Sons Ltd

Table 2.1   Types of analysers used in mass spectrometry.

Type of analyser	Symbol	Principle of separation
Electric sector	E or ESA	Kinetic energy
Magnetic sector	B	Momentum
Quadrupole	Q	m/z (trajectory stability)
Ion trap	IT	m/z (resonance frequency)
Time-of-flight	TOF	Velocity (flight time)
Fourier transform ion cyclotron resonance	FTICR	m/z (resonance frequency)
Fourier transform orbitrap	FT-OT	m/z (resonance frequency)

Triple-quadrupole and more recently hybrid instruments, for instance the quadrupole TOF instrument or the ion trap–FT ion cyclotron resonance mass spectrometer, allow one to obtain a mass spectrum resulting from the decomposition of an ion selected in the first analyser. The time-dependent decomposition of a selected ion can also be observed in ion cyclotron resonance and ion trap instruments. They allow fragments over several generations (MS^n) to be observed.

The five main characteristics for measuring the performance of a mass analyser are the mass range limit, the analysis speed, the transmission, the mass accuracy and the resolution.

The mass range determines the limit of m/z over which the mass analyser can measure ions. It is expressed in Th, or in u for an ion carrying an elementary charge, that is $z = 1$.

The analysis speed, also called the scan speed, is the rate at which the analyser measures over a particular mass range. It is expressed in mass units per second ($u\,s^{-1}$) or in mass units per millisecond ($u\,ms^{-1}$).

The transmission is the ratio of the number of ions reaching the detector and the number of ions entering the mass analyser. The transmission generally includes ion losses through other sections of the mass analyser such as electric lenses before and after the analyser. This should not be confused with the duty cycle. The duty cycle is the proportion of time during which a device or system is usefully operated. For a mass spectrometer, the duty cycle is the part of ions of a particular m/z produced in the source that are effectively analysed. The duty cycle can be expressed as a ratio or as a percentage. It is not a characteristic of the analyser but is rather a characteristic of the whole mass spectrometer. Duty cycles differ between different mass spectrometer designs. But the same mass spectrometer can have different duty cycles because they are also highly dependent on the mode of its operation. For instance, a quadrupole mass spectrometer used in selected ion monitoring mode (i.e. detecting only one specific ion) has a duty cycle of 100 %. But when this mass spectrometer is used in scan mode (i.e. scanning the analyser to detect an m/z range), the duty cycle of the spectrometer decreases according the proportion of the total observation time spent for each ion. For a quadrupole mass spectrometer scanning over 1000 amu, the duty cycle is $1/1000 = 0.1$ %. These parameters are closely associated with the sensitivity of a mass spectrometer. But the sensitivity of an instrument is better described by a factor defined as the mass spectrometer efficiency that takes into account the duty cycle, the transmission of the analyser and the efficiency of the detector.

Mass accuracy indicates the accuracy of the m/z provided by the mass analyser. It is the difference that is observed between the theoretical m/z ($m_{theoretical}$) and the measured m/z ($m_{measured}$). It can be expressed in millimass units (mmu) but is often expressed in parts

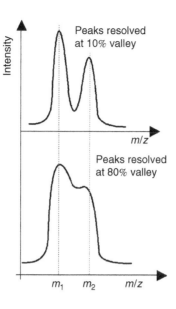

Figure 2.1
Diagram showing the concepts of
peak resolution and valley.

per million (ppm). Mass accuracy is largely linked to the stability and the resolution of the analyser. A low-resolution instrument cannot provide high accuracy. The precision obtained on the mass of the analysed sample depends also on the determination of the centroid of the peak. High mass accuracy has significant applications such as the determination of elemental composition. This will be discussed in more detail in Chapter 6 on analytical information.

Last but not least, there is the characteristic of a mass analyser concerning the resolution or its resolving power. Resolution or resolving power is the ability of a mass analyser to yield distinct signals for two ions with a small m/z difference (Figure 2.1). The exact definition of these terms is one of the more confusing subjects of mass spectrometry terminology that continues to be debated. We will use here the definitions proposed by Marshall [1]. This will be described in more details in Chapter 6.

Two peaks are considered to be resolved if the valley between them is equal to 10 % of the weaker peak intensity when using magnetic or ion cyclotron resonance (ICR) instruments and 50 % when using quadrupoles, ion trap, TOF, and so on. If Δm is the smallest mass difference for which two peaks with masses m and $m + \Delta m$ are resolved, the definition of the resolving power R is $R = m/\Delta m$. Therefore, a greater resolving power corresponds to the increased ability to distinguish ions with a smaller mass difference.

The resolving power can also be determined with an isolated peak. Indeed, the resolving power is also defined using the peak width Δm at x % of the peak height. Often x is taken to be 50 % and Δm is designated as full width at half maximum (FWHM). The relationship between the two definitions is obvious for two peaks with equal intensities. The resolution full width at x % of the peak height is equal to the resolution at $2x$ % for the valley (Figure 2.2).

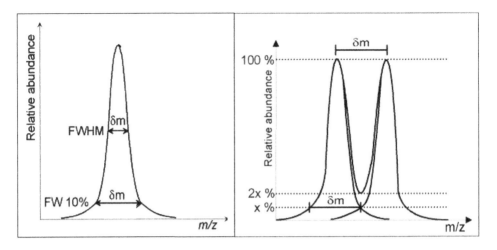

Figure 2.2
Left panel: definition of δm at a given height of the peak. Right panel: relationship between the two definitions of resolution, resolution full width at x % of the peak height or resolution at y % of the valley. As the bottom of the valley is the sum of the intensities at the cross, $y = 2x$.

It should be noted that with the FWHM definition, the resolving power needed to observe separated peaks could no longer be deduced directly from R. Indeed, as shown in Figure 2.3, when two peaks cross at 50 % height, the valley is at 100 %. Thus the resolving power needed to separate them must be higher than the one calculated using the FWHM definition. With Gaussian peak shapes, the FWHM definition of the resolving power gives values which are around double that of those obtained by 10 % valley definition.

Low resolution or high resolution is usually used to describe analysers with a resolving power that is less or greater than about 10 000 (FWHM), respectively. However, there is no exact definition of the boundary between these two terms.

The various types of mass analysers will be discussed in this chapter with a description of their principles of operation. Because each type of analyser is based on significantly different principles, each has unique properties and specifications. The main characteristics of the different analysers presented in this chapter are summarized in Table 2.2. A description of the detectors used in mass spectrometry will also be given in Chapter 3.

2.1 Quadrupole Analysers

The quadrupole analyser is a device which uses the stability of the trajectories in oscillating electric fields to separate ions according to their m/z ratios. The 2D or 3D ion traps are based on the same principle.

2.1.1 Description

Quadrupole analysers [2, 3] are made up of four rods of circular or, ideally, hyperbolic section (Figures 2.4 and 2.5). The rods must be perfectly parallel.

Table 2.2 Comparison of mass analysers.

	Quadrupole	Ion trap	TOF	TOF reflectron	Magnetic	FTICR	Orbitrap
Mass limit	4000 Th	6000 Th	>1 000 000 Th	10 000 Th	20 000 Th	30 000 Th	50 000 Th
Resolution FWHM (m/z 1000)	2000	4000	5000	20 000	100 000	500 000	100 000
Accuracy	100 ppm	100 ppm	200 ppm	10 ppm	<10 ppm	<5 ppm	<5 ppm
Ion sampling	Continuous	Pulsed	Pulsed	Pulsed	Continuous	Pulsed	Pulsed
Pressure	10^{-5} Torr	10^{-3} Torr	10^{-6} Torr	10^{-6} Torr	10^{-6} Torr	10^{-10} Torr	10^{-10} Torr
Tandem mass spectrometry	Triple quadrupoles MS/MS fragments precursors neutral loss	MS" fragments	—	PSD or TOF/TOF MS/MS fragments	Consecutive sectors MS/MS fragments precursors neutral loss	MS" fragments	—
	Low-energy collision	Low-energy collision	—	Low- or high-energy collision	High-energy collision	Low-energy collision	—

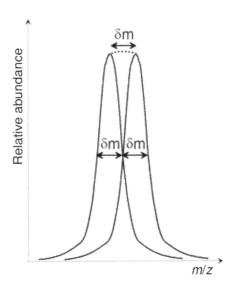

Figure 2.3
If δm is taken at half maximum height, two peaks having a mass difference equal to this width will cross at 50 % height. As the valley depth will be two times this height, it will be at 100 %. There is thus no separation.

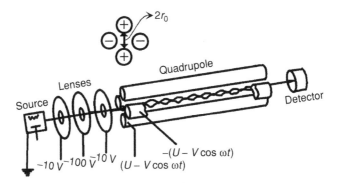

Figure 2.4
Quadrupole instrument made up of the source, the focusing lenses, the quadrupole cylindrical rods and the detector. Ideally, the rods should be hyperbolic. Reproduced (modified) from Kienitz H., Massenspektrometrie, Verlag Chemie, Weinheim, 1968, with permission.

A positive ion entering the space between the rods will be drawn towards a negative rod. If the potential changes sign before it discharges itself on this rod, the ion will change direction.

The principle of the quadrupole was described by Paul and Steinwegen [4], at Bonn University, in 1953. They started from research work on the strong focusing of ions carried

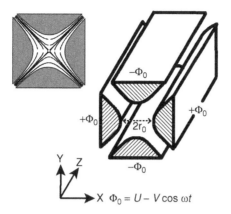

Figure 2.5
Quadrupole with hyperbolic rods and applied potentials. The equipotential lines
are represented above, on the left.

out in 1951 in Athens by the electrical engineer Christophilos. The quadrupoles have since
been developed into commercially available instruments by the work of Shoulders, Finnigan
[5] and Story.

Ions travelling along the z axis are subjected to the influence of a total electric field
made up of a quadrupolar alternative field superposed on a constant field resulting from the
application of the potentials upon the rods:

$$\Phi_0 = +(U - V \cos \omega t) \text{ and } -\Phi_0 = -(U - V \cos \omega t)$$

In this equation, Φ_0 represents the potential applied to the rods, ω the angular frequency
(in radians per second $= 2\pi v$, where v is the frequency of the RF field), U is the direct
potential and V is the 'zero-to-peak' amplitude of the RF voltage. Typically, U will vary
from 500 to 2000 V and V from 0 to 3000 V (from -3000 to $+3000$ V peak to peak).

2.1.2 Equations of Motion

The ions accelerated along the z axis enter the space between the quadrupole rods and
maintain their velocity along this axis. However, they are submitted to accelerations along
x and y that result from the forces induced by the electric fields (Figure 2.5):

$$F_x = m \frac{\mathrm{d}^2 x}{\mathrm{d}t^2} = -ze \frac{\partial \Phi}{\partial x}$$

$$F_y = m \frac{\mathrm{d}^2 y}{\mathrm{d}t^2} = -ze \frac{\partial \Phi}{\partial y}$$

Φ is a function of Φ_0:

$$\Phi_{(x,y)} = \Phi_0(x^2 - y^2)/r_0^2 = (x^2 - y^2)(U - V \cos \omega t)/r_0^2$$

Differentiating and rearranging the terms leads to the following equations of the movement (Paul equation):

$$\frac{d^2x}{dt^2} + \frac{2ze}{mr_0^2}(U - V \cos \omega t)x = 0$$

$$\frac{d^2y}{dt^2} - \frac{2ze}{mr_0^2}(U - V \cos \omega t)y = 0$$

The trajectory of an ion will be stable if the values of x and y never reach r_0, thus if it never hits the rods. To obtain the values of either x or y during the time, these equations need to be integrated. The following equation was established in 1866 by the physicist Mathieu in order to describe the propagation of waves in membranes:

$$\frac{d^2u}{d\xi^2} + (a_u - 2q_u \cos 2\xi)u = 0$$

u stands for either x or y. Comparing the preceding equations with this one, and taking into account that the potential along y has opposite sign to the one along x, the following change of variables gives to the equations of the movement the form of the Mathieu equation. First, ξ is defined as being

$$\xi = \frac{\omega t}{2} \quad \text{and} \quad \text{thus } \xi^2 = \frac{\omega^2 t^2}{4}$$

In the first term of the Paul equation, replacing t^2 by ξ^2 introduces a factor $\omega^2/4$. To compensate for this factor, the whole equation must be multiplied by the reverse, $4/\omega^2$. In the cosine term, 2ξ is equal to ωt, as needed in the Paul equations. Incorporating these changes and rearranging the terms yields the following expressions:

$$a_u = a_x = -a_y = \frac{8zeU}{m\omega^2 r_0^2} \quad \text{and} \quad q_u = q_x = -q_y = \frac{4zeV}{m\omega^2 r_0^2}$$

We will not attempt to integrate these equations [6]. We only need to recognize that they establish a relationship between the coordinates of an ion and time. As long as x and y, which determine the position of an ion from the centre of the rods, both remain less than r_0, the ion will be able to pass the quadrupole without touching the rods. Otherwise, the ion discharges itself against a rod and is not detected. Figure 2.6 shows stable and unstable trajectories in a quadrupole [7].

For a given quadrupole, r_0 is constant and $\omega = 2\pi v$ is maintained constant. U and V are the variables. For an ion of any mass, x and y can be determined during a time span as a function of U and V. Stability areas can be represented in an a_u, q_u diagram. In these areas the values of U and V are such that x and y do not reach values above or equal to r_0. Figure 2.7 represents these stability areas. The upper part of this figure represents the stability areas along the x and the y axis respectively. The overlay of these two diagrams is represented in the lower part of the figure. The regions where the ions will have stable trajectories according to both the x and y axes simultaneously are now clearly identified. Later we will consider the area A with a positive U. The area A is that currently used, with U either positive or negative. What is true for a positive U symmetrically holds true for a negative U.

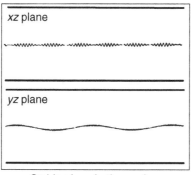

Stable along both *x* and *y*

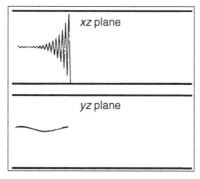

Stable along *y*, unstable along *x*

Figure 2.6
Stable and unstable trajectories of ions in a quadrupole. Reproduced (modified) from March R.E. and Hughes R.J., Quadrupole Storage Mass Spectrometry, Wiley, New York, 1989, with permission.

Considering the equations

$$a_u = \frac{8zeU}{m\omega^2 r_0^2} \quad \text{and} \quad q_u = \frac{4zeV}{m\omega^2 r_0^2}$$

we can deduce

$$U = a_u \frac{m}{z} \frac{\omega^2 r_0^2}{8e} \quad \text{and} \quad V = q_u \frac{m}{z} \frac{\omega^2 r_0^2}{4e}$$

The last terms of both the U and V equations is a constant for a given quadrupole instrument, as they operate at constant ω. We see that switching from one m/z to another results in a proportional multiplication of a_u and q_u, which means changing the scale of the drawing in U, V coordinates; thus the triangular area A will change from one mass to another, like

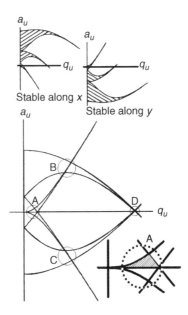

Figure 2.7
Stability areas for an ion along x or y (above) and along x and y (below); u represents either x or y. The four stability areas are labelled A to D and are circled. The area A is that used commonly in mass spectrometers and is enlarged. The direct potential part is shown for positive U (shaded) or negative U. From now on we will consider the positive area. Reproduced (modified) from March R.E. and Hughes R.J., Quadrupole Storage Mass Spectrometry, Wiley, New York, 1989, with permission.

proportional triangles. Figure 2.8 represents in a U, V diagram the areas A obtained with different masses.

We can see in this diagram that scanning along a line maintaining the U/V ratio constant allows the successive detection of the different masses. So long as the line keeps going through stability areas, then the higher the slope, the better the resolution.

If $U = 0$, there is no direct current and the resolution becomes zero. However, the value of V imposes a minimum on stable masses. Thus, if V is increased from 0 to V so as to reach slightly beyond the stability area m_1, all of the ions with masses equal to or lower than m_1 will have an unstable trajectory, and all of those with masses above m_1 will have a stable trajectory.

In practice, the highest detectable m/z ratio is about 4000 Th, and the resolution hovers around 3000. Thus, beyond 3000 u the isotope clusters are no longer clearly resolved. As

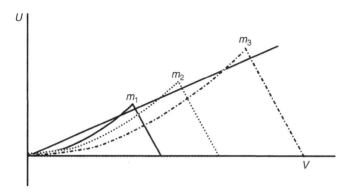

Figure 2.8
Stability areas as a function of U and V for ions with different masses ($m_1 < m_2 < m_3$). Changing U linearly as a function of V, we obtain a straight operating line that allows us to observe those ions successively. A line with a higher slope would give us a higher resolution, so long as it goes through the stability areas. Keeping $U = 0$ (no direct potential) we obtain zero resolution. All of the ions have a stable trajectory so long as V is within the limits of their stability area. Reproduced (modified) from March R.E. and Hughes R.J., Quadrupole Storage Mass Spectrometry, Wiley, New York, 1989, with permission.

is shown in Figure 2.8, adjusting the slope of the operating line allows us to increase the resolution. The resolution normally obtained is not sufficient to deduce the elementary analysis. Usually, quadrupole mass spectrometers are operated at unit resolution, that is a resolution that is sufficient to separate two peaks one mass unit apart. Quadrupoles are low-resolution instruments.

Operating at constant δm, quadrupoles require the scanning to be carried out at a uniform velocity throughout the entire mass range, as opposed to the magnetic instruments, which require an exponential scanning.

The quadrupole is a real mass-to-charge ratio analyser. It does not depend on the kinetic energy of the ions when they leave the source. The only requirements are, first, that the time for crossing the analyser is short compared with the time necessary to switch from one mass to the other, and, second, that the ions remain long enough between the rods for a few oscillations of the alternative potential to occur. This means that the kinetic energy at the source exit must range from one to a few hundred electronvolts. The weak potentials in the source allow a relatively large tolerance on the pressure. As the scan speed can easily reach $1000 \, \mathrm{Th \, s^{-1}}$ and more, this analyser is well suited to chromatographic coupling.

These quadrupoles also have the property of focusing the trajectory of the ions towards the centre of the quadrupole. Consider the diagram in Figure 2.9. The potential energy of the positive ion located at the centre of the rods of a quadrupole increases if it comes nearer a positive rod. Conversely, it decreases if it comes nearer to a negative rod. However, the alternative field actually spins the potentials, alternating the 'wells' and the 'hills'. If

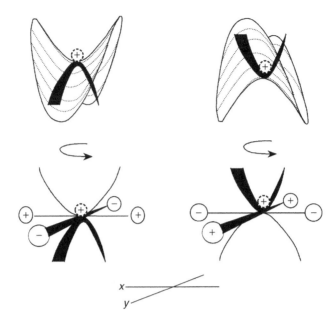

Figure 2.9
A positive ion, represented within a dotted circle, is at the
centre of a quadrupole rod, the potential signs of which are
indicated. It goes down the potential 'valley' with respect to
the negative rods, and acquires some kinetic energy in that
direction. However, the potentials quickly change, so that the
kinetic energy is converted into potential energy and the ion
goes back to the centre of the rods, as would happen for a ball
on a horse's saddle that is turned quickly. The name 'saddle
field' is an allusion to this phenomenon.

the frequency is sufficient, an ion that starts coming down the slope towards a negative
barrier is caught in the positive potential well and is thus brought back to the centre of the
quadrupole rods.

2.1.3 Ion Guide and Collision Cell

We saw that ions of all m/z ratios higher than some lower limit have a stable trajectory
when there is no direct potential as long as V is within the limit of their stability area.
When U is equal to zero (quadrupole operating in the RF only mode) all of the ions with
a mass higher than a given limit selected by adjusting the value of the RF voltage V have
a stable trajectory. But the transmission of ions with high masses suffers from their poorer
focusing. Indeed, the efficiency of focusing depends on the depth of the effective potential
well, which is inversely proportional to m/z. Consequently, ions with high m/z are weakly
focused and may be lost on the rods. Therefore, such quadrupole operating in the RF-only
mode causes all of the ions within the transmission mass range to be systematically brought
back to the centre of the rods, even if they are deflected by a collision or come close to the
rods because of their initial trajectories. The ions below this mass range are lost because

they follow an unstable trajectory, whereas the ions above this mass range are lost because they are poorly focused.

To increase the transmission of ions with high masses by a more efficient focalization, the RF voltage V is increased. Indeed, the depth of the effective potential well, which influences the efficiency of focusing, is proportional to V^2. However, the lower observable m/z limit will be higher if V is increased. There is thus a compromise to find between loss of heavy ions by lack of focusing and loss of light ions by unstable trajectory. When the mass spectrometer requires ion transmission across a wider mass range, the RF voltage V can be modulated during spectrum acquisition, providing a larger transmission window averaged over time. However, all heavy ions are poorly focused when V is low and all light ions are lost when V is high.

It is often necessary to transport ions as efficiently as possible from one point of the spectrometer to another without mass separation. For instance, ions produced in atmospheric pressure sources require to be transported across a varying pressure region to the analyser working under high vacuum. However, some phenomena lead to the loss of ions during their transport. Loss of a significant number of ions induces a low sensitivity if many of the ions formed in the source are lost before they reach the analyser. The most important event inducing the loss of ions is caused by their collision with the residual gas molecules. The trajectory of ions can also be affected by space charge effects. This effect is a consequence of the mutual repulsion between particles of like charge that causes beams or packets of charged particles to expand over time.

To obtain an effective transmission, it is necessary to reduce by focalization the loss of ions during their transport. The ion guide, also called an ion focusing device, is used for this purpose. They are designed to transport all ions efficiently and transmit them simultaneously. Thus, it is necessary to have good transmission efficiency across the largest mass range. In the same way as for the collision cells, the focusing of ions that are dispersed angularly by the collisions is important to increase the transmission of ions after the collisions.

As discussed above, the quadrupoles operating in RF-only mode have the property to focus the trajectory of ions. Therefore, they have been exploited in ion guides and collision cells. The use of quadrupoles as ion guides or ion focusing devices has been extended to other multipoles as hexapoles and octapoles. The principles of hexapole and octapole ion guides are similar to the principles applying to the quadrupole. They consist of six and eight parallel rods, respectively, placed symmetrically to form a long tube around the ion transfer axis. These rods are connected in two pairs of three or four non-adjacent rods, respectively. An RF voltage V is applied to the rods, with a polarity inverted from one rod to the next one. In this way, an effective potential, which presents a well shape, trap and focus effectively the trajectory of the ions in two dimensions along the axis of the guide. This potential, which determines if the ion travelling through the rod assembly is effectively focused, is given by the following equation:

$$U(r) = n^2 z^2 e^2 V^2 / (4mr_0^2\omega^2)(r/r_0)^{2n-2}$$

In this equation, $2n$ represents the number of rods and r is the radial distance from the centre of the rods.

As an ion moves from the centre of the multipole towards any one of the rods, the potential increases to reach a maximum at the surface of the rod. For the quadrupole ($n=2$), the potential varies as $(r/r_0)^2$, whereas the hexapole ($n=3$) and the octapole ($n=4$) have potentials that vary as $(r/r_0)^4$ and $(r/r_0)^6$, respectively. Consequently, the potential produced

Table 2.3   Main characteristics of different multipoles.

Type	Focusing power	Mass range for simultaneous transmission of ions
Quadrupole	High	Narrow
Hexapole	⇕	⇕
Octapole	Low	Wide

by the quadrupole quickly increases when the ion leaves the axis, while the octapole has a softer potential around the centre but a steeper potential close to the rods. Furthermore, for the same conditions, the maximum potential generated by an octapole has an amplitude that is four times higher than the potential generated by a quadrupole.

Owing to the shape of its potential well, the quadrupole ion guide is superior to the other multipoles in its power to focus ions towards the axis of the ion guide. Therefore, its transmission efficiency is better than the other multipoles. However, because of the weakest amplitude of its potential well, the excellent transmission efficiency of the quadrupole does not apply across a large mass range. Indeed, compared with the other multipoles, quadrupoles are characterized by the narrowest mass range for simultaneous transmission of ions. On the other hand, octapoles, which are exclusively used as ion guides, offer a lower power to focus ions but across a much larger mass range. Table 2.3 summarizes the main characteristics of these various multipoles.

It should be mentioned that the extent of the mass range for simultaneous transmission of ions is not important when the ion guide is combined with a scanning analyser. Indeed, the efficiency of transmission of the ion guide in the entire mass range of the analyser can be optimized by varying the RF voltage V during the scanning of the analyser. On the other hand, analysers such as TOF measure all the ions simultaneously. As a result, this kind of analyser requires ion guides that transmit all the ions together at the same time in the entire mass range of the analyser.

In conclusion, quadrupoles, hexapoles, octapoles and other multipoles operating in RF-only mode are able to focus efficiently ions from a mass range towards their axis without significantly affecting the axial kinetic energy of the transmitted ions. They act for a given mass range as an ion guide or ion focusing device by simultaneously transporting and transmitting these ions from one compartment of the mass spectrometer to another. As these devices work at atmospheric pressure as well as under high vacuum, they can serve to connect high- and low-pressure regions of the mass spectrometer. They can also serve as collision cells.

2.1.4 Spectrometers with Several Quadrupoles in Tandem

Figure 2.10 shows the general diagram of an instrument with three quadrupoles [8]. Quadrupole mass spectrometers are symbolized by upper case Q, and RF-only quadrupoles with a lower case q. A collision gas can be introduced into the central quadrupole at a pressure such that an ion entering the quadrupole undergoes one or several collisions.

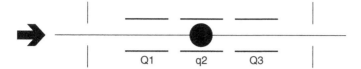

Figure 2.10
Diagram of a triple quadrupole instrument. The first and the last (Q1 and Q3) are mass spectrometers. The centre quadrupole, q2, is a collision cell made up of a quadrupole using RF only. The quadrupole mass spectrometers are symbolized by upper case Q, and the RF-only quadrupoles with a lower case q.

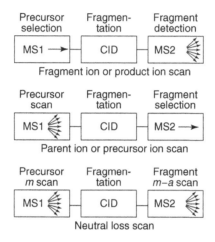

Figure 2.11
Different scan modes for a tandem mass spectrometer.

When the gas is inert, internal energy is transferred to the ion by converting a fraction of the kinetic energy into internal energy. The ion then fragments and the products are analysed by the quadrupole Q3. The kinetic energy for internal energy transfers is governed by the laws concerning collisions of a mobile species, the ion, and a static target, the collision gas. The collisions in these instruments occur at low energy. The conversion of kinetic to internal energy will be discussed in Chapter 4.

When the gas is reactive, ion–molecule reactions can be induced. The reaction products are then analysed by the quadrupole Q3.

These instruments with several analysers in series can be scanned in several ways. The most important ones are displayed in Figure 2.11.

The first scanning mode consists of selecting an ion with a chosen m/z ratio with the first spectrometer. This ion collides inside the central quadrupole and reacts or fragments. The reaction products are analysed by the second mass spectrometer. This is a 'fragment ion scan' or 'product ion scan'. This method used to be called a 'daughter scan'.

The second possibility consists of focusing the second spectrometer (Q3) on a selected ion while scanning the masses using the first spectrometer (Q1). All of the ions that produce the ion with the selected mass through reaction or fragmentation are thus detected. This method is called 'precursor scan' because the 'precursor ions' are identified. It used to be called 'parent scan'.

In the third common scan mode, both mass spectrometers are scanned together, but with a constant mass offset between the two. Thus, for a mass difference a, when an ion of mass m goes through the first mass spectrometer, detection occurs if this ion has yielded a fragment ion of mass $(m - a)$ when it leaves the collision cell. This is a 'neutral loss scan', the neutral having the mass a. For example, in chemical ionization the alcohol molecular ion loses a water molecule. Alcohols are thus detected by scanning a neutral loss of 18 mass units. On the other hand, a given mass increase can be detected if a reactive gas is introduced within the collision cell.

2.2 Ion Trap Analysers

An ion trap is a device that uses an oscillating electric field to store ions. The ion trap works by using an RF quadrupolar field that traps ions in two or three dimensions. Therefore, ion traps can be classified into two types: the 3D ion trap or the 2D ion trap.

Historically, the first ion traps were 3D ion traps. They were made up of a circular electrode, with two ellipsoid caps on the top and the bottom that creates a 3D quadrupolar field. These traps were also named quadrupole ion traps (QITs). To avoid confusion, this term should not be used but should be replaced preferably with Paul ion trap. The acronym QUISTOR derived from quadrupole ion storage is also largely used but not recommended.

Besides 3D ion traps, 2D ion traps have also been developed. They are based on a four-rod quadrupole ending in lenses that reflect ions forwards and backwards in that quadrupole. Therefore, in these 2D ion traps, which are also known as LITs, ions are confined in the radial dimension by means of a quadrupolar field and in the axial dimension by means of an electric field at the ends of the trap.

2.2.1 The 3D Ion Trap

2.2.1.1 General Principle

Paul and Steinwedel, the former being the inventor of the quadrupole analyser, described an 'ion trap' [9,10] in 1960. It was modified to a useful mass spectrometer by Stafford *et al.* [11] at the Finnigan Company. It is made up of a circular electrode, with two ellipsoid caps on the top and the bottom (Figure 2.12).

Conceptually, a Paul ion trap can be imagined as a quadrupole bent in on itself in order to form a closed loop. The inner rod is reduced to a point at the centre of the trap, the outer rod is the circular electrode, and the top and bottom rods make up the caps.

The overlapping of a direct potential with an alternative one gives a kind of '3D quadrupole' in which ions of all masses are trapped on a 3D trajectory (Figure 2.12). The inventors proposed the use of this ion trap as a mass spectrometer by applying a resonant frequency along z to expel the ions of a given mass.

In quadrupole instruments, the potentials are adjusted so that only ions with a selected mass go through the rods. The principle is different in this case. Ions of different masses are present together inside the trap, and are expelled according to their masses so as to obtain the spectrum.

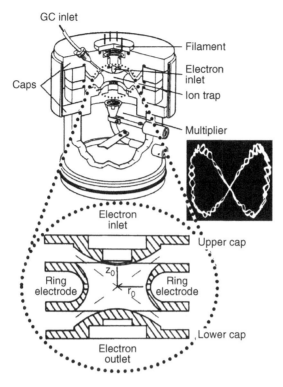

Figure 2.12
Top: a complete Paul ion trap mass spectrometer (Finnigan MAT). Bottom: detail of the trap. Right: figures obtained when injecting a fine dust of aluminium in a quadrupolar field such as is encountered in an ion trap. The particles altogether rotate along r and oscillate along z, yielding a figure-of-eight trajectory. From Finnigan MAT documentation. Reprinted, with permission.

As the ions repel each other in the trap, their trajectories expand as a function of the time. To avoid ion losses by this expansion, care has to be taken to reduce the trajectory. This is accomplished by maintaining in the trap a pressure of helium gas which removes excess energy from the ions by collision. This pressure hovers around 10^{-3} Torr (0.13 Pa). A single high-vacuum pump with a flow of about $40 \, \mathrm{l \, s^{-1}}$ is sufficient to maintain such a vacuum compared with the $250 \, \mathrm{l \, s^{-1}}$ needed for other mass spectrometers. The instrument is very simple and relatively inexpensive.

As is the case with quadrupole instruments, a potential Φ_0, the sum of a direct and an alternative potential, could be applied to the caps, and $-\Phi_0$ to the circular electrode. However, in most commercial instruments, Φ_0 is applied only to the ring electrode. In both cases, the resulting field must be seen in three dimensions.

Here again, a mathematical analysis using the Mathieu equations allows us to locate areas in which ions of given masses have a stable trajectory. These areas may be displayed in

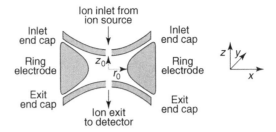

Figure 2.13
Schematic view of a 3D ion trap, and direction of
the x, y and z coordinates.

a diagram as a function of U, the direct potential, and of V, the amplitude of the alternative potential. The areas in which the ions are stable are those in which the trajectories never exceed the dimensions of the trap, z_0 and r_0. Such a diagram appears in Figure 2.18.

2.2.1.2 Theory of the Paul Ion Trap

In using the Mathieu equations to locate areas where ions of given masses have a stable trajectory, the equations are very similar to the ones used for the quadrupole. However, in the quadrupole, ion motion resulting from the potentials applied to the rods occurs in two dimensions, x and y, the z motion resulting from the kinetic energy of the ions when they enter the quadrupole field. In the Paul ion trap the motion of the ions under the influence of the applied potentials occurs in three dimensions, x, y and z. However, due to the cylindrical symmetry $x^2 + y^2 = r^2$, it can also be expressed using z, r coordinates (Figure 2.13).

The equations of the movement inside the trap are similar to those encountered for the quadrupole analyser (Section 2.1.2):

$$\beta_u = \left[a_u + \left(q_u^2/2\right)\right]^{1/2}$$

$$\frac{d^2z}{dt^2} - \frac{4ze}{m\left(r_0^2 + 2z_0^2\right)}\left(U - V\cos \omega t\right)z = 0$$

$$\frac{d^2r}{dt^2} + \frac{2ze}{m\left(r_0^2 + 2z_0^2\right)}\left(U - V\cos \omega t\right)r = 0$$

In these equations, z is used for the number of charges, to avoid confusion with the z coordinate. The general form of the Mathieu equation, whose solutions are known, is

$$\frac{d^2u}{d\xi^2} + \left(a_u - 2q_u\cos 2\xi\right)u = 0$$

u stands for either z or r. Thus, the following change of variables gives to the equations of motion the form of the Mathieu equation, where u stands for either r or z:

$$\xi = \frac{\omega t}{2}, \quad a_u = a_z = -2a_r = \frac{-16zeU}{m\left(r_0^2 + 2z_0^2\right)\omega^2}, \quad q_u = q_z = -2q_r = \frac{8zeV}{m\left(r_0^2 + 2z_0^2\right)\omega^2}$$

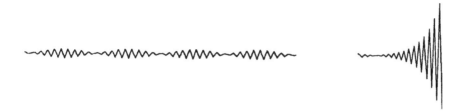

Figure 2.14
Left: a type of stable trajectory, corresponding to a purely imaginary solution of
the Mathieu equation, requiring both $\alpha = 0$ and $0 < \beta < 1$. Right: a continuously
amplified movement resulting from a not purely imaginary solution. If this occurs
along r, the ions will discharge on the wall; if along z, the ions will be expelled and
50 % will reach the detector, the others being expelled in the opposite direction.

To have a stable trajectory, the movement of the ions must be such that during this
time the coordinates never reach or exceed r_0 (r-stable) and z_0 (z-stable). The complete
integration of the Mathieu equation by the method of Floquet and Fourier requires the
use of a function $e^{(\alpha + i\beta)}$. Real solutions correspond to a continuously increasing, and thus
unstable, trajectory. Only purely imaginary solutions correspond to stable trajectories. This
requires both $\alpha = 0$ and $0 < \beta_u < 1$ (Figure 2.14).

This β_u parameter can be calculated from the q_u and a_u parameters of the Mathieu
equation. Exact values require the use of long series of terms. The following equation
allows approximate values to be obtained [12]:

$$\beta_u = \left[a_u - \frac{(a_u - 1)\, q_u^2}{2\,(a_u - 1)^2 - q_u^2} - \frac{(5a_u + 7)\, q_u^4}{32\,(a_u - 1)^3\,(a_u - 4)} - \frac{\left(9a_u^2 + 58a_u + 29\right) q_u^6}{64\,(a_u - 1)^5\,(a_u - 4)\,(a_u - 9)} \right]^{1/2}$$

A simpler approximate equation holds for q_u values lower than 0.4:

$$\beta_u = \left[a_u + (q_u^2/2) \right]^{1/2}$$

Figure 2.15 displays the iso-β lines for $\beta_u = 0$ and $\beta_u = 1$ respectively. One of the
two diagrams refers to the z coordinate and the other to the r coordinate. Remember that
$a_u = a_z = -2a_r$ and $q_u = q_z = -2q_r$. The area inside these limit values for β_u corresponds to
a, q values for a stable trajectory. However, for an ion to be stable in the ion trap, it must
have a stable trajectory along both z and r, which correspond to the overlap of the stability
areas of the two diagrams.

This overlap, as displayed in Figure 2.16, shows common areas at different places.
Commercial ion traps use the first stability region, which is circled and enlarged in the
figure.

This stability region is displayed again in Figure 2.17, where iso-β curves are drawn for
intermediate values between 0 and 1.

These areas may also be displayed in a diagram as a function of U, the direct potential,
and of V, the amplitude of the alternative potential. The areas in which the ions are stable
are those in which the trajectories never exceed the dimensions of the trap, z_0 and r_0. Such
a diagram appears in Figure 2.18.

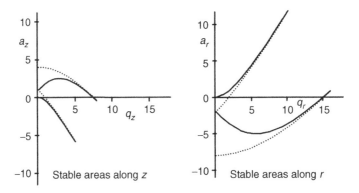

Figure 2.15
Stability diagram along r and z respectively for a 3D ion trap. The iso-β lines for $\beta_u = 0$ (solid lines) and $\beta_u = 1$ (dotted lines) are drawn. The areas inside these limits correspond to stable trajectories for the considered coordinate. They correspond to imaginary solutions of the Mathieu equation.

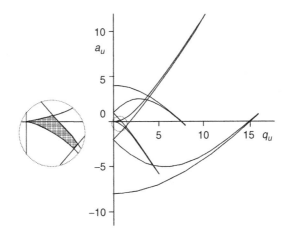

Figure 2.16
Stability areas along both r and z in a 3D ion trap. The common r and z stability area used in commercial ion traps is enlarged. This is displayed again in Figure 2.18. Redrawn and (modified) from March R.E. and Hughes R.J., Quadrupole Storage Mass Spectrometry, Wiley, New York, 1989, with permission.

Two other important parameters depending on q_z or β will be considered. First, if an RF voltage with an angular velocity $\omega = 2\pi\nu$ is applied to the trap, the ions will not oscillate at this same 'fundamental' ν frequency because of their inertia, which causes them to oscillate at a 'secular' frequency f, lower than ν, and decreasing with increasing masses. It should be noted that a_u and q_u, and thus β, are inversely proportional to the m/z ratio. The relation

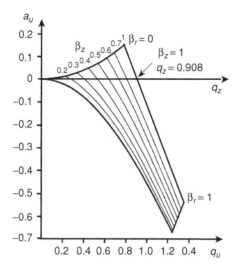

Figure 2.17
Typical stability diagram for a 3D ion trap. The value at $\beta_z = 1$ along the q_z axis is $q_z = 0.908$. At the upper apex, $a_z = 0.149\,998$ and $q_z = 0.780\,909$. Drawn with data taken from March R.E and R.J. Hughes, Quadrupole Storage Mass Spectrometry, Chemical Analysis Vol. 102, Wiley Inter-science, 1989.

between v and f along the z axis is

$$f_z = \beta_z v/2$$

As the maximum value of β for a stable trajectory is $\beta = 1$, the maximum secular frequency f_z of an ion will be half the fundamental v frequency. We will see later on that this is important for ion excitation or for resonant expulsion.

The second important parameter which is a function of q_z is the Dehmelt pseudopotential well. The trapping efficiency of ions injected in the trap can be described using the pseudopotential well given by the following equation:

$$\overline{D_z} = q_z \frac{V}{8} = \frac{eV^2}{m(r_0^2 + 2z_0^2)\omega^2}$$

2.2.1.3 Injection or Production of Ions in the Trap

The first commercially available ion traps where coupled to a gas chromatograph (GC), which provided helium for ion cooling. Ionization occurred in the trap by injection of electrons through an end cap. Some GC/MS^n ion traps still use this internal ionization. Most instruments have an external source. Figure 2.19 shows a popular ion trap instrument with an external ion source, the Finnigan LCQ. Ions produced in the source are focused

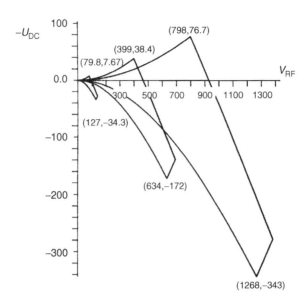

Figure 2.18
Stability areas [14] for ions simultaneously stable along
r and z. The mass-to-charge ratios of the ions displayed
are 10, 50 and 100 Th. The trap being considered has a
diameter of 1 cm and operates at a frequency of 1.1 MHz
[14]. Under such conditions, all the ions have a stable trajectory for $U = 0$ if their mass is lower than $0.108 \times V$.
For instance, a 1 Th ion has a stable trajectory up to
$V = 1/0.108 = 9.26$ V. As shown, a 50 Th ion is unstable starting from $V = 463$ V. Increasing V destabilizes
the trajectories of ions having an increasingly high mass-
to-charge ratio. Reproduced from March R.E. and R.J.
Hughes, Quadrupole Storage Mass Spectrometry, Chemical Analysis Vol. 102, Wiley Interscience, 1989 pg 199,
with permission.

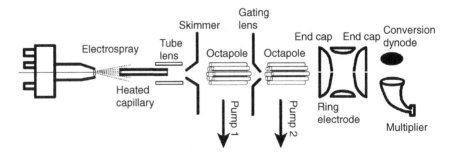

Figure 2.19
A 3D ion trap with an external ESI source (Finnigan LCQ). Ions produced by the
ESI source are focused through a skimmer and octapole lenses to the ion trap. The
gating lens is used to limit the number of ions injected in the trap, to avoid space
charge effects (see further).

through a skimmer and two RF-only octapoles. Differential pumping ensures the vacuum in the trap, while the source is at atmospheric pressure. A gating lens limits the number of injected ions to an acceptable limit. Indeed, too many ions in the trap causes a loss of resolution, and too few a loss of sensitivity. The gating lens acts as follows. For positive ions, a positive potential is applied, except during injection of ions, which occurs when a negative potential is applied. The following sequence of events is used. First, a negative potential is applied for a fixed time, and the number of ions in the trap is measured by expelling them together to the detector. From the value obtained, a gating time is calculated so as to inject the selected number of ions.

2.2.1.4 Mass Analysis of Ions in a Paul Ion Trap

After ion injection, the ions of different masses are stored together in the trap. They must now be analysed according to their mass. Most commercial traps operate by applying a fundamental RF voltage to the ring. Its frequency is constant, but its amplitude V can be varied. Additional RF voltages of selected frequencies and amplitudes can be applied to the end caps. Figure 2.20 schematically represents the wiring of such an ion trap. In this

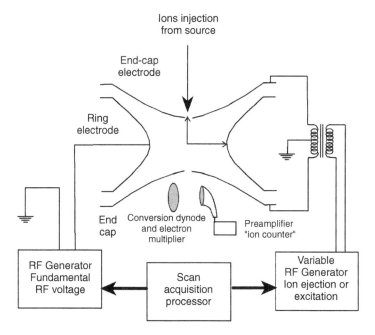

Figure 2.20
Ion trap with an RF voltage applied to the ring electrode, providing the fundamental frequency v and its associated variable amplitude V. Instead of injecting ions, electrons may be injected for internal ionization. Variable RF voltage can be applied to the end caps for ion excitation or ion ejection.

example, no DC voltage is applied. Other ion traps allow the application of DC voltages too. To explain the principle of ion analysis, we will consider an ion trap wired as in Figure 2.20.

2.2.1.5 Mass Analysis by Ion Ejection at the Stability Limit

As no DC voltage is applied, the 3D trap will be operated along the q_u axis, because in the absence of DC voltage, $a_u = 0$. As already explained, q_z is given by the following equation:

$$q_z = \frac{8zeV}{m\left(r_0^2 + 2z_0^2\right)\omega^2}$$

As the frequency is fixed, ω is a fixed value. We will consider only monocharged ions ($z = 1$): e, the elementary charge, is a constant and r_0 and z_0 are constant for a given trap; q_z will increase if V increases, and decrease if m increases. This is represented in Figure 2.21, where higher mass ions are represented by larger balls. If V is increased, all the ions will have a higher q_z value. If this value is equal to 0.908, β is equal to one, and the ion

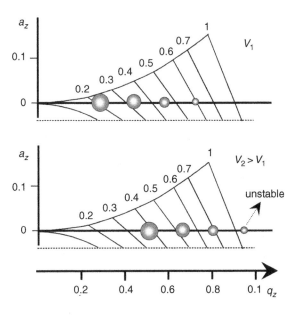

Figure 2.21
At a fixed value of the RF potential V applied to the ring electrode, heavier ions will have lower β_z values, and thus lower secular frequencies. If V is increased, β_z values increase for all the ions, and thus also the secular frequencies. In the example given, the lightest ion now has a β_z value larger than one, and is thus expelled from the 3D trap. The highest mass that can be analysed depends on the limit V value that can be applied: around 7000 to 8000 V zero to peak. For a trap having r_0 equal to 1 cm and operating at a frequency v of 1.1 MHz, the highest detectable mass-to-charge ratio is about 650 Th.

has reached its stability limit. A slight increase of V will cause this ion to have an unstable trajectory, and will be expelled from the trap in the z direction. Thus 50 % of the expelled ions will reach the detector. This allows the ions present in the trap to be analysed.

If we again consider the equation for q_z, we can write an equation with these limit values giving the maximum observable mass m_{MAX}:

$$q_z = \frac{8zeV}{m\left(r_0^2 + 2z_0^2\right)\omega^2} \qquad m_{MAX} = \frac{8ze\,8000}{0.908\,(r_0^2 + 2z_0^2)(2\pi v)^2}$$

Example of calculation

$$q_z = \frac{8zeV}{m\left(r_0^2 + 2z_0^2\right)\omega^2}$$

Let us refer to the first generation of well-known LCQs for which the parameters are as follows: $r_0 = 7.07$ mm, $z_0 = 7.83$ mm. In this equation it must be expressed in metres, and thus

$$(r_0^2 + 2z_0^2) = 1.726\ 10^{-4}\,\mathrm{m}^2$$

The RF 'zero-to-peak' voltage expressed in volts, $V_{0\rightarrow P}$, is variable up to a maximum of 8500 V.

The fundamental RF frequency v is equal to 0.76 MHz or 7.6×10^5 Hz; $\omega = 2\pi v$;

$$\omega^2 = (6.28 \times 7.6 \times 10^5)^2 = 2.278 \times 10^{13}\ \mathrm{rad\,s^{-1}}$$

$z =$ the number of charges, thus an entire number: 1 for a monocharged ion.
$e =$ the elementary charge (absolute value of the charge of one electron): 1.602×10^{-19} C.

The equation can now be written with all the constant terms:

$$q_z = \frac{8zeV}{m\left(r_0^2 + 2z_0^2\right)\omega^2} \qquad q_z = \frac{8(1.602 \times 10^{-19})}{(1.726 \times 10^{-4})(2.278 \times 10^{13})} \times \frac{zV}{m}$$

In this equation, m is expressed in **kg per ion**. If we want the classical value in atomic mass units (1 u $=$ 1 Da $= 1.66\,022 \times 10^{-27}$ kg), denoted m', we must multiply the equation by 1000 to go from kg to g, and by 6.02×10^{23} for 1 mole of ions rather than one ion. We then get the following equation, where of course the constant 0.1963 is valid only for the first series LCQ!

$$q_z = \frac{8(1.602 \times 10^{-19})(1000)(6.02 \times 10^{23})}{(1.726 \times 10^{-4})(2.278 \times 10^{13})} \times \frac{zV}{m'} = 0.1963 \times \frac{zV}{m'}$$

What is the value of q_z for an ion having a mass m' of 1500 u, monocharged ($z = 1$) at 5000 V RF amplitude? The answer is

$$q_z = 0.1963 \times \frac{(1)(5000)}{1500} = 0.6543$$

As this value of q_z is lower than 0.908 (the $\beta = 1$ stability limit), this ion will have a stable trajectory in this trap operated at 5000 V amplitude of the fundamental RF.

Thus, besides trying to increase V at higher values without arcing, the maximum observable mass can be increased by reducing the size of the trap or using a lower RF frequency v. For example, if r_0 and z_0 are reduced each to 7 mm and the RF frequency to 600 kHz, and the maximum value for V is 8000 V, the mass limit will become about 3260 Th.

2.2.1.6 Mass Analysis by Resonant Ejection

If we remember that the secular frequency at which an ion oscillates in the 3D trap is given by

$$f_z = \beta_z v/2$$

and remembering that β_z increases if q_z increases (Figure 2.17), it is possible to calculate the value of V to be applied for the ions of a given mass to oscillate at a selected frequency f_z.

If an RF voltage at frequency f_z is applied to the end caps, thus along the z axis, the ion will resonate, and the amplitude of its oscillations will increase. If the applied amplitude is sufficient, the oscillations of the ion will be so large that it will be destabilized and ejected from the trap in the z direction (Figure 2.22).

If this frequency is near to $v/2$, β_z will be near to one, it is at the stability limit ($q_z = 0.908$). If, however, the applied frequency corresponds to a lower q_z, the ions will be ejected at a lower V value. Thus, for the same maximum V value, the highest ejectable mass will be higher. This is another way to increase the mass range. For example, the Bruker Esquire instrument allows ions at $\beta_z = 2/11$ to be ejected, corresponding to $q_z = 0.25$, which allows the mass range to be extended up to 6000 Th.

It should be noted that if an ion fragments during the analysis, it is possible that its m/z ratio is such that its q_z value is higher than the resonant ejection value. If, later on, by increasing V, it reaches the stability limit ($q_z = 0.908$) and is ejected, it will then be detected at a wrong m/z, as the data system expects it to be expelled by resonance. Its apparent m/z will be higher than the true one. These 'ghost peaks' will occur more if the resonance frequency corresponds to a lower q_z value.

Figure 2.23 shows the sequence of events for a resonance ejection analysis in a pictorial way [15]. Note that in this figure the supplementary applied voltage on the end caps is designated as 'AC'.

2.2.1.7 Tandem Mass Spectrometry in the Ion Trap

There are several ways to perform tandem mass spectrometry in an ion trap. Time-dependent rather than space-dependent tandem mass spectrometry occurs in the trap. The general sequence of operations is as follows:

1. Select ions of one m/z ratio, by expelling all the others from the ion trap. This can be performed either by selecting the precursor ion at the apex of the stability diagram or by resonant expulsion of all ions except for the selected precursor.

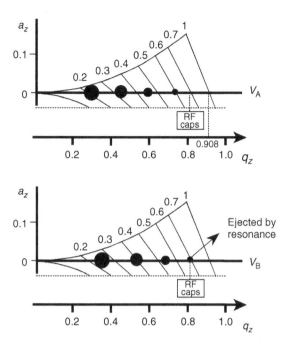

Figure 2.22
Principle of resonant ejection. Upper: ions are stored
in the 3D trap at a voltage V_A of the fundamental
RF. An additional RF is applied to the end caps corresponding to $q_z = 0.8$. On increasing V (lower panel)
ions are moved to higher q_z values. In the figure, the
smallest ion has reached $q_z = 0.8$ and is ejected by
resonance. This ejection occurs at a lower value of
V than the one needed to eject by instability at $q_z = 0.908$.

2. Let these ions fragment. Energy is provided by collisions with the helium gas, which is
   always present. This fragmentation can be improved by excitation of the selected ions
   by irradiation at their secular frequency.

3. Analyse the ions by one of the described scanning methods: stability limit or resonant
   ejection.

4. Alternatively, select a fragment in the trap, and let it fragment further. This step can be
   repeated to provide MS^n spectra.

As a first example, let us look at an MS/MS/MS or MS^3 experiment performed by
selection of ions at the stability apex. Figure 2.24 shows how correctly selecting values of
U and V allows only the ions with a given m/z ratio to be trapped. These ions fragment
in time within the trap, which then contains all of the product ions. The latter can then be
analysed by selective expulsion. Time-dependent, instead of space-dependent, MS/MS is

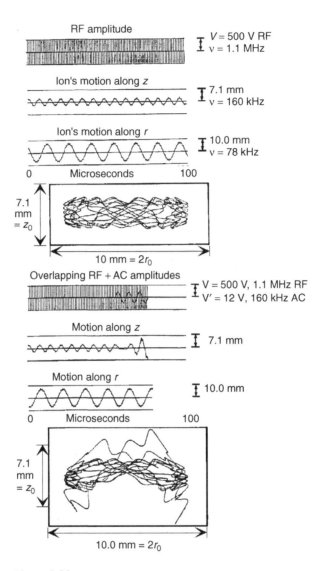

Figure 2.23
Resonant ejection of a 100 Th ion. Ions of a given mass-to-charge ratio have a characteristic oscillation frequency along both r and z. Superimposing on the fundamental RF potential an adjustable AC potential, with a frequency equal to that of the characteristic ion oscillation, transfers energy to the ion through resonance. In this example, 160 kHz is chosen. The ion trajectory along z is thus destabilized. This resonant mode allows the ejection of high-mass ions. The boxes display computer simulations of the ion trajectories. Reproduced (modified) from Cooks R.G., Glish G.L., McLuckey S.A. and Kaiser R.E., Chem. Eng. News, 69, 26, 1991, with permission.

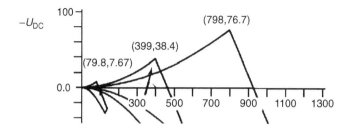

Figure 2.24
Fragment of Figure 2.18. The arrow indicates a point where
only 50 Th ions have a stable trajectory. Reproduced (modified)
from March R.E. and Hughes R.J., Quadrupole Storage Mass
Spectrometry, Wiley, New York, 1989, with permission.

thus carried out. The process can be repeated several times, selecting successive fragment
ions.

Figure 2.25 shows a typical sequence of operations used to observe third-generation
ions.

A second example describes the use of resonant ejection of ions by selected-waveform
inverse Fourier transform (SWIFT). Figure 2.26 describes an MS/MS experiment with an
instrument using RF voltages applied to the caps, but no DC voltage. In this example, the
final analysis of the fragments is performed by the stability limit method.

To fragment the ions by resonance excitation, an RF voltage must be applied to adjust
the q_z value for the selected m/z ratio to the frequency of the RF generator used. This means,
if we remember that increasing V causes the ejection of the lower mass ions, that the lower
observable m/z limit will be higher if the applied V voltage is increased. The lowest m/z
fragments are lost for this reason. In general, fragments with m/z values lower than about
20 % of the precursor ion's m/z will be lost. Thus there is an advantage to work at lower V,
which means lower q_z, to have a larger mass limit. But at higher V, the Dehmelt potential
is higher, and it is proportional to V^2 [16]:

$$\overline{D_z} = q_z \frac{V}{8} = \frac{ze V^2}{m(r_0^2 + 2z_0^2)\omega^2}$$

Figure 2.27 displays this Dehmelt potential graphically as a function of q_z, itself a function
of V.

Actually, when an ion produces fragments, they need to be efficiently trapped, which
depends on the value of the Dehmelt potential. Otherwise, parts of the fragments are lost by
lack of trapping efficiency. There is thus a compromise to find between loss of sensitivity
and loss of lower m/z fragment ions.

The gas used to slow the ions within the trap is normally helium, which has a mass of
4 Da. The ion kinetic energy amounts to only a few electronvolts. The maximum fraction
of kinetic energy convertible into internal energy is given by the equation discussed in
Chapter 4:

$$E_t = E_c \frac{m_2}{(m_1 + m_2)}$$

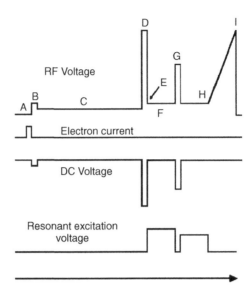

Figure 2.25
Computer-controlled sequence of operations used to carry out a typical MS/MS/MS experiment in a 3D ion trap. A: electron impact ionization of the chemical ionization gas. B: correct tuning of U (DC voltage) and V (RF voltage) allows the selection of ions with a given m/z for chemical ionization. C: 200 ms long protonation of the analyte by the chosen ion. D: the parent ion is selected. E: V is increased in order to select a mass range to observe the fragments, while keeping the surviving precursors in the trap. F: resonant excitation is used to fragment the selected precursor ions. G: one of the fragments is selected. H: resonant excitation is used to fragment this fragment. I: mass scanning to observe the second-generation fragments.

Such a weak energy is often insufficient to induce ion fragmentation within a short time span. The kinetic energy of the ions can be increased by superimposing their resonant frequency on the RF with an amplitude that is weak enough not to expel them from the trap.

In a quadrupole collision cell, the ions undergo multiple collisions. The fragments, as soon as they are formed, are reactivated by collision and can fragment further. In the ion trap, if excitation occurs by irradiation at the secular frequency of the precursor, only this ion is excited, and the product ion may be too cool to fragment further. Figure 2.28 shows an example of this behaviour [17].

In the ion trap, instead of selectively irradiating the precursor ion, broadband excitation can be used: irradiation occurs over a large range of frequencies, so that all the ions in the trap are excited, including the fragment formed.

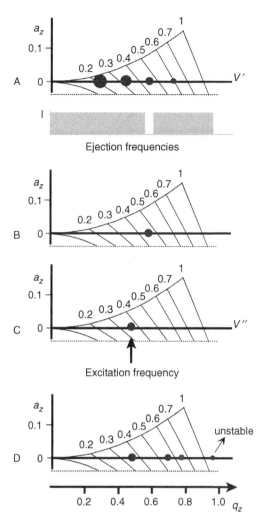

Figure 2.26
One possible sequence of events to produce an
MS/MS spectrum. A: ions of one mass-to-charge
ratio are selected by expelling all the others at
their resonance frequency applied to the caps. B:
Only ions of the selected m/z are present in the
trap. C: V is adjusted so as to bring the ion into
resonance with the excitation frequency applied
to the caps. D: ions are analysed by ejection at
the stability limit.

The ion trap allows MS^n experiments to be performed. For this purpose, an ion produced
in the source is selected and fragmented. One of the fragments is selected and fragmented
again. This process can be repeated. As an example, Figure 2.29 displays the MS^2 and MS^3
spectra obtained from permethylated oligosaccharides LNT (Gal(β1–3) GlcNac(β1–3)
Gal(β1–3) Glc) and LNnT, which differs by the 1–4 position of the first galactose

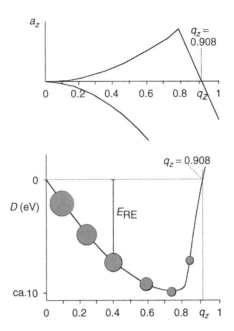

Figure 2.27
Dehmelt potential D in electronvolts (eV)
(lower graph) as a function of q_z. Heavier ions, for a given V value, have lower
Dehmelt trapping energies. An increase in
V will displace all the ions along the curve
to higher q_z values. If q_z passes the 0.908
value, the ion will be expelled. The E_{RE} line
represents the energy that has to be provided by resonant excitation in order to
expel the selected ion from the trap. The
top graph resembles the stability diagram.
The ordinate is not the same for the two
diagrams.

residue with the N-acetylglucosamine (Gal(β1–4) GlcNac(β1–3) Gal(β1–3) Glc). The
spectra show that the two different terminal disaccharides can easily be distinguished
[18]. In this paper it is also shown that these same disaccharides originating from larger
oligosaccharides yield the same fragmentation spectra, even if several fragmentation steps
are used, for instance MS^4, MS^5 or more, provided the collision energy in the last step is the
same.

2.2.1.8 Space Charge Effect

When too many ions are introduced into an ion trap, those located at the outside will act as
a shield. The field acting on the ions located at the inside will be modified, and the shape
of the stability diagram will be modified as displayed in Figure 2.30.

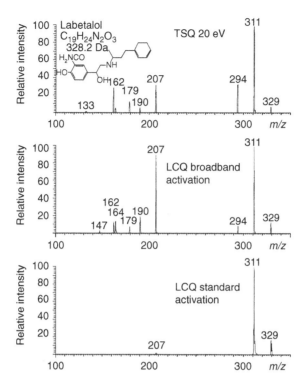

Figure 2.28
Top: fragment ion obtained from ESI protonated Labetalol molecular ion in a triple quadrupole instrument. Centre: same spectrum obtained in an ion trap with broadband excitation. All the ions, precursor or fragment, are activated by resonant irradiation. Bottom: same spectrum as obtained with an ion trap when only the precursor ion is activated. Redrawn from Senko M.W., Cunniff J.B. and Land A.P., 'Producing "Richer" Product Ion Spectra on a Quadrupole Ion Trap with Broadband Activation', Proceedings of the 46th ASMS Conference on Mass Spectrometry and Allied Topics, Orlando, Florida, p. 486, 1998, with permission.

2.2.2 The 2D Ion Trap

2.2.2.1 General Principle

The 2D ion trap, also known as the linear ion trap (LIT), is an analyser based on the four-rod quadrupole ending in lenses that repel the ions inside the rods, and thus at positive potentials for positive ions, and vice versa. In these traps, ions are confined in the radial dimension by means of a quadrupolar field and in the axial dimension by means of an electric field at the ends of the trap.

The LIT principles were recently reviewed by Douglas [19]. Once in the LIT, the ions are cooled by collision with an inert gas and fly along the z axis between the end electrodes,

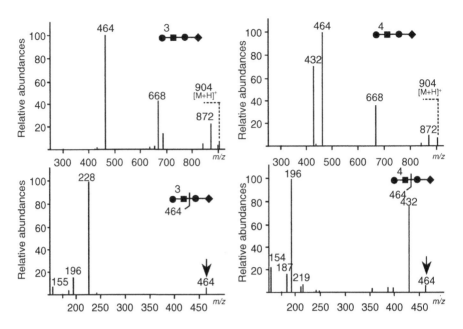

Figure 2.29

MS^2 and MS^3 mass spectra from two identical oligosaccharides except for the position of the first galactose residue on the N-acetylglucosamine. These same disaccharides originating from more complex oligosaccharides produce these same fragmentations. Reproduced (modified) from Viseux N., de Hoffmann E. and Domon B., Anal. Chem., 69, 3193–3198, 1997, with permission.

while simultaneously oscillating in the *xy* plane owing to the application of an RF-only potential on the rods. The rods are furthermore often divided into three segments. Application of an additional DC voltage to the end parts of the quadrupole also allows the ions to be trapped. This can be used either without end electrodes or together with these electrodes.

These voltages repel the ions inside the linear trap, and this repulsion is higher when the ions are closer to the ends. Ions are thus repelled towards the centre of the quadrupole if the same repelling voltages are applied at each end. Thus, the ion cloud will be squeezed at the centre of the quadrupole if the applied voltages are symmetrically applied, but can be located at closer to one end if the repelling voltage at that end is smaller.

One great advantage of LITs in comparison with Paul ion traps is a more than 10-fold higher ion trapping capacity. Furthermore, this higher trapping capacity is combined with the ability to contain many more ions before space charge effects occur owing to both a greater volume of the trap and a focusing along the central line rather than around a point. Even with 20 000 trapped ions, well resolved mass spectra can be obtained without any space charge effects in 2D ion traps, whereas more than 500 trapped ions in 3D traps induce space charge effects. Another common advantage is the higher trapping efficiency. Indeed, a trapping efficiency of more than 50 % is achieved when ions are injected into the 2D ion trap from an external source, while this trapping efficiency is only 5 % for the 3D ion trap. Both of these advantages increase the sensitivity and the dynamic range.

Ions trapped within an LIT can be mass selectively ejected either along the axis of the trap (axial ejection) or perpendicular to its axis (radial ejection). Therefore, in commercial

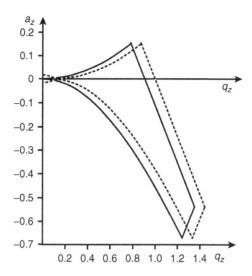

Figure 2.30
The dashed line represents the shift of the stability diagram resulting from the space charge effect. To reach the stability limit $\beta = 1$, q_z, and thus V, has to have a higher value. This could lead to an error on the mass if proper caution is not taken.

LITs two modes for the mass selective ejection of ions are used: either the ions are expelled axially using fringe field effects by applying AC voltages between the rods of the linear trap and the exit lens, or slots are hollowed out in two opposite rods and mass selective radial expulsion of ions is obtained by applying an appropriate AC voltage on these two rods.

2.2.2.2 Axial Ejection in Linear Trap

The linear trap with axial ejection was invented by Hager, from MDS Sciex, in 2002 [20]. Figure 2.31 displays a scheme of such an ion trap included in the ion path of a triple quadrupole mass spectrometer.

As shown in Figure 2.32, the ions are expelled axially using fringe field effects by applying AC voltages between the rods of the linear trap and the exit lens.

This instrument can be operated as a normal triple quadrupole with all its scan modes or as a trap in various combinations with the use of the other quadrupoles. If a slow scan rate is used to expel the ions a resolution up to 6000 FWHM can be reached by scanning at 5 Th s^{-1} using q_2 and at 100 Th s^{-1} using Q$_3$, which is at a lower pressure. As fringe field effects are used, only ions close to the exit lens are expelled. In consequence, mass selective ejection in the axial direction based on this technique is characterized by low ejection efficiency. For instance, an ejection efficiency of less than 20 % is achieved at 1 Th ms^{-1} scan rate. Different techniques have been proposed to improve the axial ejection efficiency [21], but the most promising technique for mass selective axial ejection is the technique named axial resonant excitation (AREX) [22]. Lenses are introduced between each rod of the quadrupole

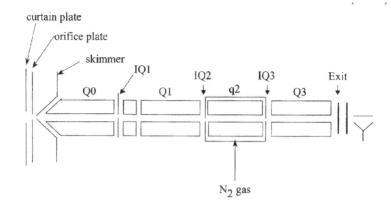

Figure 2.31
View of the linear trap included in a triple quadrupole at q2. This instrument can be operated as a regular triple quadrupole or with a trap. Reproduced from Hager J.W., Rapid Comm. Mass Spectrom., 16, 512–526, 2002, with permission.

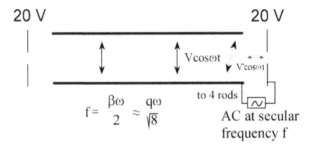

Figure 2.32
Imaged view suggesting the principle that, near the exit, the trajectory of the ions has a component along the axis at the same frequency as the radial oscillation. This can be used to expel the ions by applying an AC voltage at the same frequency between the quadrupole and the exit lens or by increasing the RF amplitude V (radial excitation).

to induce an electrostatic potential that is approximately harmonic along the central axis of the quadrupole. Inside this potential, ions with a given m/z can oscillate resonantly in the axial direction by superposing a supplemental AC field. High ejection efficiency has been observed. For instance, an ejection efficiency of more than 60 % is achieved at 10 Th ms^{-1} scan rate, three times more than with the technique using a fringe field. However, in these conditions, the mass resolution is about 1000.

2.2.2.3 Radial Ejection in Linear Trap

Radial ejection between the rods has been described [19] but never applied to commercial instruments. Ejection through slots cut in two opposite rods was described first by Senko and Schwartz from Thermo Finnigan in 2002 [23, 24].

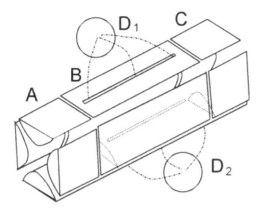

Figure 2.33
Linear trap with slots cut in two opposite rods.
Sizes are 12 mm for sections A and C and 37 mm
for B. Detectors D are placed off-line and ions
are attracted by the conversion dynodes. The
slots are 30 × 0.25 mm. Drawn according to
the data from Schwartz J.C., Senko M.W. and
Syka J.E.P., 'A Two-Dimensional Quadrupole
Ion Trap Mass Spectrometer', Proceedings of
the 50th ASMS Conference on Mass Spectrometry and Allied Topics, Orlando, Florida, 2002.

Figure 2.33 represents such a linear trap. The two detectors allow the use of all the ions expelled from the trap. Trapping efficiency is in the range 55–70 % while it is only 5 % in the Paul ion trap. Unit resolution is achieved at $16\,700\,\mathrm{Th\,s^{-1}}$ scan rate. At 27 $\mathrm{Th\,s^{-1}}$, $\Delta m = 0.05$ is observed at m/z 1520, corresponding to a resolution of 30 000 FWHM. The ion capacity is about 20 000, 40 times more than in the Paul ion trap.

The presence of the slots causes a perturbation of the RF field that can be reduced by slightly stretching the quadrupole, increasing the distance between the cut rods.

Mass selective ejection of the ions in a radial direction occurs by applying an AC voltage between the two cut rods. As for the 3D ion trap, an AC frequency corre-sponding to $q_z = 0.88$ is used. Ions of successively higher masses are brought to this q_z value by increasing V. An ejection efficiency of about 50 % is achieved at 5 $\mathrm{Th\,s^{-1}}$ scan rate [25].

Operations similar to the 3D traps can be performed, as for example to expel ions of all masses except one and observe the fragmentation, with or without ion excitation at the secular frequency. Then the fragments are analysed. This can be repeated several times for MS^n experiments. All the other operations of a 3D trap can be applied, but it also has similar limitations, for example MS/MS is limited to fragmentation scans. Thus precursor ion scan or neutral loss scan that are available with triple quadrupole instruments cannot be used on ion traps (Figure 2.11).

Segmented quadrupoles are used rather than trapping by end electrodes, to avoid pro-ducing fringe fields. To trap the ions, a DC voltage is applied to the end sections. Coupling to the source occurs through an interface comprising focusing multipoles, as described for the 3D ion trap.

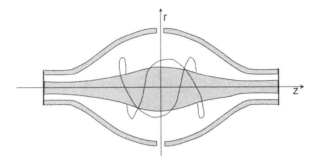

Figure 2.34
The electrostatic trap or 'orbitrap' according to the description of Makarov in [27]. An outer electrode has the shape of a barrel cut into two parts separated by a small gap. Inside is a second spindle-shaped electrode. Only DC voltages (some kilovolts) are applied. The ions turn around the central electrode while oscillating back and forth along the z axis. The resulting movement is made of intricate spirals.

2.3 The Electrostatic Trap or 'Orbitrap'

The orbitrap is an electrostatic ion trap that uses the Fourier transform to obtain mass spectra. This analyser is based on a completely new concept, proposed by Makarov and described in patents in 1996 [26] and 2004 [27], and in *Analytical Chemistry* in 2000 [28]. A third patent describes a complete instrument including an atmospheric pressure source [29]. Another article was also published with Cooks in 2005 [30]. The first commercial instrument was introduced on the market by the Thermo Electron Corporation in June 2005.

Figure 2.34 shows a face view of this trap. The external part is an electrode having the shape of a barrel cut into two equal parts with a small interval. The central electrode has a spindle shape. The maximum diameter of the central electrode is 8 mm and of the outer one 20 mm. The ions are injected tangentially through the interstice between the two parts of the external electrode (see Figure 2.35, side view K). In the first commercial instrument (2005), injection occurs off-axis through a hole in the external electrode. An electrostatic voltage of several kilovolts, negative for positive ions, is applied to the central electrode, while the outer electrode is at ground potential. The ions are injected with a kinetic energy of some kiloelectronvolts and start to oscillate in the trap in intricate spirals around the central electrode under the influence of an electrostatic field with a quadro-logarithmic potential distribution that is obtained by the DC voltage and the astute geometry of the trap.

With this instrument, at injection the outer electrode is about at ground, while the inner electrode is set at about -3200 V for positive ions. To have a reasonably circular or oval trajectory around the centre electrode, the ions need to have a kinetic energy around 1600 eV. The ions start to rotate around the inner electrode and together oscillate along the z axis.

The equation of the potential inside the trap is

$$U(r, z) = \frac{k}{2}\left(z^2 - \frac{r^2}{2}\right) + \frac{k}{2}(R_m)^2 \ln\left(\frac{r}{R_m}\right) + C$$

In this equation, r and z are cylindrical coordinates ($z = 0$ being the axis of symmetry of the field), C is a constant, k is field curvature, and R_m is the characteristic radius. The

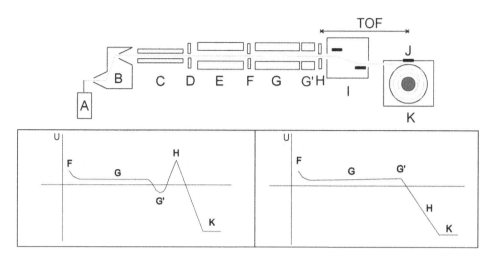

Figure 2.35
Top: simplified drawing of the complete stack of the orbitrap coupled to an electrospray source (A) at atmospheric pressure. B: lens arrangement to obtain a Z-spray, 2 mbar. C: RF quadrupole as ion guide, 10^{-2} mbar. D: Lens. E: transport RF quadrupole pumped to 10^{-5} mbar. This quadrupole may be fitted with DC voltage to select ions within a selected mass range. G and G': Segmented trapping quadrupole to store the ions to inject (10^{-5} mbar with He). F and H: lenses. I: device to deflect the ion beam prior to injection in the orbitrap. K: orbitrap side view cut at the level of the gap between the two half-barrel electrodes. The ions enter the trap, helped by the repeller electrode J, and spiral down the field to orbit around the spindle-shaped electrode. Pressure reduced to 10^{-10} mbar by the use of an additional ionic pump. Bottom: DC voltages applied during storage (left) and injection (right). Drawn according to data from Makarov A., Hardman M.E., Schwartz J.C. and Senko M., US Patent US2004108450, published June 10 2004.

first term is a quadratic field and the second the logarithmic field of a cylindrical capacitor. Inspection of this equation shows that the z coordinate appears only in one term. Thus the voltage gradient in the z direction is

$$\frac{\partial U(r, z)}{\partial z} = kz$$

An ion with mass m and total charge $q = ze$ is accelerated along z, which results from the force induced by the electric field. The electric force $-qkz$ is equal to the mass times the acceleration $m(\mathrm{d}^2 z/\mathrm{d}t^2)$:

$$F_z = m\frac{\mathrm{d}^2 z}{\mathrm{d}t^2} = -q\frac{\partial U}{\partial z} = -qkz \quad \text{or} \quad \frac{\mathrm{d}^2 z}{\mathrm{d}t^2} = -\frac{q}{m}kz \tag{2.1}$$

This equation describes a simple harmonic oscillator, like a pendulum. Introducing an energy factor E_z defined by

$$q E_z = (m/2)\left(\frac{\mathrm{d}z_0}{\mathrm{d}t}\right)^2$$

the exact solution to Equation (2.1) is

$$z(t) = z_0 \cos \omega t + \sqrt{(2E_z/k)} \sin \omega t$$

with

$$\omega = \sqrt{(q/m)k}$$

This equation shows that the frequency is directly linked to the m/q ratio and is independent of the kinetic energy of the injected ions. This is a very important property of the orbitrap. The broadband current induced by the oscillating ions is measured and converted by an FT to the individual frequencies and intensities, yielding the mass spectrum.

Figure 2.35 shows a scheme of a complete stack including an atmospheric pressure source coupling, an analytical quadrupole, a storage linear trap and the orbitrap itself. The ions are produced and focused from A to F by a classical atmospheric pressure source coupling device. The ion trap G has the particularity to allow it to accumulate the ions close to the exit to the orbitrap, adjusting the potentials on the F electrode and on the last segment G' of the trap. This allows faster injection into the orbitrap. Switching the applied voltages from those presented graphically in the lower left panel of the figure to those in the right panel allows injection in a very short time into the orbitrap at some kiloelectronvolts. From H to J, the ions fly freely except for a deflection of the beam to avoid injection of helium from the linear trap in the orbitrap. This field-free flight distance is adjusted to take account of a well-known property of the TOF instrument in which there is a focus plane at twice the length of the acceleration region, where ions of the same mass are focused successively. The distance H to J is adjusted to correspond to this focusing distance. This allows the injection of ions of one m/z at a time, focused in a short time period of some nanoseconds, for a total injection time of about 1 μs for successively higher m/z ions. This leads further to an efficient grouping of ions of the same mass, improving their detection, while dispersing ions of different masses reduce the space charge effect. As the time needed to inject ions with the same m/z is much shorter than the time of one oscillation along z, the movement of these ions will be very coherent, providing sensitive detection. The rotation around the central electrode does not have this coherence, reducing the background noise. Furthermore, this good coherence improves the mass resolution and accuracy: 150 000 FWHM resolution at m/z 600 has been demonstrated. An example of the resolution obtained with an orbitrap is shown in Figure 2.36 [31]. Another example of the use of this resolution on isotope abundance is given in Chapter 6.

Figure 2.37 shows the scheme of the first commercially available orbitrap instrument. It is somewhat different from the one previously described. First, there is an linear ion trap (LIT) that can be used for ion storage and ejection to the orbitrap by axial ejection, or independently used as a linear trap by radial ejection. It is possible to inject all the ions from the LIT, or selected ones, or product ions from the MS^n operations of the LIT. The normal acquisition cycle time in the orbitrap is 1 second.

After ejection to the orbitrap, other operations can be performed in the LIT during this cycle time. This allows one for instance to get two low-resolution spectra from the LIT while acquiring a high-resolution spectrum from the orbitrap, all in 1 second. The last lens before the LIT is a gating lens. It allows regulation of the number of ions injected, thus allowing an optimal number of ions to be injected, as well in the LIT as in the orbitrap, to optimize performance and avoid space charge effects.

The resolution is specified as 60 000 at m/z 400 when acquiring the spectrum in 1 second. The resolution increases linearly with detection time to a maximum of over 100 000. The mass accuracy with external calibration is 5 ppm and 2 ppm with internal calibration. The dynamic range is over 3000 within a spectrum and over 500 000 between spectra.

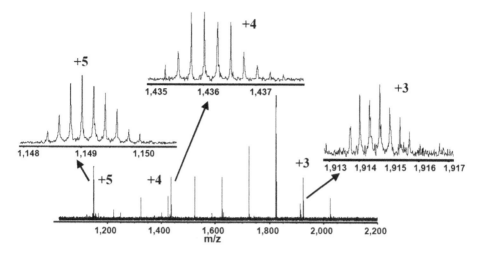

Figure 2.36
ESI mass spectrum of bovine insulin obtained with an orbitrap. The resolution at m/z 1149 is more than 60 000 FWHM. Additional peaks are due to the presence of Ultramark 1621 as internal standard. Reproduced from Hu Q., Makarov A., Noll R.J. and Cooks R.G., 'Application of the Orbitrap Mass Analyzer to Biologically Relevant Compounds', Proceedings of the 52nd ASMS Conference, Nashville, Tennessee, 2004, with permission.

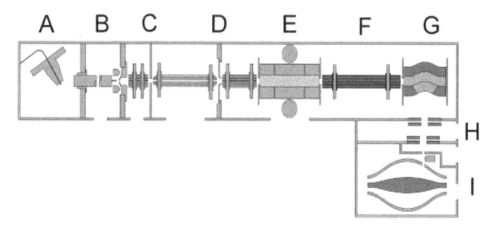

Figure 2.37
Scheme of the orbitrap instrument presented by Thermo Electron Corporation. It includes an atmospheric pressure source (A) at the left, followed by a heated capillary (B) and several multipole focusing devices (C) and a gating lens (D) followed by one more focusing octapole up to a linear trap (E) with two detectors. At the axial exit of the LIT, another focusing multipole (F) leads to a new kind of bent quadrupole (G), named 'C-trap'. This one allows storage of a bundle of ions and radial ejection toward the orbitrap. The ions enter the orbitrap laterally guided by the focusing electrodes and a repeller just at the entrance. Reproduced from Hu Q., Makarov A., Noll R.J. and Cooks R.G., 'Application of the Orbitrap Mass Analyzer to Biologically Relevant Compounds', Proceedings of the 52nd ASMS Conference, Nashville, Tennessee, 2004, with permission.

The mass range is limited by the use of the linear trap and is either m/z 50–2000 or m/z 200–4000.

The resolution at low masses is less than the one obtained with FTICR. However, the resolution of FTICR is inversely proportional to m/z whereas the resolution of the orbitrap is inversely proportional to $(m/z)^{1/2}$ and thus decreases more slowly when the mass increases. Compared with a Q-TOF instrument, the resolution is dramatically increased as well as the dynamic range.

2.4 Time-of-Flight Analysers

The concept of time-of-flight (TOF) analysers was described by Stephens in 1946 [32]. Wiley and McLaren published in 1955 the design of a linear TOF mass spectrometer which later became the first commercial instrument [33]. There has been renewed interest in these instruments since the end of the 1980s. First, progress in electronics simplified the handling of the high data flow. Second, the TOF analyser is well suited to the pulsed nature of the laser desorption ionization. The development of matrix-assisted laser desorption/ionization TOF has paved the way for new applications not only for biomolecules but also for synthetic polymers and polymer/biomolecule conjugates. TOF analysers were reviewed by Cotter [34], Guilhaus [35], Mann [36], Mamyrin [37] and Weickhardt [38]. Books on TOF mass spectrometers were published in 1994 and 1997 [39–41].

2.4.1 Linear Time-of-Flight Mass Spectrometer

Figure 2.38 displays the scheme of a linear TOF instrument. The TOF analyser separates ions, after their initial acceleration by an electric field, according to their velocities when they drift in a free-field region that is called a flight tube.

Ions are expelled from the source in bundles which are either produced by an intermittent process such as plasma or laser desorption, or expelled by a transient application of the required potentials on the source focusing lenses. These ions are then accelerated towards the flight tube by a difference of potential applied between an electrode and the extraction grid. As all the ions acquire the same kinetic energy, ions characterized by a distribution of their masses present a distribution of their velocities. When leaving the acceleration region, they enter into a field-free region where they are separated according to their velocities, before reaching the detector positioned at the other extremity of the flight tube.

Mass-to-charge ratios are determined by measuring the time that ions take to move through a field-free region between the source and the detector. Indeed, before it leaves the source, an ion with mass m and total charge $q = ze$ is accelerated by a potential V_s. It electric potential energy E_{el} is converted into kinetic energy E_k:

$$E_k = \frac{mv^2}{2} = qV_S = zeV_S = E_{el}$$

The velocity of the ion leaving the source is given by rearranging the previous equation as

$$v = (2zeV_s/m)^{1/2}$$

After initial acceleration, the ion travels in a straight line at constant velocity to the detector. The time t needed to cover the distance L before reaching the detector is given by

$$t = \frac{L}{v}$$

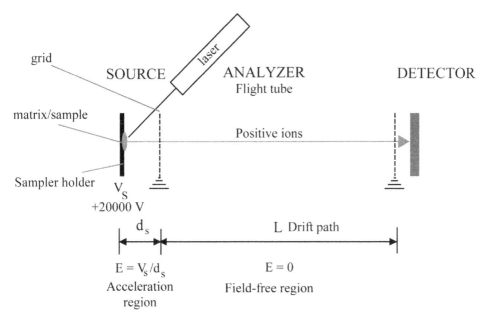

Figure 2.38
Principle of an LTOF instrument tuned to analyse positive ions produced by MALDI. After their formation during a laser pulse, ions are subject to the applied electric field. Ions are continuously accelerated and drift in a free-field region. They travel through this region with a velocity that depends on their m/z ratios. Ions are thus dispersed in time.

Replacing V by its value in the previous equation gives

$$t^2 = \frac{m}{z} \left(\frac{L^2}{2eV_s} \right)$$

This equation shows that m/z can be calculated from a measurement of t^2, the terms in parentheses being constant. This equation also shows that, all other factors being equal, the lower the mass of an ion, the faster it will reach the detector.

In principle, the upper mass range of a TOF instrument has no limit, which makes it especially suitable for soft ionization techniques. For example, samples with masses above 300 kDa have been observed by MALDI-TOF [42,43].

Another advantage of these instruments is their high transmission efficiency that leads to very high sensitivity. For example, the spectrum from 10^{-15} mol of gramicidin [44] and the detection of 100–200 attomole amounts of various proteins (cytochrome C, ribonuclease A, lysozyme and myoglobin) [45] have been obtained with TOF analysers. All the ions are produced in a short time span and temporal separation of these ions allows all of them to be directed towards the detector. Therefore, all the formed ions are in principle analysed contrary to the scanning analysers that transmit ions successively along a time scale.

The analysis speed of TOF analysers is very fast and a spectrum over a broad mass range can be obtained in micro-seconds. So, it is possible in theory to produce in 1 second several thousand TOF mass spectra over a very wide mass range. But in practice, for most of the applications, the weak number of ions detected in each individual spectrum is insufficient to provide the required precision of mass or abundance measurement. Furthermore, it is actually impossible to record all these individual spectra one by one at such a rate without exceeding the speed of data transfer and the capacity of data storage of most computers. Thus, recorded spectra are generally the addition of a number of individual spectra.

Another interesting characteristic of the TOF analyser lies in its easy mass calibration with only two reference points. As in all the mass spectrometers, the TOF mass spectrometer requires a calibration equation to relate and convert the physical property that is measured to a mass value. For the TOF spectrometer, the physical property that is measured during an analysis is the flight time of the ions. As already mentioned, the flight time of an ion is related to its mass by the following equation:

$$(m/z)^{1/2} = \left(\frac{\sqrt{2eV_s}}{L} \right) t$$

Distance L and accelerating potential V_s are constant for a given spectrometer and, thus, the terms in parentheses can be replaced with the constant A. The relationship between $m^{1/2}$ and t is linear. A constant B is added to produce a simple equation for a straight line. This constant B allows correction of the measured time zero that may not correspond exactly with the true time zero. Indeed, calibration must take into account other effects. The most basic effect is the uncertainty of the starting time of an ion. This effect can be caused by finite time delays in cables etc. Thus:

$$(m/z)^{1/2} = At + B$$

Therefore, the conversion of flight times to mass supposes a preliminary calibration with two known molecules (standards). Using their known m/z ratios and their measured flight times, this equation is solved for the two calibration constants A and B. As long as the points are not too close together, a simple two-point calibration is usually accurate.

External or internal calibration can be performed. External calibration is a method in which the calibration constants, A and B, are determined from two standards in an experiment that does not include unknown molecules. Internal calibration is a method in which the flight times of the standard and unknown ions are measured from the same spectrum providing the best possible match of experimental conditions for the three species involved. The highest degree of mass accuracy is usually achieved through internal calibration.

The resolution observed with the TOF mass spectrometer is derived from the relationship between m/z and flight time:

$$\frac{m}{z} = \left(\frac{2eV_s}{L^2} \right) t^2$$

and

$$\frac{1}{z}dm = \left(\frac{2eV_s}{L^2} \right) 2t\, dt$$

Thus

$$\frac{m}{dm} = \frac{t}{2dt}$$

Therefore, the resolution in TOF mass spectrometry is equal to

$$R = \frac{m}{\Delta m} = \frac{t}{2\Delta t} \approx \frac{L}{2\Delta x}$$

where m and t are the mass and flight time of the ion, and Δm and Δt are the peak widths measured at the 50 % level on the mass and time scales, respectively. L is the flying distance and Δx is the thickness of an ion packet approaching the detector.

As the mass resolution is proportional to the flight time and the flight path, one solution to increase the resolution of these analysers is to lengthen the flight tube. However, too long a flight tube decreases the performance of TOF analysers because of the loss of ions by scattering after collisions with gas molecules or by angular dispersion of the ion beam. It is also possible to increase the flight time by lowering the acceleration voltage. But lowering this voltage reduces the sensitivity. Therefore, the only way to have both high resolution and high sensitivity is to use a long flight tube with a length of 1 to 2 m for a higher resolution and an acceleration voltage of at least 20 kV to keep the sensitivity high.

The most important drawback of the first TOF analysers was their poor mass resolution. Mass resolution is affected by factors that create a distribution in flight times among ions with the same m/z ratio. These factors are the length of the ion formation pulse (time distribution), the size of the volume where the ions are formed (space distribution), the variation of the initial kinetic energy of the ions (kinetic energy distribution), and so on. The electronics and more particularly the digitizers, the stability of power supplies, space charge effects and mechanical precision can also affect the resolution and the precision of the time measurement.

If a pulsed sources such as MALDI seems to be well suited to the TOF analyser, the quality of its pulsed ion beam is insufficient to obtain the high resolution and high mass accuracy. This situation is substantially improved with the development of two techniques: delayed pulsed extraction and the reflectron.

2.4.2 Delayed Pulsed Extraction

To reduce the kinetic energy spread among ions with the same m/z ratio leaving the source, a time lag or delay between ion formation and extraction can be introduced. The ions are first allowed to expand into a field-free region in the source and after a certain delay (hundreds of nanoseconds to several microseconds) a voltage pulse is applied to extract the ions outside the source. This mode of operation is referred to as delayed pulsed extraction to differentiate it from continuous extraction used in conventional instruments. Delayed pulsed extraction, also known as pulsed ion extraction, pulsed extraction or dynamic extraction, is a revival of time-lag focusing, which was initially developed by Wiley and McLaren in the 1950s, shortly after the appearance of the first commercial TOF instrument.

As shown in Figure 2.39, the ions formed in the source using the continuous extraction mode are immediately extracted by a continuously applied voltage. The ions with the same m/z ratio but with different kinetic energy reach the detector at slightly different times, resulting in peak broadening. In the delayed pulsed extraction mode, the ions are initially allowed to separate according to their kinetic energy in the field-free region. For ions of the same m/z ratio, those with more energy move further towards the detector than the initially less energetic ions. The extraction pulse applied after a certain delay transmits more energy to the ions which remained for a longer time in the source. Consequently, the initially less energetic ions receive more kinetic energy and join the initially more energetic ions at the detector. So, delayed pulsed extraction corrects the energy dispersion of the ions leaving the source with the same m/z ratio and thus improves the resolution of the TOF analyser.

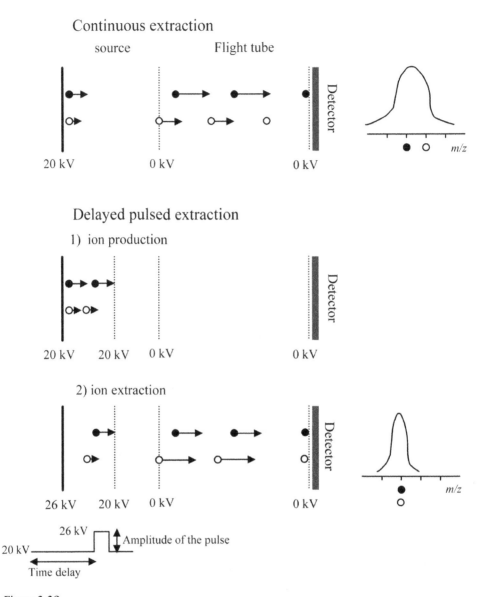

Figure 2.39
Schematic description of a continuous extraction mode and a delayed pulsed extraction mode in an linear time-of-flight mass analyser. ○ = ions of a given mass with correct kinetic energy; ● = ions of the same mass but with a kinetic energy that is too high. Delayed pulsed extraction corrects the energy dispersion of the ions leaving the source with the same m/z ratio.

The energy focusing can be accomplished by adjusting the amplitude of the pulse and the time delay between ion formation and extraction. For optimal focusing, both pulse and delay can be adjusted separately. It should be noted that the optimal pulse voltage and delay are mass dependent. Lower pulse voltages or shorter delays are required to focus ions of lower m/z ratio. In general, for a given m/z and initial velocity distribution, greater voltage pulses require shorter time delays and vice versa.

If delayed extraction increases the mass resolution without degradation of sensitivity compared with continuous extraction, it also has limitations. Indeed, delayed extraction complicates the mass calibration procedure. It can only be optimized for part of the mass range at a time and is less effective at high mass. Delayed extraction partially decouples ion production from the flight time analysis, thus improving the pulsed beam definition. However, calibration, resolution and mass accuracy are still affected by conditions in the source. For instance, in the usual axial MALDI-TOF experiments, optimum focusing conditions depend on laser pulse width and fluence, the type of sample matrix, the sample preparation method, and even the location of the laser spot on the sample.

2.4.3 Reflectrons

Another way to improve mass resolution is to use an electrostatic reflector also called a reflectron. The reflectron was proposed for the first time by Mamyrin [46]. It creates a retarding field that acts as an ion mirror by deflecting the ions and sending them back through the flight tube. The term reflectron time-of-flight (RTOF) analyser is used to differentiate it from the linear time-of-flight (LTOF) analyser. The simplest type of reflectron, which is called a single-stage reflectron, consists usually of a series equally spaced grid electrodes or more preferably ring electrodes connected through a resistive network of equal-value resistors.

The reflectron is situated behind the field-free region opposed to the ion source. The detector is positioned on the source side of the ion mirror to capture the arrival of ions after they are reflected. There are two common methods of positioning the detector. It may be coaxial with the initial direction of the ion beam. This detector has a central hole to transmit the ions leaving the source. However, the most common instrument geometry has the detector off-axis with respect to the initial direction of the ion beam. Indeed, adjusting the reflectron at a small angle in respect to the ions leaving the source allows the detector to be positioned adjacent to the ion source.

The reflectron corrects the kinetic energy dispersion of the ions leaving the source with the same m/z ratio, as shown in Figure 2.40. Indeed, ions with more kinetic energy and hence with more velocity will penetrate the reflectron more deeply than ions with lower kinetic energy. Consequently, the faster ions will spend more time in the reflectron and will reach the detector at the same time than slower ions with the same m/z. Although the reflectron increases the flight path without increasing the dimensions of the mass spectrometer, this positive effect on the resolution is of lower interest than its capability to correct the initial kinetic energy dispersion. However, the reflectron increases the mass resolution at the expense of sensitivity and introduces a mass range limitation.

If the potential of the reflectron V_R and its length D lead to an electric field in the reflectron $E = V_R/D$, an ion of charge q with a kinetic energy E_k will enter with a velocity v_{ix} and penetrate the reflectron to a depth d such that

$$d = \frac{E_k}{qE} = \frac{qV_s}{qV_R/D} = \frac{V_sD}{V_R}$$

Its speed v_x along the x axis will then be zero, and its mean velocity into the reflectron will be equal to $v_{ix}/2$. The time needed to penetrate at a distance d will be thus

$$t_0 = \frac{d}{v_{ix}/2}$$

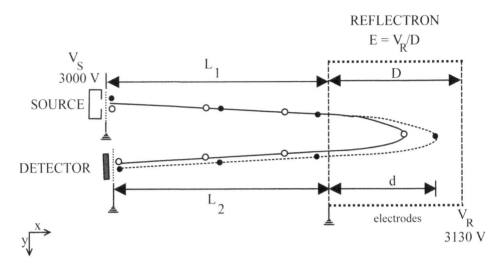

Figure 2.40
Schematic description of a TOF instrument equipped with a reflectron. ■ = ions of a given mass with correct kinetic energy; □ = ions of the same mass but with a kinetic energy that is too low. The latter reach the reflectron later, but come out with the same kinetic energy as before (see text). With properly chosen voltages, path lengths and fields, both kinds of ions reach the detector simultaneously.

Then the ion will be symmetrically repelled outside of the reflectron, so that its kinetic energy will be restored to the same absolute value as before, but the velocity will be in the opposite direction. The total flight length in the reflectron will be $2d$, and the total time t_r in the reflectron will be

$$t_r = 2t_0 = \frac{2d}{v_{ix}/2} = \frac{4d}{v_{ix}}$$

The total flight length out of the reflectron will be $L_1 + L_2$, where L_1 and L_2 are the distances covered before and after the reflectron, and the total time t out of the reflectron will be

$$t = \frac{L_1 + L_2}{v_{ix}}$$

The total flight time will be the sum of the flight time in and out of the reflectron:

$$t_{TOT} = t + t_r = (L_1 + L_2 + 4d)/v_{ix}$$

Replacing $v_{ix} = (2zV_s e/m)^{1/2}$ in the previous equation gives

$$t_{TOT}^2 = \frac{m}{z} \frac{(L_1 + L_2 + 4d)^2}{2eV_s}$$

Let us now consider two ions sharing the same mass m, one coming from the source with the correct kinetic energy E_k and the other with a kinetic energy E_k'. We define a^2 as

$$E_k'/E_k = a^2$$

The speed along the x axis during the field-free flight will be

$$E_k = \frac{mv_{ix}^2}{2} \quad v_{ix} = (2E_k/m)^{1/2} \quad v'_{ix} = (2E_ka^2/m)^{1/2} \quad v'_{ix} = av_{ix}$$

The time of field-free flight, out of the reflectron, for a path length $L_1 + L_2$ will be

$$t = L_1 + L_2/v_{ix} \quad t' = L_1 + L_2/v'_{ix} = L_1 + L_2/av_{ix}$$

and hence

$$t' = t/a$$

In the reflectron, the ions will penetrate at a depth d or d':

$$d = E_k/(qE) \quad d' = E'_k/qE = a^2E_k/qE$$

and hence

$$d' = a^2d$$

The time spent in the reflectron will be

$$t_r = 4d/v_{ix} \quad t'_r = 4d'/v'_{ix} \quad \text{or} \quad t'_r = 4a^2d/av_{ix} = 4ad/v_{ix}$$

and hence

$$t'_r = at_r$$

The total flight time for these ions sharing the same m/z but having different kinetic energies will be the sum of the flight time in and out of the reflectron:

$$t_{TOT} = t + t_r \quad \text{and} \quad t'_{TOT} = t/a + at_r$$

This means that if $a > 1$, then the ion with an excess kinetic energy will have a shorter flight time out of the reflectron (t/a), but a longer one in the reflectron (at_r). The opposite holds for $a < 1$. The variations of the flight times thus compensate each other. A correct compensation yielding the same total flight time for all ions sharing the same mass but having different kinetic energies requires choosing the proper values for E, V_s, L_1 and L_2. As far as the ions leaving the source with the correct kinetic energy spend the same time in and out of the reflectron ($t_r = t$), in others words if $4d = L_1 + L_2$, then a perfect kinetic energy focalization is obtained. A complete treatment would take into account the displacement along the y axis. In practice, the reflection angle is typically less than 2°, and this error will be small. In the figures, much larger angles have been used for clarity in the drawings.

The performance of the reflectron may be improved by using a two-stage reflectron. Indeed, to reduce the size and to improve the homogeneity of the electric field, a two-stage reflectron was proposed. In this reflectron, two successive homogeneous electric fields of different potential gradient are used. The first stage is characterized by an intense electric field responsible for the strong deceleration of the ions while the second stage is characterized by a weaker field. These two-stage reflectrons have the advantage of being more compact devices because of the strong deceleration of the ions at the first stage, but they suffer from a lower transmission.

2.4.4 Tandem Mass Spectrometry with Time-of-Flight Analyser

Tandem mass spectrometry can be performed with reflectron time-of-flight instruments. This is best achieved by combining a linear with a reflectron spectrometer. Ions after having acquired sufficient internal energy can fragment to give product ions (also called fragment ions) and neutral fragments. These fragmentations can be discriminated according to the place where they occur.

If the lifetime induced by the internal energy of an ion is greater than its flight time, this ion reaches the detector before any fragmentation has occurred. On the other hand, if the lifetime of an ion is shorter than the time spent by this ion in the source before the acceleration, this ion fragments before leaving the source. Fragmentations taking place in the source are called in-source decay (ISD) fragmentations. The precursor ion and the fragment ions have the same kinetic energy after the acceleration and thus they reach the detector separately according to their m/z. Therefore, they can be analysed as well in LTOF as in reflectron TOF. However, it is impossible to determine the precursor ions that are at the origin of these fragments.

The lifetime of an ion can be intermediate between its time spent in the source and its flight time. These ions are stable enough to leave the source but contain enough excess energy to allow their fragmentation in the field-free region of the flight tube before they reach the detector. This corresponds to the fragmentation of metastable ions. Fragmentation that occurs after the source of a TOF mass spectrometer is called post-source decay (PSD) fragmentation.

When an ion fragments after acceleration and before the entrance in the reflectron, its product ions have the same velocity as the precursor, and thus the same flight time in the absence of any field. Therefore, in an LTOF analyser, the survivor precursor ions, their fragment ions and neutral fragments reach the detector at the same time (see Figure 2.41). However, they have different kinetic energies. Indeed, if a fragment ion has the same velocity as its precursor, it has a lower kinetic energy because of its lower mass. Consequently, in a reflectron TOF analyser, these ions have different flight times that depend on their masses. The lighter fragment ions, which have a lower kinetic energy, penetrate the reflectron less deeply and spend less time inside the reflectron than the heavier fragment ions or the survivor precursor ions (see Figure 2.41). These ions thus reach the detector successively according to their m/z ratios.

If m_p is the mass of the precursor and m_f the fragment mass, their respective kinetic energies E_{kp} and E_{kf} are, since they have the same velocity v_{ix},

$$E_{kp} = \frac{m_p v_{ix}^2}{2} \quad \text{and} \quad E_{kf} = \frac{m_f v_{ix}^2}{2}$$

Thus

$$E_{kp}/E_{kf} = m_p/m_f \quad \text{or} \quad E_{kf} = E_{kp}\frac{m_f}{m_p}$$

The penetration depth d in the reflectron is given by $d = E_k/qE$. Hence, for the precursor and the fragment, the relevant equations are

$$d_p = \frac{E_{kp}}{qE} \quad \text{and} \quad d_f = \frac{E_{kf}}{qE} = \frac{E_{kp}(m_f/m_p)}{qE}$$

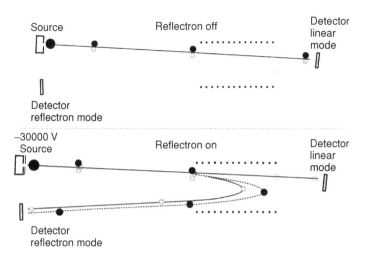

Figure 2.41
Tandem mass spectrometry with combined linear and reflectron TOF instruments. ● = bundle of ions with one given mass leaving the source; during the flight, a fraction of these ions fragments; • = Survivor ions; ⊖ fragment ions; □ = neutral fragments. Those ions fragmenting between the source and the reflectron are called PSD ions. The neutral fragments are not deflected and reach the linear detector. They will be detected provided that the parent ions have been accelerated at a sufficient kinetic energy, typically 30 keV.

Hence

$$d_f = d_p \frac{m_f}{m_p}$$

The respective flight times in the reflectron are given by

$$t_{rp} = \frac{4d_p}{v_{ix}} \quad \text{and} \quad t_{rf} = \frac{4d_f}{v_{ix}} = \frac{4d_p(m_f/m_p)}{v_{ix}} = t_{rp}(m_f/m_p)$$

This demonstrates that the time spent in the reflectron will be shorter for the fragment than for the precursor, the ratio of the respective times being equal to the mass ratio. Thus, as shown in Figure 2.41, the time of flight will be the same for the precursors and fragments in the linear mode, but will differ in the reflectron mode. When comparing the two spectra, ions observed in the reflectron spectrum but not in the linear mode result from PSD fragmentations having occurred between the source and the reflectron device. Measurement of the flight times of fragment ions correlated with a particular precursor ion allows the fragmentation of this ion to be described.

The selection of the precursor ion is obtained by a deflection gate between the source and the reflectron. Indeed, if the instrument is fitted with electrodes after the source on the ion path before the reflectron, a precursor ion can be selected to analyse its fragmentation by PSD. As shown in Figure 2.42, a potential is applied to eliminate the ions. Cutting off this potential for a short time, corresponding to the passage of the selected ion, allows this

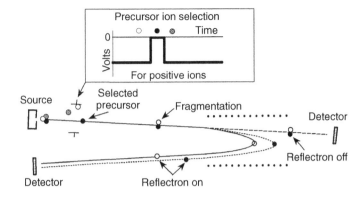

Figure 2.42
TOF spectrometer fitted with a deflection electrode for precursor
ion selection. The instrument can be operated in either the linear or
the reflectron mode. The selection resolution is about 250, which
for example gives a 4 Th window at m/z 1000.

ion to be selected. The mass of the passing ions can be determined in the linear flight mode,
while the fragments are observed in the reflectron mode.

The resolution of the selected ion on modern instruments is between 100 and 250. This
gives a window between 10 and 4 Th at m/z 1000 [47]. A system that uses two gates in series
has more recently been developed to obtain a resolution of more than 1000 [48]. When the
selected ion reaches the entrance of the first gate, the applied deflection voltage is rapidly
switched off, opening the gate and allowing the ions to travel towards the second gate. As
the selected ion passes the exit of the second gate, the voltage is switched on, closing the
gate. Therefore, the first gate rejects the lower mass ions, while the second gate rejects the
higher mass ions.

It is important to understand that the deflection gate selects not only the precursor ion but
also all the fragment ions resulting from this selected precursor ion that are produced in the
first field-free drift path of the instrument between the source and the reflectron. Indeed, as
the zone between the source and the reflectron is a field-free region, all the PSD fragments
formed in this zone have the same velocity as their precursor ions. And as selection is based
on flight time, the precursor passes between the deflecting electrodes at the same time as
its fragments.

The fragment ions have the same velocity as their precursor ions but have different
energy as a function of their mass. Therefore, the reflectron induces both their kinetic energy
focusing and time dispersion of fragment ions having different masses. It is important to
note that the optimum reflectron potential for observing the selected ion is no longer
optimum for the fragment ions and must be set up separately for each fragment ion (Figure
2.43). Therefore, the voltage on the reflectron should be adjusted to the mass of each
fragment to obtain optimum results. Indeed, with a linear field reflectron, optimal focusing
in energy is obtained when $L_1 + L_2 = 4d$, where L_1 and L_2 are the distances covered before
and after the reflectron and x is the penetration depth in the reflectron. The result is a great
dispersion of the focal points $(L_2 = 4d - L_1)$ for fragment ions of different masses, and
thus a loss of mass resolution. For example, the theobromine $m_p = 137$ u produces by loss
of CO in PSD a fragment ion $m_f = 109$ u. If the acceleration voltage $V_s = 3$ kV, $L_1 = 1$ m,

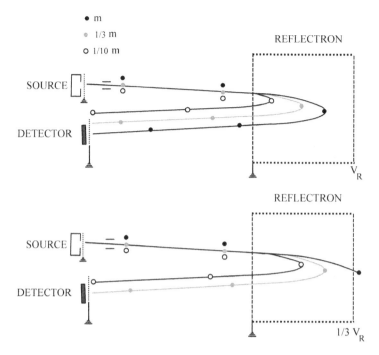

Figure 2.43
Tandem mass spectrometry by PSD, which is performed with a reflectron TOF instrument. In this method, the optimum reflectron potential for observing the precursor ion is not the optimum for the fragment ions, and must be set up separately for each fragment ion. Indeed, if the reflectron potential is optimized for the precursor ion, the fragment ions are not focused on the detector. On the other hand, if this potential is optimized for a fragment ion, the precursor ion and the heavier fragment ions travel across the reflectron whereas the lighter fragment ions are still not focused on the detector.

the electric field in the reflectron $V_R = 3120$ V and its length $D = 0.522$ m, then $L_{2p} = 1$ m and $L_{2f} = 0.59$ m. To put this situation right, it is necessary to adapt the potential of the reflectron to each mass: $V_{2R} = (m_2/m_1)V_{1R}$. Thus, the focal points and times of flight are the same for the two ions.

Although only one fragment ion is perfectly energy focused at a specific reflectron voltage, approximately 5–10 % of the full kinetic energy range of the fragment ions can be adequately focused at this voltage to produce a portion of the PSD spectrum. This requires the reflectron voltage to be systematically stepped or scanned to bring into focus other fragment ions that differ in their kinetic energies and thus to record segments of the PSD spectrum. A composite mass spectrum can be obtained by combining the individual sections of the mass spectrum that were produced from the 10–20 different reflectron potential steps that were required for observing the entire PSD spectrum.

The necessity to step the voltage applied to the reflectron to produce a complete spectrum consumes time, wastes sample and complicates the mass calibration procedure. Indeed, the acquisition of a PSD spectrum is at least 20 times longer than for a classic TOF spectrum and the main part of the sample is lost without contributing to the analysis.

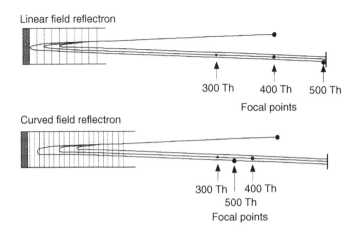

Figure 2.44
Trajectories for CID fragmentation of ions of 500 Th. The focal points in energy of the product ions of 400 and 300 Th are indicated for a linear field reflectron and for a curved field reflectron. Reproduced (modified) from Cotter R.J., Time-Of-Flight Mass Spectrometry: Instrumentation and Applications in Biological Research, ACS, Washington, DC, 1997, with permission.

To avoid this drawback, a better solution has been proposed by Cotter with the introduction of a curved field reflectron, commercially available from Shimatzu [49]. The curved field reflectron provides a retarding field that increases in a nonlinear manner. This reflectron decreases the penetration distance for the heavy ions as indicated in Figure 2.44. This then leads to a smaller dispersion of the focal points for fragment ions of different masses. So, the full kinetic energy range of the fragment ions can be adequately focused at the optimum reflectron potential for observing the selected ion. Thus, this allows the recording of product ion spectra in one step without stepping or scanning of the reflectron potential. A complete PSD ion spectrum can thus be acquired in a single step. However, the curved field reflectron causes more ion losses than the linear field reflectron.

Tandem mass spectrometry can be performed with reflectron TOF instruments using the post-source decay technique described above. However, this technique presents some limitations in mass resolution and mass accuracy. Indeed, if the resolution can reach 20 000 for the analysis of the precursor ions, it decreases to below 1000 for the fragment ions analysed by PSD. The main reason of this resolution degradation lies in the liberation of kinetic energy during the fragmentation reaction. The poor resolution of the selection of precursor ions is also another limitation of the PSD technique. Furthermore, compared with tandem mass instruments including a collision cell, the PSD technique also presents limitations through the lack of control of the ion activation conditions. Because the PSD technique is based on metastable ions, the efficiency of fragmentation is generally low. This induces a rather low quality in the fragment spectra containing a high background noise level.

To overcome many of the limitations associated with PSD, several tandem TOF mass spectrometers (TOF/TOF) have been designed. Two of these tandem TOF mass spectrometers are commercially available, one sold by Applied Biosystems and the other by Bruker. The basic principle of these two instruments is to use a short linear TOF as the first mass

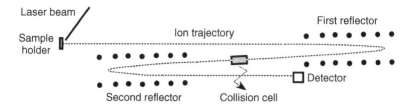

Figure 2.45
Schematic representation of a tandem mass spectrometer comprising two
RTOF analysers and, in between, a collision cell.

analyser and a reflectron TOF as the second mass analyser. These two TOF analysers are
separated by an ion deflection gate and a collision cell. The two spectrometers also have in
common the feature to reaccelerate the precursor ion and its fragments after their exit from
the collision cell. In this way, the kinetic energy spread of the selected precursor ion and
its fragment ions is reduced, and consequently this allows the recording of fragment ion
spectra in one step without scanning of the reflectron potential.

Cotter [50] has developed a compact laser desorption tandem TOF mass spectrometer
based on two reflectron TOF mass analysers (RTOF/RTOF). The instrument incorporates
two reflectron mass analysers separated by a collision cell for producing ions by collision-
induced dissociation (Figure 2.45). To allow the selection of the precursor ion, a deflection
gate is placed between the first reflectron and the collision cell. The first reflectron, which
is a single stage reflectron, provides energy focusing, while the second reflectron, which is
a curved field reflectron, provides both energy focusing and time dispersion of ions having
different masses. At the same time, another two-reflectron tandem TOF mass spectrometer
was developed by Enke [51]. However, it differs from the mass spectrometer developed by
Cotter. The second reflectron is a linear field reflectron and the fragmentation of the ions is
induced by photodissociation.

A common drawback of these two instruments is the impossibility to observe any
metastable fragmentation which occurs in the first mass analyser between the source and
the first reflectron. Because these fragment ions penetrate in the first reflectron, they do not
arrive at the deflection gate and collision cell at the same time as their precursors. Thus, they
do not contribute to the detected fragment ions and are not recorded in the fragmentation
spectrum of the selected ion. Since such metastable fragmentations are generally not minor
processes, this reduces the efficiency of these mass spectrometers.

2.4.5 Orthogonal Acceleration Time-of-Flight Instruments

TOF analysers are directly compatible with pulsed ionization techniques such as plasma or
laser desorption because they provide short, precisely defined ionization times and a small
ionization region. However, to take advantage of TOF analysers, it is interesting to combine
such powerful analysers with continuous ionization techniques. These ionization techniques
can be compatible with TOF analysers but require some adaptations to pulse the source
or to transform a continuous ion beam into a pulsed process. For instance, the coupling of
an ESI (or any other API) source with a TOF mass spectrometer is difficult, because ESI
yields a continuous ion beam, whereas the TOF system works on a pulsed process.

The most basic approach to transform a continuous ion beam into a pulsed process for use with TOF spectrometers has been by sweeping the ion beam coming out of the source over a narrow slit to the TOF spectrometer, thereby yielding an axial pulsed inlet in the analyser. Indeed, the incoming ion beam was injected along the same axis as the TOF analyser. However, most of the ions generated from the ion source were blocked by the slit and were lost, allowing only a small portion of the generated ions to be introduced into the analyser. The utilization ratio of ions or sample in these systems, also defined as the duty cycle, was very low. Another approach involves pulsed ion extraction, in which short packets of ions are periodically extracted from the source.

It is now well established that orthogonal injection, also known as orthogonal acceleration (oa), is the best technique for coupling continuous ionization sources with a TOF analyser. This technique was developed initially by O'Halloran *et al.* in the 1960s [52], but it was unknown to most researchers and was independently reinvented by different groups at the beginning of the 1990s [53, 54]. A schematic diagram of an oa-TOF spectrometer with linear mass analyser is presented in Figure 2.46. However, an orthogonal acceleration with reflectron TOF analyser is also possible.

The sample is continuously ionized in the ion source. For positive ions, the source is held at a positive voltage of V_{beam}. Ion optics focus the resulting ions into a parallel beam and direct it to the orthogonal accelerator. The beam continuously fills the first stage of the ion accelerator in the space between the plate and grid 1 (G1). Initially the voltage on the plate is at the same ground level (0 V) as the grid 1. Thus, at this point, this region is field free, so ions continue to move in their original direction. Then an injection pulse voltage of $V_{injection}$ is applied to the plate. The ions between the plate and grid 1 are pushed by the resulting electric field in a direction orthogonal to their original trajectories, and begin to fly towards the analyser. After passing through grid 1, the ions are further accelerated towards grid 2 (G2) and enter into the field-free drift region at a potential of V_{tof}, where TOF mass separation occurs.

When all ions have moved to the field-free region through grid 2, the voltage on the plate is restored to 0 V and ions from the ion source begin to fill the space between the plate and grid 1 again. Thus, during the time that the ions continue their flight in the free-field region, the orthogonal accelerator is refilled with new ion beam. One flight cycle will end when the ion with the highest m/z reaches the detector. Another flight cycle will begin by reapplying a pulse voltage to the plate. In practice, a pulsed electric field is applied to the plate at a frequency of several kilohertz. The analyser and the ion detector are similar to those used with pulsed ionization sources, apart from differences associated with a larger beam size (usually a few centimetres) that is determined by the size of the orthogonal accelerator aperture.

The direction of ion flight in the field-free region is not fully orthogonal to the direction of ions introduced. Indeed, orthogonal acceleration gives a new component of velocity to the ions that are in the ion beam. This component is orthogonal and thus independent of the velocity of the ions in the beam. But ions that are pushed in the TOF analyser also keep the velocity that they had in the ion beam. Thus, their spontaneous drift trajectories in the field-free region are inclined from the flight tube direction according to the ratio of their components of velocity in the ion beam and TOF directions. For all ions to have the same inclination, they must have the same kinetic energy in the ion beam direction when they are introduced. In other words, if ions do not have the same kinetic energy in the ion beam direction, a resulting divergence of drift trajectories is observed.

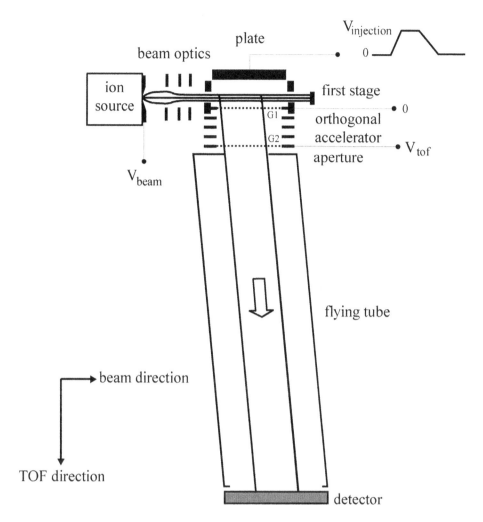

Figure 2.46
Schematic description of orthogonal acceleration with linear mass analyser. Pulses of ions are injected orthogonally from a continuous ion beam into a TOF analyser.

Orthogonal injection provides a high-efficiency interface for sampling ions from continuous beam to a TOF analyser. The TOF analyser allows simultaneous transmission of all ions and therefore all the ions formed are analysed. However, the duty cycle is far from 100 % for the oa-TOF spectrometer and it is lower than for the TOF spectrometer. This is due to the orthogonal accelerator that samples to the analyser only a part of the ions produced in the source. The duty cycle, despite this fact, is between 5 and 50 %. This is a considerable improvement over the conventional techniques described for coupling a continuous source to a TOF spectrometer.

As already mentioned, the orthogonal accelerator is filled with new ions from the ion source while the sampled ions are simultaneously analysed in the field-free region. New ions cannot be injected until the ions from the previous injection have reached the detector.

But the time required for the ion beam to fill the orthogonal accelerator is lower than the time required for the sampled ions to fly to the detector. Therefore, a part of the ions produced in the source are not pushed to the flight tube and are lost in the first stage of the orthogonal accelerator. The upper limit on the duty cycle for oa-TOF is determined by the ratio between the length L of the ions packet that is sampled and accelerated to the analyser and the distance D travelled by the ion beam between two successive injection pulses. L is determined by the size of the orthogonal accelerator aperture. D is the distance between the centres of the orthogonal accelerator aperture and the detector. Indeed, ions have the same velocity in the direction of the beam in the flight tube as in the orthogonal accelerator. Therefore, the distance travelled by the ion beam between two successive injection pulses in the first stage of the orthogonal accelerator is the same as the distance travelled by the ion in the flight tube between the orthogonal accelerator aperture and the detector.

Another important advantage of orthogonal injection is its ability to minimize the initial velocities in the TOF direction and, simultaneously, to reduce the dispersion of these velocities. This is due to the formation of a parallel ion beam with orthogonal orientation to the flight tube. This favourably affects mass resolution, mass accuracy and mass calibration stability. This makes easier to achieve resolutions of more than 10 000.

The ions produced in the source may be introduced directly into the orthogonal accelerator through beam optics that only shape and direct these ions into a parallel ion beam. As a result, the performance of this system is limited by the characteristics of the ion beam produced by the source. Therefore, this direct introduction has some limitations. One of these limitations unfavourably affects the sensitivity. Indeed, the ions usually diverge from the source. And the production of the narrow cross-section beam required for high resolution induces a significant loss of ions. Another limitation of direct introduction may lead to significant mass discrimination and comes from the velocity distribution of the ions in the beam direction. This velocity distribution leads to kinetic energy spread of the beam ions that induces mass discrimination because, as already mentioned, the directions of spontaneous drift trajectories in the field-free region are kinetic energy dependent. And ions at the extremes of the kinetic energy spread follow trajectories that do not strike the active area of the detector.

It is possible to overcome these limitations by using an RF multipole ion guide that operates at a pressure of several milliTorr. This ion guide is designed to transport efficiently the ions produced by the ion source to the orthogonal accelerator. It is a critical component of the system because it not only focuses the ions from the source, but also controls the kinetic energy of these ions by collisional cooling (also called collisional damping). This allows the high resolution and high sensitivity of the TOF analyser to be achieved. Indeed, a multipole has the property of focusing the trajectories of the ions towards its central axis, whereas collisions with the molecules of ambient gas provide both radial and axial collisional cooling of ion motion. The ions collide and gradually lose their kinetic energy to become thermalized, reducing both the energy spread and the beam diameter. The ions are then are reaccelerated to give a nearly monoenergetic beam in the axial direction towards the first stage of the orthogonal accelerator.

The ion guide produces an ion beam independently of the original parameters of the beam delivered by the source and with highly favourable properties for the oa-TOF mass spectrometer. As a result of the decoupling of the ion source and mass analyser, the performance of the instrument is independent of source conditions and this leads to improved resolution and sensitivity.

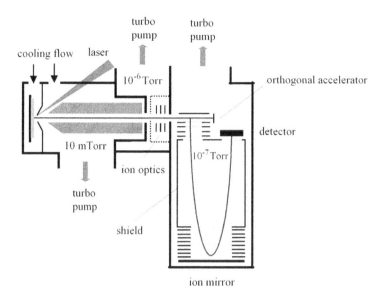

Figure 2.47
Schematic diagram of a MALDI orthogonal acceleration TOF instrument.

The oa-TOF analysers are suited for any of the different ionization techniques. An oa-TOF instrument has even been adapted with pulsed ionization techniques such as a MALDI source. The configuration of such an instrument is shown in Figure 2.47.

2.5 Magnetic and Electromagnetic Analysers

Consider an ion of mass m and charge q, accelerated in the source by a potential difference V_s. At the source outlet, its kinetic energy is

$$E_k = \frac{mv^2}{2} = qV_s \tag{2.2}$$

2.5.1 Action of the Magnetic Field

If the magnetic field has a direction that is perpendicular to the velocity of the ion, the latter is submitted to a force F_M as described in Figure 2.48. Its magnitude is given by

$$F_M = qvB$$

The ion follows a circular trajectory with a radius r so that the centrifugal force equilibrates the magnetic force

$$qvB = \frac{mv^2}{r} \quad \text{or} \quad mv = qBr \tag{2.3}$$

For every value of B, the ions with the same charge and the same momentum (mv) have a circular trajectory with a characteristic r value. Thus, the magnetic analyser selects the ions according to their momentum. However, taking into account the kinetic energy of the

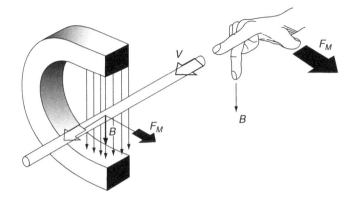

Figure 2.48
Orientation of the magnetic force on a moving ion.

ions at the source outlet leads to

$$m v^2 = 2q V_s$$

Hence

$$\frac{m}{q} = \frac{r^2 B^2}{2 V_s}$$

If the radius r is imposed by the presence of a flight tube with a fixed radius r, for a given value of B only the ions with the corresponding value of m/q go through the analyser. Changing B as a function of time allows successive observations of ions with various values of m/q. If $q = 1$ for all of the ions, the magnetic analyser selects the ions according to their mass, provided that they all have the same kinetic energy. Thus the magnetic analyser (which is fundamentally a momentum analyser) can be used as a mass analyser provided that the kinetic energy of the ions or at least their velocity is known.

Instead of positioning a guide tube and detecting the ions successively while scanning the magnetic field, it is also possible to use the characteristic that ions with the same kinetic energy but different m/q ratios have trajectories with different r values. Such ions emerge from the magnetic field at different positions. These instruments are said to be dispersive. Figure 2.54 shows two possibilities. Furthermore, Equations (2.2) and (2.3) yield

$$r = \frac{\sqrt{2m E_k}}{q B}$$

The result is that ions with identical charge and mass are dispersed by a magnetic field according to their kinetic energy. In order to avoid this dispersion, which alters the mass resolution, the kinetic energy dispersion must be controlled. This is achieved with an electrostatic analyser.

2.5.2 Electrostatic Field

Suppose a radial electrostatic field is produced by a cylindrical condenser. The trajectory is then circular and the velocity is constantly perpendicular to the field. Thus the centrifugal

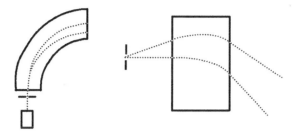

Figure 2.49
Left, energy dispersion; right, angular dispersion.

force equilibrates the electrostatic force according to the following equation, where E stands for the intensity of the electric field:

$$qE = \frac{mv^2}{r}$$

Introducing the entrance kinetic energy,

$$r = \frac{2E_k}{qE}$$

Since the trajectory is independent of the mass, the electric field is not a mass analyser, but rather a kinetic energy analyser, just as the magnetic field is a momentum analyser. Note that the electric sector separates the ions according to their kinetic energy.

2.5.3 Dispersion and Resolution

The resolving power of an analyser was defined earlier. We have seen at the beginning of this chapter how the resolution depends inversely on the dispersion at the analyser outlet. Three factors favour the dispersion, and thus the loss of resolution:

1. If the ions entering the field do not have the same kinetic energy, they follow different trajectories through the field. This is called energy dispersion (Figure 2.49).

2. If the ions entering the field follow different trajectories, this divergence may increase during the trip through the field. This is called angular dispersion (Figure 2.49).

3. The incoming ions do not originate from one point, but issue from a slit. The magnetic or electric field can only yield, at best, a picture of that slit. The picture width depends on the width of the slit and on the magnifying effect of the analyser.

2.5.3.1 Direction Focusing

An ion entering the magnetic field along a trajectory perpendicular to the field edge follows a circular trajectory as was described earlier. An ion entering at an angle α with respect to the previous perpendicular trajectory follows a circular trajectory with an identical radius and thus converges with the previous ion when emerging from the sector (Figure 2.50).

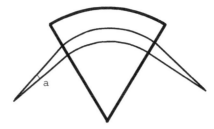

Figure 2.50
Direction focusing in a magnetic sector.

Figure 2.51
Direction focusing in an electric sector.

Choosing the correct geometry of the magnetic field (sector field) thus allows focusing of the incoming beam.

An ion entering the electric sector perpendicular to the field edge follows a curved trajectory, as was described earlier. However, if the ion trajectory at the inlet is not perpendicular to the edge, its trajectory is longer if it enters the sector closer to the outside and shorter if it enters the sector closer to the inside (Figure 2.51). Here again a suitably chosen geometry results in direction focusing.

2.5.3.2 Energy Focusing

As can be observed in Figure 2.52 when a beam of ions with different kinetic energies issues from the source, the electric and magnetic sectors produce an energy dispersion and a direction focusing. If two sectors with the same energy dispersion are oriented as is shown in Figure 2.53, the first sector dispersion energy is corrected by the second sector convergence. Double focusing instruments use this principle. A few examples appear in Figure 2.54. Rudimentary ion optics of the mass spectrograph have been described in a simple yet usable manner by Burgoyne and Hieftje [55].

2.5.4 Practical Considerations

The magnetic instrument's sources must function with potentials V_s of about 10 kV. The vacuum in the source must thus be very high so as to avoid arcing.

Classical magnets were not well suited to fast scanning, which is necessary for GC coupling, for instance, because of the hysteresis phenomenon and the magnet heating up

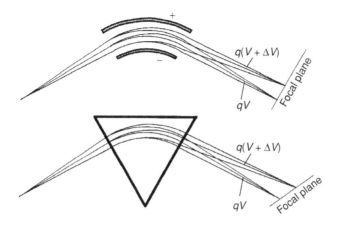

Figure 2.52
Energy dispersion: above in an electric sector and below in a
magnetic sector.

Figure 2.53
Combination of the two sectors, electrical and magnetic, shown
in Figure 2.52; the magnetic sector is turned over so as to obtain
double focusing.

by the Foucault currents induced by rapidly changing magnetic fields. Lamellar magnets
avoid such inconveniences; they have been well developed and are now widely used.

It can be shown that magnetic instruments function at constant resolution $R = m/\delta m$.
As a result, δm varies in proportion to m. In the low-mass range δm is small, while in the
high-mass range it is large. As an example, suppose the resolution is adjusted to 1000. For
an ion with mass exactly 100.0000, $\delta m = 100/1000 = 0.1$. An ion with mass 1000 yields
$\delta m = 1000/1000 = 1$. The ion with mass 100 is observed while the instrument is scanning
from mass 99.95 to 100.05. At mass 1000, this ranges from 999.5 to 1000.5. If the scanning
is carried out so as constantly to cover a mass unit within a time t, the ion with mass 100 is
observed during a time equal to $0.1t$, while that with mass 1000 is observed during $1t$. So
if the number of ions produced in the source during this time t is the same at mass 100 as
at mass 1000, then the total number of ions measured at the detector is 10 times smaller at
mass 100 than at 1000: the number of detected ions does not correspond to the number of
ions that are produced. The scanning is thus made exponential in order to correct this error
and the time spent scanning at mass 100 is 10 times longer than that at mass 1000. The
number of detected ions is then proportional to the number that is produced.

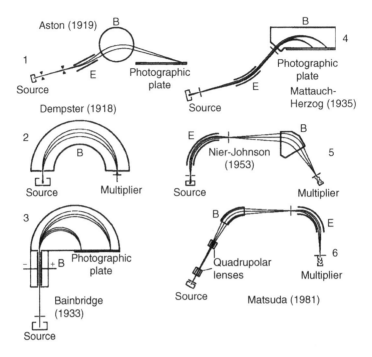

Figure 2.54
B = magnetic sector; E = electric sector. Six types of magnetic mass
spectrometers are displayed. Instruments 4, 5 and 6, with double
focusing, are most frequently used. Instrument 4 is dispersive, while
5 and 6 are scanning types. The photographic plate in 4 can be
replaced by an array detector for limited mass ranges. Reproduced
from Finnigan MAT documentation, with permission.

   An alternative technique allows one to increase artificially the mass accuracy of magnetic
spectrometers at a given resolution; this is called peak matching. This technique consists of
comparing the masses of two compounds that are simultaneously ionized in the spectrometer
source: one is unknown and its exact mass is sought; the other is a reference and its mass is
known with accuracy. This comparison is achieved by a very rapid alternative modification
of the acceleration voltage so as to focus the two ions, the intensities of the magnetic and
electric fields being kept constant. The match is perfect when the two masses' profiles
exactly overlap. If the acceleration voltages necessary for the focusing of the two ions
are known with accuracy, the mass of the unknown compound can be determined with
accuracy.
   Since sources were developed that produce high-mass ions, researchers have sought to
extend the range of mass spectrometers. In the case of magnetic instruments, the basic
equation shows how to act on the mass range:

$$\frac{m}{q} = \frac{r^2 B^2}{2V_s}$$

where V_s, the acceleration voltage in the source, can be reduced. This is a classical approach,
but it entails a loss of resolution and sensitivity.

Increasing B allows one to increase the mass range while keeping V_s constant. It is not possible to use superconducting magnets because the magnetic field cannot be scanned over a wide range, as is required for these instruments. Classical magnets allow one to reach a maximum field that depends on the nature of the alloy being used; the highest value is obtained with Permadur, a steel containing cobalt. The maximum is then 2.4 T. Finally, another technique consists of increasing the radius r. This causes a direct increase in instrument size, especially because a greater radius corresponds to a greater focal distance.

Recent instruments have attempted to overcome this inconvenience by modifying the ion trajectories and thus the focal distances. This is achieved by using two factors. First, the entrance and outlet angles of the ions can be different from the normal perpendicular to the field edge. Second, inhomogeneous fields can be used that modify the ion trajectories. For example, on a Kratos MS50RF instrument, the focal distance was reduced by 60 % using a inhomogeneous field.

2.5.5 Tandem Mass Spectrometry in Electromagnetic Analysers

We saw that some ions decompose during the flight inside the analyser. These metastable or collision-induced fragmentations can occur in different parts of the instrument. The fragmentations that occur inside the analyser are normally not observable. However, those that occur in the field-free regions can be observed under proper experimental conditions and yield interesting information. The first region located between the source and the first analyser is called the first field-free region. The second field free region is located between the first and the second analysers, and so on. They can have dimensions ranging from a few centimetres to a metre. The region that is most appropriate to the type of study being carried out is chosen: ion structure, reaction mechanism, thermochemical determinations, ion–molecule reactions, and so on.

Instruments with combined magnetic and electric analysers can be assembled according to either of two configurations. The electric sector is located either in front of the magnetic sector, which is the most frequent case, or behind it. The magnetic sector is labelled B and the electric sector is labelled E. The first configuration is called EB (or also 'Nier–Johnson').

The geometry is less frequently a BE one. It is sometimes called a 'reverse Nier-Johnson' or 'reverse geometry'. This geometry allows an easier analysis of the ion kinetic energy (IKE or MIKE spectrometry).

MS/MS is possible with a double sector instrument even though it cannot exactly be considered as made up of two mass spectrometers hooked in series. A technique called linked scan is used. It consists of a simultaneous scan of the E and B sectors according to a mathematical relationship depending on the system geometry, on the region under study and on the type of information that is looked for. These different types of scans are listed in Table 2.4.

We will now examine the most common types of scans.

2.5.5.1 Detection of Metastable Ions in Normal Scan

All of the ions have the same kinetic energy at the source outlet. If an ion dissociates in a field-free region, the fragments have approximately the same velocity as the precursor ion and thus have different kinetic energies from that of the precursor ion. Since the electric sector sorts the ions according to their kinetic energy, all of the metastable ions formed

Table 2.4 Types of scans for electromagnetic analysers. FFR1:
first field-free region; FFR2: second field-free region.

EB configuration		
Product ion spectrum	FFR1	B/E constant
Precursor ion spectrum	FFR1	B^2/E constant
Neutral loss spectrum	FFR1	$(B^2/E^2)(1-E)$ constant
BE configuration		
Product ion spectrum	FFR1	B/E constant
	FFR2	E scan (MIKE)
Precursor ion spectrum	FFR1	B^2/E constant
	FFR2	B^2E constant
Neutral loss spectrum	FFR1	$(B^2/E^2)(1-E)$ constant
	FFR2	$B^2(1-E)$ constant

before the electric sector are not observed in the normal spectrum. Thus the metastable ions produced in the first field-free region of an EB spectrometer or in BE instruments are not detected in a normal scan. Only EB instruments produce spectra that show the metastable ions formed between the two sectors. At the electric sector outlet the ion with a mass m_p has a velocity

$$v = \sqrt{\frac{2qV_s}{m_p}}$$

The fragment ion with mass m_f has the same velocity, and thus a momentum

$$m_f v = m_f \sqrt{\frac{2qV_s}{m_p}}$$

Since the focusing condition in the magnetic sector is $Bq = mv/r$, we have, for the precursor with mass m_p:

$$B_p = \sqrt{\frac{m_p}{q}} \frac{1}{r} \sqrt{2V_s}$$

and for the fragment with mass m_f:

$$B_f = \frac{m_f}{qr} \sqrt{\frac{2qV_s}{m_p}}$$

which corresponds to an apparent measured mass m^*:

$$B_f = \sqrt{\frac{m^*}{qr}} \sqrt{2V_s} = \frac{m_f}{qr} \sqrt{\frac{2qV_s}{m_p}} \quad \text{which gives } m^* = \frac{m_f^{\,2}}{m_p}$$

The ion issuing from the metastable fragmentation shows up at the apparent mass m^*, linked to m_p and m_f through the relation $m^* = m_f^2/m_p$. The kinetic energy released during the fragmentation brings up a dispersion in velocity, thus an alteration in the resolution.

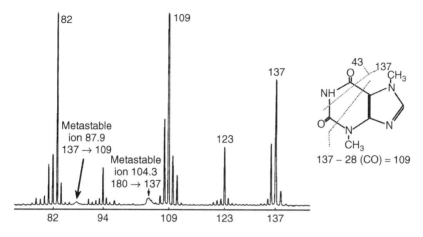

Figure 2.55
Example of a theobromine spectrum obtained with a magnetic instrument using an EB configuration. The signal detected at an apparent m/z 87.9 comes from the metastable fragmentation of the ion at 137 Da, which loses 28 Da and yields the 109 Da fragment. Similarly, the signal at apparent m/z 104.3 comes from the metastable fragmentation $180 \rightarrow 137$ Da. Reproduced from Kratos documentation, with permission.

Figure 2.55 shows as an example the spectrum of theobromine, the metastable fragmentation of which is shown in the scheme.

2.5.5.2 BE Instruments and MIKE Spectroscopy

Mass analysed ion kinetic energy (MIKE) is the simplest observation method for metastable or collision-induced ions. It requires an instrument with a reverse BE geometry.

At the source outlet, every ion, if we consider only the singly charged ones, has a kinetic energy expressed by the following equation:

$$e V_s = \frac{m v^2}{2}$$

This kinetic energy is constant throughout the flight.

If an ion dissociates between the magnetic and electric sectors, the fragments have, as a first approximation, the same velocity as the precursor ion.

Consider an ion M with mass m_p yielding a fragment F with mass m_f. The fragment has the velocity v_p of the ion M. The kinetic energies are, respectively,

$$E_{kp} = e V_s = \frac{m_p v_p^2}{2}$$

$$E_{kf} = \frac{m_f v_p^2}{2}$$

We therefore deduce

$$\frac{E_{kp}}{E_{kf}} = \frac{m_p}{m_f}$$

The ions going through the electric sector must obey the condition

$$eE = \frac{mv^2}{r}$$

Normally, the value of E is the same for all the primary ions, because they all have the same kinetic energy. The metastable ions are then eliminated: they are not observed in the normal spectrum taken with a BE instrument. However, if we modify the value of the field E so that

$$eE_p = \frac{m_p v_p^2}{r} \quad \text{and} \quad eE_f = \frac{m_f v_p^2}{r}$$

then the value E_f of the metastable ion field is observed. Thus,

$$\frac{E_p}{E_f} = \frac{m_p}{m_f}$$

Knowing the ratio E_p/E_f and the value of m_p allows us to determine the mass m_f of the fragment that is observed. The result is MIKE.

Normally, it is sufficient to isolate a beam of precursor ions by the magnetic sector (constant B value) and to scan the electric sector. As opposed to the other linked scans, the MIKE experiments exploit the independent use of the magnetic and electric sectors.

A first important use for MIKE is the determination of the fragment filiations that are observed in the spectrum. Such information is, of course, important for establishing fragmentation mechanisms.

The analysis we carried out supposes that the fragment ion retained the velocity of the precursor. This is true only if the fragmentation occurs without any release of kinetic energy. Usually the fragmentation goes along with a conversion of part of the internal energy into kinetic energy of the fragments. If the reaction releases kinetic energy E_f, the fragment then has a kinetic energy ranging from $(E_{kf} + E_f)$ and $(E_{kf} - E_f)$, depending on whether the orientation of the precursor ion induced an increase or a decrease in the velocity. This variation results in a widening of the peak compared with the width due to the normal dispersion of velocities and to the instrument resolution.

The MIKE technique thus allows a direct measurement of the kinetic energy that is released during the fragmentation between the magnetic and electric sectors, in a BE configuration instrument, by scanning the electric sector. The kinetic energy T released during the fragmentation of a metastable ion can also be measured by scanning the acceleration potential high-voltage scan. Both of these methods are shown in Figure 2.56 [56].

2.5.5.3 The B/E = Constant Scan

Consider an ion that fragments between the source and the first analyser. The fragment always has the velocity of the precursor. If the precursor is focused in the magnetic sector for a field B_p such that

$$eB_p = \frac{m_p v_p}{r}$$

the fragment ion is focused for

$$eB_f = \frac{m_f v_p}{r}$$

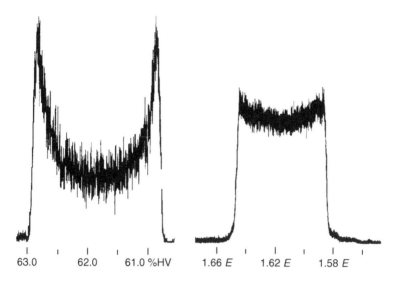

63.0   62.0   61.0 %HV   1.66 *E*   1.62 *E*   1.58 *E*

Figure 2.56
Widening of metastable peaks analysed by high-voltage scan (left) and
electric sector scan MIKES (right) and resulting from the kinetic energy
that is released during the fragmentation. Reproduced from Cooks R.G.,
Beynon J.H., Caprioli R.N. and Lester G.R., 'Metastable Ions', Elsevier,
New York, 1973, with permission.

Hence

$$\frac{B_\mathrm{p}}{B_\mathrm{f}} = \frac{m_\mathrm{p}}{m_\mathrm{f}}$$

For the electric sector, these conditions are

$$eE_\mathrm{p} = \frac{m_\mathrm{p}v_\mathrm{p}^2}{r'}$$

$$eE_\mathrm{f} = \frac{m_\mathrm{f}v_\mathrm{p}^2}{r'}$$

$$\frac{E_\mathrm{p}}{E_\mathrm{f}} = \frac{m_\mathrm{p}}{m_\mathrm{f}} = \frac{B_\mathrm{p}}{B_\mathrm{f}}$$

or

$$\frac{B_\mathrm{p}}{E_\mathrm{p}} = \frac{B_\mathrm{f}}{E_\mathrm{f}} = \text{contant}$$

The measurement principle is the following: both of the instrument sectors are first
focused on the precursor normal focusing values. The magnetic and electric fields are then
simultaneously scanned so as to respect the condition that $B/E = \text{constant}$ by reducing the
values of B and E with respect to their initial values. All of the fragments from the selected
precursor are thus successively detected. This is called the B/E linked scan. Note that this
technique is useful both for BE and for EB geometry instruments. The mass resolution
is better than for the MIKE technique, as double focusing is still available. However, as
the energy dispersion of the ions remains relatively high, especially if a collision gas is
used, only a weak resolution is possible in order to have sufficient sensitivity. The effective

resolution is about 350, which allows one to reach up to a 350 Da mass with a 'unit' or better resolution. This type of scan does not yield information on the kinetic energy that is released, since ions with equal masses but with different kinetic energies are focused upon one point.

2.5.5.4 The $B^2/E = Constant$ Scan

The fragment and precursor ions produced in the source carry the same kinetic energy ($m_p v_p^2 = m_f v_f^2$). However, the fragment ions formed in the first field-free region have a velocity that is identical with that of their precursors issued from the source ($v_p = v_f'$).

Given E_f and B_f, the respective values of the electric and magnetic fields allowing the transmission of fragment ions produced in the source

$$e E_f = \frac{m_f v_f^2}{r'} \quad \text{and} \quad e B_f = \frac{m_f v_f}{r}$$

and E_f' and B_f' the values of the electric and the magnetic fields allowing the transmission of the fragment ions produced in the first field-free region from the same precursors,

$$e E_f' = \frac{m_f v_f'^2}{r'} \quad \text{and} \quad e B_f' = \frac{m_f v_f'}{r}$$

the following relationships can be established:

$$\frac{E_f}{E_f'} = \frac{m_f v_f^2}{m_f v_f'^2} = \frac{m_p v_p^2}{m_f v_p^2} = \frac{m_p}{m_f}$$

$$\frac{B_f}{B_f'} = \frac{m_f v_f}{m_f v_f'} = \frac{v_f}{v_p} = \sqrt{\frac{m_p}{m_f}}$$

Hence

$$\frac{B_f^2}{E_f} = \frac{B_f'^2}{E_f'} = \text{constant}$$

In fact, the initial values of the magnetic and electric fields are those allowing the detection of selected fragment ions issuing from the source. Then the two sectors are scanned simultaneously, while keeping B^2/E constant.

Another scanning technique, the defocusing voltage scan, allows the determination of precursor ions from a given fragment.

2.5.5.5 Accelerating Voltage Scan or Defocusing

If an ion fragments between the source outlet and the first analyser, the product ion has the same velocity as the precursor, instead of its own velocity, and thus is not detected. However, if the precursor is accelerated to a velocity equal to that needed by the product ion to be focused, then the product ion is detected:

$$e V_s = \frac{m_p v_p^2}{2} = \frac{m_f v_f^2}{2}$$

In order for the product ion of mass m_f issuing from the precursor of mass m_p to be focused, the precursor ion has to be accelerated, before its dissociation, by a potential difference V' such that its velocity is v_f. We thus have

$$eV'_s = \frac{m_p v_f^2}{2}$$

Dividing the last two equations, we obtain:

$$\frac{V'_s}{V_s} = \frac{m_p}{m_f}$$

In order to exploit the method, we proceed as follows. First, the instrument is entirely focused on an ion issuing from the precursors we want to identify, with a source potential V_s. The potential V_s is then progressively increased. The fragment ion is 'defocused' if it is produced inside the source, but detected again every time the potential V_s' corresponds to a precursor of this ion. The mass m_p of the precursor is easily calculated:

$$m_p = \left(\frac{V'_s}{V_s}\right) m_f$$

It is thus possible to detect and analyse the mass of all of the precursors of a given fragment, so long as they yield a metastable or collision-induced fragmentation.

2.5.5.6 Constant Neutral Loss Scan: $B^2(1 - E)/E^2 = $ Constant Scan

Some groups give rise to the loss of typical neutrals. Consider the alcohols as an example, as they easily lose H_2O. The detection, in a spectrum, of all the fragments that lose $18\,Da$ allows the detection of those with a high probability of containing one of the hydroxyl groups of the starting molecule.

During a chromatographic analysis, the detection of compounds that give rise to losses of $18\,Da$ leads to highly selective detection of the chromatographic peaks of compounds containing a hydroxyl group. This technique is called neutral loss scanning.

Given m_n as the mass of a neutral loss, then starting from a precursor with mass m_p, the fragments with a mass m_f equal to $m_p - m_n$ have to be focused.

The focusing condition in the electric sector is written in the same way as for the analysis of the product ions:

$$\frac{m_f}{m_p} = \frac{E_f}{E_p} = E' = 1 - \frac{m_n}{m_p}$$

$$m_p = \frac{m_n}{(1 - E')}$$

The focusing condition in the magnetic sector is again that the B value corresponds to an ion with an apparent mass m^* such that

$$m^* = \frac{m_f^2}{m_p} = m_f E' = (m_p - m_n)E'$$

$$= \left(\frac{m_n}{(1 - E') - m_n}\right) E' = \frac{m_n E'^2}{(1 - E')}$$

$$\frac{m^*}{q} = \left(\frac{m_n E'^2}{(1 - E')q}\right) = \frac{r^2 B_f^2}{2V_s}$$

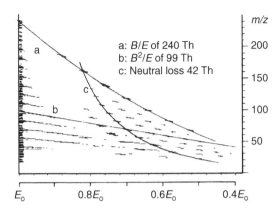

Figure 2.57
Metastable map obtained by scanning B for different
values of E using a heptadecane sample. Reproduced
from Kratos documentation, with permission.

Grouping the constant terms on the right, we obtain

$$\frac{B_f^2 \left(1 - E'\right)}{E'^2} = \frac{2 m_n V_s}{q r^2} = \text{constant}$$

If the scan is carried out while respecting the condition deriving from this equation, the fragment ion formed between the source and the analyser is passed on by the two sectors only if it differs by a constant mass m_n from its precursor.

2.5.5.7 The Metastable Map B = f(E)

A map of the detected ions can be drawn as a plot of B as a function of E. In order to obtain experimental points, E is kept constant and B is scanned. Then E is increased and a new scan of B is carried out. Repeating this process allows one to cover all of the possible combinations of E and B within a chosen range. Such a map contains simultaneously any function of B and of E, as is shown in Figure 2.57.

2.5.5.8 Instruments with More Than Two Sectors

We can imagine combining more than two sectors, thereby obtaining an increasing number of possibilities for MS/MS analyses. Instruments with three and four sectors in some configurations are commercially available.

The combination of two electric sectors and two magnetic sectors within an EBEB configuration, for example, allows, in theory, the high-resolution analysis of both the precursors and the fragments. However, the ions produced during the fragmentation have the velocity of the corresponding precursors. If a collision chamber is located between the two mass spectrometers and if the ions are not slowed, then the situation at the inlet of the second analyser is the same as that prevailing for the fragments formed between the source and the analyser in single analyser instruments. The focusing condition was shown earlier.

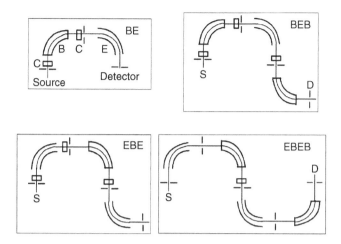

Figure 2.58
Common combinations of electric (E) and magnetic (B) sectors
and collision cells (C).

It calls for scanning while keeping the B/E ratio constant and equal to the value that this ratio normally has for the precursor ion.

The second instrument must thus scan according to a function that depends on the precursor selected in the first analyser. In practice, this can only be carried out through automatic computer control.

The second possible solution consists in slowing down the ions issuing from the first spectrometer, in achieving low-energy collisions and then in reaccelerating the ions so that they all have the same kinetic energy. The second analyser then functions as a normal spectrometer. However, the greater energy dispersion brings about a resolution loss of the second spectrometer. Figure 2.58 shows examples of such configurations.

2.6 Ion Cyclotron Resonance and Fourier Transform Mass Spectrometry

2.6.1 General Principle

We saw how the trajectories of ions are curved in a magnetic field. If the ion velocity is low and if the field is intense, the radius of the trajectory becomes small. The ion can thus be 'trapped' on a circular trajectory in the magnetic field: this is the principle of the ion cyclotron or Penning trap.

Suppose that an ion is injected into a magnetic field B with a velocity v. The equations are

$$\text{Centripetal force}: \quad F = qvB$$

$$\text{Centrifugal force}: \quad F' = \frac{mv^2}{r}$$

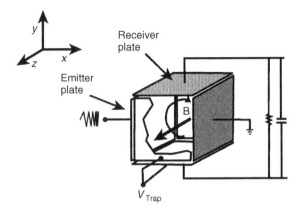

Figure 2.59
Diagram of an ion cyclotron resonance instrument. The
magnetic field is oriented along the z axis. Ions are injected in the trap along the z axis. They are trapped along
this axis by a trapping voltage, typically 1 V, applied to
the front and back plates. In the xy plane, they rotate
around the z axis owing to cyclotron motion and move
back and forth along the z axis, between the electrostatic trapping plates. The sense of rotation indicated is
for positive ions. Negative ions will orbit in the opposite
sense.

The ion stabilizes on a trajectory resulting from the balance of these two forces:

$$qvB = \frac{mv^2}{r} \quad \text{or} \quad qB = \frac{mv}{r}$$

The ion completes a circular trajectory of $2\pi r$ with a frequency

$$v = \frac{v}{2\pi r}$$

Thus the angular velocity ω is equal to

$$\omega_c = 2\pi v = \frac{v}{r} = \frac{q}{m}B$$

As a result of this equation, the frequency and the angular velocity depend on the ratio
$(q/m)B$, and are thus independent of the velocity. However, the radius of the trajectory
increases, for a given ion, in proportion to the velocity. If the radius becomes larger than
that of the cell, the ion is expelled.

In practice, the ions are injected into a box (Figure 2.59) a few centimetres along its side,
located in a magnetic field of 3 to 9.4 T produced by a superconducting magnet. For a 3 T
field, the cyclotron frequency is 1.65 MHz at 28 Th and 11.5 kHz at 4000 Th. The frequency
range is thus very large. Currently (2007) magnets giving a 15 T maximum field are used.
Magnets of 24 T have been tested at the National High Magnetic Field Laboratory. [57]

The relationship between the frequency and the mass shows that determining the mass
in this case consists of determining the frequency. The latter can be measured according to

several methods which are classified into one of two categories: those based on observing isolated frequencies and those using complex waves and Fourier transforms.

2.6.2 Ion Cyclotron Resonance

The first application of ion cyclotron resonance (ICR) to mass spectrometry is due to Sommer. [58]

Irradiating with an electromagnetic wave that has the same frequency as an ion in the cyclotron allows resonance absorption of this wave. The energy that is thus transferred to the ion increases its kinetic energy, which causes an increase in the radius of the trajectory. The 'image current' that is induced by the ions circulating in the cell wall perpendicularly to the trajectory of the ions can be measured. In this case an ion excitation phase, targeting only ions with a given mass so as to have them fly close to the wall, alternates with a detection phase.

To be detected, ions of a given mass must circulate as tight packets in their orbits. Ions of the same mass excited to the same energy will be in the same orbit and rotate with the same frequency. If, however, they are located anywhere on the orbit, when one ion passes close to one of the detecting plate then statistically there will be another ion of the same mass passing close to the opposite detecting plate. The resulting induced current will be null. To avoid this, ions have to be excited in a very short time, so that they are all grouped together in the orbit, and thus in phase.

2.6.3 Fourier Transform Mass Spectrometry

Fourier transform mass spectrometry (FTMS) was first described by Comisarow and Marshall in 1974 [59, 60] and was reviewed by Amster [61] in 1996 and by Marshall *et al.* [62] in 1998. This technique consists of simultaneously exciting all of the ions present in the cyclotron by a rapid scan of a large frequency range within a time span of about 1 µs. This induces a trajectory that comes close to the wall perpendicular to the orbit and also puts the ions in phase. This allows transformation of the complex wave detected as a time-dependent function into a frequency-dependent intensity function through a Fourier transform (FT), as shown in Figures 2.60 and 2.61.

Ions of each mass have their characteristic cyclotron frequency. It can be demonstrated that ions excited by an AC irradiation at their own frequency and with the same energy, thus the same V_0 potential, applied during the same time T_{exc}, will have an orbit with the same radius, and with an appropriate radius will all pass close to the detection plate:

$$r = \frac{V_0 T_{exc}}{B_0}$$

This equation, demonstrated in [62], is indeed independent of the m/z ratio. Thus broadband excitation will bring all the ions onto the same radius, but at frequencies depending on their m/z ratio, provided that the voltage is the same at each frequency. This can be best performed by applying a waveform calculated by the inverse Fourier transform, namely SWIFT [63].

As usual for a technique based on Fourier transform, the resolution depends on the observation time, which is linked with the disappearance of the detected signal (relaxation time). Here the disappearance of the signal mainly results from the ions being slowed by

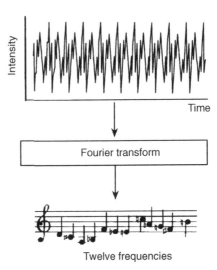

Figure 2.60
Principle of the Fourier transform: a
sound signal whose intensity is measured
as a time-dependent function is made up
of many frequencies superposed one over
the other, each with its own intensity. The
Fourier transform allows one to find the
individual frequencies and their intensities. Reproduced (modified) from Finnigan MAT documentation, with permission.

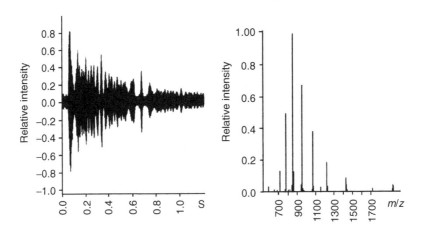

Figure 2.61
Signal intensity as a function of time is transformed, through a Fourier transform, into intensity as a function of frequency, and hence into an intensity
to m/z relationship. Redrawn from Amster I.J., J. Mass. Spectrom., 31, 1325–
1337, 1996, with permission.

the residual ion–molecule collisions. In order to achieve high resolution, a very high cell vacuum is necessary (about 10^{-5} Pa), which is an important limitation to this technique.

The FT requires a considerable amount of calculation, as the sampling velocity must be at least twice as great as the highest frequency that has to be measured (Nyquist theorem). The data flow to be treated is very large and needs appropriate computers. This inconvenience is impossible to overcome in the case of coupling with a chromatographic technique, unless we only want the spectra of a very limited number of components.

Comparing FTMS with Fourier transform nuclear magnetic resonance (FTNMR), we first notice how the frequency range to be covered here is very large. Second, relaxation in NMR is invariably linked with the interaction among liquid-phase or solid-phase molecules. In the gas phase, relaxation depends on the vacuum and on the stability of the ions being observed. If the vacuum is not sufficient, collisions slow the ions and their movement becomes incoherent. The observation of an ion is also limited to its lifetime.

FTMS allows one to achieve time spans of about 1 s per spectrum, such as in other mass spectrometric methods.

These instruments are sensitive enough to detect about 10 ions in the cell. However, the number of ions in the cell cannot exceed 10^6, as the repulsion between the ions scatters them considerably. The dynamic range is thus limited to about 10^5.

Techniques based on cyclotron resonance are also interesting because they allow the observation of ions over long time spans. This allows the study of slow fragmentations that are not observable in classical mass spectrometry, and also equilibria between ionic species and ion–molecule reactions.

The possibility of selectively eliminating ions from the cell through intense irradiation at resonance frequencies, and of eventually keeping only ions of a single mass within the cell, offers possibilities for the high-resolution study of ion reactions. For example, a gas-phase acidity scale was obtained with this method [64].

Using an ESI source coupled to an ICR FTMS system, Smith et al. [65] were able to observe an ion with a mass of 5×10^6 Da, with 2610 charge units. One ion was isolated in the cell and allowed to discharge with a collision gas. Figure 2.62 shows that the ion discharges in a quantified way. The number of charges can be deduced from the mass shifts that are observed.

2.6.3.1 Additional Features of Modern FTMS

In modern FTICR instruments, an ion of mass m and charge q is subjected to an axial homogenous magnetic field B and a quadrupolar electric field E derived from a potential $V = V_0(z^2 - \rho^2/2)/2d^2$, with $d^2 = (z_0^2 + \rho_0^2/2)/2$, where ρ_0 is the polar coordinate in the xy plane.

The motion of this ion results [66] from the superposition of an axial oscillation due to the trapping voltage V_0 on the end plates separated by a distance d with frequency $\omega_z = (qV_0/md^2)^{1/2}$, a cyclotron motion with frequency $\omega_c = qB/m$ and a magnetron motion with frequency $\omega_m = \omega_z^2/2\omega_c$ It can be shown that

$$\omega_m \ll \omega_z \ll \omega_c$$

so that the magnetron motion (Figure 2.63) has a radius much larger than the cyclotron motion.

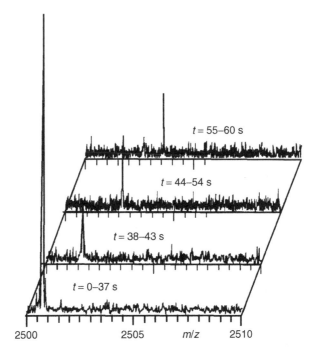

Figure 2.62
One multiply charged ion, produced in an ESI source, is
isolated in an ICR FTMS cell. During this time it discharges
by collision with a neutral gas in a quantified way, proving
that it is indeed an isolated ion. From the observed masses,
the number of charge can be determined, as explained for
the ESI source. Reproduced from Smith R.D., Cheng X.,
Bruce J.E., Hofstadler S.A. and Anderson G.A., Nature,
369, 137, 1994, with permission.

Collisional damping, reducing the energy of the ion by collision with an inert gas,
reduces the amplitude of the axial and the cyclotron motions while increasing the radius of
the magnetron motion, and hence the ions are lost on the wall of the cell. In order to prevent
this inconvenience, methods have been devised for axialization of the motion. One efficient
method is the superposition of an azimuthal electric quadrupolar field with frequency ω_c
which converts the magnetron motion into collision-damped cyclotron motion [67]. To
achieve the collision damping, helium gas is injected for a short time to raise the pressure
to about 10^{-2} Pa. This gives a spectacular increase in resolution, sensitivity and selectivity
[68]. A resolution of 1 770 000 has been obtained for leucine enkephalin [69].

There is another advantage [70] of the high resolution so obtained: for the multiply charged ions produced by ESI, the $^{12}C/^{13}C$ isotopic peaks must exhibit unit mass
spacing so that the number of them in a unit m/z ratio must represent the number of
charges. A theoretical intensity distribution may be compared with the experimental
distribution. Figure 2.64 displays a spectrum of ubiquitin from McLafferty's laboratory
[71].

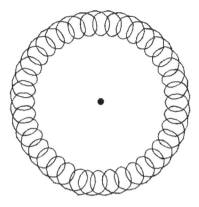

Figure 2.63
Cyclotron and magnetron motion. The ions move in an orbit known as the cyclotron movement. But that orbit itself turns around a centre, represented by a black dot, in a magnetron movement, responsible for a loss of resolution. The magnetron motion is the more important, the larger the ion, and thus causes mainly loss of resolution at higher masses. Application of a quadrupolar RF field allows this magnetron movement to be suppressed.

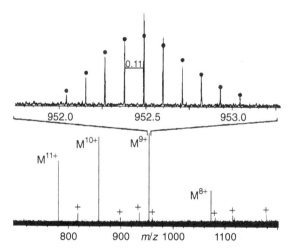

Figure 2.64
High-resolution FT ICR mass spectrum of ubiquitin. The upper trace displays the detail of the peak at 952.5 Th. The distance of 0.11 Th between the isotopic peaks allows one to deduce that the charge is 9. This spectrum was obtained with 5 amol of ubiquitin. Reproduced from Kelleher N.L., Senko M.W., Little D., O'Connor P.B. and McLafferty F.W., J. Am. Soc. Mass Spectrom., 6, 220, 1995, with permission.

2.6.4 MS^n in ICR/FTMS Instruments

In an ICR cell, ions exposed to an axial magnetic field B and having a charge ze have a cyclotron motion with frequency

$$\omega_c = \frac{zeB}{m}$$

Thus, monocharged ions will have, for a given value of B depending on the instrument, a frequency inversely related to their mass. They can be detected by the current they induce in the wall of the cell, in a non-destructive way. This is a distinct feature of ICR which is important in MS/MS experiments. By applying an AC electric field perpendicular to the axis, it is possible to excite ions, or to expel ions by discharge on the wall. This allows all ions to be expelled except those having a selected mass. These ions are then allowed to fragment during a short time. Sometimes a collision gas is introduced by a pulse to induce fragmentation by collision, and excitation by irradiation at the cyclotron frequency is also used. However, sustained excitation at the resonance frequency results in larger cyclotron radii, and fragment ions are then produced with non-zero magnetron radii. It is now common practice to use sustained off-resonance irradiation (SORI) [72, 73]. This results in ions being alternatively accelerated and decelerated, limiting the cyclotron radius. The fragments are then produced close to the centre. Fragments are detected while they are formed in a non-destructive way. They can be detected repetitively, increasing sensitivity and resolution. The selection and fragmentation process can then be repeated, providing MS^n capability, without reloading ions from the source as the detection is not destructive. This contrasts with the situation in ion traps, where the detection of ions empties the trap, which then has to be reloaded from the source.

However, the magnetic field has no focusing properties. Over time, there is an off-axis displacement of the centre of the ion cyclotron orbit, known as magnetron radial expansion. Reaxialization can be performed using the focusing properties of a quadrupolar RF electric field. Combined with collision damping, this allows reversion of the magnetron radial expansion and greatly improves the resolution and the sensitivity, especially by reducing the loss of ions during MS/MS experiments. High resolution can be obtained on both precursor *and* fragment ions.

2.7 Hybrid Instruments

Some mass spectrometers combine several types of analysers. The most common ones include two or more of the following analysers: electromagnetic with configurations EB or BE, quadrupoles (Q), ion traps (ITs) with Paul ion traps or linear ion traps (LITs), time-of-flight (TOF), ion cyclotron resonance (ICR) or orbitrap (OT). These are named hybrid instruments. The aim of a hybrid instrument is to combine the strengths of each analyser while avoiding the combination of their weaknesses. Thus, better performances are obtained with a hybrid instrument than with isolated analysers. Hybrids are symbolized by combinations of the abbreviations indicated in the order that the ions travel through the analysers.

Initially, most combinations of electromagnetic sectors with quadrupoles have been proposed. The first hybrids described were instruments of type BEqQ but several other

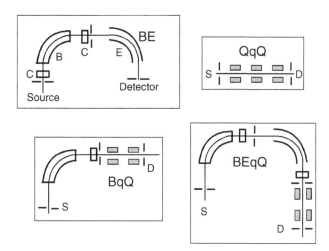

Figure 2.65
Common combinations of electric (E) and magnetic (B) sectors, quadrupoles (Q) and collision cells (C). The lower case q indicates a focusing quadrupole.

types of hybrids have been built over the years. Indeed, almost all possible combinations of two different types of analysers have been reported, and many are commercially available. We will describe some important hybrid instruments below.

2.7.1 Electromagnetic Analysers Coupled to Quadrupoles or Ion Traps

As already mentioned, the first hybrid instruments result from the combination of an electromagnetic analyser with a quadrupole [74, 75]. They are most frequently made up of an electromagnetic instrument in front of a quadrupole analyser. They generally include between the two analysers a quadrupolar RF-only collision cell and thus correspond to the BEqQ or the EBqQ configurations. Figure 2.65 gives examples of such configurations. Ions can be analysed at high resolution in the magnetic instrument and then with low resolution in the quadrupole part. Two options are available for the two analysers, which respect the requirements of each type of analyser regarding kinetic energy. Indeed, the kinetic energy must be in the kilovolt range in the electromagnetic part and tens of volts in the quadrupoles.

The first option consists of slowing all of the ions when they come out of the electromagnetic analyser, which is easy because at this stage they all have the same kinetic energy. Ions with a low energy are then fragmented through collisions in the first quadrupole q and analysed in the second quadrupole Q. The second option consists of achieving high-energy collisions in a collision cell located at the exit of the magnetic analyser, before entering the first quadrupole. The fragments are then slowed to a kinetic energy that is compatible with the quadrupoles. However, in this case, all the fragments have the same velocity as their precursors, and thus have different kinetic energies. The kinetic energies of the fragments are indeed equal to a fraction m_1/m_2 of the kinetic energy of their precursors. The voltage applied on the electrode to slow the ions before the quadrupoles must be a function of the

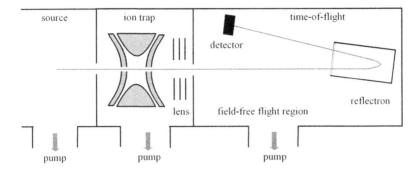

Figure 2.66
Scheme of an IT TOF hybrid mass spectrometer including a Paul IT analyser
and TOF analyser equipped with a reflectron.

ratio of the mass m_1 of the precursor focused at the exit of the electromagnetic part and
of the mass m_2 of the fragment to be analysed in the quadrupole. Furthermore, the BEqQ
or EBqQ instruments can be used for MS^3 experiments. Indeed, other scan modes allow
MS/MS in the electromagnetic analyser using MIKE or linked scans. In combination with
the quadrupole part, this allows various MS^3 experiments to be performed.

Similar hybrid instruments in which the quadrupole analyser is replaced by an IT have
also been described, for instance in a BE IT configuration [76]. Similar corrections of the
kinetic energy must be applied. Such instruments are interesting for two reasons. First, MS^n
can be performed with precursor ions selected at high resolution. Second, precursor ions
can be accumulated in the IT over a longer time period, allowing very high sensitivity to
be reached. Here again, combinations with linked scans are possible.

2.7.2 Ion Trap Analyser Combined with Time-of-Flight or Ion Cyclotron Resonance

Several hybrid instruments have more recently been proposed that combine TOF and IT
analysers in the IT TOF configuration [77–79]. The scheme of such an instrument is given
in Figure 2.66.

The IT is used to accumulate ions and to perform ion selection and activation in MS/MS
experiments before analysis in the TOF analyser. All the ions accumulated in the trap are
then ejected in the RTOF analyser. Therefore, the TOF analyser is used for mass analysis
instead of the classical ion ejection methods used with ITs, namely mass selective ion ejection at the stability limit or resonant ejection. In comparison with TOF instruments, higher
sensitivity is achieved by ion accumulation in the IT. In comparison with IT instruments,
the analysis by TOF reduces the time as the TOF analyser allows faster mass analysis,
extends the mass range, and gives a better resolution and much better accuracy.

An IT analyser has also been coupled to an ICR FTMS instrument, yielding a hybrid
instrument in the IT ICR configuration. This hybrid instrument gives high sensitivity at
the attomole level, a high resolution of 100 000 FWHM at 1 s scan rate and a high mass
accuracy of 1 to 2 ppm with external calibration at 1 scan per second. A similar hybrid
instrument in which the ICR analyser is replaced by an orbitrap analyser has also been

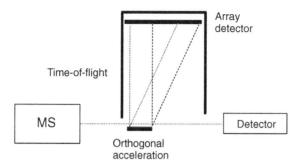

Figure 2.67
Principle of the combination of a mass spectrometer with an oa-TOF spectrometer. Ions coming from the mass spectrometer are directed to the detector. When a pulse voltage is applied to the orthogonal acceleration repeller, the ions are analysed by the TOF instrument.

reported. This instrument has main specifications similar to the IT ICR instrument. These orbitrap hybrids are described in Section 2.3.

2.7.3 Hybrids Including Time-of-Flight with Orthogonal Acceleration

The advantages of oa-TOF analysers with continuous ion beam sources have been exploited since the 1990s in hybrid instruments that employ this analyser as the second mass analyser. The first mass analyser is coupled to the TOF in such a way that acceleration of the ions in the TOF occurs perpendicular to the initial trajectory of the ions. The TOF analyser can be advantageous as the second stage of the instrument because its capability to transmit simultaneously all ions leads to a useful increase of sensitivity. Indeed, the TOF analyser has the advantage of detecting all ions simultaneously to a good precision, while scanning analysers detect ions successively during the time. Consequently, such a combination has clear speed and sensitivity advantages, even when a broad mass range is analysed. This is valid in both MS and MS/MS mode.

As already mentioned, when ions fragment during their flight in a field-free region, with or without collision activation, the precursor and fragments have the same velocity and reach the detector simultaneously. If, however, they are accelerated perpendicularly to their trajectories and so pushed in the TOF mass analyser, ions with different masses will arrive at different times and will be detected separately according to their respective m/z [80]. However, owing to their initial velocities, they will reach the array detector at different places.

Figure 2.67 displays the principle of such an instrument. After the first analyser, the ions are focused into a parallel beam. They are then directed towards the oa-TOF analyser where they enter continuously. Initially no voltage is applied to the repeller. Thus, at this point, this region is field free, so ions continue to move in their original direction. Then an injection pulse voltage is applied to the repeller. This induces an electric field that is perpendicular to the trajectories of the ions. The ions are then pushed by the resulting electric field in a direction orthogonal to their original trajectories, and begin to fly towards the analyser.

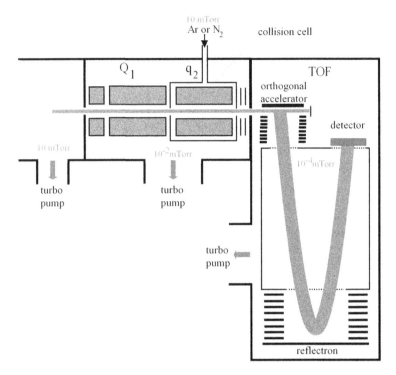

Figure 2.68
Scheme of a hybrid mass spectrometer including a quadrupole analyser,
a quadrupolar collision cell and an orthogonal acceleration time-of-flight
analyser.

Before entry into the flight tube where mass separation occurs, the ions are further
accelerated and acquire their final energy of several kiloelectronvolts. When all ions are in
the flight tube, the voltage on the repeller is no longer applied and ions from the ion source
again begin to fill the orthogonal accelerator. Thus, during the time that the ions continue
their flight in the flight tube, the orthogonal accelerator is refilled with new ion beam. One
flight cycle will end when the ion with the highest m/z reaches the detector. Another flight
cycle will begin by reapplying a pulse voltage to the repeller. In practice, a pulsed electric
field with a frequency of a few kilohertz is used. As the beam of ions is parallel, the spread
of initial velocities in the TOF direction is minimized. Furthermore, the width of the
ion beam gives a spatial spread that can be easily corrected by refocalization in the TOF
analyser.

The first hybrid instrument to take advantage of orthogonal injection combines an
electromagnetic analyser with an oa-TOF instrument [81,82]. However, the most successful
type of hybrid instruments in which an oa-TOF has been integrated combines a quadrupole
analyser with a TOF instrument in a Q TOF configuration. These instruments are powerful
and robust with unique performances [83–85]. They give high sensitivity in the attomole
range, a resolution of 10 000 FWHM allowing the assignment of the number of charges of
multiply charged ions, a mass range extended to about m/z 20 000, and a fair mass accuracy

of about 5 to 10 ppm. The rapid success of this type of hybrid instrument is due to the attractive combination of the simplicity of the quadrupole and the high performance of the TOF. The high sensitivity and high mass accuracy are available in both MS and tandem (MS/MS) modes.

As shown in Figure 2.68, the most common of these instruments include a quadrupole analyser Q1 and a quadrupolar collision cell q2, followed by an oa-TOF. They thus have the QqTOF configuration. This instrument can be described as a triple quadrupole where the last quadrupole is replaced by an oa-TOF, or as the addition of a quadrupole analyser and a collision cell to a TOF analyser. In some commercial instruments, the q2 quadrupole is replaced by an hexapole, but the principle remains the same.

In MS mode, the two quadrupoles Q1 and q2 are in RF-only mode, and they act only as ion guides, but are limited in their mass range. The TOF analyses all the ions that have been orthogonally accelerated and acts as the only mass analyser. The resulting spectra support the good performances of the TOF analyser regarding resolution and mass accuracy, but with a mass range limited by the transmission of the quadrupoles. The q2 quadrupole may or not contain a collision gas. If it contains a gas at sufficiently low pressure and the ions have a low kinetic energy, the energy transmitted will remain lower than the fragmentation threshold but the ions will lose kinetic energy in the radial and axial directions. This will improve the resolution and the sensitivity in the TOF analyser by reducing the flight time distribution of ions with a given mass value. Altogether in these instruments, the velocity spread of the ions is strongly reduced. Furthermore, the beam is strongly focused in the axial direction, thus also reducing the spatial spread at injection time. The beam is then much better prepared for injection in the TOF than those issuing from a directly connected ion source. Furthermore, any continuous ion source, by opposition to pulsed ones, can be connected. However, the extended mass range, over about m/z 20 000, is lost.

In principle, it is also possible to use the scanning capability of Q1, using the TOF part only as a total ion current detector. However, this scan mode is not used, as all the above-mentioned advantages of using the TOF analyser are lost.

In MS/MS mode, the ion filter capability of Q1 is used to transmit and to select only precursor ions of interest. These ions are then accelerated by a potential difference that is between 20 and 200 eV before they enter into the collision cell q2 where they undergo fragmentation induced by collision with neutral gas molecules (usually argon or nitrogen). The resulting fragment ions and the remaining parent ions continue to undergo collisions before being analysed in the TOF. Consequently, these ions lose kinetic energy in the radial and axial directions. In other words, this not only focuses the ions, but also controls their kinetic energy by collisional cooling. This step is even more important in QqTOF instruments than it is in triple quadrupoles because the TOF analyser is much more sensitive to the kinetic energy spread of the incoming ions than is Q3 in a triple quadrupole instrument.

In comparison with a tandem TOF/TOF instrument, in this hybrid instrument the mass range of the selected precursor ion is reduced by the limitations of the quadrupole analyser to about m/z 4000, though ion focusing and transmission are much easier. On the other hand, compared with a triple quadrupole, the TOF analyser offers better resolution and better mass accuracy. Sensitivity is increased with the TOF if spectra are measured over a broad mass range owing to the ability of this analyser to analyse all the ions almost together. Consequently, this sensitivity advantage is lost for selected reaction monitoring or multiple reaction monitoring. For the same reason, the advantage is not met for all of the MS/MS scan modes but only for the product ion scan. Precursor ion scan or neutral loss scan cannot be performed with a QTOF, but some similar measurements are possible without the gain in sensitivity.

References

1. Marshall, A.G., Hendrickson, C.L. and Shi, S.D.H. (2002) *Anal. Chem.*, **74**, 253A–9A.
2. Ferguson, R.E., McKulloh, K.E. and Rosenstock H.M. (1965) *J. Chem. Phys.*, **42**, 100.
3. Kienitz, H. (1968) *Massenspektrometrie*, Verlag Chemie, Weinheim.
4. Paul, W. and Steinwedel, H.S. (1953) *Z. Naturforsch.*, **8a**, 448.
5. Finnigan, R.E. (1994) *Anal. Chem.*, **66**, 969A.
6. A partial treatment of this problem can be found in March, R.E. and Hughes, R.J. (1989) *Quadrupole Storage Mass Spectrometry*, John Wiley & Sons, Inc., New York. For a detailed treatment, see also the work of the mathematician R. Campbell, *Théorie Générale de l'Equation de Mathieu*, Masson, Paris, 1955.
7. March, R.E. and Hughes, R.J. (1989) *Quadrupole Storage Mass Spectrometry*, John Wiley & Sons, Inc., New York.
8. Yost, R.A. and Enke, C.G. (1979) *Anal. Chem.*, **51**, 1251A.
9. Paul, W. and Steinwedel, H.S. (1960) US Patent, 2939952.
10. Todd, J.F.J. (1991) *Mass Spectrom. Rev.*, **10**, 3.
11. Stafford, G.C., Kelley, P.E., Syka, J.E. *et al.* (1984) *Int. J. Mass Spectrom. Ion Processes*, **60**, 85.
12. March, R.E. and Hughes, R.J. (1989) *Quadrupole Storage Mass Spectrometry*, John Wiley & Sons, Inc., New York, p. 200.
13. March, R.E. and Hughes, R.J. (1989) *Quadrupole Storage Mass Spectrometry*, Chemical Analysis vol. **102**, John Wiley & Sons, Inc., New York.
14. March, R.E. and Hughes, R.J., *Quadrupole Storage Mass Spectrometry*, John Wiley & Sons, Inc., New York, 1989, p. 199.
15. Cooks, R.G., Glish, G.L., McLuckey, S.A. and Kaiser, R.E. (1991) *Chem. Eng. News*, **69**, March 25, 26.
16. Major, F.G. and Dehmelt, H.G. (1968) *Phys. Rev.*, **170**, 91.
17. Senko, M.W., Cunniff, J.B. and Land, A.P. (1998) Producing 'richer' production spectra on a quadrupole ion trap with broadband activation. *Proceedings of the 46th ASMS Conference on Mass Spectrometry and Allied Topics, Orlando, Florida*, p. 486.
18. Viseux, N., de Hoffmann, E. and Domon, B. (1997) Structural assignment of permethylated oligosaccharide subunits using sequential tandem mass spectrometry. *Anal. Chem.*, **69**, 3193–8.
19. Douglas, D.J., Frank, A.J. and Mao, D. (2005) Linear ion traps in mass spectrometry. *Mass Spectrom. Rev.*, **24**, 1–29.
20. Hager, J.W. (2002) A new linear ion trap mass spectrometer. *Rapid Commun. Mass Spectrom.*, **16**, 512–26.
21. Moradian, A. and Douglas, D.J. (2005) Axial ion ejection from linear quadrupoles with added octopole field. *Proceedings of the 53rd ASMS Conference on Mass Spectrometry and Allied Topics, San Antonio, Texas*, June.
22. Hashimoto, Y., Hasegawa, H., Baba, T. and Waki, I. (2006) Mass selective ejection by axial resonant excitation from a linear ion trap. *J. Am. Soc. Mass Spectrom.*, **17**, 685–90.
23. Schwartz, J.C., Senko, M.W. and Syka, J.E.P. (2002) A two-dimensional quadrupole ion trap mass spectrometer. *Proceedings of the 50th ASMS Conference on Mass Spectrometry and Allied Topics, Orlando, Florida*, June.

24. Senko, M.W. and Schwartz, J.C. (2002) Trapping efficiency measured in a 2D ion trap mass spectrometer. *Proceedings of the 50th ASMS Conference on Mass Spectrometry and Allied Topics, Orlando, Florida*, June.

25. Schwartz, J.C., Senko, M.W. and Syka, J.E.P. (2002) A two-dimensional quadrupole ion trap mass spectrometer. *J. Am. Soc. Mass Spectrom.*, **13**, 659–69.

26. Makarov, A. (1999) Mass spectrometer. US Patent, 5886346.

27. Makarov, A., Hardman, M.E., Schwartz, J.C. and Senko, M. (2004) US Patent, 2004108450.

28. Makarov, A. (2000) Electrostatic axially harmonic orbital trapping: high-performance technique of mass analysis. *Anal. Chem.*, **72**, 1156–62.

29. Makarov, A., Hardman, M.E., Schwartz, J.C. and Senko, M. (2004) US Patent, 2004108450.

30. Hu, Q., Noll, R.J., Li, H. *et al.* (2005) The orbitrap: a new mass spectrometer. *J. Mass Spectrom.*, **40**, 430–43.

31. Hu, Q., Makarov, A., Noll, R.J. and Cooks, R.G. (2004) Application of the orbitrap mass analyser to biologically relevant compounds. *Proceedings of the 52nd ASMS conference, Nashville, Tennessee*, May 23–27.

32. Stephens, W. (1946) *Phys. Rev.*, **69**, 691.

33. Wiley, W.C. and McLaren, J.B. (1955) *Rev. Sci. Instrum.*, **16**, 1150.

34. Cotter, R.J. (1999) *Anal. Chem.*, **71**, 445A.

35. Guilhaus, M. (1995) *J. Mass Spectrom.*, **30**, 1519.

36. Mann, M. and Talbo, G. (1996) Developments in matrix-assisted laser desorption/ionization peptide mass spectrometry. *Curr. Opin Biotechnol.*, **7**, 11–9.

37. Mamyrin, B.A. (2001) *Int. J. Mass Spectrom.*, **206**, 251–66.

38. Weickhardt, C., Moritz, F. and Grotemeyer, J. (1996) Time-of-flight mass spectrometry: state-of-the-art in chemical analysis and molecular science. *Mass Spectrom. Rev.*, **15**(3), 139–62.

39. Cotter, R.J. (ed.) (1994) *Time-of-Flight Mass Spectrometry*, ACS Symposium Series, No. 549), American Chemical Society, Washington, DC.

40. Schlag, E.W. (ed.) (1994) *Time-Of-Flight Mass Spectrometry and Its Applications*, Elsevier Science, Amsterdam.

41. Cotter, R.J. (1997) *Time-Of-Flight Mass Spectrometry : Instrumentation and Applications in Biological Research*, American Chemical Society, Washington, DC.

42. Imrie, D.C., Pentney, J.M. and Cottrell, J.S. (1995) *Rapid Commun. Mass Spectrom.*, **9**, 1293.

43. Moniatte, M., vanderGoot, F.G., Buckley, J.T. *et al.* (1996) Characterisation of the heptameric pore-forming complex of the Aeromonas toxin aerolysin using MALDI-TOF mass spectrometry. *FEBS Lett.*, **384**(3), 269–72.

44. Lange, W., Greifendorf, D., Van Leyen, D. *et al.* (1986) *Springer Proc. Phys.*, **9**, 67.

45. Onnerfjord, P., Nilsson, J., Wallman, L. *et al.* (1998) Picoliter sample preparation in MALDI-TOF MS using a micromachined silicon flow-through dispenser. *Anal. Chem.*, **70**(22) 4755–60.

46. Mamyrin, B.A., Karataev, V.I., Schmikk, D.V. and Zagulin, V.A. (1973) *Sov. Phys. – JETP*, **37**, 4.

47. Spengler, B.J. (1997) *J. Mass Spectrom.*, **38**, 1019.

48. Piyadasa, C.K.G., Hakansson, P., Ariyaratne, T.R. and Barofsky, D.F. (1998) *Rapid Commun. Mass Spectrom.*, **12**, 1655.

49. Cornish, T.J. and Cotter, R.J. (1993) *Rapid Commun. Mass Spectrom.*, **7**, 1037.

50. Cornish, T.J. and Cotter, R.J. (1993) *Anal. Chem.*, **65**, 1043.

51. Seeterlin, M.A., Vlasak, P.R., Beussman, D.J. *et al.* (1993) *J. Am. Chem. Soc.*, **4**, 751.

52. O'Halloran, G.J., Fluegge, R.A., Betts, J.F. and Everett, W.L. (1964) Technical Documentary Report No. ASD-TDR-62–644, The Bendix Corporation, Research Laboratory Division, Southfield, MI.

53. Dovonof, A.F., Chernushevich, I.V. and Laiko, V.V., (1991) *Proceedings of 12th International Mass Spectrometry Conference, 26–30 August, Amsterdam, the Netherlands*, p. 153.

54. Dawson, J.H.J. and Guilhaus, M. (1989) *Rapid Commun. Mass Spectrom.*, **3**, 155–9.

55. Burgoyne, T.W. and Hieftje, G.M. (1996) An introduction to ion optics for the mass spectrograph. *Mass Spectrom. Rev.*, **15**(4), 241–59.

56. Cooks, R.G., Beynon, J.H., Caprioli, R.N. and Lester, G.R. (1973) *Metastable Ions*, Elsevier, New York.

57. Shi, S.D.H., Drader, J.J., Hendrickson, C.L. and Marshall, A.G. (1999) Fourier transform ion cyclotron resonance mass spectrometry in a high homogeneity 25 tesla resistive magnet. *J. Am. Soc. Mass Spectrom.*, **10**, 265–8.

58. Sommer, H., Thomas, H.A. and Hipple, J.A. (1949) *Phys. Rev.*, **76**, 1877.

59. Comisarow, M.B. and Marshall, A.G. (1974) Fourier transform ion cyclotron resonance spectroscopy. *Chem. Phys. Lett.*, **25**, 282–3.

60. Comisarow, M.B. and Marshall, A.G. (1974) Frequency sweep Fourier transform ion cyclotron resonance spectroscopy. *Chem. Phys. Lett.*, **26**, 489–90.

61. Amster, I.J. (1996) Fourier transform mass spectrometry. *J. Mass Spectrom.*, **31**, 1325–37.

62. Marshall, A.G., Hendrickson, C.L. and Jackson, G.S. (1998) Fourier transform ion cyclotron resonance mass spectrometry: a primer. *Mass Spectrom. Rev.*, **17**, 1–35.

63. Marshall, A.G., Wang, T.C.L. and Ricca, T.L. (1985) *J. Am. Chem. Soc.*, **107**, 7893.

64. Bowers, M.T. (ed.) (1979) *Gas Phase Ion Chemistry*, vol. **2**, Academic Press, New York.

65. Smith, R.D., Cheng, X., Bruce, J.E. *et al.* (1994) *Nature*, **369**, 137.

66. Brown, L.S. and Gabrielse, G. (1986) *Rev. Mod. Phys.*, **58**, 233.

67. Bollen, G., Moore, R.B., Savarde, G. and Stolzenberg, H. (1990) *Appl. Phys.*, **68**, 4355.

68. Guan, S., Wahl, M.C., Wood, T.D. and Marshall, A.G. (1993) *Anal. Chem.*, **65**, 1753; Scheikhard, L., Guan, S. and Marshall, A.G. (1992) *Int. J. Mass Spectrom. Ion Processes*, **120**, 71.

69. Guan, S. and Marshall, A.G. (1993) *Rapid Commun. Mass Spectrom.*, **7**, 857.

70. Beu, S.C., Senko, M.W., Quiinn, J.P. and McLafferty, F.W. (1993) *J. Am. Soc. Mass Spectrom.*, **4**, 190.

71. Kelleher, N.L., Senko, M.W., Little, D. *et al.* (1995) *J. Am. Soc. Mass Spectrom.*, **6**, 220.

72. Gauthier, J.W., Trautman, T.R. and Jacobson, D.B. (1991) Sustained off-resonance irradiation for collision-activated dissociation involving Fourier transform mass spectrometry. Collision-activated dissociation technique that emulates infrared multiphoton dissociation. *Anal. Chim. Acta*, **246**, 211–25.

73. Amster, I.J. (1996) Fourier transform mass spectrometry. *J. Mass Spectrom.*, **31**, 1325–37.

74. Glish, G.L., McLuckey, S.A., McBay, E.H. and Bertram, L.K. (1986) *Int. J. Mass Spectrom. Ion Processes*, **70**, 321.

75. Schoen, A.E., Amy, J.W., Ciupek, J.D. *et al.* (1985) *Int. J. Mass Spectrom. Ion Processes*, **65**, 125.

76. Loo, J.A. and Muenster, H. (1999) *Rapid Commun. Mass Spectrom.*, **13**, 54.

77. Waldren, R.M. and Todd, J.F.J. (1979) *Int. J. Mass Spectrom. Ion Phys.*, **29**, 314.

78. Michael, S.M., Chien, B.M. and Lubman, D.M. (1993) *Anal. Chem.*, **65**, 2614.

79. Doroshenko, V.M. and Cotter, R.J. (1998) *J. Mass Spectrom.*, **33**, 305.

80. Guilhaus, M., Selby, D. and Mlynski, V. (2000) *Mass Spectrom. Rev.*, **19**, 65.

81. Batemen, R.H., Green, M.R., Scott, G. and Clayton, E. (1995) *Rapid Commun. Mass Spectrom.*, **9**, 1227.

82. Strobel, F.H., Solouki, T., White, M.A. and Russell, D.H. (1991) *J. Am. Soc. Mass Spectrom.*, **2**, 91.
83. Morris, H.R., Paxton, T., Dell, A. *et al.* (1996) *Rapid Commun. Mass Spectrom.*, **10**, 889.
84. Shevchenko, A., Chernushevich, I., Ens, W. *et al.* (1997) *Rapid Commun. Mass Spectrom.*, **11**, 1015.
85. Chernushevich, I.V., Loboda, A.V. and Thomson, B.A. (2001) *J. Mass Spectrom.*, **36**, 849.

3
Detectors and Computers

3.1 Detectors

The ions pass through the mass analyser and are then detected and transformed into a usable signal by a detector. Detectors are able to generate from the incident ions an electric current that is proportional to their abundance. Detectors used in mass spectrometry were reviewed in 2005 [1]. The most common types of ion detectors are described below. Their specifications and their features are also discussed.

Several types of detectors presently exist. The choice of detector depends on the design of the instrument and the analytical applications that will be performed. A variety of approaches are used to detect ions. However, detection of ions is always based on their charge, their mass or their velocity. Some detectors (Faraday cup) are based on the measurement of direct charge current that is produced when an ion hits a surface and is neutralized. Others (electron multipliers or electro-optical ion detectors) are based on the kinetic energy transfer of incident ions by collision with a surface that in turn generates secondary electrons, which are further amplified to give an electronic current. As described in Chapter 2, detection of ions in FTICR or orbitrap (OT) mass spectrometers is characteristically different from these detections. In this case, the detector consists of a pair of metal plates within the mass analyser region close to the ion trajectories. Ions are detected by the image current that they produce in a circuit connecting the plates.

Because the number of ions leaving the mass analyser at a particular instant is generally quite small, significant amplification is often necessary to obtain a usable signal. Indeed, 10 incident ions per second at the detector corresponds to an electric current of 1.6×10^{-18} A. In consequence, subsequent amplification by a conventional electronic amplifier is required. Furthermore, with the exception of Faraday cup and image current detection, the other detectors multiply the intensity of the signal by a cascade effect.

Ion detectors can be divided into two classes. Some detectors are made to count ions of a single mass at a time and therefore they detect the arrival of all ions sequentially at one point (point ion collectors). Others detectors, such as photographic plates, image current detectors or array detectors, have the ability to count multiple masses and detect the arrival of all ions simultaneously along a plane (array collectors).

These detectors are effective for most of the applications in mass spectrometry. Nevertheless, if a detector must ideally be free of any discrimination effect, its efficiency generally decreases when the mass of the ion increases. This induces limitations for the detection of high-mass ions and can compromise a quantitative analysis from these data because the signal decreases exponentially with increasing mass. On the other hand, progress in mass spectrometry has led to the advent of entirely new ionization sources and analysers that allow the study of analytes with very high molecular mass. For these reasons, the

Mass Spectrometry: Principles and Applications, Third Edition Edmond de Hoffmann and Vincent Stroobant

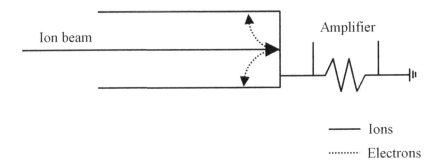

Figure 3.1
Schematic diagram of a Faraday cup.

development of new detectors which eliminate these limitations is required. These new classes of detectors such as the charge or inductive detector [2] or cryogenic detector [3] are under development. They are based on physical principles that differ from those used for the current detectors and their efficiencies are the same for all the detected ions and are unrelated to their masses. They have near 100 % efficiency for very large slow ions. Indeed, the inductive detector simply produces a signal by inducing a current on a plate generated by a moving charged ion. Its efficiency is related to the charge of the ion but is independent of its mass and its kinetic energy. In the same way, the cryogenic detector is a kinetic energy sensitive calorimetric detector operating at low temperatures.

3.1.1 Photographic Plate

The first mass spectrometers used photographic plates located behind the analyser as detectors. Ions sharing the same m/z ratio all reach the plate at the same place and the position of the spots allows the determination of their m/z values after calibration. The darkness of the spots gives an approximate value of their relative abundance. This detector, which allows simultaneous detection over a large m/z range, has been used for many years but is obsolete today.

3.1.2 Faraday Cup

A Faraday cup is made of a metal cup or cylinder with a small orifice. It is connected to the ground through a resistor, as illustrated in Figure 3.1. Ions reach the inside of the cylinder and are neutralized by either accepting or donating electrons as they strike the walls. This leads to a current through the resistor. The discharge current is then amplified and detected. It provides a measure of ion abundance.

Because the charge associated with an electron leaving the wall of the detector is identical to the arrival of a positive ion at this detector, secondary electrons that are emitted when an ion strikes the wall of the detector are an important source of errors if they are not suppressed. In consequence, the accuracy of this detector can be improved by preventing the escape of reflected ions and ejected secondary electrons. Various devices have been used to capture ions efficiently and to minimize secondary electron losses. For instance, the cup is coated with carbon because it produces few secondary ions. The shape of the cup and the use of a weak magnetic field prevent also any secondary electrons produced inside to exit.

The disadvantages of this simple and robust detector are its low sensitivity and its slow response time. Indeed, the sensitivity of such detectors is limited by the noise of the amplifiers. Furthermore, this detector is not well adapted to ion currents that are not stable in the same time as during the scanning of the analyser because of its slow response time. These detectors are nevertheless very precise because the charge on the cylinder is independent of the mass, the speed and the energy of the detected ions.

The Faraday cup was widely used in the beginning of mass spectrometry but all the characteristics of this detector mean that it is now generally used in the measurement of highly precise ratios of specific ion species as in isotopic ratio mass spectrometry (IRMS) or in accelerator mass spectrometry (AMS). To obtain a highly accurate ratio in such relative abundance measurements, the intensities of the two stable beams of specific ions are measured simultaneously with two Faraday cups.

3.1.3 Electron Multipliers

At present (2006), the most widely used ion detector in mass spectrometry is the electron multiplier (EM). In this detector, ions from the analyser are accelerated to a high velocity in order to enhance detection efficiency. This is achieved by holding an electrode called a conversion dynode at a high potential from ± 3 to ± 30 kV, opposite to the charge on the detected ions. A positive or negative ion striking the conversion dynode causes the emission of several secondary particles. These secondary particles can include positive ions, negative ions, electrons and neutrals. When positive ions strike the negative high-voltage conversion dynode, the secondary particles of interest are negative ions and electrons. When negative ions strike the positive high-voltage conversion dynode, the secondary particles of interest are positive ions. These secondary particles are converted to electrons at the first dynode. These are then amplified by a cascade effect in the electron multiplier to produce a current. The electron multipliers may be of either the discrete dynode or the continuous dynode type (channeltron, microchannel plate or microsphere plate).

The discrete dynode electron multiplier is made up of a series of 12 to 20 dynodes that have good secondary emission properties. As shown in Figure 3.2, these dynodes are held at decreasing negative potentials by a chain of resistors. The first dynode is held at a high negative potential from -1 to -5 kV, whereas the output of the multiplier remains at ground potential. Secondary particles generated from the conversion dynode strike the first dynode surface causing an emission of secondary electrons. These electrons are then accelerated to the next dynode because it is held at a lower potential. They strike the second dynode causing the emission of more electrons. This process continues as the secondary electrons travel towards the ground potential. Thus a cascade of electrons is created and the final flow of electrons provides an electric current at the end of the electron multiplier that is then increased by conventional electronic amplification.

There is another design of electron multiplier for which the discrete dynodes are replaced by one continuous dynode. A type of continuous-dynode electron multipliers (CDEM), which is called a channeltron, is made from a lead-doped glass with a curved tube shape that has good secondary emission properties (Figure 3.3). As the walls of the tube have

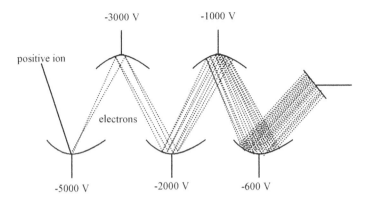

Figure 3.2
Schematic diagram of electron multiplier. The first dynode is a conversion dynode to convert ions into electrons.

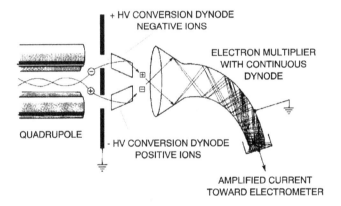

Figure 3.3
Continuous dynode electron multiplier, also known as the channeltron. ○, incident ions; □, secondary particles. Reproduced (modified) from Finnigan MAT documentation, with permission.

a uniform electric resistance, a voltage applied between the two extremities of the tube will therefore produce a continuous accelerating field along its length. Secondary particles from the conversion dynode collide with the curved inner wall at the detector entrance and produce secondary electrons, which are then accelerated by the field towards the exit of the tube. These electrons pass further into the electron multiplier, again striking the wall, causing the emission of more and more electrons. Thus a cascade of electrons is created and finally a metal anode collects the stream of secondary electrons at the detector exit and the current is measured

The amplifying power is the product of the conversion factor (number of secondary particles emitted by the conversion dynode for one incoming ion) and the multiplying factor of the continuous dynode electron multiplier. It can reach 10^7 with a wide linear dynamic range (10^4–10^6). Their lifetime is limited to 1 or 2 years because of surface

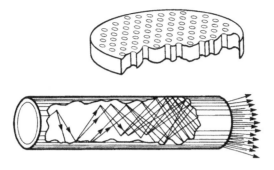

Figure 3.4
Cross-section of a microchannel plate and electron
multiplication within a channel. Reproduced from
Galileo documentation, with permission.

contamination from the ions or from a relatively poor vacuum. Their high amplification
and their fast response time allow their use with rapid scanning of the analyser. The
conversion factor is highly dependent on the impact velocity of the detected ions and on
their nature (mass, charge and structure), so these detectors are not as precise as Faraday
cups. Since the detection efficiency on all electron multipliers is highly dependent on ion
velocity, they are characterized by the mass discrimination effect for ions with constant
energy. Because of their slower velocity, large ions produce fewer secondary electrons
and thus the efficiency decreases when the mass of the ion increases. However, as already
mentioned, the conversion dynodes at high voltages reduce the mass discrimination effect
and serve to increase signal intensity and therefore sensitivity because they accelerate ions
to a high velocity in order to enhance detection efficiency. Conversion dynodes are thus
very useful for detecting high-mass ions, especially with analysers delivering ions at low
kinetic energy, such as quadrupoles or ion traps.

Another type of continuous dynode electron multipliers is the microchannel plate (MCP)
detector. It is a plate in which parallel cylindrical channels have been drilled. The channel
diameter ranges from 4 to 25 μm with a centre-to-centre distance ranging from 6 to 32 μm
and a few millimetres in length (Figure 3.4). The plate input side is kept at a negative
potential of about 1 kV compared with the output side.

Electron multiplication is ensured by a semiconductor substance covering each channel
and giving off secondary electrons. Curved channels prevent the acceleration of positive
ions towards the input side. Two plates can also be connected herringbone-wise or three
plates can be connected following a Z shape, as shown in Figure 3.5.

The snowball effect within a channel can multiply the number of electrons by 10^5. A
plate allows an amplification of 10^2–10^4, whereas by using several plates the amplification
can reach 10^8. This detector is characterized by a very fast response time because the
secondary electron path inside the channel is very short. In consequence, it is well suited to
TOF analysers, which need precise arrival times and narrow pulse widths. Furthermore, the
large detection area of the microchannel plate allows the detection of large ion beams from
the analyser without additional focalization. However, the microchannel plate detectors
have some disadvantages. They are fragile, sensitive to air and their large microchannel
plates are expensive.

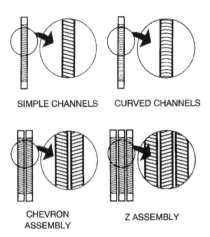

SIMPLE CHANNELS   CURVED CHANNELS

CHEVRON   Z ASSEMBLY
ASSEMBLY

Figure 3.5
Microchannel plate types and connec-
tions. Reproduced from Galileo documentation, with permission.

At the output side, a single metal anode collects the stream of secondary electrons of all the channels and the current is measured as in other types of electron multipliers. In that case, all the microchannels are then connected between them, so as to act as a large single detector.

The microchannel plate detector can, however, also work with a metal anode that gathers the stream of secondary electrons at every channel exit. To avoid any confusion, the term array detector is preferably used to describe a microchannel plate where every microchannel remains as an individual ion-detecting element. This array detector acts as electronic photoplates. Indeed, it resembles that of a photographic plate: ions with different m/z ratios reach different spots and may be counted at the same time during the analysis. The advantage of array detectors is that analyser scanning is not necessary and therefore sensitivity is improved because simultaneous detection of ions implies that more ions are collected, and this greater efficiency leads to lower limits of detection than for other detectors.

A newer and less expensive alternative to the microchannel plate is the microsphere plate (MSP). As illustrated in Figure 3.6, this electron multiplier consists of glass beads with diameters from 20 to 100 μm that are sintered to form a thin plate with a thickness of 0.7 mm. This plate is porous with irregularly shaped channels between the planar faces. The surfaces of the beads are covered with an electron emissive material and the two sides of the plate are coated to make them conductive. The operating principle of this electron multiplier is similar to that of the microchannel plate. A potential difference of between 1.5 and 3.5 kV is applied across the plate, with the output side of the plate at the more positive potential. When particles hit the input side of the microsphere plate, they produce secondary electrons. These electrons are then accelerated by the electric field through the porous plate and collide with other beads. Secondary electron multiplication in the gaps occurs and finally a large number of secondary electrons are emitted from the output side of the plate.

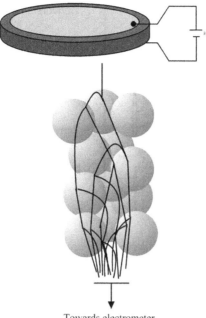

Towards electrometer

Figure 3.6
Microsphere plate detector and electron
multiplication through the porous plate.
Reproduced (modified) from El-Mul Tech-
nologies Ltd documentation, with permission.

The microsphere plate offers some advantages over the microchannel plate. It is less expensive and its gain of 10^6–10^7 is higher. This higher amplification is due to the fact that nearly the entire surface of the input side is active and therefore emits secondary electrons that will be accelerated onto and through the plate to give the final signal. In comparison, the surface of the microchannel plate between the microchannels, which corresponds to about 50 % of the entire surface, is inactive.

3.1.4 Electro-Optical Ion Detectors

The electro-optical ion detector (EOID) combines ion and photon detection devices. This type of detector operates by converting ions to electrons and then to photons. The most common electro-optical ion detector is called the Daly detector. As shown in Figure 3.7, this type of detector is made up of two conversion dynodes, a scintillation or phosphorescent screen and a photomultiplier. This device allows the detection of both positive and negative ions. As for the electron multipliers, ions from the analyser strike a dynode. In the positive mode, ions are accelerated towards the dynode that carries a negative potential, whereas in the negative mode, ions are accelerated towards the positive dynode. Secondary electrons

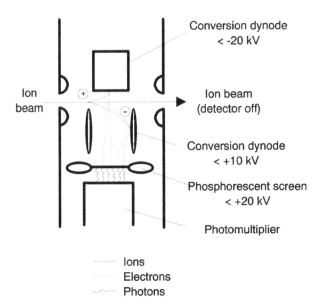

Figure 3.7
Schematic diagram of a Daly detector.

that are given off are then accelerated towards the phosphorescent screen that emits photons as a result. These photons are detected by the photomultiplier that is positioned behind the phosphorescent screen and is converted into an electric current that will be amplified. The phosphorescent screen surface is covered with a thin layer of aluminium conductor to avoid the formation of a charge that would prevent new electrons from reaching it. The electro-optical ion detector, although relatively complex because it requires multiple conversions (ion to electron to photon to electron), has some advantages. Its lifetime is longer than the lifetime of electron multipliers because the photomultiplier is sealed in glass and held under vacuum. This prevents contamination and allows the detector to maintain its performance for a considerably longer period than conventional electron multipliers. In addition, it has a fast response time and a similar sensitivity to electron multipliers with an amplification ranges from 10^4 to 10^5.

Another electro-optical ion detector, which is called the electro-optical array detector, allows the simultaneous measurement of ions spatially separated along the focal plane of the mass spectrometer. It combines the microchannel plate and Daly detector. The ions are converted in a microchannel plate into electrons that are amplified. The released secondary electrons finally strike a phosphorescent screen that emits photons. These photons are then detected with a photodiode array or CCD detector. This array detector acts as electronic photoplates.

3.2 Computers

This section describes briefly the functioning of a computer dedicated to mass spectrometry. Applications in GC/MS and HPLC/MS will appear in Chapter 5.

3.2.1 Functions

A computer dedicated to mass spectrometry is usually capable of three basic operations: (i) control of the mass spectrometer; (ii) acquisition and processing of data from the mass spectrometer; and (iii) interpretation of data.

The computer can control the mass spectrometer by introducing the values and variations of different parameters. As the computer treats digital data, whereas the mass spectrometer produces and receives analogue data, an interface is necessary to convert one type of data into another.

A computer dedicated to mass spectrometry registers the data given out by the mass spectrometer and converts them either into values of masses and peak intensities, or into total ionic current, temperatures, acceleration potential values, and so on. It is also capable of data processing. It allows calculation of average spectra and subtraction of one spectrum from another in order to eliminate the background noise or simply to emphasize the differences between two spectra.

The computer can also calculate the possible compositions of ions of a given mass, taking into account only the elements in the molecular formula. For example, if a compound is known to contain only CHON as elements, a fragment detected at m/z 39 can have only C_2HN or C_3H_3 as its elemental composition. More examples can be found in Appendix 5. The computer can also calculate for a given molecular formula the theoretical relative abundances within the isotopic cluster for comparison with experimental values. It can also compare the spectra that it observes with a library of spectra.

3.2.2 Instrumentation

The interface receives, conditions and digitizes the analogue signals (those given by a beam, an ionic current, temperatures, voltages) output by the mass spectrometer and sends them to the computer. It acts an analogue-to-digital converter (ADC).

The dynamic range of analogue values that are measured increases with the number of bits (with 24 bits, it ranges from 0 to $2^{24} - 1$); the sampling velocity determines the number of measures per peak that is possible.

A digital-to-analogue converter (DAC) transmits to the mass spectrometer the analogue voltages that are needed for the functioning and that are determined by the computer according to the values given by the operator.

The computer furnishes the capacity for calculation and controls the operation sequence required by a program. It transfers the data to the central memory. The software transmits the instructions typed on the keyboard and converts them into machine language.

3.2.3 Data Acquisition

The data acquisition system is designed to digitize electrical signals from the detector and transfer them to the data system in a compatible format. Different types of data acquisition systems are used in mass spectrometry because ion signals are measured and recorded by detectors that can be operated in the analogue mode or in the pulse counting mode. The most common detection systems operate in analogue mode with an ADC. The detector furnishes an analogue electrical signal that is amplified. The background noise is reduced

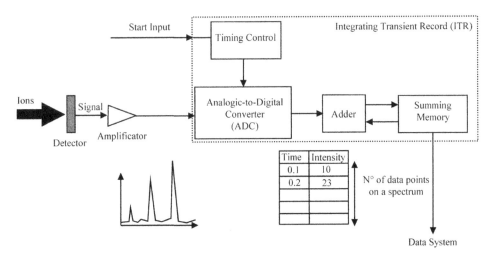

Figure 3.8
Principle of TOF mass spectrum acquisition with ITR in analogue mode. Data from JEOL documentation.

by a filter that cuts off high frequencies. The data acquisition system converts the signal from the amplifier with its ADC to a digital value, which is sent to the memory and finally forwarded to the data system.

The volume of information forwarded from the ADC to the memory is unrelated to the number of ions detected, but proportional to the data points on one spectrum. Thus, the ADC sampling frequency that is needed for the data acquisition depends on the scan speed (mass range per unit time) of the mass spectrometer. Indeed, if a higher scan speed is needed, or if a higher resolution is used, the sampling frequency must be increased because the number of data points per unit of time increases. For instance, quadrupole instruments cover a range of 1000 mass units within 1 s, that is 1 mass unit within 1 ms. Furthermore, to identify the peak position with high accuracy, it is desirable to have at least 10 data points for every peak. Thus, the sampling frequency must be at least equal to 10 kHz. In other words, the data acquisition system measures and digitizes the current from the detector 10 000 times per second. The scan speed of TOF mass spectrometers can easily reach 50 mass units within 1 μs, that is 1 mass unit within 20 ns. The sampling frequency of these instruments must be at least equal to 500 MHz.

For some mass spectrometers, such as TOF mass spectrometers, for reasons of data transfer speed and data storage capacity, the data acquisition system needs to accumulate data for a period of time in the summing memory, and forward the accumulated data to the data system. Each spectrum is added to the sum of the previous spectra so that a continuous summation process takes place. This type of data acquisition system is called digital signal averaging (DSA) or integrating transient recorder (ITR). Figure 3.8 illustrates the principle of mass spectrum acquisition with this type of system.

With an 8-bit ADC, the intensity of the signal from the detector is converted into a numerical value that ranges from 0 to 255. To improve this small dynamic range, slow scanning instruments use typically 16- or 24-bit ADCs corresponding to intensity values from 0 to 65 535 or 0 to 16 777 215, respectively. Another possibility with fast scanning

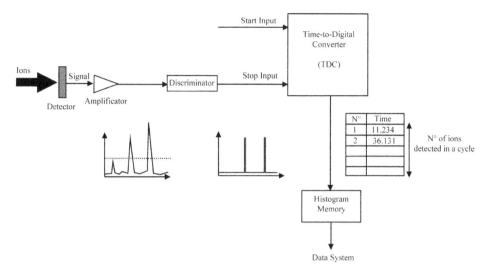

Figure 3.9
Principle of TOF mass spectrum acquisition with TDC in pulse counting mode. Data from
JEOL documentation.

instruments is to sum several single spectra. Thus, this type of detection system has a
wide dynamic range. However, the signal distortion and electrical noise from the detection
system directly affect the spectral quality. The width of the signal from the detector and
amplifier is also reflected in the data, resulting in a lower apparent resolution. Furthermore,
it is not well suited for single ion detection because of the background noise associated
with analogue detection.

Another type of detection system used in the TOF mass spectrometer operates in pulse
counting mode with a time-to-digital converter (TDC). Each ion that strikes the detector
generates a pulse of electrons. This pulse from the detector is amplified and forwarded to
the discriminator. The pulse exceeding a certain threshold selected by the discriminator
is used to trigger a timing pulse that is sent to the TDC. The TDC counts the time from
the start pulse to each of the stop pulses, and the resulting list is stored in the histogram
memory.

The volume of information forwarded from the TDC to the histogram memory is relatively small, since the size of this list acquired from a single flight cycle is proportional to the
number of ions detected in that flight cycle. The histogram memory normally accumulates
data over the course of the acquisition period and forwards these data to the data system
as a spectrum, which is a histogram of the detected events from several thousand cycles.
For one flight cycle, the *y* axis of the spectral data represents only two values: 0 if no ion
is detected and 1 if the signal given by the detector reaches a certain threshold. The basic
principle of the TOF mass spectrum recorded with a TDC is presented in Figure 3.9.

Since the TDC is a time counting device, when two or more ions arrive at the detector
simultaneously in one flight cycle, the system counts them as one ion. In the same way,
when two ions arrive at the detector in sequence within a certain interval, the system does
not count the latter ion because the TDC is unable to register another count during the
period after each ion event (counting dead time). Consequently, the dynamic range of this

detection system is limited and mass measurement accuracy is affected. Indeed, the result is that intense mass peaks become distorted by depletion of the top and right side of the peak. Thus, the height of the mass spectrum peak is decreased and the peak position is shifted to the left.

These saturation effects, resulting from the loss of ions due to the TDC dead time, can be statistically corrected by applying a correction factor [4]. However, there is no effective correction when the quantity of ions increases and several ions arrive at the detector simultaneously. In conclusion, detection systems that operate in pulse counting mode are well suited to detect small quantities of ions by accumulation for a long period of time and when detection of individual ion events is important in order to obtain a good signal-to-noise ratio.

3.2.4 Data Conversion

A peak position is given by a time value (or a Hall voltage value in a magnetic analyser or a voltage value in a quadrupole), which must be converted into a mass value. This supposes a preliminary calibration with known products (PFK, $(CsI)_n$, etc.). The masses furnished by the calibrating product are stored in the computer. A relationship of the type $m = ax + b$ is calibrated with two known peaks and is checked or corrected with known peaks located in other mass areas.

The peak position can be taken as the position of the largest digital sample or can be calculated by the centroid:

$$t_c = \sum V_n t_n \Big/ \sum V_n$$

The peak intensity is given either by the maximum voltage V_{max}, or by the peak area.

3.2.5 Data Reduction

The computer can order the printer or the screen to furnish spectra in table or graph form, supposing a normalization to 100 of the base peak intensity or of a fraction of this intensity if the detection of weak peaks is sought. It can also order the production of time-dependent data (total ion current, temperatures, voltages).

In addition, it can also print a chromatogram, a variation with time of the intensities registered at a few given masses. It can suggest elementary compositions for the peaks.

3.2.6 Library Search

The computer can hold a database where the main peaks of known products are stored. The spectra obtained by electron ionization alone are reproducible enough to be useful.

Two types of library search have been developed. The first type, called forward search, compares the new spectrum with the spectra stored in the library and looks for the best match of the spectra. The second type, called reverse search, checks for the possible presence in the new spectra of a spectrum chosen in the library.

A filtering that eliminates the majority of the library spectra as being far too different from the new spectrum precedes the search. The latter may identify the new spectrum or suggest possible structures. Figure 3.10 shows an example of the result of a library search.

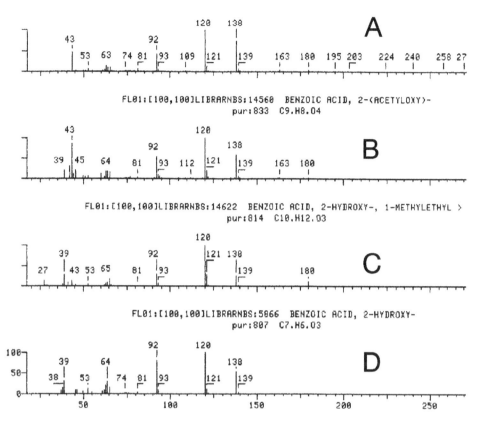

Figure 3.10
Example of a library search. A: spectrum of an aspirin sample; B to D: library spectra identified by the computer.

Products having isomeric structures most often have almost identical spectra. Moreover, totally different products may accidentally have very similar spectra. Identification mistakes are thus frequent. A systematic recourse to another criterion is necessary, for example the retention time.

References

1. Koppenaal, D.W., Barinaga, C.J., Denton, M.B. *et al.* (2005) Mass spectrometry detectors. *Anal. Chem.*, 419A–27A.
2. Schultz, J.C., Hack, C.A. and Benner, W.H. (1997) *J. Am. Soc. Mass Spectrom.*, **9**, 305–13.
3. Hilton, G.C., Martinis, J.M., Wollman, D.A. *et al.* (1998) *Nature*, **391**, 672–5.
4. Coates, P.B. (1992) *Rev. Sci. Instrum.*, **63**, 2082.

4

Tandem Mass Spectrometry

Tandem mass spectrometry, abbreviated MS/MS, is any general method involving at least two stages of mass analysis, either in conjunction with a dissociation process or in a chemical reaction that causes a change in the mass or charge of an ion [1–3].

In the most common tandem mass spectrometry experiment a first analyser is used to isolate a precursor ion, which then undergoes spontaneously or by some activation a fragmentation to yield product ions and neutral fragments:

$$m_p^+ \longrightarrow m_f^+ + m_n$$

A second spectrometer analyses the product ions. The principle is illustrated in Figure 4.1. The product ions spectrum will not display isotope peaks if the selected precursor m/z contains only one isotope for each atomic species, which most often will be the case.

It is possible to increase the number of steps: select ions of a first mass, then select ions of a second mass from the obtained fragments, and finally analyse the fragments of these last selected ions. This is labelled as an MS/MS/MS or MS^3 experiment. The number of steps can be increased to yield MS^n experiments (where n refers to the number of generations of ions being analysed).

4.1 Tandem Mass Spectrometry in Space or in Time

Basically, a tandem mass spectrometer can be conceived in two ways: performing tandem mass spectrometry in space by the coupling of two physically distinct instruments, or in time by performing an appropriate sequence of events in an ion storage device. Thus there are two main categories of instruments that allow tandem mass spectrometry experiments: tandem mass spectrometers in space or in time.

Common space instruments have two mass analysers, allowing MS/MS experiments to be performed. A frequently used instrument of this type uses quadrupoles as analysers. The QqQ configuration indicates an instrument with three quadrupoles where the second one, indicated by a lower case q, is the reaction region. It operates in RF-only mode and thus acts like a lens for all the ions. Other instruments combine electric and magnetic sectors (E and B) or E, B and qQ, thus electric and magnetic sectors and quadrupoles. TOF instruments with a reflectron, or a combination of a quadrupole with a TOF instrument, are also used.

To obtain higher order MS^n spectra requires n analysers to be combined, increasing the complexity and thus also the cost of the spectrometer. Furthermore, successive analysers

Mass Spectrometry: Principles and Applications, Third Edition Edmond de Hoffmann and Vincent Stroobant
© Copyright 2007, John Wiley & Sons Ltd

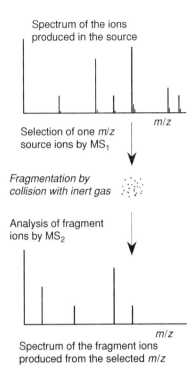

Figure 4.1
Principle of MS/MS: an ion M_1 is selected by the first spectrometer MS_1, fragmented through collision, and the fragments are analysed by the second spectrometer, MS_2. Thus ions with a selected m/z value, observed in a standard source spectrum, can be chosen and fragmented so as to obtain their product ion spectrum.

can be arranged in any number, but because the fraction of ions transmitted at each step is low, the practical maximum is three or four analysers in the case of beam instruments.

Besides this spatial separation method using successive analysers, tandem mass spectrometry can also be achieved through time separation with a few analysers such as ion traps, orbitrap and FTICR, programmed so that the different steps are successively carried out in the same instrument. This method was described in the case of the ion traps and FTMS in Chapter 2. The maximum practical number of steps for these instruments is seven to eight. In these instruments the proportion of ions transmitted is high, but at each step the mass of the fragments becomes lower and lower.

A significant difference between the two types of trapping instruments is that in the ion trap mass spectrometers, ions are expelled from the trap to be analysed. Hence they can be observed only once at the end of the process. In FTMS, they can be observed nondestructively, and hence measured at each step in the sequential fragmentation process.

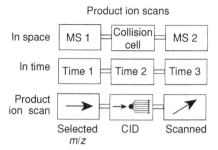

Figure 4.2
Comparison of a product ion scan performed by a space-based and a time-based instrument.

Figure 4.2 represents a schematic comparison a simple MS/MS product ion scan performed by either a 'space'- or a 'time'-based instrument.

Nomenclature: the main names and acronyms used in MS/MS are listed below [4,5]

Molecular ion: ion formed by the addition or the removal of one or several electrons to or from the sample molecule.

Ion of the molecular species: ion originating from the molecule by the abstraction of a proton $[M - H]^-$ or by a hydride abstraction $[M - H]^+$, or by interaction with a proton or a cation to form a protonated molecule $[M + H]^+$ or a cationized molecule $[M + Cat]^+$. These ions allow the molecular weight to be deduced. The ambiguous and obsolete terms 'quasimolecular ion' or 'pseudomolecular ion' should be avoided.

Adduct ion: ion formed by direct combination of two separate molecular entities, usually an ion and a neutral molecule, in such a way that there is change in connectivity promoted by intermolecular binding, but no loss of atoms within the precursor entities.

Cluster ion: ion formed by a multi-component atomic or molecular assembly of one or more ions with atoms, ions or molecules.

Precursor ion (used to be parent ion): any ion undergoing either a fragmentation or a charge change.

Product ion (used to be daughter ion): ion resulting from the above reaction.

Fragment ion: ion resulting from the fragmentation of a precursor ion.

Neutral loss: fragment lost as a neutral species.

CA   Collisional Activation
CAD   Collision-Activated Dissociation
CID   Collision-Induced Dissociation
DADI   Direct Analysis of Daughter Ion (obsolete name for MIKES)

_____ *Continued on page 192* ___

Continued from page 191

MIKES	Mass-analysed Ion Kinetic Energy Spectrometry
CAR	Collision-Activated Reaction
E	Electric sector
B	Magnetic sector
Q	Quadrupole
q	RF-Only Quadrupole as Collision Cell
N	Target Gas in a Collision
NR	Neutralization Reionization
N_fR	Neutral Fragment Reionization

Source spectrum: spectrum in which only ions from the source are recorded.

4.2 Tandem Mass Spectrometry Scan Modes

The four main scan modes available using tandem mass spectrometry are represented in Figure 4.3. Many other MS/MS scan modes are possible. We will focus on fragmentations that occur with an inert collision gas. MS/MS needs a computer system able to control all the experimental factors and to record the results.

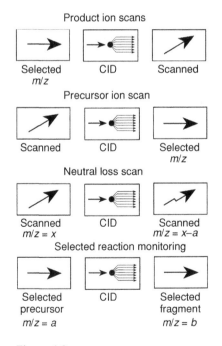

Product ion scans

Selected   CID   Scanned
m/z

Precursor ion scan

Scanned   CID   Selected
m/z

Neutral loss scan

Scanned   CID   Scanned
m/z = x   m/z = x–a

Selected reaction monitoring

Selected   CID   Selected
precursor   fragment
m/z = a   m/z = b

Figure 4.3
Main processes in tandem mass spectrometry (MS/MS). CID stands for collision-induced dissociation, as occurs when an inert gas is present in the collision cell.

1. Product ion scan (daughter scan) consists of selecting a precursor ion (or parent ion) of a chosen mass-to-charge ratio and determining all of the product ions (daughter ions) resulting from collision-induced-dissociation CID. If reactive gas is used in the collision cell, collision-activated-reaction CAR products are observed. When only fragment ions are produced, this scan mode is also referred to as 'fragment ion scan'.

2. Precursor ion scan (parent scan) consists of choosing a product ion (or daughter ion) and determining the precursor ions (or parent ions). This method is called 'precursor scan' because the 'precursor ions' are identified. This scan mode cannot be performed with time-based mass spectrometers. This scan requires the focusing of the second spectrometer on a selected ion while scanning the masses using the first spectrometer. All of the precursor ions that produce ions with the selected mass through reactions or fragmentations thus are detected.

3. Neutral loss scan consists of selecting a neutral fragment and detecting all the fragmentations leading to the loss of that neutral. As in the case of the precursor ion scan, this scan mode is not available with time-based mass spectrometers. This scan requires that both mass spectrometers are scanned together, but with a constant mass offset between the two. Thus, for a mass difference a, when an ion of mass m goes through the first mass spectrometer, detection occurs if this ion has produced a fragment ion of mass $(m - a)$ when it leaves the collision cell.

4. Selected reaction monitoring (SRM) consists of selecting a fragmentation reaction. For this scan, both the first and second analysers are focused on selected masses. There is thus no scan. The method is analogous to selected ion monitoring in standard mass spectrometry. But here the ions selected by the first mass analyser are only detected if they produce a given fragment, by a selected reaction. The absence of scanning allows one to focus on the precursor and fragment ions over longer times, increasing the sensitivity as for selected ion monitoring, but this sensitivity is now associated with a high increase in selectivity.

A symbolism to describe the various scan modes is represented in Figure 4.4 [6].

Note that precursor or neutral loss scans are not possible in time separation analysers. The product ion scan is the only one available directly with these mass spectrometers. However, with these instruments, the process can be easily repeated over several ion

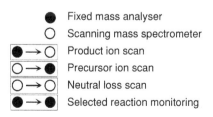

Figure 4.4
Symbolism proposed by Cooks *et al.* for the easy representation of various scan modes.

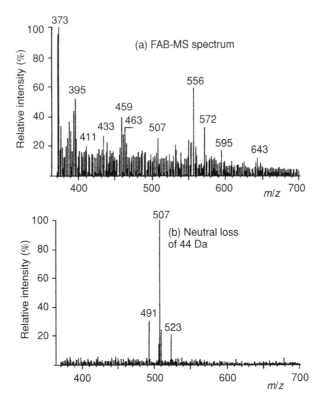

Figure 4.5
(a) Source FAB spectrum of the bile acids in a urine sample.
(b) Neutral loss scan of 44 Da measured immediately after
spectrum A. Althrough only about 5 % of the ions are transmitted, the signal-to-noise ratio is considerably increased.
The ion at m/z 507 is now well detected, and the m/z 491
and 523 ions, not detected in (a), are now visible. Reprinted
(modified) from Libert R., Hermans D., Draye J.P., Van
Hoof F., Sokal E. and de Hoffmann E., Clin. Chem., 37,
2102, 1991, with permission.

generations. Thus time-based instruments easily allow MS^n product ion spectra to be
obtained.

As the number of transmitted ions decreases at every step because of the ionic species
trajectory lengths and the presence of a collision gas, the practical maximum number of
steps is three or four in the case of tandem mass spectrometry in space, and seven or eight
for tandem mass spectrometry in time. However, chemical noise may decrease more than
the ionic signal, so often the signal-to-noise ratio increases in addition to the sensitivity.
This principle is illustrated in Figure 4.5 displaying the analysis of abnormal bile acids
in urines [7]. While only a few per cent of the ions are transmitted in the 44 Da neutral
loss scan spectrum, the signal-to-noise ratio is strongly increased. This allows a very good
detection of the m/z 507 ions, and displays ions corresponding to two analogous bile acids

containing one less hydroxyl group at m/z 491 and one more at m/z 523. These last ions are completely buried in the noise in the source FAB spectrum.

4.3 Collision-Activated Decomposition or Collision-Induced Dissociation

Tandem mass spectrometry requires the fragmentation of precursor ions selected by the first analyser in order to allow the second analyser to analyse the product ions. According to the Warhaftig diagram (see Chapter 7), the ions leaving the source can be classified into three categories. The first category of ions, with a lifetime greater than 10^{-6} s, reaches the detector before any fragmentation has occurred. The second category, with a lifetime smaller than 10^{-7} s, fragments before leaving the source and only the fragments are detected. The last category, called metastable ions, have an intermediate lifetime. These ions are stable enough to be selected by the first analyser while containing enough excess energy to allow their fragmentation before they reach the second analyser. The probability associated with this phenomenon is relatively low (1 %) because their number is small and they spend a very short time in the reaction region. If precursor ions undergo a collisional activation (CA), that is an increase in their internal energy that induces their decomposition, then the situation described above is much improved. This collision-activated decomposition (CAD) technique, also known as collision-induced dissociation (CID), allows one to increase the number of precursor ions that fragment in the reaction region and also the number of fragmentation pathways. The structural analysis thus becomes easier.

An overall view shows the CID process as a sequence of two steps. The first step is very fast (10^{-14} to 10^{-16} s) and corresponds to the collision between the ion and the target when a fraction of the ion translational energy is converted into internal energy, bringing the ion into an excited state. The second step is the unimolecular decomposition of the activated ion. The collision yield then depends on the activated precursor ion decomposition probability according to the theory of quasi-equilibrium or RRKM. This theory is explained elsewhere. Let us recall that it is based on four suppositions:

1. The ion dissociation time is long with respect to the formation time and the excitation time.

2. The dissociation rate is low with respect to the rate of the redistribution of the excitation energy among all of the ion internal modes.

3. The ion achieves an internal equilibrium condition where the energy is distributed with an equal probability among all of the internal modes. Considering that an ion with N non-linear atoms has $3N - 6$ vibration modes, it is easy to understand how the collision yield decreases in a manner inversely proportional to the ion mass.

4. The dissociation products that are observed result from a series of competitive and consecutive reactions.

As a consequence of these suppositions, CID is an 'ergodic' ion activation method, which allows a redistribution of the energy in the vibrational modes of the ion because the dissociation rate is slower than the rate of energy randomization. In these conditions, the fragmentation pathways depend on the amount of energy deposited and not on the method of ion activation used. As the energy is distributed with an equal probability among all of the internal modes of the ion, this leads preferentially to cleavage sites at the weakest bonds. For the same reason, molecules with more atoms will need more energy, or more time, to dissociate. Modern ionization techniques allow the production of ions with high-molecular-weight compounds. This is why other ion activation methods that are being developed now increase the energy transfer [8].

Several methods exist that activate the ions through collisions. The most common one consists of colliding the low- or high-energy accelerated ions with gas molecules as immobile targets. To achieve collisional activation in MS/MS instruments with spatially separate analysers, a collision cell is placed between the two mass analysers. This cell often corresponds simply to a small chamber with entrance and egress apertures and contains an inert target gas at a pressure sufficient for collisions with ions to occur. In MS/MS instruments based on time-separated mass analysis steps, an inert gas is simply introduced into the ICR or the ion trap. In some elaborations of these experiments, it is introduced only into a certain region or at a particular time in the operating sequence using a pulsed valve. An article describing the history of this technique [9] and the fundamental aspects are detailed in the literature [10–12].

4.3.1 Collision Energy Conversion to Internal Energy

The kinetic energy for internal energy transfers is governed by the laws concerning collisions of a mobile species (the ion) and a static target (the collision gas).

Consider an ion of mass m_1 hitting a target of mass m_2. If $\mathbf{R}$ represents the position vectors in the laboratory reference frame O, and $\mathbf{r}$ represents the position vectors in the reference frame linked to the centre of gravity G, we see on the following diagram that

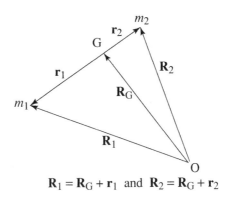

$$\mathbf{R}_1 = \mathbf{R}_G + \mathbf{r}_1 \quad \text{and} \quad \mathbf{R}_2 = \mathbf{R}_G + \mathbf{r}_2$$

Defining the centre of gravity as $(m_1 + m_2)\mathbf{R}_G = m_1\mathbf{R}_1 + m_2\mathbf{R}_2$ then

$$\mathbf{r}_1 = m_2(\mathbf{R}_1 - \mathbf{R}_2)/(m_1 + m_2) \quad \text{and} \quad \mathbf{r}_2 = -m_1(\mathbf{R}_1 - \mathbf{R}_2)/(m_1 + m_2)$$

Differentiating with respect to t,

$$\mathbf{u}_1 = m_2\mathbf{g}/(m_1 + m_2) \quad \text{and} \quad \mathbf{u}_2 = -m_1\mathbf{g}/(m_1 + m_2)$$

with

$$\mathbf{g} = \mathbf{v}_1 - \mathbf{v}_2 \quad \text{and} \quad m_1\mathbf{u}_1 + m_2\mathbf{u}_2 = 0$$

According to the law of conservation of momentum, the kinetic energy of a rapid particle colliding with a static target ($\mathbf{v}_2 = 0$) cannot be entirely converted into internal energy. The kinetic energy available for conversion, E_t, termed 'relative kinetic energy', is actually the kinetic energy in the centre of mass reference frame. Indeed, in this G frame the momentum is equal to zero. We thus obtain

$$E_t = (m_1u_1^2 + m_2u_2^2)/2 = \mu g^2/2$$

with

$$\mu = m_1m_2/(m_1 + m_2)$$

Thus

$$E_t = E_k m_2/(m_1 + m_2) \quad \text{where} \quad E_k = m_1v_1^2/2$$

The kinetic energy is conserved in an *elastic* collision

$$\mu g^2/2 = \mu g'^2/2$$

In an *inelastic* collision, a part Q of the kinetic energy, at a maximum equal to E_t, is converted into internal energy:

$$\mu g^2/2 = Q + \mu g'^2/2$$

Thus, the conservation of energy and of momentum imply that only a fraction of the translational energy is converted into internal energy under inelastic conditions. This energy fraction is given by the following equation:

$$E_{cm} = E_{lab}\frac{M_t}{M_i + M_t}$$

where M_i is the ion mass; M_t the target gas mass; E_{lab} the ion kinetic energy in the laboratory frame of reference; and E_{cm} the maximum energy fraction converted into internal energy. Consequently, an increase in the ion kinetic energy or in the target gas mass increases the energy available for the conversion. This energy decreases as a function of $1/M_i$.

For instance, a $100\,u$ ion with a kinetic energy of $10\,eV$ colliding with argon (atomic mass 40) has a maximum increase in its internal energy amounting to

$$10\frac{40}{40 + 100} = 2.86\,eV$$

At low collision energy this maximum will be almost reached, but at high collision energy, that is in the range of kiloelectronvolts, a fraction of this maximum will be converted to internal energy. Always keep in mind that 1 eV per ion is equal to around $100\,kJ\,mol^{-1}$.

A measure of the average amount of kinetic energy converted into internal energy in the collision process was obtained by Harrison and Lin [13]. They observed that the fragmentation of the n-butylbenzene molecular ion gives two fragments, $C_7H_7^+$ and $C_7H_8^{\bullet+}$, and that the ratio of their intensities increases with increasing amount of internal energy of the parent ion $C_{10}H_{14}^{\bullet+}$ measured by charge exchange. Nacson and Harrison [14], using similar techniques, concluded that 87 % of the available collision energy is converted into internal energy for collision of 60 eV n-octylbenzene with an N_2 target, while the efficiency decreases to 33 % for n-butylbenzene with an Ar target.

In practice, two collision regimes should be distinguished: low energy, in the range of 1–100 eV, as occurring in quadrupole, ICR or ion trap instruments, and high energy, several thousand electronvolts, as common for magnetic and TOF instruments.

For theoretically comparable energy exchange, different fragmentation patterns are observed for low and high collision energy. Generally speaking, one might state that high-energy spectra give simpler, more clear-cut fragmentations, while low-energy ones lead to spectra with more diverse fragmentation pathways, often including more rearrangements.

4.3.2 High-Energy Collision (keV)

Although the various types of analysers (electromagnetic, quadrupolar, TOF, ion trap and ICR) allow tandem mass spectrometry, only electromagnetic, TOF or hybrid instruments in which ions have very high translational energy can function at high energy. In these instruments, a collision cell is placed between two analysers. The instrument configuration that is most favourable to CID studies at high energy is that which offers the best resolution both of the precursor ions and of the product ions, namely EBEB or BEEB. An alternative may be found in the use of triple sector instruments such as EBE and BEB or double sector instruments such as EB or BE. Each of these configurations sacrifices at least the resolution of the precursor ions or of the product ions.

The technique consists of introducing a gas into the collision cell as a molecular beam directed perpendicularly to the ion beam accelerated by a potential of a few kilovolts towards an outlet where a pump removes it from the instrument so as to preserve a vacuum compatible with the use of electromagnetic analysers. This collision cell is located at a focal point to avoid losses. It is better to float the collision cell electrically to distinguish the metastable from the CID products.

If, for example, an ion with m/z 100 dissociates to yield an ion with m/z 60 with a source at 8 kV and with grounded analysers, the product ions have an energy of 4.8 keV, whether they are formed through collision or through metastable dissociation. However, if the collision cell is floated at 6 keV, the parent ions undergoing CID have a kinetic energy of 2 keV and the product ions thus have an energy of 1.2 keV plus 6 keV acceleration when leaving the collision cell, that is 7.2 keV. The kinetic energy difference allows one to distinguish the ions formed through the two types of reaction because the ions issuing from metastable decomposition have an energy of 4.8 keV [15].

At high energy, the ion excitation is mostly electronic [16]. The internal kinetic energy conversion occurs most efficiently when the collision interaction time and the internal period of the mode that undergoes the excitation are comparable [17]. Hence, in the case of an ion with mass 1000 Da having an energy of 8 kV, the value for the interaction time with a target a few angstroms wide is close to 10^{-15} s. This time corresponds to the vertical electronic transition. The energy thus acquired is then redistributed in the form of vibrational energy. A bond cleavage may result.

Helium is the most common target gas used in CID studies at high energy. In fact, it minimizes the dissociation reaction, that is the neutralization of the precursor ion and the deviation of the product ions beyond the angle allowed for the focalization of ions into the second sector of the spectrometer. However, helium is not very efficient in transferring internal energy, so using a heavier gas such as argon or xenon allows the collision yield to be increased.

The internal energy that is acquired during the ions' stay in the collision cell is on average 1–3 eV with an energy dispersion that can reach 15 eV [18,19]. Only a small fraction of the energy available is actually converted. Let us recall that the maximum energy that can be converted in a system where an ion with mass 1000 Da has an energy of 4 keV and where the collision gas is helium is 16 eV. In addition, only collisions with an angular deviation of less than 1° are observable through this technique. Mechanisms producing a great angular dispersion of the products are thus rarely observed.

4.3.3 Low-Energy Collision (Between 1 and 100 eV)

Low-energy CID spectra are measured using triple quadrupole, ion trap, ICR or hybrid instruments. For tandem mass spectrometers in space, the collision chamber is most often a quadrupole in the RF mode only, which allows one to focus the ions that are angularly dispersed by the collision. The pressure difference between the collision cell and the rest of the analyser is obtained through differential pumping.

At low energy, the ions' excitation energy is mostly of a vibrational nature [20] because the interaction time between an ion with mass 200 Da at an energy of 30 eV with a target of a few angstroms is about 10^{-14} s, corresponding approximately to the bonds' vibration period.

The nature of the collision gas is more important than it is for the high-energy collisions. Normally, heavier gases such as argon, xenon or krypton are preferred because they allow the transfer of more energy.

The energy that is deposited is slightly lower than at high energy but has a weaker dispersion. However, the collision yields are extremely high when compared with the energy available. This is due on the one hand to the great focusing characteristic of RF-only quadrupoles and on the other to the length of the collision chamber that allows multiple collisions. Let us recall that the maximum energy that can be converted in such a system where an ion with m/z 1000 has a kinetic energy of 20 eV and where the collision gas is xenon is 2.3 eV.

4.4 Other Methods of Ion Activation

One of the main inconveniences with CID is the limitation of the energy transferred to an ion and thus the limitation of its degree of fragmentation. MS/MS is used more and more to fragment large molecules. When the ions are large, the energy is distributed on a greater number of bonds. The result is a slower reaction rate of the fragmentation. Furthermore, a collision gas must be introduced into the instrument, compromising the vacuum. In order to avoid these inconveniences, other ion activation methods were developed [21].

Another activation method, which is performed without gas, uses collisions with a solid surface [22, 23]. This method is called surface-induced dissociation (SID). In practice, this technique typically involves collisions of precursor ions on a metallic surface at an approximate 45° angle of incidence. The products of SID depend on the energy of the ion and on the nature of both the ion and the surface. In addition to dissociation of the impacting ions, other reactions can be observed. At a collision energy below 100 eV, dissociation is in competition with the formation of ion–surface reaction products, whereas at a collision energy of several hundred electronvolts the sputtering of materials absorbed on the surface dominates dissociation. This method minimizes the pumping that is necessary because no gas is introduced into the collision cell. Moreover, the interaction efficiency is 100 % and it is able to transfer large energy (7–100 eV) with an energy distribution that is narrower than that observed in gas collision activation. When a non-metallic surface is used, charge accumulates on the surface and repels other ions. With a metallic surface, discharge of the ions is in competition with dissociation and reduces the sensitivity. Better results are obtained when a non-conducting substance is adsorbed on the metal or when the metal is coated with a thin layer of a non-metal. An example is the use of an interfacial monolayer of alkyl-thiols on gold.

Two other ion activation methods were developed to replace the gas molecules as targets by laser beams (photodissociation or infrared multiphoton dissociation IRMPD) or by electron beams (electron capture dissociation ECD). These two methods can be applied to ions that are trapped during their excitations by photons or electrons, respectively. Thus, they are most often used with ion trap or ICR analysers because the residence time and the interaction time are longer.

The photodissociation uses a laser that is directed through a window to irradiate the interior of the analyser. The mechanism of fragmentation by photodissociation involves the absorption by the ion of one or more photons. As each photon is absorbed, the ion increases its internal energy. The energy accumulates in the various vibration modes, and finally it is sufficient to provoke dissociation resulting in gas-phase fragments of the ion. As in CID, the photodissociation is an 'ergodic' ion activation method because the energy is added slowly and is redistributed in the vibrational modes of the ion before its fragmentation.

The photodissociation has some advantages over the other ion activation methods. Indeed, this method is selective because only the ions that absorb at the wavelength of the incident radiation can be photodissociated. Furthermore, the amount of energy that is acquired by photodissociation is well defined and the distribution of the internal energy is narrow.

Different lasers have been used for photodissociation. Historically, UV and visible lasers were the first to be used. But more recently, the use of IR lasers has become more common. Compared with UV lasers, IR lasers are rather non-selective and are of lower energies. Consequently, when the capture of only one UV photon provides enough energy to induce dissociation of ions, the energy received from multiple IR photon absorption is needed to activate and dissociate ions. IRMPD is the ion activation method based on the capture of several IR photons. It uses a continuous CO_2 laser of 10.6 μm with a power of 25–50 W. The amount of energy acquired by an ion can be varied by the length of the laser interaction with the ions, typically in the range of 10–300 ms. In the case of the FTICR analyser, the main advantage of the IRMPD method over CID is a much faster MS/MS acquisition time because it does not require a pulsed collision gas to excite the ions.

Electron capture dissociation (ECD) has recently been developed as an alternative activation method and is now widely used [24, 25]. The ECD activation method is applied to multiply charged positive ions submitted to a beam of low energy produced by an emitter cathode. Low-energy electrons (<0.2 eV) must be used to have a sufficient capture cross-section. When multiply charged positive ions are irradiated with low-energy electrons, electron capture occurs with charge state reduction to produce a radical positive ion that can then dissociate.

The mechanism of dissociation by ECD involves electron capture by the multiply charged ion. The 5–7 eV of recombination energy induces the increase of its internal energy that allows its dissociation. This activation process is very short and it is assumed to occur much faster than 10^{-14} s. Because bond dissociation occurs faster than a typical bond vibration, ECD allows only the direct bond cleavage without prior randomization of the internal energy. Consequently, in contrast to the most commonly used activation methods such as CID and IRMPD, ECD has been described as a 'non-ergodic' process. This explains why this activation method is well suited to the fragmentation of large ions. Bond cleavage can be induced by addition of a low energy because it occurs before the randomization of this energy over a great number of vibration modes of a large ion.

The dissociation obtained by ECD involves the fragmentation of an odd-electron cation. Furthermore, from the non-ergodic nature of ECD, it is not generally the weakest bonds that are preferentially the cleavage sites but the dissociation of the ion that is restricted to positive charge sites where the electron is captured. As a result, fragmentation pathways are governed by radical ion chemistry and they can be different to all other activation methods. Thus, ECD is a complementary activation method to CID and IRMPD. However, from the relatively narrow diameter of the electron beam compared with the size of the volume where the ions are trapped, only a small number of ions are activated instantaneously. Consequently, the overall efficiency of ECD is typically lower than that obtained with CID or IRMPD and longer interaction times are needed.

It is difficult to perform the ECD method with an ion trap analyser because the strong electric field applied to the trap affects the movement of the electrons in the trap. In consequence, electrons cannot maintain their thermal energy and finally they are expelled from the trap. To overcome this limitation, an ECD-like activation method for use with an ion trap has been developed [26]. This method, which is called electron transfer dissociation (ETD), uses gas–phase ion/ion chemistry to transfer an electron from singly charged anthracene anions to multiply charged ions. The mechanism of this method and the observed fragmentation pathways are analogous to those observed in ECD. A summary of the main characteristics of the different activation methods used in mass spectrometry is given in Table 4.1.

Table 4.1   Different ion activation methods used in mass spectrometry.

Activation methods	Energy	Instruments
CID	Low	QqQ, IT, QqTOF, QqIT, ICR
	High	TOF, magnetic
SID	Low	QqQ, IT, ICR, BqQ
	High	TOF
Photodissociation or IRMPD	Low	ICR, IT
ECD	Low	ICR
ETD	Low	IT

4.5 Reactions Studied in MS/MS

The type of reaction produced by the collision with a neutral molecule N, which is most widely studied in tandem mass spectrometry, corresponds to a mass change. It is illustrated by the following equations that describe covalent dissociation by ejection of neutral species:

$$m_p^{n+} + N \longrightarrow m_f^{n+} + m_n + N$$

If the target gas is a chemically reactive species, an association reaction may occur, which can yield an adduct ion with a mass greater than that of the precursor:

$$m_p^+ + m_n \longrightarrow (m_p + m_n)^+$$

These two types of reactions also occur with negative ions.

The reaction of fullerene radical cation with helium, to produce endohedral $He@C_{60}$, is a particular example of the last type of reaction, and is displayed in Figure 4.6 [27]:

$$C_{60}^{\bullet+} + He \longrightarrow He@C_{60}^{\bullet+}$$

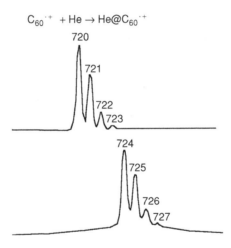

Figure 4.6
Fullerene C_{60} gives by electron ionization the spectrum at the top. The m/z 720 to 723 ions result from the ^{13}C isotopes. If these fullerene ions are allowed to collide at high energy with ^4He, the spectrum at the bottom is observed. It results from endohedral addition of He 'in' the fullerene. The endohedral nature has been demonstrated by neutralization–reionization of the m/z 724 ion. With data from Weiske Th., Wong Th., Kratschmer W., Terlouw J.K. and Scwarz H., Angew. Chem. Int. Ed. Engl., 31, 183–185, 1992.

Another type of reaction produced through collision is the charge variation:

1. Charge exchange:

$$m_p^{\bullet+} + N \longrightarrow m_p + N^{\bullet+}$$
$$m_p^{\bullet-} + N \longrightarrow m_p + N^{\bullet-}$$

An example [28] is $Ne^{\bullet+} + Pb \rightarrow Pb^{\bullet+} + Ne$

2. Partial charge transfer:

$$m_p^{2+} + N \longrightarrow m_p^{\bullet+} + N^{\bullet+}$$

An example [29] is $C_6H_6^{2+} + C_6H_6 \rightarrow 2C_6H_6^{\bullet+}$

3. Ionization, charge stripping or charge inversion (normally observed at collision energies of about 1 keV):

$$m_p + N \longrightarrow m_p^{\bullet+} + N + e^-$$
$$m_p^{\bullet+} + N \longrightarrow m_p^{2+} + N + e^-$$
$$m_p^- + N \longrightarrow m_p^+ + N + 2e^-$$

Examples are

(1) [30] $C_6H_5CH_3 + He \longrightarrow C_6H_5CH_3^{\bullet+} + He + e^-$

(2) [31] $Ar^{\bullet+} + Ar \longrightarrow Ar^{2+} + Ar^* + e^-$

(3) [32] lactic acid $(M - H)^- + 2\,SF_6 \rightarrow (M - H)^+ + 2\,SF_6^{\bullet-}$

As all of these processes, by necessity, can only result from electronic excitation, this suggests that collision activation under these conditions leads at least in part to electronic excitation. However, especially for larger molecules, the mechanisms of collisional activation are more likely to involve vibrational excitation. In this case, it may occur through impulsive collision of the target atom or molecule with a selected atom or group of atoms in the region of the collision site.

The charge change can be combined with a mass change such as in the reaction

$$m_p^- + N \longrightarrow m_f^+ + m_n + N + 2e^-$$

which corresponds to dissociative charge stripping and

$$m_p^{2+} + N \longrightarrow m_{f1}^+ + m_{f2}^+ + N$$

which corresponds to Coulomb explosion with charge separation.

The type of reaction that is observed depends on the kinetic energy of the precursor ion when it collides with N. When $N = O_2$, the neutral ionization reaction occurs readily:

$$m_n + O_2 \longrightarrow m_n^{\bullet+} + O_2^{\bullet-}$$

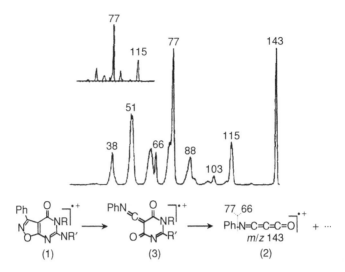

Figure 4.7
Neutralisation–reionization of the fragment (N_fR) at m/z 143
from the compound (1). The inset shows the CID spectrum. The
NR spectrum is the same as the CID one, except for m/z 66. This
shows that this m/z 66 ion is produced during the neutralization
process and reionized. Both spectra confirm the (2) structure for
the m/z 143 fragment. It is possible that structure (3) is an intermediate in this process. Reproduced (modified) from Flammang
R., Laurent S., Flammang-Barbieux M. and Wentrup C., Rapid
Comm. Mass Spectrom., 6, 667, 1992, with permission.

This allows the ionization of neutrals with kiloelectronvolt kinetic energy that can either
be formed through neutralization of an ion by a charge exchange reaction or be produced
by a fragmentation. The acronyms NR are used for neutralization reionization and N_fR for
neutral fragment reionization [33–35].

Positive ion neutralization is performed using collisions with molecules whose ionization
energy is low. Ammonia is often used for this purpose:

$$m_n^{\bullet+} + NH_3 \longrightarrow m_n + NH_3^{\bullet+}$$

A typical neutralization–reionization experiment combines the following steps: collision
with ammonia to neutralize the ions, elimination of all charged species by passing between
the plates of a condenser; and reionization by collision with oxygen.

Figure 4.7 gives an example of such a neutralization–reionization reaction [36].

4.6 Tandem Mass Spectrometry Applications

MS/MS applications are plentiful, for example in elucidation of structure, determination
of fragmentation mechanisms, determination of elementary compositions, applications to
high-selectivity and high-sensitivity analysis, observation of ion–molecule reactions and
thermochemical data determination (kinetic method). We will examine them in detail
through a few examples.

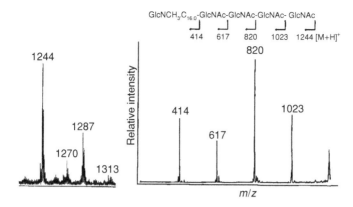

Figure 4.8
Structure elucidation by MS/MS of a nodulation factor. Left: *m/z* values observed by FAB from a chromatographic fraction of nodulation factors. Right: the *m/z* 1244 ion is selected as precursor and fragmented by *B/E* linked scan in a magnetic instrument. The clear-cut fragmentation at the glycosylic bonds allows straightforward sequence assignment as shown by the fragmentation scheme. Reproduced (modified) from Mergaert P., D'Haeze W., Geelen D., Prome D., Van Montagu M., Geremia R., Prome J.C. and Holsters M., J. Biol. Chem., 49, 29217–29223, 1995, with permission.

4.6.1 Structure Elucidation

Tandem mass spectrometry allows more structural information to be obtained on a particular ionic species, because the used ionization method yields relatively few structurally diagnostic fragments, or because its fragmentation is obscured by the presence of other compounds in the mixture introduced in the source, or because it is obscured by other ions generated from the matrix in the course of ionization.

An example of structure elucidation by tandem mass spectrometry is displayed in Figure 4.8. The FAB mass spectrum of an isolated fraction containing several nodulation factors appears at different *m/z* values [37]. One of them, *m/z* 1244, is selected and fragmented using *B/E* linked scan on a magnetic instrument. The observed fragment ions suggest the structure shown, corresponding to features known for this class of compounds, that is an oligosaccharide bearing at one end an alkyl chain bound to an amino sugar.

The analogy to a chromatographic separation coupled to a mass spectrometer should be mentioned. In this case, the FAB mass spectrum is analogous to a chromatographic trace displaying compounds of the mixture. The *B/E* linked scan yields the mass spectrum of one of these compounds. While chromatography gives a separation requiring a long time, the separation according to the mass is almost instantaneous, providing much shorter analysis time.

Tandem mass spectrometry also allows the determination of isomers and diastereoisomers. As an example, glycosylmonophosphopolyisoprenols [38,39] display in the negative ion FAB mode abundant $(M - H)^-$ ions. When these ions are fragmented under low-energy collisions, an abundant fragment corresponding to the phospholipid anion is observed,

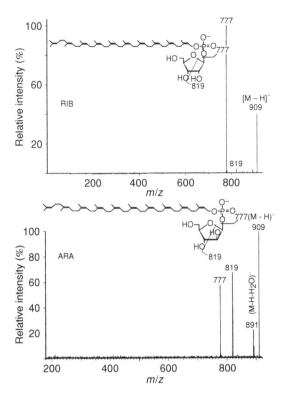

Figure 4.9
FAB CID at low collision energy of the $(M-H)^-$ anion from the decaprenylphosphates of β-D-ribose (top) and β-D-arabinose (bottom) respectively. The *trans* configuration of the 2-OH and phosphate groups in β-D-ribose favours the formation of the phospholipid fragment. Reproduced (modified) from Wolucka B. and de Hoffmann E., J. Biol. Chem., 270, 20151, 1995, with permission.

together with fragments across the sugar ring. This fragmentation is very sensitive to the stereochemistry, as illustrated in Figure 4.9. Decaprenylphospho-D-ribose displays the decaprenylphosphate anion as almost the sole fragment, while the analogous D-arabinose derivatives also produce fragments resulting from cleavage across the sugar ring.

From a study of several compounds, it appears that a 2-hydroxyl *trans* to the phosphate strongly favours the formation of the phospholipid anion. The ribose derivative differs from the arabinose derivative only by the *trans* configuration of the 2-hydroxy group with respect to the phosphate.

An example of elucidation of fragmentation mechanisms by tandem mass spectrometry is displayed in Figure 4.10 which shows the mass spectrum of *p-t*-butylphenol obtained through electron ionization. An important fragment is observed at m/z 107, which corresponds to a loss of 43 Da. We can consider either the loss of a methyl radical followed by the elimination of CO, or the loss of a C_3H_7 fragment after a rearrangement in one or several steps. Spectrum A gives a partial answer: selecting the ion of m/z 135 as a precursor, we

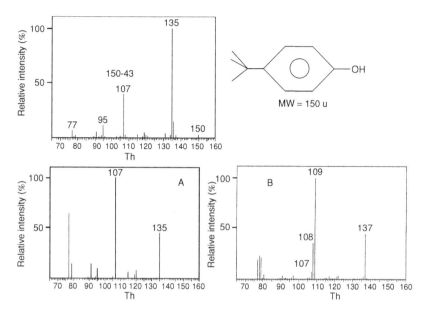

Figure 4.10
Top: electron ionization spectrum of *p-t*-butylphenol. Spectrum A: fragments of the precursor ion 135 Th. This spectrum shows that the ion of *m/z* 107 results from a loss of 28 Da after the loss of a methyl by the molecular ion. Spectrum B proves that the oxygen atom is not lost during this fragmentation: the loss of 28 Da is thus a loss of C_2H_4, not of CO.

observe that it produces the fragment of 107 Th. Thus the first step is the loss of a methyl that produces the ion of 135 Th. The subsequent loss of 28 Da may result from a neutral CO or C_2H_4. In order to determine this, we select 137 Th as a precursor, an ion corresponding to 135 Th with an ^{18}O isotope or with two ^{13}C. Spectrum B shows that the ion produced is displaced from 107 to 109 Th: the oxygen atom is preserved in the fragment. The fragments that are observed at 108 and 107 Th in this last spectrum are due to the contribution of ^{13}C. A complete calculation [40] shows that the observed masses correspond to the loss of a neutral with composition C_2H_4. We can thus conclude that the molecular ion loses a methyl, rearranges and then loses C_2H_4.

4.6.2 Selective Detection of Target Compound Class

The following example explains the development of a rapid selective analysis method for the components of a complex mixture based on MS/MS. The aim is to find an easy way of detecting a class of compounds, the carnitines, in biological fluids. This compound participates in carrying fatty acids with various chain lengths through cell membranes; it contributes to eliminating excess fatty acids in urines. It is therefore of interest to detect them in order to arrive at a diagnosis.

Elaborating a selective detection method starts with a study of the fragmentation reactions. Ions that are typical of carnitine are looked for, independently of the nature of the fatty acid. Figure 4.11 shows the product ion spectrum of the molecular ion of octanoylcarnitine.

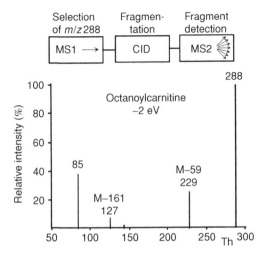

Figure 4.11
Spectrum of the ions produced by a collision-induced dissociation of the pseudomolecular ion of octanoylcarnitine. This spectrum is obtained by selecting the 288 Th ion by means of the first analyser, fragmenting it through collision in the collision cell and analysing the fragments using the second spectrometer.

The fragmentation schemes deduced from this spectrum, and confirmed by the CID spectra of the fragments, and also by the product ion spectra of other carnitines, are shown in Figure 4.12. Referring to these fragmentation schemes, it can be seen that precursor scans of either m/z 144 and 85 are good candidates. Indeed, these fragments do not contain parts of the fatty acids, and should thus not be sensitive to their nature. The fragment ions at m/z 229 (M − 59) and 127 (M − 161), acylium ion obtained after loss of the carnitine moiety, contain the fatty acid chain, and thus are not good candidates for precursor ion scans that would detect all the carnitine conjugates. However, the neutrals lost in these fragmentations contain a part (59 Da) or the complete (161 Da) carnitine moiety. Neutral loss scan of either 59 or 161 Da thus should allow the selective detection of all the carnitine conjugates.

We conclude that we could selectively detect the carnitine conjugates by looking for the precursors of the 85 and 144 Th fragments, or by detecting the losses of 59 or 161 Da neutrals.

An example of an application is given in Figure 4.13. The first spectrum is the FAB spectrum, sometimes called the source spectrum because it detects the ions formed in the source. The second spectrum is that obtained by selectively detecting the precursors of m/z 85. We see that the carnitine conjugates clearly dominate.

This first sample came from the urine of a patient suffering from a short-chain fatty acid metabolism defect (short-chain acylCoA dehydrogenase deficiency). As a comparison, the spectrum of a urine sample from a patient suffering from a medium-chain fatty acid metabolism defect (medium-chain acylCoA dehydrogenase deficiency) is displayed in Figure 4.14. Other examples of selectivity obtained by MS/MS are given in Chapter 5.

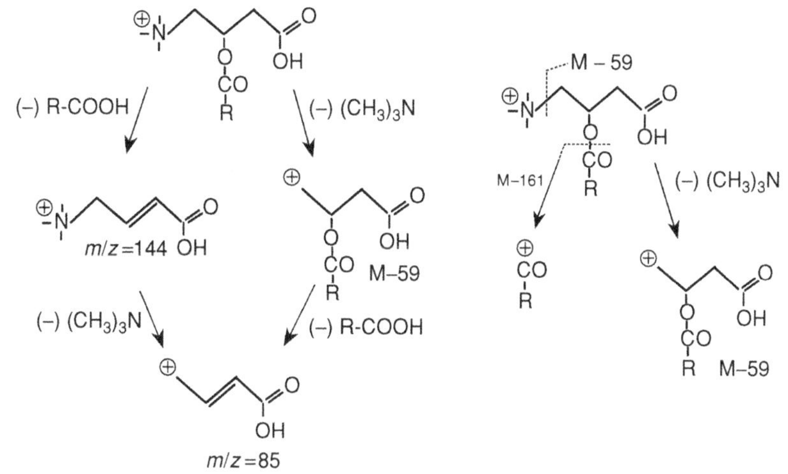

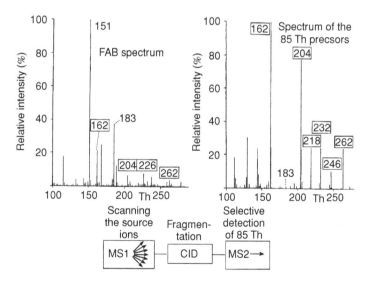

Figure 4.12
Fragmentation diagram of carnitines leading to 144 and 85 Th fragment ions, and to ions resulting from the losses of 59 and 161 Da neutrals. All these fragmentations are typical of all the carnitines, whatever the nature of the fatty acid chain R.

Figure 4.13
Left: FAB spectrum of a crude biological sample containing carnitine conjugates. Right: the spectrum of the same sample obtained by detecting the precursors of the 85 Th fragment. At 162 Th, carnitine; at 204 Th, acetylcarnitine; at 218 Th, propionyl; and so on.

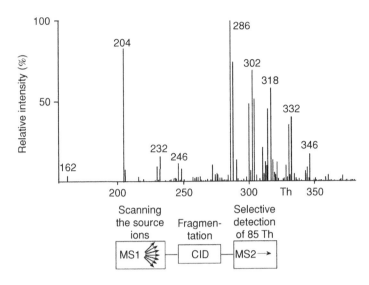

Figure 4.14
Spectrum of the precursors of the ion of 85 Th as in Figure 4.13, but
from a sample containing longer chain carnitines: 288 Th, octanoylcarnitine, and so on.

This general methodology for the selective detection of compounds or compound classes
can be combined with a chromatographic separation, allowing selective detection of one
compound in a complex mixture. Furthermore, as already discussed at the beginning of this
chapter, tandem mass spectrometry also allows the signal-to-noise ratio to be substantially
improved in the detection of selected compounds or compound classes.

The possibility to obtain both high selectivity and high sensitivity by tandem mass
spectrometry is largely used in the pharmaceutical industry to monitor and quantify a
selected compound in pharmacokinetics studies, which is the field that uses the largest
number of tandem mass spectrometers.

4.6.3 Ion–Molecule Reaction

If a reactant gas is introduced into the collision cell, ion–molecule collisions can lead to the
observation of gas-phase reactions. Tandem-in-time instruments facilitate the observation
of ion–molecule reactions. Reaction times can be extended over appropriate time periods,
typically as long as several seconds. It is also possible to vary easily the reactant ion energy.
The evolution of the reaction can be followed as a function of time, and equilibrium can be
observed. This allows the determination of kinetic and thermodynamic parameters, and has
allowed for example the determination of basicity and acidity scales in the gas phase. In
tandem-in-space instruments, the time allowed for reaction will be short and can be varied
over only a limited range. Moreover, it is difficult to achieve the very low collision energies
that promote exothermic ion–molecule reactions. Nevertheless, product ion spectra arising
from ion–molecule reactions can be recorded. These spectra can be an alternative to CID
to characterize ions.

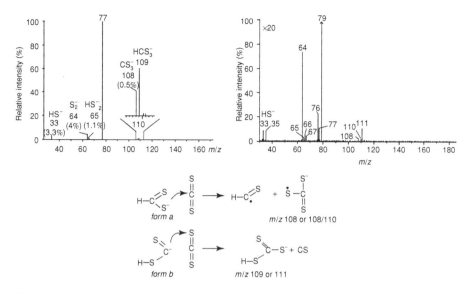

Figure 4.15
Top left: spectrum obtained when HCS_2^- ions formed in the source by chemical ionization of CS_2 are selected with the first quadrupole and allowed to collide in the collision cell filled with neutral CS_2. Top right: spectrum obtained when $HC^{32}S^{34}S^-$ at 79 Th is selected instead of m/z 77 $HC^{32}S_2^-$ as for the first spectrum. The observed ions and their isotope shifts can be explained by the mechanism displayed. Reproduced (modified) from Gimbert Y., Arnaud R., Tabet J.C. and de Hoffmann E., J. Phys. Chem., 102, 3732–3737, 1998, with permission.

For a simple example of an ion–molecule reaction, Figure 4.15 displays the spectrum obtained when HCS_2^- ions are selected by the first quadrupole of a triple quadrupole instrument, QqQ, and allowed to collide with CS_2 molecules in the central quadrupole collision cell q [41].

The mechanism displayed in Figure 4.15 explains the observed ions produced as well the isotope pattern observed when 79 Th $HC^{32}S^{34}S^-$ is selected by the first quadrupole, rather than 77 Th $HC^{32}S_2^-$. Indeed, the first reaction produces $HCS_3^{\bullet-}$ which contains one sulfur atom originating from the $HC^{32}S_2^-$ reactant. When $HC^{32}S^{34}S^-$ is selected, only one out of the two sulfur atoms will be included in this $HCS_3^{\bullet-}$ product ion, which is dedoubled to 108 and 110 Th in equal abundances. The second reaction produces the ions at m/z 109 which contains both sulfur atoms from the reacting ion from the source. It is thus shifted entirely to m/z 111 when $HC^{32}S^{34}S^-$ is selected as the reactant ion.

4.6.4 The Kinetic Method

The kinetic method [42, 43] is a relative method for thermochemical data determination which is based on measurement of the rates of competitive dissociations of mass-selected cluster ions. This method was introduced by Cooks [44] for proton affinity determination. Later, an extension of this method was proposed by Fenselau [45].

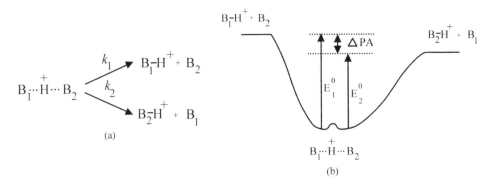

Figure 4.16
Proton affinity determination by the kinetic method. (a) This method is based on competitive
dissociation of heterodimer clusters. (b) Potential energy diagram for proton-bound dimer
dissociation.

For proton affinity determination, the kinetic method involves the formation of the proton
bound heterodimer between the two bases whose affinities are to be compared. By tandem
mass spectrometry, the appropriate cluster ion $[B_1HB_2]^+$ is selected and its spontaneous or
collisional dissociation is observed. As shown in Figure 4.16, the competitive dissociation
leading to the two protonated monomers is analysed and the relative abundances of the
monomers $[B_1H]^+$ and $[B_2H]^+$ are measured. From these abundances, the relative proton
affinities of the two bases B_1 and B_2 can be calculated and the proton affinity of one of the
two bases can be determined, if the proton affinity of the other is known.

The kinetic method is based on the following equation:

$$\ln(k_1/k_2) = \ln(Q_1^{\ddagger}/Q_2^{\ddagger}) - (E_1^0 - E_2^0)/RT$$

where the k_x are the rate constants, $Q_x^{\ddagger}$ are the partition functions of the transition state,
E_x^0 are the activation energies for the reactions, R is the Boltzmann constant, T is the dimer
temperature and the subscripts refer to channel x.

The above equation can be simplified to give an equation that shows the relation of the
proton affinity (PA) with the rate constant ratio of the decomposition reaction:

$$\ln(k_1/k_2) = \Delta PA/RT$$

This equation is verified by the assumption that both the differences in the entropy changes
for the competitive dissociation channels and the reverse activation barriers for the dissociation of the proton-bonded dimer are negligible or similar. In these conditions, $Q_1^{\ddagger} = Q_2^{\ddagger}$
and $E_1^0 - E_2^0 = PA_2 - PA_1 = \Delta PA$. The entropy effects are generally comparable when
the competing dissociations involve species that are similar in structure.

The rate constant ratio of the competitive decomposition reaction is equal to the ratio of
the two fragment ion intensities (I_x) if the comparison is made for competitive dissociation
of a weakly proton dimer having no other decomposition channel. In these conditions,

$k_1/k_2 = I_1/I_2$ and thus the above equation can be modified to allow the evaluation of the difference in proton affinities of B_1 and B_2:

$$\ln(I_1/I_2) = \Delta PA/RT$$

The kinetic method provides an alternative to equilibrium measurements for the determination of gas–phase thermochemical properties. It has been applied more and more in thermochemical data determination mainly because of its ability to measure very small energy differences and its simplicity. Indeed, it can be executed easily on any tandem mass spectrometer. Furthermore, this method is sensitive and is applicable with impure compounds. Its applications are broad, covering thermochemical properties in the gas phase such as proton affinity [46], electron affinity [47], metal ion affinity [48], ionization energy [49], acidity [50] or basicity [51]. In addition to the determination of thermochemical data, the kinetic method has also been applied in structural and chemical analysis such as chiral distinctions. This method is able to distinguish enantiomers and to measure precisely enantiomeric ratios [52].

References

1. Cooks, R.G., Beynon, J.H., Caprioli, R.M. and Lester, G.R. (1973) *Metastable Ions*, Elsevier, Amsterdam.
2. McLafferty, F.W. (ed.) (1983) *Tandem Mass Spectrometry*, John Wiley & Sons, Inc., New York.
3. Busch, K.L., Glish, G.L. and McLuckey, S.A. (1988) *Mass Spectrometry/Mass Spectrometry: Techniques and Applications of Tandem Mass Spectrometry*, VCH, New York.
4. Price, P. (1991) *J. Am. Soc. Mass Spectrom.*, **2**, 336.
5. Glish, G. (1991) *J. Am. Soc. Mass Spectrom.*, **2**, 349.
6. Schwartz, J.C., Wade, A.P., Enke, C.G. and Cooks, R.G. (1990) Systematic delineation of scan modes in multidimensional mass spectrometry. *Anal. Chem.*, **62**, 1809–18.
7. Libert, R., Hermans, D., Draye, J.P. *et al.* (1991) *Clin. Chem.*, **37**, 2102.
8. Laskin, J. and Futrell, J.H. (2005) Activation of large ions in FTICR mass spectrometry. *Mass Spetrom. Rev.*, **24** (2), 135–67.
9. Cooks, R.G. (1995) *J. Mass Spectrom.*, **30**, 1215.
10. de Hoffmann, E. (1996) Tandem mass spectrometry: a primer. *J. Mass Spectrom.*, **31**, 129–37.
11. Shukla, A.K. and Futrell, J.H. (2000) Tandem mass spectrometry: dissociation of ions by collisional activation. *J. Mass Spectrom.*, **35**, 1069.
12. Jennings, K.R. (2000) The changing impact of the collision-induced decomposition of ions in mass specrometry. *Int. J. Mas Spectrom.*, **200**, 479.
13. Harrison, A.G. and Lin, M.S. (1983) *Int. J. Mass Spectrom. Ion Phys.*, **51**, 353.
14. Nacson, S. and Harrison, A.G. (1985) *Int. J. Mass Spectrom. Ion Processes*, **63**, 325.
15. Busch, K.L., Glish, G.L. and McLuckey, S.A. (1988) *Mass Spectrometry/Mass Spectrometry: Techniques and Applications of Tandem Mass Spectrometry*, VCH, New York, p. 50.
16. Yamaoka, H., Dong, P. and Durup, J. (1969) *J. Chem. Phys.*, **51**, 3465.
17. Beynon, J.H., Boyd, R.K. and Brenton, A.G. (1986) *Adv. Mass Spectrom.*, **10A**, 437.
18. Neumann, G.M. and Derrick, P.J. (1984) *Org. Mass Spectrom.*, **19**, 165.

19. Wysocki, V.H., Kenttamaa, H.I. and Cooks, R.G. (1985) *Int. J. Mass Spectrom. Ion Processes*, **75**, 181.
20. Schwartz, R.N., Slawsky, Z.I. and Herzfeld, K.F. (1952) *J. Chem. Phys.*, **20**, 1591.
21. Sleno, L. and Volmer, D.A. (2004) Ion activation methods for tandem mass spectrometry. *J. Mass Spectrom.*, **39**, 1091–1112.
22. Mabud, M.A., DeKrey, M.J. and Cooks, R.G. (1985) *Int. J. Mass Spectrom. Ion Processes*, **75**, 285.
23. Laskin, J., Denisov, E.V., Shukla, A.K. *et al.* (2002) *Anal. Chem.*, **74**, 3255.
24. Zubarev, R.A., Kelleher, N.L. and McLafferty, F.W. (1998) Electron capture dissociation of multiply charged protein cations. A nonergodic process. *J. Am. Chem. Soc.*, **120**, 3265–6.
25. Zubarev, R.A., Haselmann, K.F., Budnik, B. *et al.* (2002) Towards an understanding of the mechanism of electron capture dissociation: a historical perspective and modern ideas. *Eur. J. Mass Spectrom.*, **8**, 337.
26. Syka, J.E.P., Coon, J.J., Schroeder, M.J. *et al.* (2004) Peptide and protein sequence analysis by electron transfer dissociation mass spectrometry. *Proc. Natl. Acad Sci. USA*, **101**, 9528–33.
27. Weiske, T., Wong, T., Krätschmer, W. *et al.* (1992) The neutralization of HeC60 in the gas phase: compelling evidence for the existence of an endohedral structure for He@C60. *Angew. Chem. Int. Ed. Engl.*, **31** (2), 183–5.
28. Gran, W.H. and Duffenbach, O.S. (1937) *Phys. Rev.*, **51**, 804.
29. Mathur, B.P., Burger, E.M., Boswick, D.E. and Moran, T.F. (1981) *Org. Mass Spectrom.*, **16**, 92.
30. McLafferty, F.W., Todd, P.J., McGilvery, D.C. and Balwin, M.A. (1980) *J. Am. Chem. Soc.*, **102**, 3360.
31. Asr, T., Beynon, J.H. and Cooks, R.G. (1972) *J. Am. Chem. Soc.*, **94**, 6611.
32. Douglas, D.J. and Shushan, B. (1982) *Org. Mass Spectrom.*, **17**, 198.
33. McLafferty, F.W. (1992) *Int. J. Mass Spectrom. Ion Processes*, **118/119**, 221.
34. Goldberg, N. and Schwarz, H., *Acc. Chem. Res.*, **27**, 347.
35. Zagorevskii, D.V. and Holmes, J.L. (1999) Neutralization-reionization mass spectrometry applied to organometallic and coordination chemistry. *Mass Spectrom. Rev.*, **18** (2), 87–118.
36. Flammang, R., Laurent, S., Flammang-Barbieux, M. and Wentrup, C. (1992) *Rapid Commun. Mass Spectrom.*, **6**, 667.
37. Mergaert, P., D'Haeze, W., Geelen, D. *et al.* (1995) Biosynthesis of azorhizobium caulinodans nod factors. *J. Biol. Chem.*, **49**, 29217–23.
38. Wolucka, B., McNeil, M.R., de Hoffmann, E. *et al.* (1994) *J. Biol. Chem.*, **269**, 23328.
39. Wolucka, B. and de Hoffmann, E. (1995) *J. Biol. Chem.*, **270**, 20151.
40. de Hoffmann, E. and Auriel, M. (1989) *Org. Mass Spectrom.*, **24**, 748.
41. Gimbert, Y., Arnaud, R., Tabet, J.C. and de Hoffmann, E. (1998) Gas phase reaction of neutral carbon disulfide with its hydride adduct anion: tandem mass spectrometry and theoretical studies. *J. Phys. Chem.*, **102**, 3732–7.
42. Cooks, R.G., Patrick, J.S., Kotiaho, T. and McLuckey, S.A. (1994) *Mass Spectrom. Rev.*, **13**, 287.
43. Cooks, R.G. and Wong, P.S.H. (1998) *Acc. Chem. Res.*, **31**, 379.
44. Cooks, R.G. and Kruger, T.L. (1977) *J. Am. Chem. Soc.*, **99**, 1279.
45. Chen, X., Wu, Z. and Fenselau, C. (1993) *J. Am. Chem. Soc.*, **115**, 4844.
46. McLuckey, S.A., Cameron, D. and Cooks, R.G. (1981) *J. Am. Chem. Soc.*, **103**, 1313.
47. Chen, G., Cooks, R.G., Corpuz, E. and Scott, L.T. (1996) *J. Am. Soc. Mass Spectrom.*, **7**, 619.

48. Yang, S.S., Chen, G., Ma, S. *et al.* (1995) *J. Mass Spectrom.*, **30**, 807.
49. Wong, P.S.H., Shuguang, M. and Cooks, R.G. (1996) *Anal. Chem.*, **68**, 4254.
50. Wright, L.G., McLuckey, S.A., Cooks, R.G. and Wood, K.V. (1982) *Int. J. Mass Spectrom. Ion Processes*, **42**, 115.
51. Liquori, A., Napoli, A. and Sindona, G. (1994) *Rapid Commun. Mass Spectrom.*, **8**, 89.
52. Tao, W.A., Zhang, D.X., Wang, F. *et al.* (1999) *Anal. Chem.*, **71**, 4427.

5

Mass Spectrometry/ Chromatography Coupling

In order to analyse a complex mixture, for example natural products, a separation technique – gas chromatography (GC), liquid chromatography (LC) or capillary electrophoresis (CE) – is coupled with the mass spectrometer. The separated products must be introduced one after the other into the spectrometer, either in the gaseous state for GC/MS or in solution for LC/MS and CE/MS. This can occur in two ways: the eluting compound is collected and analysed off-line; or the chromatograph is connected directly to the mass spectrometer and the mass spectra are acquired while the compounds of the mixture are eluted. The latter method operates on-line. Reviews on the coupling of separation techniques with mass spectrometry have been published in the last few years [1–4].

The most obvious advantage drawn from coupling a separation technique with mass spectrometry consists of obtaining a spectrum used for identifying the isolated product. However, this is not the only goal that can be reached. The ideal detector should:

1. Have no alteration of the chromatographic resolution, which means not producing within the detector a mixture of the products separated before the detection.

2. Have the highest possible sensitivity.

3. Be universal, which means to detect all of the eluted products.

4. Furnish the maximum structural information possible, at best to allow positive identification of all the eluted compounds.

5. Be selective, i.e. to allow the identification of target products in a mixture. This characteristic is automatically present if item 4 is verified.

6. Give out a signal proportional to the concentration.

7. Have a constant, or at least predictable, response factor.

8. Have as low as possible cost/performance ratio.

9. Not be harmful to the product.

10. Not produce artefacts.

11. Allow the deconvolution of chromatographic peaks, i.e. the decomposition of unresolved peaks into constituents.

Mass Spectrometry: Principles and Applications, Third Edition Edmond de Hoffmann and Vincent Stroobant
© Copyright 2007, John Wiley & Sons Ltd

The importance of this last characteristic comes out of the statistical analysis of the probability of having two products within one chromatographic peak. Consider a mixture of N constituents. We will define the resolution, R, or the peak capacity in chromatography as being the maximum number of distinct compounds practically distinguishable. So, if a chromatogram is registered on a sheet of paper 1 m long and if we can consider that two compounds are detected separately when the distance between the peak centroids is at least 1 mm, we will be able to detect a maximum of 1000 compounds. This example supposes a constant peak width, which is very rare. According to our definition, we will say that the resolution $R = 1000$. If p is the probability of not having more than one compound per peak, it can be shown that these values are linked by the equation [5, 6]:

$$N = (2R \ln 1/p)^{1/2}$$

Let us calculate the maximum number N of compounds acceptable if we require $p = 90\%$, i.e. 9 chances out of 10 to have only one compound per peak, in the case of a resolution of 1000. We find $N = 14$. Of these 14 peaks, statistically 1 out of 10 will contain two unseparated compounds. This example shows that the statistical probability of having two compounds per peak is very high. Figure 5.1 shows these results graphically, from a more elaborate statistical analysis [7].

This statistical result might surprise users of 'classical' chromatographic methods, but it does correspond to reality for those who use GC/MS, LC/MS or CE/MS.

5.1 Elution Chromatography Coupling Techniques

The separation methods furnish a given flow of eluate (liquid or gaseous) generally under atmospheric pressure. One way or another, the eluted substances must find their way into the mass spectrometer source, where a high vacuum is necessary. We saw in the Introduction that the mean free path could be evaluated according to the following equation, where the pressure p is expressed in Pa and the distance L is expressed in centimeters:

$$L = \frac{0.66}{p}$$

A pressure of about 10^{-4} Torr (0.01 Pa or 100 nbar or 10^{-7} atm) ensures a mean free path of about 50 cm, and constitutes the upper limit of tolerable pressure for the source. Under this pressure, 1 cm^3 of gas under atmospheric pressure becomes 10^7 cm^3. A flow of 1 cm^3 min^{-1} thus requires a pumping flow of 10^7 cm^3 min^{-1} under this pressure, i.e. 166 l s^{-1}; we thus obtain an approximation of the pumping capacity required by this chromatographic flow.

If the eluent is liquid, 1 mol will yield about 24 l of gas under atmospheric pressure. To gain an idea, suppose the eluent is water and its flow rate is 0.1 cm^3 min^{-1}. A gas flow of $(0.1/18) \times 25\,000 = 139$ cm^3 min^{-1} under atmospheric pressure will be observed, which would require a pump flow at the source of $139 \times 166 = 23\,000$ l s^{-1}. This flow is exceedingly high: the eluate from a liquid chromatograph cannot be entirely evaporated in the source. We will examine the existing solutions to this problem.

In fact the actual capacities of pumps in mass spectrometry range from 50 to 1000 l s^{-1}, so that the maximum gas flow hovers around 5 cm^3 min^{-1} under 1 atm pressure.

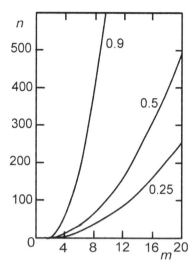

Figure 5.1
Graphical representation of the probability that there be more than one compound per peak. The peak capacity is along the n axis. The m axis is the number of acceptable constituents that have the probability indicated next to each curve of having only one compound per peak. The statistical analysis used to produce this graph is slightly different from that leading to the equation in the text. Redrawn from Martin M., Herman D.P. and Guiochon G., Anal. Chem., 58, 2200, 1986, with permission.

5.1.1 Gas Chromatography/Mass Spectrometry

Packed columns yield a carrier gas flow of about $20\,cm^3\,min^{-1}$, whereas capillary columns achieve only about $1\,cm^3\,min^{-1}$.

In mass spectrometry, filled columns yield a higher flow than is tolerated in the source, whereas the capillary columns fall within the acceptable values, which simplifies the gas flow problem.

The sample concentration in the carrier gas is also an important variable. If Q is the quantity in micrograms of a compound with molecular mass M injected into the column, d is the gas flow rate at the exit of the column in $cm^3\,min^{-1}$ and l is the peak mid-height width expressed in seconds, a good approximation for the average concentration is the ratio between the injected quantity converted into a corresponding gas volume and the volume of the carrier gas that passes in a length of time l. As the gas volume of 1 mol under atmospheric pressure is about $25\,000\,cm^3$ at ordinary temperatures, the volume V_e

corresponding to Q/M μmol injected sample is

$$V_e = \left(\frac{Q}{M}\right) \times 25\,000 \times 10^{-6}$$

Hence, for 10 μg injected and a molecular weight of 100 Da, we have 0.0025 cm^3. If the chromatographic peak width at mid-height is 10 s, the volume V_p of carrier gas for a flow rate of 20 cm^3 min^{-1} will be

$$V_p = \left(\frac{20}{60}\right) \times 10 = 3\,\text{cm}^3$$

We see that, under these conditions, the eluted sample represents less than one-thousandth of the total volume. This volume ratio is independent of the pressure and temperature. Note that an increase in the chromatographic resolution means a smaller width for the peak, and thus an improvement in this ratio. It is clearly more favourable with a capillary column. Obviously, improving this ratio before introduction into the mass spectrometer is very interesting. The ionization of the carrier gas must also be avoided, as it induces focusing difficulties and, in sources with high potential differences, discharges. Consequently, helium is generally chosen.

Two types of coupling interface between the gas-phase chromatograph and the mass spectrometer will be examined: open coupling and direct coupling [8, 9].

Interfaces that allow enrichment of the carrier gas with the eluted substance also exist. They are important if packed columns are used. However, they are now rarely used and will not be described here.

5.1.1.1 Open Coupling

An open-split coupling is shown in Figure 5.2. The chromatographic column leads to a T-shaped tube that contains a smaller diameter tube. A platinum or deactivated fused-silica capillary also leads to this tube and goes into the mass spectrometer source. The capillary is kept in a vacuum-sealed device and is heated in order to avoid condensation. The T-shaped tube is closed at both extremities but is not sealed, so that the pressure remains equal to atmospheric pressure. A helium current passes through it in order to avoid any contact with atmospheric oxygen which could oxidize the eluted compounds. The tube that enters the mass spectrometer has a diameter and a length that are chosen to furnish into the source a flow close to the acceptable maximum with respect to the gaseous conductance of this source and the pumping capacity. Thus, a capillary of 0.15 mm in diameter and 50 cm long

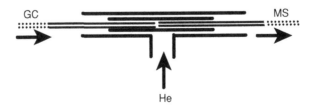

Figure 5.2
Open-split coupling. The helium current protects the eluent from air oxidation.

heated to 250 °C carries 2.5 ml min^{-1} of eluted gas into the source. This is sufficient in practice to pump everything that comes out of a capillary column.

This coupling does not cause any enrichment of the eluted compounds. It allows work under the usual chromatographic conditions, one end of the column being under atmospheric pressure. It does not require any special setting and changing the column is very easy. In principle, this system can be used with any type of column so long as no enrichment is necessary.

5.1.1.2 Direct Coupling

This coupling consists of having the capillary column directly enter the spectrometer source by a set of vacuum-sealed joints. Its yield can only be 100 %. The pumping is not problematic, for the capillary is necessarily very long. A length of at least 15 m is necessary for a column with an inside diameter of 0.25 mm.

Compared with the open coupling, this has the inconvenience of not allowing the elimination of the solvent and making the change of the column complicated. This last inconvenience can be limited by connecting one end of the chromatographic column to a glass capillary that enters the source through a Teflon pad pierced with an opening such that the junction is sealed.

The chromatography is carried out between atmospheric pressure at the injector and a vacuum at the other end of the column, provided the column is long enough. We saw earlier that no major inconvenience results from this, except for the comparison of chromatographic traces.

5.1.2 Liquid Chromatography/Mass Spectrometry

The coupling of liquid chromatography is more delicate because gas-phase ions must be produced for mass spectrometry. Liquid chromatography normally is used for compounds that are not volatile and are not suitable for gas chromatography.

The problem is made even more difficult by the need to eliminate the elution solvent. Suppose, for example, that this solvent is water and that the column being used has a very small diameter that allows the flow rate to be limited to 0.1 ml min^{-1}. This corresponds to 0.1 g min^{-1} of water or 5.6 mmol min^{-1}, which yields a flow rate of 135 cm^3 min^{-1} of gas at atmospheric pressure, which is still too much to be injected in a source under vacuum.

Different methods are used to tackle these problems [10–13]. Some of these coupling methods, such as moving-belt coupling or the particle beam (PB) interface, are based on the selective vaporization of the elution solvent before it enters the spectrometer source. Other methods such as direct liquid introduction (DLI) [14] or continuous flow FAB (CF-FAB) rely on reducing the flow of the liquid that is introduced into the interface in order to obtain a flow that can be directly pumped into the source. In order to achieve this it must be reduced to one-twentieth of the value calculated above, that is 5 µl min^{-1}. These flows are obtained from HPLC capillary columns or from a flow split at the outlet of classical HPLC columns. Finally, a series of HPLC/MS coupling methods such as thermospray (TSP), electrospray (ESI), atmospheric pressure chemical ionization (APCI) and atmospheric pressure photoionization (APPI) can tolerate flow rates of about 1 ml min^{-1} without requiring a flow split. Introducing the eluent entirely into the interface increases the detection sensitivity of these methods. ESI can accept flow rates from 10 nl min^{-1} levels to

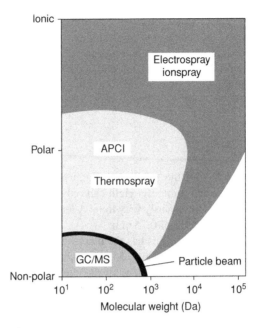

Figure 5.3
Application range of different coupling methods.

$0.2\,ml\,min^{-1}$. As it is more dependent on concentration than on the total quantity injected, there is little advantage of using it at high flow rates. Figure 5.3 shows the application range of different chromatography coupling methods (except DLI and CF-FAB) as a function of the mass and nature of the eluted compound. Figure 5.4 shows an example of an application of HPLC/APCI coupling [15].

5.1.2.1 Particle Beam Interface Coupling

The particle beam interface [16, 17] is a device that allows the solvent to be separated rapidly and efficiently from the eluted molecules issuing from a liquid chromatographic column. The chromatographic eluate is pumped through the capillary up to a glass concentric nebulizer (Figure 5.5). The eluate then is transformed into a cloud of droplets that are scattered by a helium concentric flow. The spray passes through a desolvation chamber, the walls of which are heated and the pressure of which is slightly lower than atmospheric pressure. During this travel the droplets undergo partial desolvation and yield slightly solvated eluted compound droplets.

   The desolvation chamber is linked with a double stage pumping molecular beam separator. When it leaves this evaporation chamber the mixture of helium, solvent vapour and particles undergoes a supersonic expansion into the first pumping stage (about 10 Torr). A high-speed gas beam containing particles of the eluted molecules results from this. As the mass difference between the gas molecules and the particles is important, the particles diffuse less rapidly from the centre of the beam towards the periphery. The separation is achieved by skimming the peripheral layers of the beam with a skimmer.

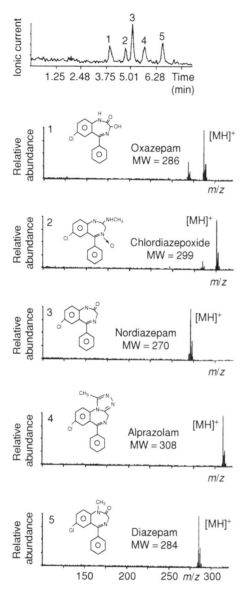

Figure 5.4
HPLC/APCI separation of a mixture of five benzodiazepines. A 1.2 ml min⁻¹ CH₃CN–CH₃OH–H₂O (40:25:35 v/v) flow is used for the analysis. Reproduced (modified) from Huang E.C., Wachs T., Conboy J.J. and Henion J.D., Anal. Chem., 62, 713A, 1990, with permission.

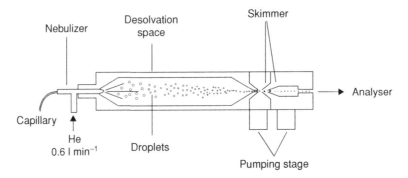

Figure 5.5
Diagram of a particle beam interface.

The skimmed solvent vapour and the helium are mechanically pumped out of the apparatus. This process is repeated in the second pumping stage (about 500 mTorr). Finally, the flow enriched in particles consisting of a narrow particle beam with a diameter smaller than 100 nm is sent into the mass spectrometer source without disturbing the vacuum. In the EI or CI mode, the particles injected into the source are rapidly vaporized before ionization. In the FAB mode, the particle beam is directed onto an FAB nozzle covered by a matrix. Thus the particles collide with the matrix surface and are trapped in it.

5.1.2.2 Continuous Flow FAB Coupling

The continuous flow FAB coupling invented by Caprioli et al. [18, 19] consists of linking the end of a capillary chromatographic column with the end of an FAB nozzle by a capillary passing through the introduction nozzle.

From 1 to 5 % glycerol is added to the chromatographic solvent. The flow rate varies from 1 to 5 μl min^{-1}, which is pumped easily. The solvent evaporates but the glycerol stays at the nozzle surface with the eluted compounds and serves as a FAB matrix. Figure 5.6 shows a diagram of such a coupling. Figure 5.7 gives an application example [20].

5.1.2.3 Coupling Using Atmospheric Pressure Ionization Sources

ESI, APCI and APPI sources were described in Chapter 1. As these ionization methods require the analysed sample in solution, they are ideal for coupling liquid chromatography with mass spectrometry. Modern sources are versatile. These sources can tolerate flow rates up to 1 ml min^{-1}, allowing an easy coupling to a classical HPLC system. With minor modifications, they can also generate results with flow rates down to 10 μl min^{-1}. APCI and APPI are usually used at higher flow rates than ESI. Optimum conditions are at a few hundred microlitres per minute. However APPI is still well suited for flow rates down to 0.1 ml min^{-1}, where APCI sensitivity is reduced. For ESI, a flow rate at 1 ml min^{-1} or even higher is technically possible, but may cause a reduction in the signal-to-noise ratio. The best sensitivity is achieved at low flow rate and the tolerated flow can go down to about 100 nl min^{-1}, with the nano-ESI version, allowing the coupling with HPLC using capillary or nano columns. Capillary electrophoresis can be coupled too, as will be discussed next.

As is the case for most sources, APCI and APPI ionization sources are sensitive to the total quantity of sample injected and not to its concentration. So, these ionization

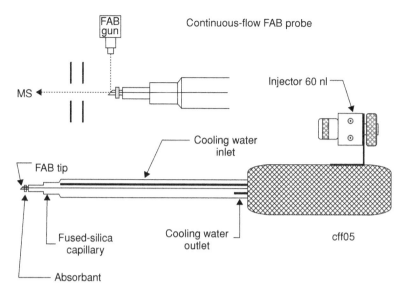

Figure 5.6
Diagram of a continuous flow FAB coupling system. The HPLC capillary column made of fused silica enters through the nozzle of an introduction FAB probe. The solvent evaporates and the glycerol remains as a FAB matrix.

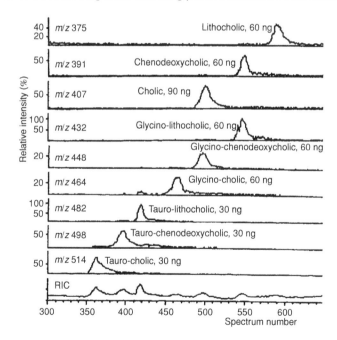

Figure 5.7
Example of an application of continuous flow FAB and capillary column liquid chromatography. One microlitre containing the indicated quantities of bile salts was injected. The total chromatography time is 11 min. Methanol gradient in water from 50 to 100 %, plus 1 % glycerol.

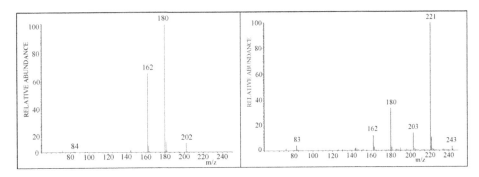

Figure 5.8
Spectra obtained from underivatized glucosamine (M) with two elution solvents. Left: 50/50 water/methanol. Only peaks from the glucosamine are observed: 180 $[M + H]^+$, 202 $[M + Na]^+$, 162 $[M - H_2O]^+$. Right: 50/50 water/acetonitrile (Acn). The same glucosamine peaks are observed at m/z 180 and 162, but abundant peaks are observed at 203 $[M - H_2O + Acn + H]^+$, 221 $[M + Acn + H]^+$, 243 $[M + Acn + Na]^+$ and 83 $[2 Acn + H]^+$. Private communication from ing. A. Spote.

techniques are mass, not concentration, sensitive detectors. On the other hand, ESI is a concentration sensitive detector. Thus, there is no advantage in using high flow rates. This feature explains why ESI coupled with nano and capillary HPLC gives a large gain in sensitivity. The sensitivity of the ESI source to concentration, and not to the total quantity of sample injected in the source as is the case for most other sources, has already been discussed in Chapter 1 (Section 1.11.3).

Using a mass spectrometer as a detector puts some constraints on the methods used. These constraints are different from what is to be considered with other detectors. Indeed, the methods used when the HPLC is coupled with a UV detector are not always compatible with these various ionization sources. Significant changes to LC methods are often required to adapt them to HPLC/MS even if modern sources are more tolerant and require less and less adjustments. Buffers and pH modifiers should be preferentially volatile. Operating the mass spectrometer with non-volatile buffers such as phosphate is technically possible, but the salt deposit in the source will have to be removed periodically. It is recommended to use formic acid, acetic acid, ammoniac, triethylamine, carbonates, ammonium formate, ammonium acetate, ammonium carbonate. Buffer concentration must be kept as low as possible. Ion pairing reagents and surfactants must be avoided because they perturb the ionization process and induce ion suppression. It should not be forgotten that ESI does not tolerate high salt concentrations or non-polar solvents. In the same way, trifluoroacetic acid largely decreases the analyte response.

The choice of the elution solvent can have a marked effect on the spectrum obtained. As an example, Figure 5.8 displays the spectra obtained from underivatized glucosamine when 50/50 water/methanol is used, or 50/50 water/acetonitrile. The spectrum with acetonitrile shows that this solvent can compete with glucosamine for the proton, while methanol is not sufficiently basic. Better sensitivity and linearity is obtained with methanol than with acetonitrile, even if actually acetonitrile gives a better chromatographic performance.

ESI coupling can be applied to organic as well as to inorganic compounds. As an example, cisplatin (I) is a drug used in cancer therapy. By hydrolysis, chlorine atoms can be replaced by a hydroxyl group, yielding toxic metabolites. Figure 5.9 describes the HPLC/MS and HPLC/MS/MS analysis of cisplatin and its mono and dihydroxy analogues

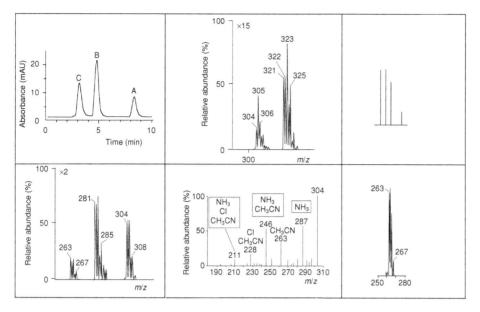

Figure 5.9
The HPLC ESI/MS and MS/MS analysis of cisplatin and its hydroxo metabolites. Top left: HPLC trace using porous graphite column. Top centre: ESI spectrum from HPLC peak A compound. Ion at m/z 321 corresponds to the sodium adduct of cisplatin. Top right: relative abundances of the platinum isotopes, without the contribution of chlorine atoms. Bottom left: spectrum corresponding to HPLC peak B. The m/z observed is difficult to interpret. However, the bottom centre spectrum displays the MS/MS fragments from m/z 304. It clearly shows that the hydroxy group of monohydroxo-cisplatin has been replaced by an acetonitrile molecule. Bottom right: molecular mass region of the spectrum of peak C. Ion at m/z 263 corresponds to protonated dihydroxo-cisplatin. Redrawn (modified) from Ehrsson H.C., Wallin I.B., Andersson A. S. and Edlund P.O., Anal. Chem., 67, 3608–3611, 1995, with permission.

[21]. The platinum has a complicated isotope pattern as shown. The spectrum corresponding to the peak A in the HPLC trace can be interpreted easily as corresponding to cisplatin. Indeed, its molecular weight calculated from the first isotopes of both platinum and chlorine is $194 + (2 \times 35) + (2 \times 17) = 298$ u. Adding 23 u for a sodium ion adduct leads to the observed m/z 321 peak, accompanied by the expected isotopes.

The spectrum from the compound of the HPLC peak B (Figure 5.9) is not straightforward to interpret. However, the MS/MS fragmentation of m/z 304 shows that acetonitrile (41 u) is incorporated and has replaced the hydroxyl group of monohydroxo-cisplatin. Indeed, the calculated mass $194 + 35 + (2 \times 17) + 41 = 304$ corresponds to the observed one. The spectrum of HPLC peak C is easy to interpret. It corresponds to protonated dihydroxo-cisplatin.

H₃N Cl	H₃N OH	H₃N OH
Pt	Pt	Pt
H₃N Cl	H₃N Cl	H₃N OH
A: Cisplatin	B: Monohydroxo-cisplatin	C: Dihydroxo-cisplatin

5.1.3 Capillary Electrophoresis/Mass Spectrometry

Capillary electrophoresis (CE) is a family of related techniques that employ narrow-bore (20–200 µm internal diameter) capillaries to perform separation of both large and small analytes. These separations are obtained by the use of high voltages that generate respectively electro-osmotic and electrophoretic flow of buffer solutions and ionic species. The coupling of capillary electrophoresis can be performed using ESI. The nanospray version yields the best results. Indeed, the flows produced by capillary electrophoresis are typically in the range of some hundreds of nanolitres per minute. This requires a dilution of the sample to adjust the flow before introduction into the conventional ESI source. The main advantages of capillary electrophoresis, namely speed, high resolution and small volumes of samples, are saved by the use of nano-ESI/MS. This separation technique allows analysis in aqueous solutions that are complementary to classical techniques such as HPLC/MS. The scheme of such a coupling is represented in Figure 5.10.

The interfacing of CE with ESI mass spectrometry has progressed substantially in recent years. On-line CE/MS has been widely used for both qualitative and quantitative analysis of many chemically diverse molecules. This method becomes a useful and sensitive analytical tool for the separation, quantification and identification of biological, therapeutic, environmental and other important classes of chemical analytes. The developments of CE/MS have been reviewed [22–26].

Figure 5.11 displays the analysis of a mixture of peptides from ovalbumin obtained by CE/MS coupled to ESI. The spectrum shown is a mean of the spectra acquired during the elution of the indicated broad peak. It corresponds to a mixture of doubly or triply charged ions of several glycopeptides [27]. The displayed structures were actually deduced from MS/MS fragmentation spectra of these multiply charged ions.

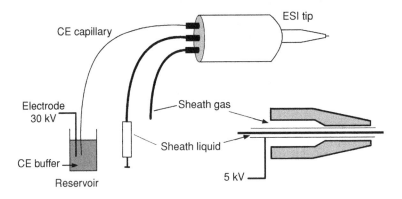

Figure 5.10
Scheme of a capillary zone electrophoresis coupling device to a mass spectrometer.

5.2 Chromatography Data Acquisition Modes

No matter which ionization technique or chromatographic method is used, three acquisition modes exist: scanning, selected-ion monitoring (SIM) (not to be mistaken with SIMS, which means secondary ion mass spectrometry) and selected-reaction monitoring (SRM).

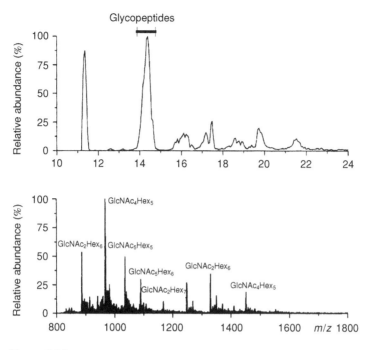

Figure 5.11
Chromatographic trace obtained by capillary electrophoresis and spectrum of the mixture of doubly or triply charged glycopeptides eluted together in the peak. Reproduced (modified) from Kelly J.F., Ramaley L. and Thibault P., Anal. Chem., 69, 51–60, 1997, with permission.

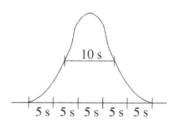

In the scan mode, complete spectra are repeatedly measured between two extreme masses. Suppose that the time width of the chromatographic peak is 10 s at mid-height. As is shown schematically in the diagram, at least one spectrum must be measured every 5 s in order to ascertain that one of them lies entirely within the chromatographic peak. Suppose the scan covers masses 50–550 Da at low resolution. A short calculation shows that 0.01 s per mass (500 in 5 s) can be used. All of the ions reaching the detector during that time span are counted. Increasing this time span increases the sensitivity, as the total number of counted ions increases. To increase the sensitivity, either the range of masses scanned is decreased, or the scan time is increased. In the first case, analytical information may be lost, and in the second case, a good mass spectrum may not be obtained, or two eluted compounds in the same peak may not be deconvoluted (for deconvolution, see below).

If the analysis aims at detecting target compounds, with known spectral characteristics, with a maximum sensitivity, SIM is useful. Thus, if we choose to detect a given compound by monitoring three characteristic fragments, we switch the analyser rapidly from one mass to another. The sensitivity gain can be enormous. In fact, integrating over three masses during 5 s means that 1.66 s per mass is available instead of 0.01 s as was the case before.

The SRM acquisition mode allows one to obtain a sensitivity and selectivity gain with respect to SIM. The detection of selected reactions, based on the decomposition reactions of ions that are characteristic of the compounds to be analysed, requires the use of a tandem mass spectrometric instrument. In order to carry out this type of analysis, the spectrometer is set so as to let through only the ions produced by a decomposition reaction in the chosen reaction region: for example, the first spectrometer selects the precursor ion with an m_p^+/z ratio that is characteristic of the compound to be detected, while the second spectrometer selects the fragment ion with an m_f^+/z ratio resulting from the characteristic decomposition reaction of the compound to be analysed, $m_p^+ \rightarrow m_f^+ + m_n$, that occurs between the two analysers.

The selectivity gain results from the fact that the fragmentation reaction implies two different characteristics of the compound under study. In fact, ions must meet two conditions in order to reach the detector: the m/z ratios both of the product ions and of the precursor ions must satisfy the selected values. The sensitivity gain results from the greater signal-to-noise ratio that is characteristic of tandem mass spectrometry; the gain is high in spite of the weak ion transmission through the instrument.

Figure 5.12 gives an example of SRM. The problem is to detect a contaminant in a fuel sample. In the total ion reconstructed chromatogram, which corresponds to a standard GC trace, the compound cannot be seen at its retention time. If, however, selected fragmentations are monitored through detection of MS/MS product ions, the only peak observed is that of the impurity, at the expected retention time. The instrument used here is a GCQ ion trap (Finnigan MAT).

5.3 Data Recording and Treatment

One can no longer imagine performing GC/MS or HPLC/MS without an on-line data system. Such a system includes an acquisition processor, a magnetic recording device and a computer with its accessories.

5.3.1 Data Recording

The spectrometer provides two series of data as a function of time: the number of detected ions and, simultaneously, a physical value indicating the mass of these ions. This can be, for example, the value of the magnetic field measured by a Hall probe in the case of a magnetic instrument. The ions of every mass appear with a certain distribution over a time period, as shown in the sketch. The surface under the curve is proportional to the number of detected ions, while the value of the magnetic field at the centroid of the peak is an indicator of the ion mass. This centroid and mass must therefore be determined. In order to achieve

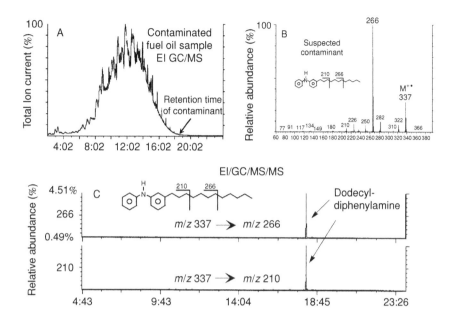

Figure 5.12
(A) Total ion chromatographic trace of a fuel oil. (B) Spectrum of the sought contaminant. (C) SRM spectrum of the two main reactions: the sought compound is the only one detected. Reprinted (modified) from Finnigan MAT documentation, with permission.

this, the acquisition processor accumulates the signal related to the number of ions during a short time with respect to the peak width. Usually, it samples eight times within the peak width.

As an illustration, consider this example. The spectrometer covers a range of 500 mass units within 1 s, that is 1 mass within 2 ms. During that time, the processor must carry out about eight measurements of the number of ions; 0.25 ms is allotted for every sample. In other words, it must measure 4000 samples per second. Its sampling frequency is said to be 4 kHz.

In order to achieve this, the current from the ion detector goes through a resistance 4000 times a second, and the acquisition processor reads the potential difference at the resistance ends, which is proportional to the number of detected ions, and digitizes it. This value goes to the y axis of the mass spectrum. This signal is then associated with the simultaneous reading of the mass indicator, which yields the x axis value. This gives a profile spectrum. To convert it into a bar graph, a suitable algorithm allows the processor to determine the limits of the peak and its centroid. The sum of the values read within these limits during the condenser discharges is proportional to the number of ions, while the interpolation of the indicator value at the centroid yields the ion mass. A bar graph spectrum is obtained. This is illustrated in Figure 5.13.

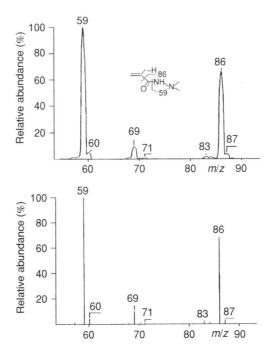

Figure 5.13
The same spectrum displayed on top as a profile
of ion abundances and below as a bar graph of the
centroids of the peaks.

If a larger mass range is scanned per unit time, or if a higher resolution is used, the
sampling speed must be increased, as the number of data points per unit time increases.

Another important characteristic of the acquisition processor is its dynamic range, which
is linked in part to the signal digitization possibilities. An example illustrates this. An
ion detector can typically detect between one and a million ions reaching the detector
simultaneously. Its own dynamic range, that is the ratio of the largest to the smallest
detectable signal, is thus equal to 10^6. If the ADC uses 16 bits, a usual value, it can yield
numerical values between 1 and $2^{16} - 1$, that is 65 535. Its dynamic range is thus much
lower than that of the detector itself. This problem can be overcome by reading different
value ranges alternately.

5.3.2 Instrument Control and Treatment of Results

With commercially available programs, the computer can carry out the following operations.

5.3.2.1 Data Acquisition Management

The acquisition program allows the operator to choose the scan mode or the SIM mode, the
range of masses that are scanned, either low or high resolution, and so on. The acquisition
processor is set according to the data given by the operator concerning the analysis to be
carried out. Most recent instruments are entirely electronically controlled from parameters

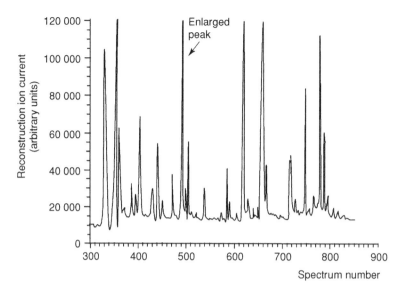

Figure 5.14
Chromatographic trace reconstructed by summing the intensities of all of the detected masses in every spectrum, and used as arbitrary units along the *y* axis. The *x* axis is the spectrum number. This is an analysis of volatile acids as trimethylsilylated derivatives from the urine of a patient suffering a metabolic disorder. If asked, the computer can indicate the retention time of every peak. The peak that is pointed out is enlarged in Figure 5.15.

provided by the operator. Instruments with automatic injectors can thus carry out several successive chromatographic analyses without any intervention by the operator.

Expert systems are available on recent instruments (from about 1990). They allow one to program the instrument tuning modifications or changes in the type of measurement to be carried out, based on a computer investigation of the data. For example, 'if an ion with a given *m/z* value is detected at a set retention time, then measure the spectrum in a negative ion mode, then come back to starting mode'. These possibilities are now revolutionizing mass spectrometry.

5.3.2.2 Interpretation of Results

An interactive program that lets the operator intervene and modify the parameters at any time allows the investigation of the chromatographic results. Usually, it allows the following operations:

1. To show the reconstructed ion chromatogram (RIC) from the sum of the intensities of all the detected ions. Some ions present in the background noise can sometimes be excluded. The spectrum numbers corresponding to the chromatographic peaks are given on the *x* axis. Figure 5.14 shows an example. Retention times may be indicated.

2. The operator can use magnifying effects on this chromatographic trace to select part of the chromatographic trace and enlarge it over the whole screen, or amplify the

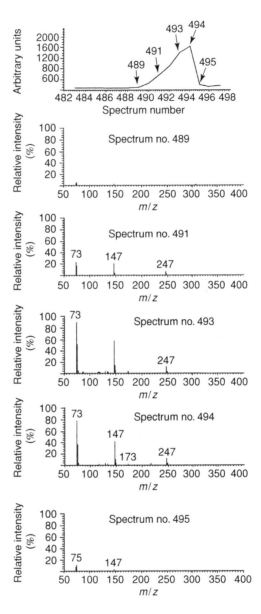

Figure 5.15
Enlarged picture of the chromatographic peak
pointed out in Figure 5.14 and a few spectra
measured during the elution time of that chromatographic peak. The TMS derivative of suc-
cinic acid can be recognized.

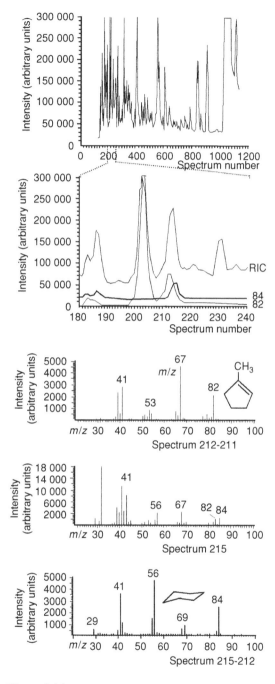

Figure 5.16
Deconvolution of a chromatographic peak containing two compounds. Spectrum 212, where the background noise was subtracted, is different from 215.
The two spectra are obtained by the appropriate subtraction.

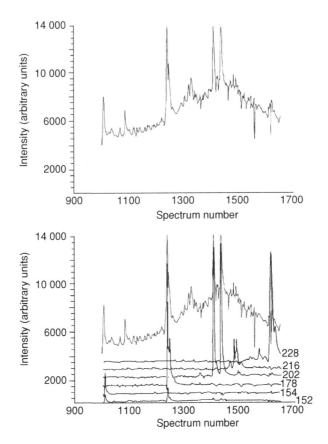

Figure 5.17
Top: chromatogram of compounds sampled over a discharge. The complexity of the mixture entails an envelope of the baseline. Bottom: selective search for polycyclic aromatics, which are important pollution agents. The following are detected: 228, benzanthracenes; 216, benzofluorenes; 202, fluoranthene and pyrene; 178, anthracene and phenanthrene. The next two are only weakly detected: acenaphthene and acenaphthylene, 154 and 152. These identifications must be confirmed by the retention times and by complete mass spectra.

chromatographic trace vertically to emphasize the low-intensity peaks. Figure 5.15 shows an enlarged version of the peak pointed out in the chromatogram in Figure 5.14 and every spectrum recorded during the elution of that peak. Only spectra 493 and 494 are of good quality.

3. To select one or several spectra to appear on the screen. Two spectra can thus be compared, one measured at the beginning of the elution of a chromatographic peak and the other at the end. If they are identical, the probability that the peak contains only one compound is high. However, if they are different, the peak contains more than one compound. Figure 5.16 shows an example of this. The spectra indicate the presence of two compounds with m/z 82 and 84. A spectrum can also be selected within

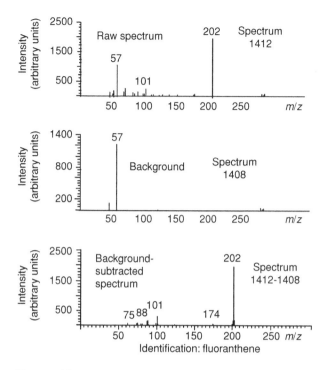

Figure 5.18
Same analysis as in Figure 5.17. Confirmation of fluoranthene identification by the full spectrum, after background subtraction.

the chromatographic peak and another, representing the background noise, at the peak edge. The second spectrum is then subtracted from the first.

4. To draw mass chromatograms. This consists of drawing a graph of the abundance of ions of a selected m/z value, or of a chosen sum of ion abundances, as a function of time. This process allows one to look through a chromatograph for the possible presence of known compounds by choosing the masses of typical ions. Target compounds can thus be pointed out within a chromatographic envelope where they cannot be identified solely from their retention times. Figure 5.17 shows an example.

The definitive identification will be the result of a combination of retention time, ion chromatogram and, as shown in Figure 5.18, the complete spectra.

Figure 5.19 shows the analysis of a hydrocarbon mixture obtained from the irradiation-induced dimerization of a mixture of n-$C_{10}H_{22}$ and n-$C_{10}D_{22}$ [28] as an example. The mixture is produced from the dehydrodimerization of hydrocarbons $C_{20}H_{42}$, $C_{20}H_{21}D_{21}$ and $C_{20}D_{42}$. The dimerization produces 15 eicosane isomers for every isotopic composition, that is 45 isomers in total, some of which also show diastereoisomers. These isomers are observed through chemical ionization with methane, which induces the abstraction of a hydride or of D^-. Hydrocarbons $C_{20}H_{42}$ are observed at 281 Th as $(M - H)^+$, $C_{20}H_{21}D_{21}$ at 302 or 301 Th corresponding to $(M - H)^+$ or $(M - D)^+$, and $C_{20}D_{42}$ at 322 Th as $(M - D)^+$. All of the 45 isomers are observed

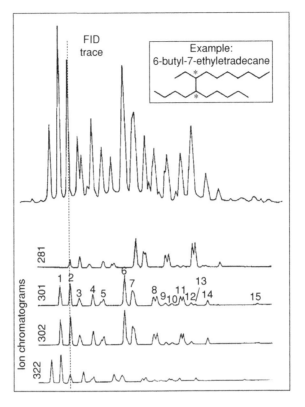

Figure 5.19
Top: GC trace with a flame ionization detector (FID).
One of the isomers is shown as an example. It has
two asymmetric carbon atoms. Bottom: ion chromatograms corresponding to $(C_{20}H_{42} - H)^+$ at 281 Th,
$(C_{20}H_{21}D_{21} - D)^+$ at 301 Th or $-H$ at 302 Th, and
$(C_{20}D_{42} - D)^+$ at 322 Th. Reprinted (modified) from de
Hoffmann E., Baudson Th. and Tilquin B., Journal High
Resolut. Chromatogr. Chromatogr. Commun., 10, 153,
1987, with permission.

in these ion chromatograms, while the FID trace, even though obtained at a very high
resolution, shows only 20 peaks.

5. To detect target compounds that are not observable in the chromatographic trace. Figure
    5.20 illustrates this. According to the retention time, TMS-phenol (TMS, trimethylsilyl)
    should appear in spectrum 342, but it is not visible on the trace covering this region.
    The following trace shows the intensity relative to two ions typical of TMS-phenol:
    151 (the most intense) and 166 (the molecular ion). Thus the phenol shows up! Its
    identification is confirmed by the spectra shown in Figure 5.21.

Such uses of mass chromatograms end up increasing the dynamic range of chromatography. They can also be used to increase the effective resolution, as is shown in the example
in Figure 5.19.

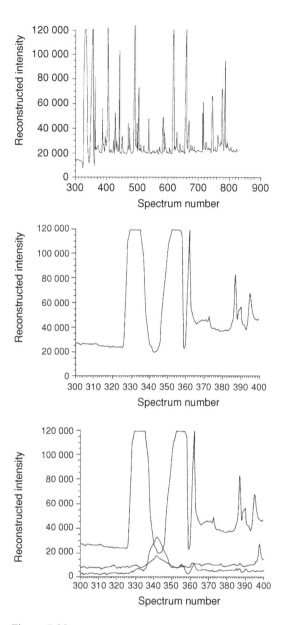

Figure 5.20
Top: complete chromatographic trace of volatile acids from the urine of a patient suffering from Wilson's disease. Middle: Trace showing part of the chromatograph ranging from spectra 300 to 400. Bottom: Trace showing the chromatograms of m/z 151 and 166 that are typical of TMS–phenol and thus indicate its presence.

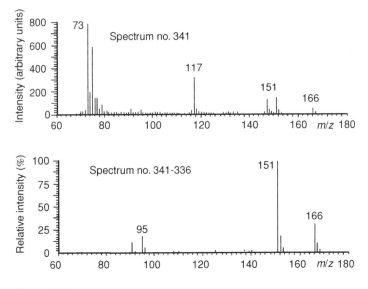

Figure 5.21
Proof of the presence of TMS–phenol that is invisible in the valley separating the two major peaks. The second spectrum results from the subtraction of the background noise.

5.3.2.3 Other Programs

The other programs that are usually available are described below.

Individual programs allow one to draw spectra with various formats, or to carry out comparisons of spectra, or to draw complete or mass chromatograms with two or three dimensions with various formats.

A spectrum subtraction program allows the substraction of one spectrum from another in order to eliminate the background noise or simply to emphasize differences between two spectra. An example is given in Figure 5.21.

Library search programs allow direct searches to determine whether or not the spectrum that is obtained is present in the collection, or to carry out inverse searches to determine if the spectrum of a given compound in a collection occurs among those recorded during a chromatographic analysis, for example. An example of a library search is given at the end of Chapter 3.

The elemental composition for a given mass can be calculated by limiting oneself to chemically acceptable formulae. Thus, for example, C_2H_5O and CHO_2 are acceptable for mass 45 Da but C_3H_9 is not. The calculation can be carried out with low or high resolution. In the first case, the number of possible formulae rapidly becomes too high, and is of interest only if limits can be ascribed to the number of atoms of each species.

The isotopic abundances for a given formula can be calculated and compared with experimental values.

Finally, utility programs allow one to extract some spectra from an analysis, to erase others, and so on.

References

1. Kostiainen, R., Kotiaho, T., Kuuranne, T. and Auriola, S. (2003) *J. Mass Spectrom.*, **38**, 357.
2. Gelpi, E. (2002) *J. Mass Spectrom.*, **37**, 241.
3. Klampfl, C.W. (2004) *J. Chromatogr. A*, **1040**, 131.
4. Mastovska, K. and Lehotay, S.J. (2003) *J. Chromatogr. A*, **1000**, 153.
5. Davis, J.M. and Giddings, J.C. (1983) *Anal. Chem.*, **55**, 418.
6. Hirschferd, T. (1980) *Anal. Chem.*, **52**, 297A.
7. Martin, M., Herman, D.P. and Guiochon, G. (1986) *Anal. Chem.*, **58**, 2200.
8. Kitson, F.G., Larsen, B.S. and McEwen, C.N. (1996) *Gas Chromatography and Mass Spectrometry: A Practical Guide*, Academic Press, New York.
9. McMaster, M.C. and McMaster, C. (1998) *GC/MS: A Practical User's Guide*, John Wiley & Sons, Inc., New York.
10. Ardey, R.E. (2003) *Liquid Chromatography–Mass Spectrometry: An Introduction*, John Wiley & Sons, Inc., New York.
11. Ardrey, B. (ed.) (1993) *Liquid Chromatography/Mass Spectrometry*, VCH, New York.
12. Willoughby, R., Sheehan, E. and Mitrovitch, S. (1998) *A Global View of LC/MS: How to Solve Your Most Challenging Analytical Problems*, Global View Publishing, Pittsburgh.
13. Niessen, W.M.A. (1998) *Liquid Chromatography–Mass Spectrometry*, Marcel Dekker, New York.
14. Arpino, P.J., Baldwin, M.A. and McLafferty, F.W. (1974) *Biomed. Mass Spectrom.*, **1**, 80.
15. Huang, E.C., Wachs, T., Conboy, J.J. and Henion, J.D. (1990) *Anal. Chem.*, **62**, 713A.
16. Greene, F.T. (1975) *Proceedings of the 23rd ASMS Conference*, Houston, TX, p. 695.
17. Wiloughby, R.C. and Browner, R.F. (1984) *Anal. Chem.*, **56**, 2626.
18. Caprioli, R.M., Fan, T. and Cottrell, J.S. (1986) *Anal. Chem.*, **58**, 2949.
19. Caprioli, R.M. (1990) *Continuous Flow Fast Atom Bombardment Mass Spectrometry*, John Wiley & Sons, Inc., New York.
20. Van Hoof, F., Libert, R., Hermans, D. *et al.* (1990) *Adrenoleukodystrophy and Other Peroxysomal Disorders* (eds G. Uziel, R. Wanders and M. Cappa), Excerpta Medica, Amsterdam, p. 52.
21. Ehrsson, H.C., Wallin, I.B., Andersson, A.S. and Edlund, P.O. (1995) Cisplatin, transplatin, and their hydrated complexes: separation and identification using porous graphitic carbon and electrospray ionisation mass spectrometry. *Anal. Chem.*, **67**, 3608–11.
22. Niessen, W.M.A., Tjaden, U.R. and Van Dergreef (1993) Capillary electrophoresis mass spectrometry. *J. Chromatogr.*, **636** (1), 3–19.
23. Issaq, H.J., Janini, G.M., Chan, K.C. and El Rassi, L. (1995) *Adv. Chromatogr.*, **35**, 101.
24. Tomlinson, A.J., Guzman, N.A. and Naylor, S. (1995) Enhancement of concentration limits of detection in CE and CEMS: a review of on-line sample extraction, cleanup, analyte preconcentration, and microreactor technology. *J. Capillary Electrophor.*, **2** (6), 247–66.
25. Figeys, D. and Aebersold, R. (1998) High sensitivity analysis of proteins and peptides by capillary electrophoresis tandem mass spectrometry: recent developments in technology and applications. *Electrophoresis*, **19** (6), 885–92.
26. Banks, J.F. (1997) Recent advances in capillary electrophoresis electrospray mass spectrometry. *Electrophoresis*, **18** (12–13), 2255–66.
27. Kelly, J. F., Ramaley, L. and Thibault, P. (1997) Capillary zone electrophoresis-electrospray mass spectrometry at submicroliter flow rates: practical considerations and analytical performance. *Anal. Chem*, **69** (1), 51–60.
28. de Hoffmann, E., Baudson, T. and Tilquin, B. (1987) *J. High Resolut. Chromatogr. Chromatogr. Commun.*, **10**, 153.

6
Analytical Information

In order to analyse the spectrum of an unknown compound, we must proceed by steps, rather like in the quiz game – 'man or woman?', 'young or old?', and so on – asking increasingly precise questions, until the structure is deduced or until we give up.

First the origin of the sample, its history, is taken into account. This often allows the elimination of some hypotheses or narrowing of the research field. For example, a side product does not contain nitrogen if none of the reactants and none of the solvents contained it.

6.1 Mass Spectrometry Spectral Collections

The spectrum is examined first. Before starting any interpretation, it is strongly recommended that a computer or a manual library search is performed to check whether this spectrum belongs to an existing collection. Identification of an unknown compound in this way depends directly on the quality and comprehensiveness of the collection used. However, only libraries of electron ionization spectra are efficient. Other ionization techniques yield spectra that are much too dependent on the instruments and experimental conditions.

If the spectrum is found, the research is over. However, we must take into account the fact that close isomers often have identical mass spectra, and that sometimes very similar mass spectra belong to different compounds. We cannot blindly trust identification from a comparison.

Three main spectra collections now exist. The first is the NIST/EPA/NIH mass spectral database, which contains 190 000 spectra of 163 000 compounds [1]. This original collection of spectra and related information is produced by the National Institute of Standards and Technology (NIST) with the assistance of expert advisors from the Environmental Protection Agency (EPA) and National Institutes of Health (NIH). This library is available on CD-ROM for personal computers with integrated tools for GC/MS deconvolution, mass spectra interpretation and chemical substructure identification. This US government publication is very cheap and of very high quality. This library is widely spread in many commercial mass spectrometers. Mass spectra for over 15 000 compounds are accessible on-line [2].

McLafferty and Stauffer published in *The Wiley Registry of Mass Spectral Data* a collection of about 380 000 spectra of over 200 000 compounds [3]. It is the largest and most comprehensive library of reference spectra that contains over 180 000 searchable structures and over 2 million chemical names. The eighth edition of *The Wiley Registry of Mass Spectral Data* is available in electronic format and is compatible with the software of most instrument manufacturers and NIST MSSearch. Furthermore, it is accompanied by an interpretation help program called 'Probability Based Matching' (PBM).

Mass Spectrometry: Principles and Applications, Third Edition Edmond de Hoffmann and Vincent Stroobant
© Copyright 2007, John Wiley & Sons Ltd

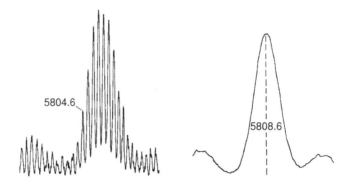

Figure 6.1
FAB spectrum of human insulin isotopic cluster ($C_{257}H_{383}$-$N_{65}O_{77}S_6$) at 6000 resolution (left) and at 500 resolution (right).

The *Eight-Peak Index of Mass Spectra* published by the Mass Spectrometry Data Centre of the Royal Society of Chemistry is a popular printed index of mass spectral data that now contains some 81 000 spectra of over 65 000 different compounds [4]. These spectra are published in the shape of lists of the eight main peaks. The complete data are sorted in three different ways to allow easy identification of unknown compounds: by (i) molecular weight subindexed on molecular formula, (ii) molecular weight subindexed on m/z value and (iii) m/z value of the two most intense ions.

Smaller specialized libraries of mass spectrometry also exist in hardcopy or electronic format: main polluting agents [5], environmental contaminants [6,7], drugs and metabolites [8], pharmaceutical products [9], and so on.

If the spectrum is not described in a library, the information contained in the spectrum is used. The mass spectrum furnishes various types of analytical information. The first is the molecular mass. The ionization techniques that are now accessible have made this information available for almost all compounds.

The molecular mass measured by mass spectrometry can correspond to the average mass calculated using the average atomic weight of each element of the molecule or to the monoisotopic mass calculated using the 'exact mass' of the most abundant isotope for each element. The type of mass measured by mass spectrometry depends largely on the resolution of the analyser. Indeed, if the instrument is unable to resolve the isotopes, the various peaks in the isotopic cluster combine and form a single peak that spreads over several masses. Thus, the mass determined by the instrument corresponds to the average mass. Sometimes it may be difficult to determine even the average mass with reasonable accuracy. On the other hand, if the resolution is high enough to distinguish the different peaks in the isotopic clusters, the mass determined by the instrument corresponds to the calculated monoisotopic mass. Furthermore, high-resolution mass spectrometry leads to very narrow peaks, allowing for greater accuracy. Figure 6.1 shows an example of how resolution affects the observed mass.

The second type of analytical information is the elemental composition. Indeed, measurement of the mass with sufficient accuracy provides an unequivocal determination of the total elemental composition. This method is often referred to as 'high resolution'.

However, this information is not always available. Other elements of the spectrum provide more information, even if it is only fragmentary, on the elemental composition: isotopic abundance, mass of small fragments or of lost neutrals.

Once the crude formula has been established, or is partially known, we try to define the structure from the fragmentation. This is the subject of the following chapters.

6.2 High Resolution

We will use here the definitions proposed by Marshall [10]:

- *Mass peak width* (Δm at 50 % height): FWHM of mass spectral peak.

- *Mass resolving power* ($m/\Delta m_{50\%}$): The observed mass divided by the mass peak width at 50 % height for a well-isolated single mass spectral peak.

- *Mass resolution* (either $m_2 - m_1$ in Da or $(m_2 - m_1)/m_1$ in ppm): The smallest difference between equal-magnitude peaks such that the valley between them is a specified fraction of either peak height.

- *Mass precision*: Root-mean-square deviation in a large number of repeated measurements.

- *Mass accuracy*: The difference between measured and actual mass.

- *Mass defect*: The difference between 'exact' and nominal mass.

Figure 6.2 illustrates the difference between precision and accuracy.

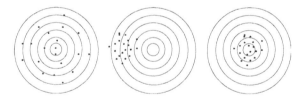

Figure 6.2
Left: low precision, but fair accuracy as the points are well distributed around the target. Centre: better precision, but bad accuracy. Right: same precision as centre, but much better accuracy.

Resolution and instrumental mass resolving power have already been discussed in Chapter 2 and shown in Figures 2.1–2.3. High-resolution instruments, including double focusing, FTICR and orbitrap mass spectrometers, are capable in principle of measuring the mass of an ion and its associated isotope peaks with sufficient accuracy to allow the determination of its elemental composition. This is possible because each element has a slightly different characteristic mass defect. The accurate mass measurement shows the total mass defect and provides a means to determine its elemental composition. However, the accuracy requirements in measuring mass and thus the resolution requirements increase very rapidly as the mass increases. Indeed, as the number of atoms increases, the number

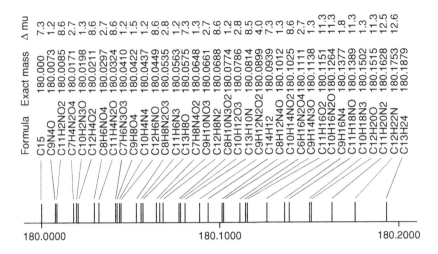

Figure 6.3
Exact masses and corresponding formulae for various possible ions of m/z 180 containing only carbon, hydrogen, nitrogen and oxygen atoms in limited number (C_{6-15}, H_{0-24}, N_{0-4} and O_{0-4}).

of possible combinations also increases, with mass differences that become increasingly small. For example, $C_{24}H_{19}N$ and $C_{21}H_{23}NS$, with masses of 321.1517 and 321.1551 u respectively, require an accuracy of 11 ppm for their distinction and a resolving power $m/\Delta m_{50\%}$ better than 94 400 for their separation. Only some commercial electromagnetic instruments, FTICR and orbitrap permit such a resolution. However, with measurement conditions that require scans of less than 0.3 s, such as with the coupling of a gas or liquid chromatograph, a resolving power of 10 000 is hardly achieved by the best instruments. For example, the resolving power of the orbitrap varies with the scan time from 7500 at 0.3 s to 60 000 at 1 s and to a maximum of 100 000 at about 2 s per full scan at m/z 500.

The elemental compositions of fragments with masses less than 100 u are easy to obtain and much information concerning the composition of the molecular ion can be deduced, at least if the spectrum is from a high-purity compound. Moreover, knowing the elemental composition of the fragments is of valuable help in elucidating the structure and understanding the fragmentation mechanism.

High resolving power also allows an increase in the selectivity of detection in analyses aimed at screening known target compounds.

Another example is illustrated in Figure 6.3. The possible compositions encompassing C_{6-15}, H_{0-24}, N_{0-4} and O_{0-4} are considered at the mass 180 u. There are no compositions with accurate masses closer together than 0.0012 u. Thus, for a measurement at m/z 180, an accuracy of 6 ppm would be required to eliminate all the possibilities and thus to define a particular elemental composition, within the fixed limits for the elements. The resolving power to separate two peaks with this mass difference would be more than 150 000. However, if only one pure compound is present, a lower resolution already allows the determination of the mass with this accuracy as will be discussed below. It should also be noted that if more elements have to be considered, for instance the introduction of sulfur atoms, the number of formulae dramatically increases and thus also the resolution or the mass accuracy needed to separate them or to distinguish between them.

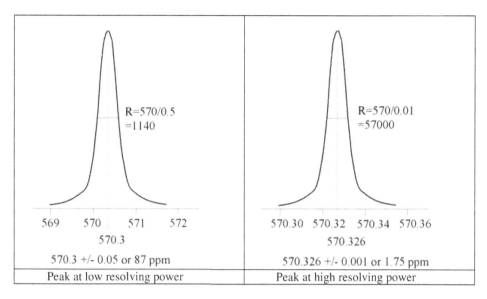

Figure 6.4
The centroid can be estimated with an accuracy of about a tenth of the width at half maximum. The reader can prove it using a ruler. Comparing detection at low (left) and high (right) resolving power clearly shows that at a higher resolution a better accuracy can be obtained. The peak displayed is a good-quality one: low background noise, good symmetry. Several factors can make the centroid less accurate, as shown in the next figure.

When the accurate mass measurement is used for confirmation of the formula or for elemental composition analysis, all the candidates fitting the experimentally determined value and its reportable uncertainty must be considered. Setting fixed acceptable error limits for mass measurement is not correct.

Sentences are often found meaning explicitly or implicitly '5 ppm accuracy is sufficient to deduce the elemental composition'. This is absolutely not true and is at the origin of many exaggerations. This value comes from a rule of the American Chemical Society that states: 'For most new compounds, HRMS data accurate within 5 ppm or combustion elemental analysis accurate within 0.4 % should be reported to support the molecular formula assignment' [11]. Thus, this rule does not tell that 5 ppm accuracy can be used to deduce an elemental composition, but that it can be used in *support of a proposed formula*, but not as a proof of that formula.

Second, the 'accuracy' notion has drifted. If a peak has a width Δm, the centroid at the maximum of the peak can be estimated with a lower divergence than $\Delta m/2$, as shown in Figure 6.4. Qualitatively, a divergence of $\Delta m/10$ can be achieved if the peak is of high quality. Figure 6.4 also shows that a spectrum of the same quality acquired with a higher resolving power leads to better accuracy in the estimation of the mass.

It is important to realize the difference between accurate mass determination and resolving power. Accurate mass can be determined at a quite low resolution but requires an isolated peak, and thus applies safely only on a high-quality spectrum of one pure compound. Risks of having mixed peaks, that is peaks of compounds having the same nominal mass but different elemental compositions, with pure compounds still exist for the

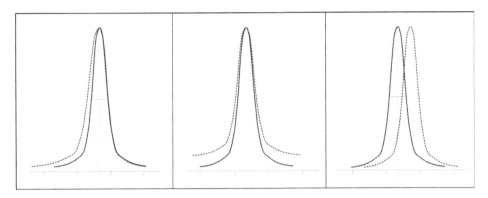

Figure 6.5
Dashed lines: experimental peak. Some problems alter the accuracy of the mass determination at the centroid. Actually, the experimental peak may contain several of these alterations. Furthermore, peaks with low noise on the signal are used here.

fragments. High resolving power not only increases the accuracy but also considerably reduces the risk of having mixed peaks for mixtures or fragment ions.

Another very important requirement for elucidating elemental composition is high-quality detection of the isotopic peaks, in both relative abundances and mass accuracies. This requires low background noise and a very good signal-to-noise ratio, or in other words good ion statistics or good dynamic range.

There are several causes of errors in the determination of an accurate mass, illustrated in Figure 6.5, from the centroid of the peak:

1. The peak is not symmetric. A first reason can be a defect of the spectrometer. In this case all the peaks in the same mass region should have a similar distortion. This can in some cases be corrected through the calibration. If not, a mathematical treatment of the peak can be performed if available in the software, or, better, another mass spectrometer can be used! Another reason may be that the peak actually contains more than one component with similar masses. This should only be observed on this peak. Purification of the sample or better direct chromatographic separation can be tried if it is believed that this is due to an impurity. Otherwise a higher resolving power should be used to separate the two masses.

2. There is an obvious background that results in the broadening of the peak, mainly near the bottom, and alters the accuracy of the centroid. Furthermore, background noise will reduce the detection limit, and make centroid determination on small peaks, for example isotopic peaks, difficult or impossible. This reduces the dynamic range, thus the ratio of abundance between the highest peak and the lowest one still well detected, leading to bad isotope detection. The background noise can result from a dirty source or instrument or from bleeding of a chromatographic column, and the remedy is obvious. Or it can result from poor ion statistics. The remedy can be to increase the number of ions injected, if the spectrometer has a sufficient ion capacity. The danger is to provoke space charge effects that will strongly reduce the resolution. Another possibility is to

increase the number of spectra averaged, if the acquisition time is not critical, as in chromatographic coupling. Otherwise there is no remedy except to use a better mass spectrometer.

3. With some instruments, the instability of the analyser is such that there is a drift in mass on relatively short times, say less than 1 day. When this drift is large, the use of an internal standard or a dual source to inject ions alternately from the sample and from the standard is needed. The first solution increases the number of ions in the analyser, and thus the importance of the space charge effect. The second increases the acquisition time. Frequent recalibration of the spectrometer may also be required. If the instrument is stable, a lock mass can be used either from a compound added to the sample or from a known ion from the solvent or the background.

6.2.1 Information at Different Resolving Powers

Different resolving powers of the mass spectrometer allow various levels of information to be obtained. Remember that for all mass spectrometers the resolving power does not vary in the same way for different masses:

- Quadrupoles and ion traps are instruments with constant bandwidth. This means that if the FWHM of a monocharged peak is 0.5 u at m/z 40, it will also be 0.5 u at m/z 1000. But the resolving power at m/z 40 is $40/0.5 = 80$, while at m/z 1000 it is $1000/0.5 = 2000$.

- TOF and magnetic analysers have a resolution that remains constant when the mass increases. If the resolving power FWHM is 20 000 at m/z 40, the bandwidth $\Delta m_{50\%}$ will be $40/20\,000 = 1/500$ or 0.002 u. As the mass accuracy will be reasonably a tenth of this value, it will be 0.0002 u or $0.0002/40 = 5 \times 10^{-6}$ or 5 ppm. At m/z 4000 these figures become $4000/20\,000 = 0.2$ u and the mass accuracy $0.02/4000 = 5$ ppm, thus constant. It should be noted that for the purpose of elemental composition determination this accuracy is sufficient at m/z 40 but by far not at m/z 4000.

- FTICR: at constant detection time the resolving power R is inversely proportional to m/z. Thus, if $R = 10^7$ at m/z 100 it will be at 10^6 at m/z 1000.

- Orbitrap: at constant detection time the resolving power R is inversely proportional to the square root of m/z. Thus, if R is 100 000 at m/z 100, at m/z 1000 it will be $100\,000(100/1000)^{1/2} = 31\,646$.

To illustrate the use of resolving power we refer to an example given in [10]. Ubiquitin ($C_{378}H_{630}N_{105}O_{118}S$) has a monoisotopic mass of 8560.6254 u for the neutral molecule. The spectra of interest are displayed in Figure 6.6.

In ESI, 1.007 825 u must be added for each attached proton. Thus, the 9+ charge state has a monoisotopic mass of $(8560.6254 + (9 \times 1.007\,825)) = 8569.6958$ u, and will appear at m/z $8569.6958/9 = 952.1884$. For the 8+ charge state, the monoisotopic mass will be $8560.6254 + (8 \times 1.007\,825) = 8568.6880$, and will appear at m/z $8568.6880/8 = 1071.0860$, giving a difference of m/z $1071.0860 - 952.1884 = 118.8976$ between these two charge states. To separate them a resolving power of more than

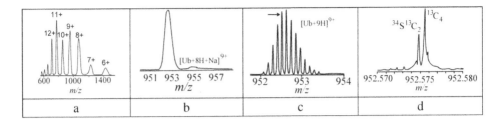

Figure 6.6
Increasing the resolving power for the detection of peaks from ubiquitin ($C_{378}H_{630}N_{105}O_{118}S$).
Redrawn from data in Marshall A.G., Hendrickson C.L. and Shi S.D.H., Anal. Chem., 74,
253A–259A, 2002.

$1071.0860/118.976 = 9$, say 20, will be sufficient. Thus, observing the charge states only
can be done at very low resolution (Figure 6.6a).

At this charge state it often happens that a sodium adduct is formed, so one proton is
replaced by one sodium ion. The mass difference will be mass of Na minus mass of H,
that is $(22.989\,768 - 1.007\,825) = 21.9819$, or divided by the charge, giving a difference of
m/z $21.9819/9 = 2.4424$ and will be located at m/z 954.6308. Thus in order to observe this
sodium adduct, m/z 952.2 must be separated from m/z 954.63. The required resolving power
is $952/2.4424 = 391.6$. Remember that by using the FWHM definition of the resolving
power, two peaks at exactly the required resolving power are not separated (see Chapter 2,
Figure 2.3). Thus the minimum resolution needed using the FWHM definition will be about
1000, as can be estimated by looking at the peaks in Figure 6.6b.

Each of the observed peaks in a spectrum is composed of several isotopic peaks. Between
two successive isotopic peaks, the rounded mass difference will be m/z $1/9 = 0.11$, and
the minimum required resolving power $1071/0.11 = 9730$. If the used resolving power
corresponds to the FWHM definition of the resolving power, there will be no separation at
this value. To get a 10 % valley separation, the resolving power with the FWHM definition
would be about 20 000. See spectrum Figure 6.6c. Then, the charge state can be confirmed
with higher accuracy. Note that this resolution is mandatory if we apply it on the MS/MS
fragments of a multiply charged ion, because there is no other way to measure the charge
state (see Chapter 1, Figure 1.25).

Each observed isotopic peak in spectrum c in Figure 6.6 is also composed of several isotope combinations. For example, the peak marked by an arrow in panel c is
shown at high resolution in panel d, where two main peaks appear. According to the
observed accurate masses, the first one contains one ^{34}S associated with two ^{13}C, and
the second contains four ^{13}C atoms. Indeed the accurate mass of ^{34}S is 33.967 080,
that of ^{32}S 31.972 070, and that of ^{13}C 13.003 355. Thus compared with a molecule
containing only the main isotopes of the elements, the shift in mass due to four ^{13}C
is $4 \times (13.003\,355 - 12.000\,000) = 4.0134$ u. The shift due to two ^{13}C and one ^{34}S is
$2 \times (13.003\,355 - 12.000\,000) + 1 \times (33.967\,080 - 31.972\,070) = 4.0034$.

The difference in shift between the two formulae is $4.0134 - 4.0034 = 0.0100$ u. Taking
into account the charge state, the m/z difference to measure is $0.01/9 = 0.0011$ u. The
required resolving power is $952.6/0.0011 = 875\,340$. In 2006, only a high-field FTICR is
capable of such a resolution.

6.2.2 Determination of the Elemental Composition

Even with very good accuracy for the determined mass, the elemental composition cannot generally be determined only from the accurate mass except at very low masses, say less than 100 Da. Restrictions must be imposed from other information about the elemental composition: for example, a synthetic compound cannot contain an element not used during synthesis; some elements have typical isotope distributions; and so on. The next sections will look at the available information.

It should also be stressed that all the atomic masses are experimentally determined relative masses. The arbitrary reference is the mass of ^{12}C and the mass unit is defined as one-twelfth of the mass of ^{12}C. However, the Avogadro number is only known with some error margin, and thus the mass of one ^{12}C is not known exactly. An 'exact' mass exists of course, but cannot be determined with the techniques available in 2006. There is no analytical mass spectrometer able to determine directly the mass of an ion: a calibration needs to be done with compounds having accurately known masses, but these are never exact.

Different brands of mass spectrometers use different tables of accurate atomic masses, and thus different calibration tables. This is why high-resolution experimental data from one mass spectrometer cannot be used with the software of another mass spectrometer if they do not use the same references.

6.3 Isotopic Abundances

Most elements appear in nature as isotope mixtures. Thus, natural carbon is a mixture of 98.90 % of isotope ^{12}C and 1.10 % of isotope ^{13}C. Table 6.1 lists the abundances of a few elements that are important in organic chemistry. In Appendix 4, the masses and the isotopic abundances are given for all the isotopes except the rare earths.

These isotopes are responsible for the peaks in the mass spectrum appearing as isotopic clusters that are characteristic of the elemental composition. They provide important analytical data. Indeed, even without exact mass measurement, the possibilities for elemental composition determination can often be restricted by using isotopic abundance data. For example, the fragments $C_{10}H_{20}$ and $C_8H_{12}O_2$, both with a nominal mass of 140 u, produce peaks at mass 141 with 11 and 8.8 %, respectively, of the abundance at mass 140 u. This is the result of a different statistical probability of having ^{13}C isotopes. These two elemental compositions can thus be distinguished in a mass analysis.

However, the actual possibilities are limited. Measurement artefacts can modify such percentages. In the case of fragments of heavier molecules, the observed intensity can result from several fragments with different compositions, all present in the isotopic cluster, such as $C_{10}H_{21}$ at mass 141 in the example above.

In principle, this risk does not exist for the molecular peak. In fact, the contribution of a peak with composition $(M + H)^+$ associated to a molecular $M^{\bullet+}$ peak should never be excluded. This artefact can have high abundance in the case of compounds such as amides in electron ionization.

However, some atoms have isotopic compositions that are more obvious. This is the case for chlorine, bromine, selenium and sulfur, for example (Figure 6.7).

Table 6.1   Isotopic abundances.

| Isotope | Relative abundance (%) | Mass (u) | Mean atomic mass[a] | |
			Calculated	Measured
^1H	99.985	1.007 825	1.007 976	1.007 94
^2H	0.015	2.014 0		
^{12}C	98.90	12.000 000	12.011 036	12.011 1
^{13}C	1.10	13.003 355		
^{14}N	99.63	14.003 074	14.006 762	14.006 74
^{15}N	0.37	15.000 108		
^{16}O	99.76	15.994 915	15.999 324	15.999 43
^{17}O	0.04	16.999 131		
^{18}O	0.20	17.999 160		
^{19}F	100	18.998 403	18.998 403	18.998 4
^{23}Na	100	22.989 767	22.989 767	22.989 76
^{31}P	100	30.973 762	30.973 762	30.973 76
^{32}S	95.02	31.972 070	32.064 385	32.066 6
^{33}S	0.75	32.971 456		
^{34}S	4.21	33.967 866		
^{36}S	0.02	35.967 080		
^{35}Cl	75.77	34.968 852	35.452 737	35.452 79
^{37}Cl	24.23	36.965 903		
^{39}K	93.2581	38.963 707	39.098 299	39.098 31
^{40}K	0.0117	39.963 999		
^{41}K	6.7302	40.961 825		
^{79}Br	50.69	78.918 336	79.903 526	79.904 1
^{81}Br	49.31	80.916 289		

[a] Mean value for the natural mixture of isotopes.

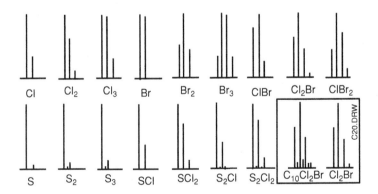

Figure 6.7
Useful isotope combinations in mass spectrometry. Isotopes of other atoms that are possibly associated must always be taken into account, as is shown in the framed section.

The relative abundances of isotopes in a molecule or in a fragment result from their statistical distribution. Let us consider an example in order to explain the calculation of isotopic clusters. What are the relative intensities of the isotopic peaks accompanying the molecular peak of CS_2? Sulfur shows up as a mixture of three main isotopes with nominal masses 32, 33 and 34. These three isotopes occupy two possible positions in the CS_2 molecule. The total number of possible combinations is 3^2, that is 9:

$^{32}S\ ^{32}S$	Total mass: 64, one combination.
$^{32}S\ ^{33}S$ or $^{33}S\ ^{32}S$	Total mass: 65, two combinations.
$^{32}S\ ^{34}S$ or $^{34}S\ ^{32}S$	Total mass: 66, two combinations.
$^{33}S\ ^{33}S$	Total mass: 66, one combination.
$^{33}S\ ^{34}S$ or $^{34}S\ ^{33}S$	Total mass: 67, two combinations.
$^{34}S\ ^{34}S$	Total mass: 68, one combination.
	Total: nine combinations.

Consider now the probability of observing each mass. Mass 64 results from the single possible combination of two isotope 32 atoms. This isotope amounts to 95.02 % in nature, so the probability of having two such atoms simultaneously is $(0.9503)^2$, that is 90.31 %.

Mass 65 results from two possible combinations of isotopes 32 and 33, which amount to 95.03 and 0.75 %, respectively. Each combination has a probability of occurring that is equal to $(0.9503) \times (0.0075) = 0.72$ %, which must be multiplied by the number of combinations: $0.72 \times 2 = 1.42$ %.

Mass 66 can occur in either of two ways: either two isotopes 33, or one isotope 32 and one 34. $^{33}S_2$ has a probability of $(0.0075)^2 = 0.005\,62$ %, in one combination, and $^{32}S^{34}S$ has a probability of $(0.9503) \times (0.0422) = 4.01$ %, which must be multiplied by two for the number of possible combinations, that is 8.02 %. In total, the probability of observing mass 66 is equal to $8.02 + 0.00562 = 8.026$ %.

Mass 67 has a probability equal to $(2 \times 0.0075) \times (0.0422) = 0.063$ %, and mass 68 has $(0.0422)^2 = 0.1781$ %. In summary:

Mass	%	% of the predominant peak
64	90.31	100
65	1.42	1.572 4
66	8.026	8.889
67	0.063	0.069 75
68	0.1781	0.197 2

We now have the possible combinations of two sulfur atoms, and the relative abundances that are calculated are those which would be observed in the case of a fragment with composition S_2. We must now combine each of these with the carbon isotopes. As there is only one carbon atom in CS_2, we have 98.90 % of isotope 12 and 1.10 % of isotope 13. We must now look for all the possible combinations of both of the following isotope series, with their probabilities, while observing that the total mass is the sum of the masses and the probability is the product of the probabilities. Finally, we sum the probabilities of all the ions with equal masses:

S_2		C	
Mass	%	Mass	%
64	90.31	12	98.90
65	1.42	13	1.10
66	8.026		
67	0.063		
68	0.1781		

Then, with Mass S_2 + Mass C = Total mass, and Probability S_2 × Probability C = Total probability:

Mass S_2	Mass C	Total mass	Prob. S_2	Prob. C	Total probability
64	12	76	90.31	98.90	89.317
64	13	77	90.31	1.10	0.993 41
65	12	77	1.42	98.90	1.404
65	13	78	1.42	1.10	0.015 6
66	12	78	8.026	98.90	7.938
66	13	79	8.026	1.10	0.088
67	12	79	0.063	98.90	0.062
67	13	80	0.063	1.10	0.000 69
68	12	80	0.1781	98.90	0.176
68	13	81	0.1781	1.10	0.002
			Total probability %		99.996 6

Gathering the identical masses, we obtain the abundances of the CS_2 peaks:

Mass	Total %	% of the predominant peak
76	89.31	100
77	2.3974	2.684
78	7.9536	8.9056
79	0.152	0.17
80	0.177	0.198
81	0.0019	0.002

If the molecule contains another element, we should combine its masses and abundances anew.

Note that this calculation does not take the sulfur isotope of mass 36 into account; it is 0.017 % abundant and affects the result for the carbon sulfide of mass 80 u.

Mathematical methods for the calculation of theoretical relative abundances within the isotopic cluster, for comparison with experiment, usually rely on expansion of the polynomial expression based on an extension of the binomial probability distribution [12, 13]. Indeed, for an element with x isotopes and relative abundances $I_1, I_2, \ldots, I_x$, and for

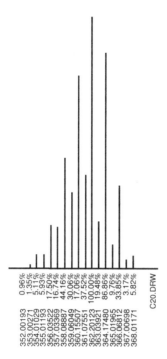

Figure 6.8
Isotopic cluster of $C_9H_{16}Se_3$ molecular ion, calculated by Hsu's program. The peaks below 1 % are omitted. The numbers indicate the exact mass and the relative intensity (%).

n atoms of this element in the molecule, the problem consists of calculating, for every element, the terms of the following polynomial expression:

$$(I_1 + I_2 + \cdots + I_x)^n$$

The coefficient of the jth term then corresponds to the number of possible combinations that match a given distribution; the term itself corresponds to the probability of this combination. When considering several different types of atoms, we calculate the products for each term in these polynomials. We then match the right mass to each of these combinations, and we gather the identical masses.

Figure 6.8 shows an example of a calculation based on a method developed by Hsu and carried out on a personal computer. Some isotope distribution calculators are also available on-line on the Internet [14, 15].

Another method has been developed for calculating isotope distributions starting from a molecular formula and elemental isotopic abundances [16, 17]. This method uses Fourier transforms to do the multiple convolutions required to determine molecular isotope distributions and calculates ultrahigh-resolution distributions over a limited mass range. Because discrete Fourier transforms can be calculated very efficiently, this new way of looking at

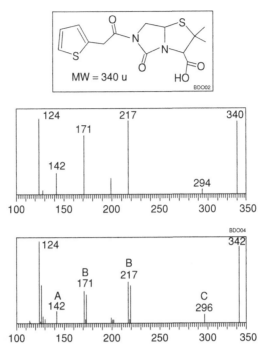

Figure 6.9
Product ion spectra obtained from the fragmentation of the molecular ion and of the corresponding (M + 2) isotopic precursor shown above. (A)
Unshifted ion, thus sulfurless. (B) Dedoubled ions,
containing one of the two sulfur atoms. (C) Ions
shifted two mass units: they contain both sulfur
atoms.

the problem has significant practical implications. Specifically, this method for calculating
isotope distributions is very fast, accurate and economical in its use of computer memory
and thus can be applied to extremely large molecules.

Even more information from isotopic abundances can be obtained by tandem mass spectrometry (MS/MS), as is shown in the following example. The molecule shown in Figure
6.9 contains two sulfur atoms and its molecular peak at 340 Th is thus accompanied by an
isotopic peak at 342 Th. Its relative intensity is $2 \times 4.21 = 8.42\,\%$, if the participation of
other elements' isotopes is neglected. The spectra of the ions derived from the fragmentation of the molecular ion and of the isotopic ion (M + 2) are shown. The (M + 2) ion
contains one ^{34}S and one ^{32}S sulfur atom. If both sulfur atoms are kept in a fragment, then
this fragment also contains one ^{34}S and one ^{32}S sulfur atom. It is shifted by 2 mass units. If
it contains only one of the two sulfur atoms, the probability that the atom being retained is
^{34}S is the same as the probability that it is ^{32}S; the peak is therefore dedoubled. Finally, if
the fragment does not contain sulfur atoms, it shows up at the same mass in both spectra.
Such an approach therefore provides information on the elemental composition.

With instruments of high resolving power, still more information on isotopes can be
obtained as the isotopes do not have the same mass difference towards the main isotope

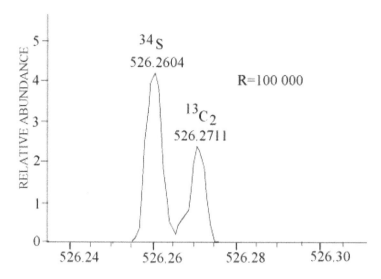

Figure 6.10
A + 2 of protonated MRFA peptide measured with a resolving power
of 100 000 with an orbitrap mass spectrometer. The peaks due to the
indicated isotope compositions are clearly separated and measured with
correct relative abundances. Redrawn, with permission from Thermo-
Finnigan documentation.

[18]. For instance, the peptide MRFA contains one sulfur atom. The main isotope peak is
at m/z 524.265. As shown in Figure 6.10, at m/z 526 two accurate mass peaks appear at
m/z 526.2604 and 526.2711. These peaks are due to the presence of one ^{34}S atom or two
^{13}C atoms, respectively.

6.4 Low-Mass Fragments and Lost Neutrals

The molecular ion fragments and produces ions and neutrals that are not observed in the
spectrum. The mass of the neutral product can be deduced from the difference between the
mass of the parent ion and that of the observed ionic fragment.

One can deduce much information concerning the elemental composition from the
masses of the neutrals or of the low-mass fragments.

In the case of fragments containing only carbon and hydrogen, masses range from 36
(C_3) to 43 (C_3H_7) u for three carbon atoms. For two atoms the limits are 24 and 29 u, for
four atoms 48 and 57 u, and for five the minimum is 60 u, for C_5, which of course is very
unlikely! We thus observe that ions or neutrals ranging from 30 to 35 u, for example, must
necessarily contain atoms other than carbon and hydrogen.

Some masses can only have one reasonable formula, such as 20 u, which corresponds
only to HF.

In the case of lost neutral fragments, it is important to take into account the fact that
a fragment can result from several successive steps starting from the molecular ion. For
example, the spectra of bile salts often contain an ion with mass $(M - 20)$ u although

they contain no fluorine. This fragment is actually the result of successive losses $(M - H_2O - H_2)^+$.

This information can be combined with that furnished by the isotopic abundances. Mass 47 u may correspond to the crude formula CH_3O_2 or CH_3S. In the latter case, however, this ion must be accompanied by another with mass 49 u amounting to 4.2 % of its intensity and caused by ^{34}S.

All mass spectrometric data systems allow the calculation of the possible compositions of fragments with given masses while limiting the possibilities to the elements contained in the molecular formula.

6.5 Number of Rings or Unsaturations

When the elemental analysis of a molecule or of a fragment is known, the number of rings or unsaturations can be calculated. Even though this does not hold just for mass spectrometry, let us recall the principle.

An aliphatic hydrocarbon has a formula C_nH_{2n+2}. Each ring or unsaturation that is present in a hydrocarbon decreases the number of hydrogen atoms by two units. Let x be the number of hydrogen atoms observed; then, if N_i is the number of rings and unsaturations, we have

$$x = 2n + 2 - 2N_i \quad \text{or} \quad N_i = \frac{2n + 2 - x}{2}$$

For example, benzene has formula C_6H_6:

$$N_i = \frac{((2 \times 6 + 2) - 6)}{2} = 4$$

Benzene indeed contains one ring and three unsaturations.

The presence of an oxygen or a sulfur atom in the molecule does not modify these numbers, as we can see when comparing CH_4 with CH_3OH and CH_3SH, or CH_3CH_3 with CH_3OCH_3 and CH_3SCH_3.

Each halogen atom 'replaces' a hydrogen atom and decreases the number of hydrogen atoms by an equal number, as is shown by CH_4 and CH_2Cl_2.

Each nitrogen or phosphorus atom increases the number of hydrogen atoms of a compound by one unit: compare CH_4 with CH_3NH_2 and CH_3PH_2 as an example.

Let n be the number of carbon atoms, and nX and nN be those of halogens and nitrogen (or phosphorus), respectively. The following rule can be deduced, where N_i is the number of rings or unsaturations and x is the number of hydrogen atoms that are found experimentally:

$$N_i = \frac{(2n + 2 + nN - nX) - x}{2}$$

In mass spectrometry, this equation can only be strictly applied to the molecular ion. Protonation, deprotonation, fragmentation and other modifications change the rule. For instance, a fragment ion often results from a bond cleavage, so that the theoretical number of hydrogen atoms in a saturated fragment is decreased by one unit. Thus, for example, the saturated ion CH_3^+ contains a hydrogen atom less than the molecule CH_4. This is easily understood as the equation above leads to half a whole value for N_i. We then only need to subtract half of that value in order to obtain the correct number. In the same way, protonated molecules will have an increment of 0.5, and deprotonated a decrement of 0.5.

6.6 Mass and Electron Parities, Closed-Shell Ions and Open-Shell Ions

6.6.1 Electron Parity

Common molecules have an even number of electrons. Stable radicals are rare exceptions, such as NO. In classical chemistry, we most often meet active species that are ions with an even number of electrons, or radicals, an uncharged species with an odd number of electrons. In mass spectrometry, we observe ions with an even number of electrons, but we also often meet radical ions, a species uncommon in solution chemistry and having specific characteristics.

6.6.2 Mass Parity

The atomic masses used in common chemical calculations are based on averages resulting from mixtures of isotopes. In mass spectrometry, the calculation is based on the mass of the predominant isotope of each element. As the isotopes are separated in the spectrometer, we always face several peaks with different masses, and with intensity ratios defined as described earlier. Thus, for example, dichloromethane has a classical molecular mass equal to $12.01 + 2 \times 1.00 + 2 \times 35.45 = 84.91$ Da. The molecular mass in mass spectrometry is (if mass defects are neglected) $12 + 2 + 2 \times 35 = 84$ u. Several isotopic peaks are observed in the spectrum, the second most important being observed at m/z 86 with an intensity equal to 64.8 % of that of the m/z 84 peak.

Organic molecules are normally made up of atoms of C, H, N, O, S, P and halogens, and we limit the following discussion to these elements. Molecular masses that are considered here are calculated by using the value of the atomic mass of the predominant isotope of each element, as usual in mass spectrometry.

The nitrogen rule requires that the molecular mass is always even when the number of nitrogen atoms is even or zero. This results from the fact that nitrogen has a different mass parity and valence electrons parity: mass 14 u, five peripheral electrons. Both of these parities are identical in the case of any other atom. It should be noted that this holds only if we consider the mass of the predominant isotope. Thus, the 'chemical' mass of bromine is 80 u, an even number, but its predominant isotope is that of mass 79 u, an odd mass. In the same way, isotopically labelled compounds do not always obey this rule.

Another approach to the problem consists of saying that adding one nitrogen atom in a molecular formula entails also adding one hydrogen to the 'saturated' formula. However, nitrogen has an even mass, and the number of new atoms with a unit mass is one. Thus, for example, CH_3CH_3 has a molecular mass of 30 u, while CH_3NHCH_3 has a mass of 45 u, an odd number.

6.6.3 Relationship Between Mass and Electron Parity

First, consider a molecule that contains no nitrogen and is ionized through electron ionization. The ionization process consists of expelling one electron in order to produce a radical cation:

$$M + e^- \longrightarrow M^{\bullet+} + 2e^-$$

At the beginning the molecule M has an even mass because, hypothetically, it contains no nitrogen, and has an even number of electrons. After ionization, the mass has not changed, but the number of electrons has decreased by one unit, and has become odd. We have obtained a radical cation, which is represented by a dot combined with a plus sign: $M^{\bullet+}$.

If this molecule M is ionized through chemical ionization, we obtain a protonated molecule, $(M + H)^+$. The protonated molecule has an odd mass: the mass, which is even, increased by one because of the new proton. The number of the ion electrons is the same as that of the neutral molecule, that is even. It is a normal cation, not a radical cation. The same deductions apply to negative ions.

An odd number of nitrogen atoms brings about an odd molecular mass in daltons such as is defined in mass spectrometry: NH_3 17, CH_3NH_2 31, and so on. Thus, in the case of an odd number of nitrogens, the earlier rule must be inverted: for the ion, the mass parity is the same as the electron parity.

From an analytical point of view, we see that an odd molecular mass, based on the predominant isotopes, indicates the presence of an odd number of nitrogen atoms.

Figure 6.11 shows the spectrum of butanol brought down to its five most intense ions. The following rule is confirmed:

> When there are no or an even number of nitrogen atoms, any ion with an even mass has an odd number of electrons and is a radical cation or radical anion. Any ion with an odd mass, in turn, has an even number of electrons and is a cation or an anion. The reverse holds true for an odd number of nitrogen atoms. There is no exception within the C, H, N, O, S, P, alkali metal and halogen elements.

6.7 Quantitative Data

The goal of quantitative analysis in mass spectrometry is to correlate the intensity of the signals with the quantity of the compound present in the sample. Several quantitative analysis methods using mass spectrometry have been developed and many applications using these methods have been described [19].

6.7.1 Specificity

The degree of specificity of a quantitative analysis using mass spectrometry depends on how the spectrometer is used, and even more on the signal that is used during the correlation. For example, we can use the ion current as a signal to determine the concentration of the compound that is studied as long as there is no interference with other substances (in

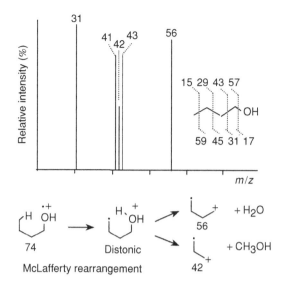

Figure 6.11
Mass spectrum of butanol limited to its five most abundant fragments. As is shown in the scheme, the ions resulting from the direct cleavage of a single bond in the molecular ion lead to odd mass fragments. These ions have a vacant valence and are thus cations with an even number of electrons, such as CH_3^+, $CH_3CH_2^+$, and the lost neutral is a radical: there is always a vacant valence. However, a rearrangement followed by the cleavage of a bond leads to a neutral molecule and to a radical cation, and thus to an odd number of electrons, but an even mass. Ions resulting from the rearrangement have a charge site separated from the radical site and are termed 'distonic ions'.

other words, as long as the substance being analysed is pure or separated beforehand using chromatography).

Suppose that we know exactly the structure and the mass spectrum of the substance whose concentration we want to determine; we can use as a signal one or several ions specific to this substance. It is important to choose these ions from among those with high masses (higher than 200 Th if possible). In fact, the significance of a signal depends on its ionic abundance and on its mass. It is logical to expect intense peaks corresponding to high masses to be more typical of the molecule than any low-mass ion.

An in-depth study of mass spectra of over 29 000 different compounds containing only the elements C, H, N, O, F, Si, P, S, Cl, Br and I showed that, on average, the probability of the presence of a peak in a spectrum halves every 100 Th [20]. Thus the probability of having in a mixture compounds with ions typical of the same m/z decreases as the mass increases.

Another specificity criterion lies in the link between the intensity ratio of these characteristic ions and that of their respective abundance in the mass spectrum of the pure substance.

In fact, the probability of finding in a mixture two compounds with several identical characteristic masses in the portion of the spectrum above 100 Th becomes almost zero, unless these two compounds are isomers. Several characteristic ions are necessarily used when the quantitative analysis of complex biological mixtures containing many molecules from the same family is undertaken.

Many methods exist that improve the specificity of quantitative analysis using mass spectrometry. These methods can be classified into either of two categories: those that act upon the sample and those that act on the spectrometer.

The first type of method that increases the specificity is based on a simple preliminary purification which can be an extraction into a solvent that is less polar, an acid–base separation, and so on, in order to eliminate all possible interference. Another possibility is to use different characteristic ions with higher masses after having formed a higher mass derivative from the compound. A further approach consists of making the spectrometer more selective, for example by increasing the resolution, or by resorting to a different ionization technique, or by applying another data acquisition mode such as SRM, which was discussed in Chapters 4 and 5.

All of these techniques can, of course, be used simultaneously with chromatographic separation techniques. Methods that allow an improvement in the mass spectrometer specificity can thus be worthwhile alternatives to long procedures for sample preparation. However, quantitative analysis carried out directly on neat samples may undergo a matrix effect characterized by a variation in the sample response because of the effect of the matrix on the abundance of ions within the source.

6.7.2 Sensitivity and Detection Limit

Sensitivity in mass spectrometry is defined as the ratio of the ionic current change to the sample change in the source. The recommended unit is $C \, \mu g^{-1}$. It is important that the relevant experimental conditions corresponding to sensitivity measurement should always be stated.

The detection limit should be differentiated from sensitivity. It is the smallest sample quantity that yields a signal that can be distinguished from the background noise (generally a signal equal to 10 times the background noise). It should be noted that this minimum quantity is not enough to obtain an interpretable mass spectrum. Figure 6.12 shows an example detection limit obtained using mass spectrometry when measuring dihydrocholesterol. In these experimental conditions, the detection limit is lower than 1 picogram but 1 nanogram of the compound is necessary to obtain an interpretable mass spectrum.

The limit of detection depends considerably on the abundance of the ionic species that is measured. The more abundant the measured ionic species is with respect to all of the ions derived from the analysed molecule, the higher is the limit of detection, as shown in Figure 6.12. The goal is thus to produce a signal that is as intense as possible. Several methods allow this goal to be reached, such as the modification of the ionization conditions, reversal into the negative mode, the use of other softer ionization techniques or the derivative of the sample, in order to increase the number of ions produced in the source or to reduce their fragmentation.

Another factor influencing the sensitivity corresponds to the length of time of signal integration. Of course, the longer that time is, the more intense is the signal. The data

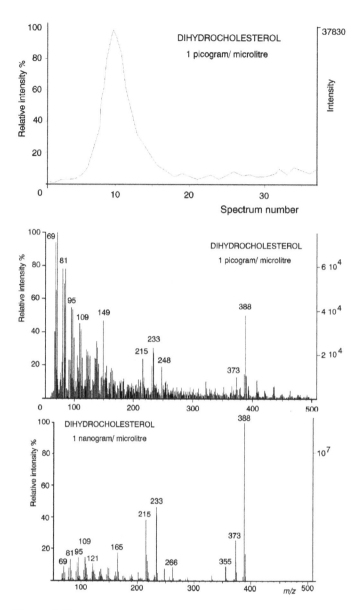

Figure 6.12
Detection limit obtained in mass spectrometry in the case of dihydrocholesterol.

acquisition mode thus influences the sensitivity by this factor. Three acquisition modes
exist: the scan, SIM and SRM modes. These three modes are cited in ascending order of
their effect on the sensitivity.

As a reminder, the scan consists of measuring complete spectra between two limit
masses several times. The detection of selected ions consists of tuning the analyser so as
to focus on to the detector only those ions with a specific m/z ratio. If several ions with

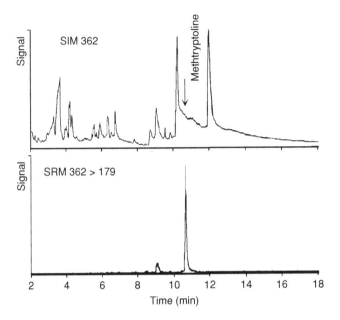

Figure 6.13
Detection of the heptafluorobutyryl derivative of methtrypto-
line using GC. Top: SIM of the 362 Th negative ion. Bottom:
SRM of the 362→179 Th fragmentation. Reprinted (modified)
from Yost R.A., Adv Mass Spectrom, 10B, 1479, 1985, with
permission.

different directly m/z ratios have to be detected, the analyser goes rapidly from one mass to
another. This method is clearly more sensitive than the scan mode as the length of time of
signal integration is greater and the increase in the resulting signal-to-noise ratio allows a
sensitivity gain of up to a 1000-fold.

The detection of selected reactions is a technique that requires tandem mass spectrometry,
which makes the SRM even more sensitive and more selective than the SIM. In order to
carry out this type of acquisition, the instrument is tuned to transmit only the ions derived
from a fragmentation reaction in the chosen reaction region. The sensitivity gain obtained
by using this method compared with SIM is due to the increase in the signal-to-noise ratio
that is characteristic of tandem mass spectrometry. Figure 6.13 shows the sensitivity gain
and the selectivity gain of SRM compared with SIM [21].

The increased sensitivity of the SIM and SRM data acquisition results in lower flexibility.
The full-scan mode furnishes complete data so new structural information can be extracted
in order to answer questions that may arise after the initial experiment. This is not possible
for data acquired in the SIM or SRM modes.

6.7.3 External Standard Method

The role of a standard is to determine the mathematical relationship, in the concentration
range to be measured, between the selected signal intensities and the mixture composition.

This external standard method consists of preparing a synthetic sample containing a known quantity of the molecule to be measured (M_{ste}), then introducing a precise volume of this solution into the spectrometer and recording the intensity of the response signal (I_{ste}).

Then, without any modification of the analytical conditions, an equal volume of the solution containing the molecule to be quantified (M_x) is introduced into the spectrometer and the intensity of its response signal (I_x) is measured. Since the volumes that are introduced are equal, there is a proportionality between the response intensities and the quantities as long as the response signal intensity remains linear with respect to the concentration and as long as the signal intensity is zero at zero concentration. That is,

$$M_x = I_x \times \frac{M_{ste}}{I_{ste}}$$

or $M_x = I_x \times RF_x$ if the response factor is defined as

$$RF_x = \frac{M_{ste}}{I_{ste}}$$

In electron ionization, the response is normally linear with respect to the concentration over a wide range, often six orders of magnitude. This is not true for the other ionization techniques because of the influence which the sample quantity can have on the number of ions that is produced and on the fragmentation, and thus on the production yield of the various ionic species. For instance, in the chemical ionization mode the formation of adducts appearing at higher sample pressures changes the relative intensities in the spectrum.

Thus, verification by a calibration curve is necessary. This calibration curve allows one to calculate the quantities of a compound to be measured in the unknown samples, to confirm the method specificity and also to define its sensitivity. In order to do this, equal volumes of a series of synthetic samples containing an increasing quantity of the molecule to be measured are introduced into the mass spectrometer and the intensity of their response signal is recorded.

This allows the determination of the mathematical relationship, in the concentration range to be measured, between the selected signal intensity (I_x) and the quantity of the molecule to be measured (M_x) that is present in the mixture. Ideally this relationship should correspond to the equation of a straight line with a slope equal to one in order to ensure maximum precision. Figure 6.14 shows an example of a quantitative analysis of phenobarbital within a mixture using GC/MS.

6.7.4 Sources of Error

Experimental procedures for quantitative mass spectrometric analysis usually involve several steps. The final error results from the accumulation of the errors in each step, some steps in the procedure being higher error sources than others. A separation can be made between the errors ascribable to the spectrometer and its data treatment on the one hand and the errors resulting from the sample handling on the other.

Normally, the sample handling errors are higher than those due to the mass spectrometer and thus make up the greater part of the final error. Handling errors can be numerous, such as the error in measuring a sample volume and the error in the introduction into the spectrometer.

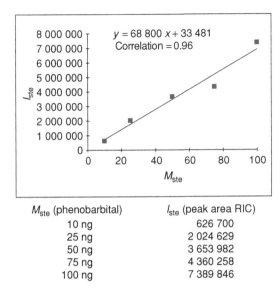

M_{ste} (phenobarbital)	I_{ste} (peak area RIC)
10 ng	626 700
25 ng	2 024 629
50 ng	3 653 982
75 ng	4 360 258
100 ng	7 389 846

Figure 6.14
Calibration curve obtained by external standardiza-
tion of phenobarbital present within a mixture, in
order to quantify it.

The errors due to the mass spectrometer are also numerous, such as the variation in the source conditions and the instability of the mass scale. For statistical reasons, every measurement of a signal intensity carries a minimal intrinsic error. This error is inversely proportional to the square root of the number of ions detected for that signal. In order to optimize the reproducibility or the precision of the measurements, a maximum number of ions must thus be detected for every ionic species.

All of these error sources, other than the minimal intrinsic error, can be reduced by using the internal standard method. The absolute measurement of the signal is replaced by the measurement of the signal ratio for the molecule that is measured and for the internal standard. The same compound can play the role of the internal standard for the quantification of various compounds within the mixture.

For an evaluation of error sources in quantitative GC/MS determinations, the reader is referred to the paper by Claeys *et al.* [22].

6.7.5 Internal Standard Method

This method is based on a comparison of the intensities of the signal corresponding to the product that has to be quantified with the one of a reference compound called the internal standard. This method allows the elimination of various error sources other than the minimal intrinsic error due to statistical reasons. In fact, if we choose as an internal standard a molecule with chemical and physical properties as close as possible to the properties of the molecule to be measured, the latter and the internal standard undergo the same loss in the extraction steps and in the derivative or the same errors in the introduction of the sample into the mass spectrometer, when the source conditions are varied. As both

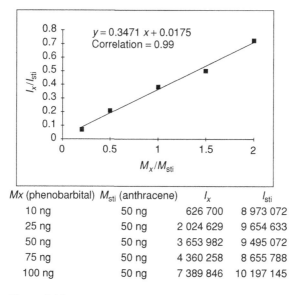

Mx (phenobarbital)	M_{sti} (anthracene)	I_x	I_{sti}
10 ng	50 ng	626 700	8 973 072
25 ng	50 ng	2 024 629	9 654 633
50 ng	50 ng	3 653 982	9 495 072
75 ng	50 ng	4 360 258	8 655 788
100 ng	50 ng	7 389 846	10 197 145

Figure 6.15
Calibration curve for the quantitative analysis of pheno-barbital using anthracene as an internal standard.

compounds undergo the same losses and the same errors, their ratio remains unchanged during the procedure. Knowing the quantity of the internal standard that is added from the start and the relative proportion of the quantity of both compounds allows these losses and errors to be neglected. It is important to add the internal standard as early as possible in the procedure, in order to obtain the maximum precision.

The method consists first of carrying out measurements on synthetic samples containing the same known quantity of the internal standard and increasing quantities of the compound to be measured. With these results a calibration curve is constructed. This allows a mathematical relationship to be obtained between the intensities of the signals corresponding to the compound to be analysed and the internal standard (I_x/I_{sti}) and the quantity of compound present in the sample (M_x). As a reminder, maximum precision is obtained if the relationship corresponds to the equation of a straight line with a slope equal to one.

The measurements are then carried out on the unknown samples that had a constant quantity of internal standard added to them before they were treated according to the experimental procedure. The quantity of compound in each unknown sample can be measured using the calibration curve. Figure 6.15 shows how useful an internal standard can be for measuring purposes. This figure uses once again the example of the quantitative analysis of phenobarbital using an internal standard, anthracene.

The internal standard should show physical and chemical properties that are as close as possible to those of the molecule that has to be measured. It must be pure, absent from the sample and, of course, inert towards the compounds in the sample. The internal standards can be classified into three categories: structural analogues that are labelled with stable isotopes, structural homologues and compounds from the same chemical family. These various types of internal standards are classified here in descending order according to their usefulness and their price. In fact, the starting material for labelled compounds is fairly cheap, but most require many steps in their total synthesis and are thus very expensive. In the case of standards corresponding to structural homologues or in the case of compounds belonging to the same chemical family, the ions that are used must have masses differing from that of the compound that must be measured if direct introduction is used. However, if the introduction is carried out by chromatographic coupling and if the retention time of the compound is different from that of the internal standard, then the ions can have identical masses.

It is also worth mentioning that, depending upon the nature of the internal standard, weighted or non-weighted linear regression analysis to construct the calibration curve should be considered [22].

6.7.6 Isotopic Dilution Method

Isotope dilution mass spectrometry (IDMS) can be considered as a special case of the internal standard method: the internal standard that is used is an isotopomer of the compound to be measured, for example a deuterated derivative. Note that an internal standard is necessary for every compound to be measured. This internal standard is as close as possible to perfection since the only property that distinguishes it from the compound to be measured is a slight mass difference, except for some phenomena that involve the labelled atoms, such as the isotopic effect. In that case, we have an absolute reference, that is the response coefficients of the compound and of the standard are identical. This method is often used to establish standard concentrations. The basic theory of this method rests on the analogy between the relative abundance of isotopes and their probability of occurrence [23].

The method consists of examining the spectrum of the compound that must be measured in order to select an intense characteristic peak which is used to measure the analyte. A known, exact quantity of labelled internal standard is added to the sample with an unknown concentration. When the labelled internal standard is added, the peak corresponding to the characteristic peak is moved to a different position in the spectrum, according to the number and nature of the atoms that were used in the labelling. The ratio of these two signal intensities is used to measure their relative proportion.

Suppose that there are N atoms or molecules in the mixture that yield a peak which is characteristic of the mass m. Suppose also that there are M atoms or molecules of the labelled compound in the substance which yield a characteristic peak for the mass $m + n = o$, where n corresponds to the mass displacement caused by the introduction of isotopes into the molecule.

The ratio R_{mo} of the ion intensities over the masses m and o is given by

$$R_{mo} = \frac{N P_m + M Q_m}{N P_o + M Q_o}$$

where P_m, P_o, Q_m and Q_o represent the relative isotopic abundances normalized over the isotopes for the natural product and the labelled product with masses m and o respectively. Since P_m, P_o, Q_m, Q_o and M are known, and R_{mo} can be deduced from the spectrum of the mixture containing the natural compound and the labelled compound, then the value of N can be precisely calculated (Figure 6.16).

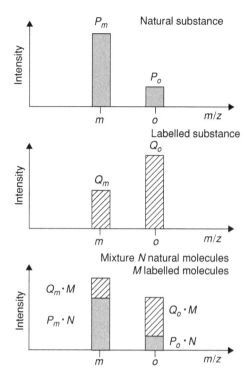

Figure 6.16
Principle of isotopic dilution illustrated by a
molecule having two isotopes of mass m and
o.

If A represents the Avogadro number, x and y represent the quantity of natural product
and the quantity of labelled product added to the sample, respectively, and E and F represent
the molecular masses of the natural product and of the labelled product, respectively, then
the equation is

$$R_{mo} = \frac{(Ax/E)P_m + (Ay/F)Q_m}{(Ax/E)P_o + (Ay/F)Q_o}$$

or, simply

$$R_{mo} = \frac{(x/y \times P_m/E) + Q_m/F}{(x/y \times P_o/E) + Q_o/F}$$

This mathematical equation corresponds to the equation of a curve. The straight line
represents only a special case. The nature of the calibration curve that is obtained depends
on the normalized relative isotopic abundances and thus on the nature of the isotopes that
are introduced, on the increase in the molecular weight and on how enriched the labelled
compound is.

If P_o and Q_m are zero (there is no interference between the natural compound at high mass o and the labelled compound at low mass m), then the equation becomes

$$R_{mo} = \frac{x}{y} \times \frac{P_m}{Q_o} \times \frac{F}{E}$$

This equation corresponds to the equation of a straight line passing through the origin but with a slope that can be other than one.

Hence, in theory, the calibration curve for a certain analysis can be calculated without any reference to experiments on synthetic samples containing increasing quantities of the compound to be measured. If the isotopic abundances are known with precision, the method allows one to determine the exact quantity of the substance to be measured without having to set up a calibration curve. In practice, disagreements between the theoretical and practical calibration curves are possible.

Quantitative applications using a direct inlet probe are analogous to those using a GC or an LC inlet, except for one important difference: the separation carried out before introducing the sample when using chromatographic coupling. This allows one to increase the specificity of measurement by the acquisition of an extra datum which is the retention time. The lack of resolution of the direct inlet mode causes peaks to overlap and reduces the sensitivity. This can negatively influence the precision of the measurement.

References

1.   http://www.nist.gov/srd/nist1.htm (21 March 2007).
2.   http://www.webbook.nist.gov/chemistry (21 March 2007).
3.   McLafferty, F.W. and Stauffer, D.B. (2006) *The Wiley Registry of Mass Spectral Data*, 8th edn, John Wiley & Sons, Inc., Hoboken, NJ.
4.   *Eight Peak Index of Mass Spectra*, vol. **3**, Royal Society of Chemistry, Cambridge, 1991.
5.   Middelditch, B.S., Missler, S.R. and Hines, H.B. (1981) *Mass Spectrometry of Priority Pollutants*, Plenum Press, New York.
6.   Stemmler, E.A. and Hites, R.A. (1988) *Electron Capture Negative Ion Mass Spectra of Environmental Contaminants and Related Compouds*, VCH, New York.
7.   Hites, R.A. (1992) *Handbook of Mass Spectra of Environmental Contaminants*, Lewis, Boca Raton, FL.
8.   Pfleger, K., Maurer, H.H. and Weber, A. (1992) *Mass Spectral and GC Data of Drugs, Poisons, Pesticides, Pollutants and Their Metabolites*, VCH, Weinheim.
9.   Ardrey, R.E., Allen, A.R., Bal, T.S. and Moffat, A.C. (1985) *Pharmaceutical Mass Spectra*, Pharmaceutical Press, London.
10. Marshall, A.G., Hendrickson, C.L. and Shi, S.D.H. (2002) *Anal. Chem.*, **74**, 253A–9A.
11. Guidelines for authors. *J. Org. Chem.* **70** (1) (2005).
12. Yergey, J.A. (1982) *Int. J. Mass Spectrom. Ion Processes*, **52**, 337.
13. Hsu, C.S. (1984) Diophantine approach to isotopic abundance calculations. *Anal. Chem.*, **56**, 1356.
14. http://www.sisweb.com/mstools.htm (21 March 2007).
15. http://www.winter.group.shef.ac.uk/chemputer/isotopes.html (21 March 2007).
16. Rockwood, A.L. VanOrden, S.L. and Smith, R.D. (1995) Rapid calculation of isotope distributions. *Anal. Chem.*, **67** (15), 2699–704.

17. Rockwood, A.L. and VanOrden, S.L. (1996) Ultrahigh-speed calculation of isotope distributions. *Anal. Chem.*, **68** (13), 2027–30.
18. Makarov, A., Denisov, E., Kholomeev, A. *et al.* (2006) Performance evaluation of a hybrid linear ion trap/orbitrap mass spectrometer. *Anal. Chem.*, **78** (7), 2113–20.
19. Millard, B.J. (1978) *Quantitative Mass Spectrometry*, Heyden, London.
20. McLafferty, F.W. and Venkataraghavan, R. (1982) *Mass Spectral Correlations*, American Chemical Society, Washington, DC.
21. Yost, R.A. (1985) *Adv. Mass Spectrom.*, **10B**, 1479.
22. Claeys, M., Markey, S.P. and Maenhaut, W. (1977) *Biomed. Mass Spectrom.*, **4**, 122.
23. Pickup, J.F. and McPherson, K. (1976) *Anal. Chem.*, **48**, 1885.

7

Fragmentation Reactions

Most of the chemical reactions occur in the condensed phase or in the gas phase under conditions such that the number of intermolecular collisions during the reaction time is enormous. Internal energy is quickly distributed by these collisions over all the molecules according to the Maxwell–Boltzmann distribution curve.

7.1 Electron Ionization and Fragmentation Rates

Using conventional electron ionization mass spectrometry, everything occurs in a vacuum such that any collision is highly unlikely. The molecule receives energy from an electron beam and is ionized into a radical cation. The ion that is thus formed is subjected to an electric field that directs it towards the analyser. For instance, suppose a singly charged 100 u mass ion is accelerated by a 1000 V potential difference in the source. This ion has a mass of

$$100 \times 1.66 \times 10^{-27}\,\text{kg} = 1.66 \times 10^{-25}\,\text{kg}$$

At the source outlet, its kinetic energy is

$$\frac{mv^2}{2} = (1000\,\text{V}) \times (1.6 \times 10^{-19}\,\text{C}) = 1.6 \times 10^{-16}\,\text{J}$$

The square of its speed is then

$$v^2 = \frac{(2 \times 1.6 \times 10^{-16})}{(1.66 \times 10^{-25})} = 1.93 \times 10^9\,\text{J kg}^{-1}$$

Hence

$$v = 4.39 \times 10^4\,\text{m s}^{-1} = 1.58 \times 10^5\,\text{km h}^{-1}$$

Thus the time spent in the source ranges from 10^{-6} to 10^{-7} s, and in the analyser about one or two orders of magnitude longer.

Let us remember also that the ionization is usually carried out with a 70 eV electron beam, and that 1 eV corresponds to 96.48 kJ mol^{-1}. The ion excess energy can be several electronvolts. In the condensed phase, the cooling of an excited ion or molecule results from the collisions or from the emission of photons. Under high vacuum, only the latter possibility remains. The time delays necessary for cooling by radiation can be measured

---

Mass Spectrometry: Principles and Applications, Third Edition Edmond de Hoffmann and Vincent Stroobant
© Copyright 2007, John Wiley & Sons Ltd

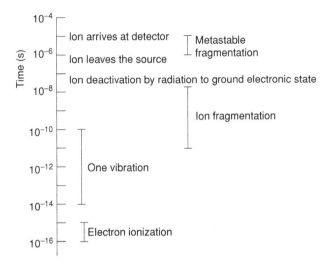

Figure 7.1
In electron ionization, the ionization occurs in a very short
time. The energy is redistributed through the vibrations in
the ion. After 10^{-8} s (one-hundredth of a microsecond), the
ion deactivates to the ground state if it did not fragment. This
occurs well before the ion leaves the source. This explains why
electron ionization is reproducible from one spectrometer to
another.

[1]. In the case of electronic excitations, the radiation occurs in the UV or in the visible
range and typically occurs after 10^{-8} s. The rotation and vibration excitations emit in the IR
range after times from 1 ms to a few seconds. Figure 7.1 summarizes the time events after
electron ionization. The ionization occurs in a very short time. The energy is redistributed
within a time corresponding to a few vibrations, that is less than 10^{-10} s and then the ion
may or not fragment. After about 10^{-8} s, the ion will go back to the ground electronic state
by radiation and will almost no longer fragment. This occurs well before the ion leaves the
source and explains why electron ionization spectra are reproducible from one instrument
to another.

Here is the problem: a great quantity of energy is transferred to isolated molecules. The
notion of temperature, the statistical distribution of energy, loses all meaning. Also, in a
few microseconds, the whole process is over. The reaction has or has not occurred, and the
products are detected.

The reactions that are observed are unimolecular fragmentations. Recombinations are
impossible since the collisions, under the usual conditions, are non-existent. The rearrangements and the opening of rings are detectable only if they are followed by a fragmentation.
These reactions thus depend only on the structure and on the energy contained within the
molecule.

Both of these characteristics, isolated unimolecular reaction and very short time, imply
that the products of these reactions are the kinetic products. Hence explanations based
on the stability of the fragment ions may be misleading. However, many fragmentation

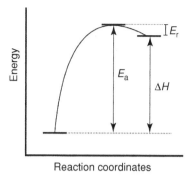

Figure 7.2
The recombination activation energy E_r of a radical with a cation is low. As is shown, this means $\Delta H = E_a - E_r \approx E_a$.

reactions correspond to zero- or low-energy reverse reactions, such as the recombination of a radical with a cation. According to the principle of micro-reversibility, the fragmentation activation energy is equal to the endothermicity (Figure 7.2). In this case, the reasoning based on thermodynamics is often correct.

In thermal reactions, the reaction rate depends on the activation–deactivation equilibrium and on the decomposition rate of the activated complex:

$$M + N \underset{k_{-1}}{\overset{k_1}{\rightleftharpoons}} M^* + N$$

$$M^* \xrightarrow{k_2} P_1 + P_2$$

Values of k_2 are difficult to deduce for thermal reactions in the condensed phase because the rates measured are always combinations of these three steps. In the gas phase, however, k_2 is directly measurable, but the values are not directly transferable to the reactions in the liquid phase, since the reacting species in the gas phase are not solvated.

7.2 Quasi-Equilibrium and RRKM Theory

Two almost identical theories explaining the phenomena observed in the case of unimolecular reactions in the gas phase at high vacuum were proposed in 1952. One of them, the quasi-equilibrium theory (QET), was suggested by Rosenstock *et al.* [2] and applies to mass spectrometry. The other is named after the initials of its authors, RRKM, standing for Rice, Rampsberger, Kassel and Marcus [3], and deals with neutral molecules.

Both are based on assumptions and postulates. The first assumption is that the rotation, vibration, translation and electronic movements are independent of each other. The second assumption states that the movement of the nuclei can be expressed by classical mechanics. However, some quantum mechanical corrections are used.

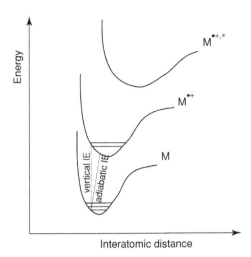

Figure 7.3
Morse curves for electron ionization. A vertical transition is observed if the interatomic distances have no time to adjust. The opposite is represented by the oblique line. Ionization can also lead to higher energies, including the excited electronic state.

The first postulate states that all the microscopic states are equally probable. In other words, all of the degrees of freedom participate in the energy distribution with the same probability. The second postulate states that the system can be described by movements on a multidimensional surface, and that a border surface exists that separates the reactants from the products. This surface can only be crossed in one direction: any reactant that crosses the transition state is irreversibly transformed into products.

The electron impact ionization of a molecule M to give a molecular ion in the ground and excited electronic states (respectively $M^{\bullet +}$ and $M^{\bullet +*}$) occurs over a very short time. An electron accelerated by a 10 V potential difference has a speed equal to 1.88×10^8 cm s^{-1}. It thus flies a distance of 1.88 Å, or 1.88×10^{-8} cm, in 10^{-16} s. This time is the interaction time of an ionizing electron with a molecule. The ionization must occur within that period. This is verified experimentally. This ionization is thus a vertical transition. The process is much more rapid than the time of one vibration, which is about 10^{-14} s in the case of the fastest ones. The distances between atoms thus do not change during the ionization.

This vertical ionization requires a higher ionization energy than the adiabatic process, as shown in Figure 7.3. Note that adiabatic in this case does not have the same meaning as in thermodynamics. The total energy transferred to the ion can, of course, be higher than that of the vertical transition. Indeed, the increasing energy of the electrons that are used raises the probability of vertical transitions that are more endothermic and thus favours the formation of ions of higher internal energy. However, the internal energy of the formed ions cannot be superior to a threshold corresponding approximately to 15 eV. Generally, the molecular ions of excited electronic states do not survive in the source because they lead to fragmentation or they return to their ground electronic state by emission of a photon. The molecular ions that leave the source to form stable or metastable ions thus contain generally a weak internal energy.

Electronic energy is unable by itself to lead to fragmentation of the molecular ion. The fragmentation requires a previous internal conversion of electronic energy to vibrational energy.

After the ionization, the energy distributes itself over the various degrees of freedom in a statistical fashion. The fast exchange of internal energy is done not only between the various degrees of freedom of the same electronic state but also between all the degrees of freedom of all the electronic states. These exchanges lead to the conversion of electronic energy acquired during ionization into vibrational and rotational energy of the ground electronic state of the molecular ion $M^{\bullet+}$. Consequently, the excited electronic states of the ionized molecule may be populated initially but relaxation to the ground electronic state occurs prior to their fragmentations. Thus, the fragmentation processes of the molecular ion are induced from this same electronic state in which vibrational and rotational energy are accumulated. As the energy distribution is much faster than the reaction, the energy will be distributed in a statistical fashion on the various degrees of freedom. It can be shown experimentally that the statistical energy distribution is carried out within a time span corresponding to a few vibrations, that is less than 10^{-10} s. Note that this time span is very short with respect to the time spent in the spectrometer source, at least 10^{-7} s.

As soon as an oscillator contains more energy than a certain E_0 value that is characteristic of it, this oscillator becomes the reaction coordinate and the molecule undergoes the fragmentation reaction. Moreover, this reaction occurs faster if the internal energy distributed over this oscillator in excess of E_0 is higher.

Both theories lead to the following expression for the rate constant [4,5]:

$$k(E) = \frac{1}{h} \frac{Z^{\ddagger}}{Z^*} \frac{P^{\ddagger}(E - E_0)}{\rho_E}$$

where h is Planck's constant, Z is the partition function for adiabatic degrees of freedom, $\ddagger$ refers to the activated complex, * refers to the active ionic species, $P^{\ddagger}(E - E_0)$ represents the total number of states corresponding to the activated complexes between energies zero and $E - E_0$ and ρ_E represents the density of states of the ion at energy E.

The following simplified equation is also used. It yields results that are flawed with important errors when it is applied to low or high energies.

$$k(E) = v \left(\frac{E - E_0}{E} \right)^{n-1}$$

where n is the number of vibrational degrees of freedom, v is the frequency factor, E is the internal energy of the ion and E_0 is the transition energy. Structural modifications in a molecule bring about variations of v and E_0; n also varies from one molecule to another. The v factor is an inverse function of the activated complex steric requirements; ΔS^* is all the weaker when the activated complex is highly ordered:

$$v = \frac{E - E_0}{h} \cdot e^{\Delta S^*/R}$$

Thus, referring to the spectrum in Figure 6.11, butanol can fragment, for example, according to the two following paths:

The first reaction can occur thanks to a transition state of any conformation, also called a loose complex. However, the second reaction must go through a transition state with a definite conformation called a tight complex. This brings about a reduction in the activation entropy for the second reaction. The growth of the rate constant with respect to energy is thus greater in the case of the first reaction than in the second. However, the activation energy of the second reaction is weaker than that of the first as the bond cleavages are partially compensated for by the formation of new bonds in the transition state.

As opposed to thermal reactions, the distribution of energy among the ions is not regular. Figure 7.4 represents a Warhaftig diagram deduced from these theories. An increment in the energy of the ionizing electrons induces a displacement of the energy distribution curve towards higher energies.

On the curve of rate constants as a function of energy, three limits are indicated by dotted lines: line 1 shows a rate constant lower than $10^6 \, s^{-1}$ and corresponds to an average lifetime higher than $10^{-6} \, s$. The corresponding ions reach the detector intact, without any fragmentation. Those corresponding to rate constants higher than $10^7 \, s^{-1}$ (at line 2 on the graph) fragment before leaving the source: only their fragments are detected. Those located between these limits fragment after leaving the source, but before reaching the detector. These metastable ions are detected only in peculiar circumstances. The dotted line 3 shows the intersection between both rate curves: on the left, the reaction with rearrangement is faster; on the right, the opposite is true.

As is shown by these curves, the proportion of ions that decompose during the time of flight is small. Most of the ions are either stable within the measurement time or dissociated within the source: the game is over before leaving the source. As a result, modifications of the TOF within the instruments through changes in the acceleration potential within the source or by switching from one instrument to another do not modify the appearance of the spectrum substantially. Thus spectra measured by magnetic instruments where the acceleration voltages hover around a few kilovolts are not very different from those obtained using quadrupole instruments where the acceleration voltages are up to a few dozen volts, or from those obtained with ICR instruments where the time spent by the ions in the instrument can be long.

This justifies the fact that spectral libraries used for comparative identifications are based on electron ionization. This method furnishes the spectra that are most easily compared.

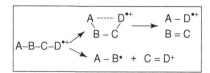

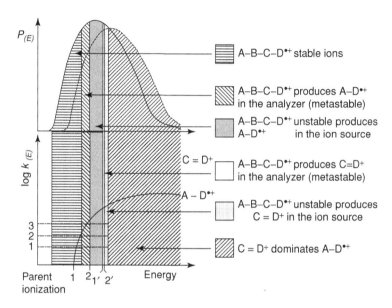

Figure 7.4
Warhaftig diagram. The x axis is the internal energy of the ions. The y axis of the top diagram, P, is the proportion of ions with the energy represented on the x axis. Point 1 indicates the activation energy of the first fragmentation reaction. An energy 2 is necessary, however, for the ion to decompose fast enough for the fragment to be observable: a rate constant higher than that indicated by the horizontal dotted line 2.

7.3 Ionization and Appearance Energies

The ionization of a neutral molecule requires a minimum energy called the ionization energy (IE) or the ionization potential (IP). Mass spectrometers allow one to determine this ionization energy. The principle consists of increasing the energy of the ionizing electrons in the source and observing which minimum energy allows one to observe the molecular ion. In practice, analytical mass spectrometers yield an energy dispersion that is much too wide for a precise determination. Instruments dedicated to this goal yield results in agreement with those obtained by photoionization.

The appearance of a fragment ion occurs from an energy including the ionization energy of the neutral molecule, the activation energy and the kinetic shift. The molecular ion dissociates into fragments only if it contains an excess energy great enough to allow decomposition; that is, enough energy to overcome the activation barrier. The corresponding energy is called appearance energy (AE) or appearance potential (AP) and corresponds to E_0. This is sufficient for observing the product ion provided it is observed over a long time,

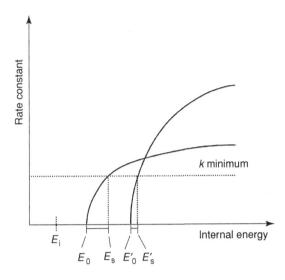

Figure 7.5
E_i: ionization energy of the precursor ion. E_0: the ion contains enough excess energy in order to fragment but the reaction is not fast enough to allow observation of the fragment. E_s: the excess energy is enough for the ion to dissociate before leaving the source. The difference $E_0 - E_s$ is called the kinetic shift.

as is possible, for example, in FTMS. In classical instruments the fragments are observed after a very short time of about 10^{-8} s, as we have seen. The parent excess energy must be enough to ensure a dissociation rate of that magnitude. Figure 7.5 illustrates this principle.

7.4 Fragmentation Reactions of Positive Ions

The terminology and the symbolism suggested by McLafferty [6, 7] are used here as they have become universal. Moreover, the name McLafferty is associated with a rearrangement that we will discuss later.

7.4.1 Fragmentation of Odd-Electron Cations or Radical Cations ($OE^{\bullet+}$)

In radical cations, the charge is delocalized over the whole molecule. The most favoured radical and charge sites in the molecular ion are assumed to arise from the loss of the electron of lowest ionization energy in the molecule. Consequently, when we write fragmentation reactions on paper, we represent the charge as localized on the site with the weakest ionization energy. As the following order $n- > \pi- > \sigma$-electrons is observed for ionization, the heteroatoms with weak ionization energies carry the charge preferentially. The symbol $^{\bullet+}$ at the end of the molecules means an odd-electron ion without specifying the localization of either the radical or the charge site. However, use of either $^{\bullet}$ or $^{+}$ within the molecule implies localization of the radical or the charge.

Positive charge and radical sites are electron-deficient sites. Cleavage reaction initiated by the positive charge site involves attraction of an electron pair because positive charge corresponds to the loss of an electron. The movement of an electron pair induces heterolytic cleavage with migration of the charge site. Cleavage reaction initiated by the radical site arises from its strong tendency for electron pairing. An odd electron is donated to form a new bond whereas the transfer of a single electron induces homolytic cleavage with migration of the site of the unpaired electron.

Thus, the charge and the radical sites induce different cleavage reactions. The indication i and α involve all types of reactions initiated respectively at a charge or a radical site.

The fragmentation of these radical cations without any rearrangement or without any cleavage of an even number of bonds such as occurs in rings necessarily leads to an even-electron, or closed shell, ion and to a neutral radical. The parity rules were discussed in Section 6.6.

7.4.1.1 Direct Dissociation (σ)

The expulsion of an electron from a σ bond can bring about the direct dissociation of the latter. We then speak of 'σ fragmentations'. One of the fragments keeps the charge, while the other is a radical:

$$R-R' + e^- \longrightarrow R-R'^{\,\bullet+} + 2e^-$$

$$R-R'^{\,\bullet+} \Big\langle \begin{array}{l} R^{\bullet} + R'^{+} \\ R^{+} + R'^{\,\bullet} \end{array}$$

According to *Stevenson's rule*, if two charged fragments are in competition to produce a neutral radical by electron attachment, the radical having the highest ionization energy will be produced. The other ion, whose corresponding neutral radical has a lower ionization energy, will hold its charge and will thus be the observed fragment. Indeed, this reaction can be considered as a competition between two cations to carry away the electron:

$$R^+ \cdots e^- \cdots R'^+$$

This rule is illustrated by the two examples in Figure 7.6. However, Stevenson's rule must only be applied to the *competitive* formation of fragment ions. Further dissociation or additional formation by other pathways could respectively decrease or increase the abundance of the fragment ions and lead to erroneous interpretations.

7.4.1.2 Cleavage of a Bond Adjacent to a Heteroatom (i)

The bond adjacent to a heteroatom can be broken by a *charge-site-initiated reaction*, that is by attraction of an electron pair from this bond, and we talk about an induced cleavage (i):

$$R-CH_2-\overset{\displaystyle\frown}{Y}{}^{+}-R' \xrightarrow{\ i\ } R-CH_2{}^+ + {}^{\bullet}Y-R'$$

This cleavage of the bond adjacent to the heteroatom can be seen as a direct dissociation assisted by inductive electron withdrawal due to the difference in electronegativity, but it occurs after the ionization.

In principle, Stevenson's rule still applies. By this rule, the cleavage of the adjacent bond could involve radical migration and charge retention if the ionization energy of ${}^{\bullet}YR'$ is less

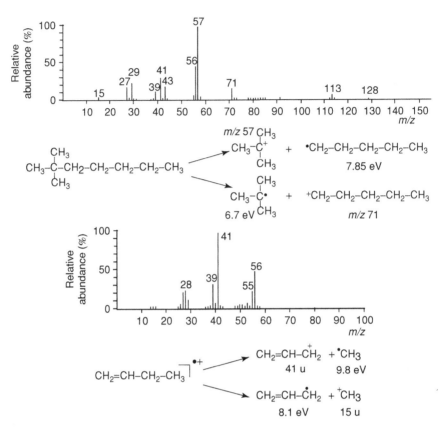

Figure 7.6
The ionization energy of the *t*-butyl radical is lower than that of the pentyl radical. Thus, the *t*-butyl ion is preferentially observed in the top spectrum. The same holds true for the comparison between the allyl and methyl radicals.

than that of $RCH_2{}^{\bullet}$. This counterpart reaction is classified as a special case of a σ bond cleavage and can be written as follows:

$$R-CH_2-\overset{\bullet+}{Y}-R' \xrightarrow{\sigma} R-CH_2^{\bullet} + {}^+Y-R'$$

7.4.1.3 Cleavage of the Alpha Bond

The alpha bond to the radical cation site can be broken by a *radical-site-initiated reaction*, that is by a transfer of the unpaired electron to form a new bond to an adjacent atom (α atom) with concomitant cleavage of another bond of this atom. The new bond compensates energetically for the cleaved bond. In this case we speak about a 'radical-site-initiated α fragmentation':

$$R-CH_2-\overset{\bullet+}{Y}-R' \xrightarrow{\alpha} R^{\bullet} + CH_2=\overset{+}{Y}-R'$$

A half arrow indicates the displacement of a single electron. Note that when the heteroatom corresponds to an oxygen atom, the positively charged oxygen in this last cation is isoelectronic with nitrogen and thus forms three covalent bonds.

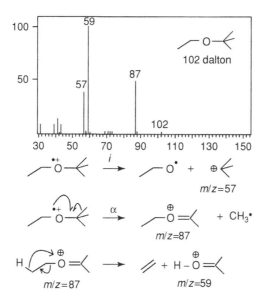

Figure 7.7
Fragmentation of *t*-butyl ethyl ether.

Sometimes, several competitive cleavages are possible. Among the different possibilities, the loss of the radical of the highest ionization energy is generally observed. However, if several alkyl chains can be lost as radicals, the loss of the longest chain is favoured.

By Stevenson's rule, the cleavage of the alpha bond could involve radical retention and charge migration if the ionization energy of $^{\bullet}CH_2YR'$ is more than of $R^{\bullet}$. This counterpart reaction is classified as a special case of a σ bond cleavage and can be written as follows:

$$R-CH_2-\overset{\bullet+}{Y}-R' \xrightarrow{\ \sigma\ } \overset{+}{R} + {}^{\bullet}CH_2-Y-R'$$

The following spectra illustrate these different mechanisms. In the mass spectrum of *t*-butyl ethyl ether presented in Figure 7.7, the fragmentation of the adjacent bond is observed because the *t*-butyl ion is very stable. The high electronegativity of the oxygen allows this fragment to carry away the electron; the ion with m/z 45 ($CH_3CH_2O^+$) is not observed. The other fragmentation path, α cleavage, leads to the loss of one methyl at m/z 87. It is immediately followed by a rearrangement which brings about the loss of ethylene and leads to the ion with m/z 59.

As a comparison, Figure 7.8 shows the mass spectrum of ethyl 2-butyl ether. The α fragmentation induced by the radical site has two possibilities for losing a methyl group, and only one for the loss of an ethyl group, giving the 87 and 73 Th fragments respectively. We see that the loss of the larger hydrocarbon radical is preferred. It is followed by the loss of an ethylene molecule via a hydrogen rearrangement, which dominates the spectrum at 45 Th. In the same way, the loss of a methyl followed by the elimination of ethylene leads to a 59 Th fragment. The cleavage of the adjacent bond which leads to the butyl cation (57 Th) also is observed.

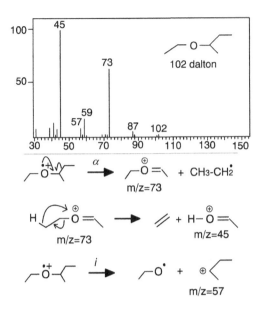

Figure 7.8
Spectrum of ethyl 2-butyl ether.

Figure 7.9 displays two examples where, respectively, the cleavage of the α bond initiated by the radical and cleavage of the adjacent bond initiated by the charge are in competition with their counterpart reactions.

7.4.1.4 Competition Between the Cleavages of the Adjacent and the Alpha Bonds

The cleavage of the adjacent bond occurs more easily if the heteroatom is a large atom. In the case of neighbouring atoms, the more electronegative one leads more easily to the cleavage of the adjacent bond. The α cleavage becomes predominant for electron donors. The following order is observed:

$$Br, Cl < R^\bullet, \pi \, bond, S, O < N$$

Thus the halogens preferentially cause the loss of the radical $X^\bullet$ through a cleavage of the adjacent bond, whereas the amines preferentially lose a radical through a cleavage of the alpha bond.

As an example compare the spectra of butylamine and of butanethiol presented in Figure 7.10. In the case of butylamine, the ion that is most intense corresponds to a radical-initiated α cleavage leading to the $CH_2 = NH_2^+$ ion at 30 Th:

$$\text{\raisebox{0.5ex}{\quad}} \overset{\alpha}{\longrightarrow} CH_2{=}\overset{\oplus}{N}H_2 + \text{\raisebox{0.5ex}{\quad}}$$
$$m/z = 30$$

However, in the case of butanethiol, the main fragmentation occurs through a cleavage of the adjacent bond and corresponds to the loss of an $HS^\bullet$ radical and the most intense ion is

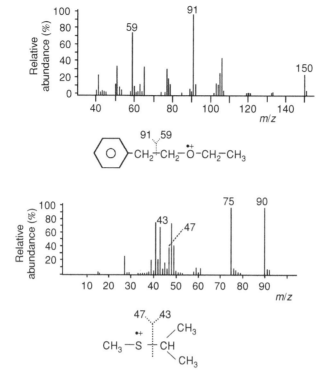

Figure 7.9
Top: in this spectrum, the radical-site-initiated α cleavage is observed at *m/z* 59. The α bond cleavage with charge migration is also observed at *m/z* 91. Bottom: in this spectrum, the adjacent bond cleavage initiated by the charge is observed at *m/z* 43. The counterpart reaction corresponding to the adjacent bond cleavage with charge retention is also observed at *m/z* 47.

the butyl cation at 57 Th. The radical-initiated α cleavage leading to the $CH_2 = SH^+$ ion at 47 Th is also observed but is clearly less important.

7.4.1.5 Fragmentation of Radical Cations with Rearrangement

The rearrangements that can occur are very numerous and often make the interpretation of the spectra very difficult. However, some are frequent, very specific and well understood.

Thus the McLafferty rearrangement consists of the transfer of a hydrogen atom to a radical cation site using a six-atom ring as an intermediate. The radical cation that results now has a radical site far away from the cation site: it is a *distonic radical cation*. This rearrangement is then followed by either a radical-site-induced or a charge-site-induced fragmentation, yielding in both cases a neutral molecule and a new radical cation. In the absence of nitrogen, these ions have an even mass and are easily detected in the spectrum.

Figure 7.11 shows the spectrum of 2-hexanone. For ketones both of the paths, i and α, lead to identical products except for the charge. Once again the cation corresponding

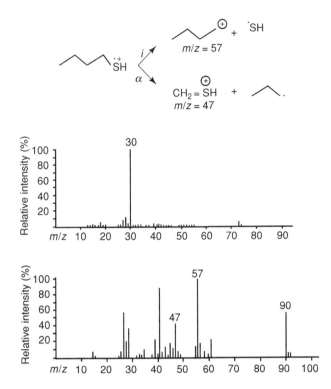

Figure 7.10
Spectra of (top) butylamine and (bottom) butanethiol.

to the radical with the lowest ionization energy is predominant. In this case the enol is better stabilized by the resonance with the lone pair on the oxygen. Note that the radical cation with an odd number of electrons appears at the even mass-to-charge ratio of 58 Th, in agreement with the parity rules.

The predominant peak at m/z 43 in Figure 7.11 corresponds to the alpha cleavage initiated by the radical:

$$\text{structure} \longrightarrow \text{structure} \cdot + CH_3-\overset{\oplus}{C}O$$
$$m/z = 43$$

7.4.2 Fragmentation of Cations with an Even Number of Electrons (EE⁺)

Using electron ionization, the molecular radical cation is formed in the source. This radical cation fragments into a radical and a cation with an even number of electrons or, through rearrangements or multiple steps, into a neutral molecule and a new odd-electron cation. The latter are often easily recognized in the spectrum because their mass is even in the absence of a nitrogen atom.

The soft ionization techniques such as FAB, TSP, CI ESI, MALDI, APCI, and so on produce molecular species with an even number of electrons, most often by the addition

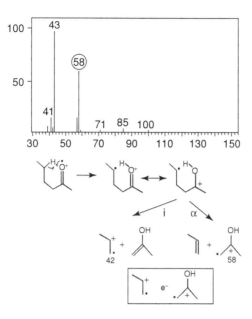

Figure 7.11
McLafferty rearrangement through a six-atom
ring intermediate. Figure 6.11 shows another
example and details the rules linking the mass
and electron parities.

or abstraction of a proton. As opposed to radical cations, which are met with almost
exclusively in mass spectrometry, the even cations are common in 'classical' chemistry,
and their reactions seem much more familiar to the chemist. These molecular species
are generally more stable than the radical cations produced by electron ionization. From
an analytical point of view, the spectra are much simpler. However, compared with EI,
the rearrangements are more frequent and more varied. Usually these spectra yield less
information and are more difficult to interpret than the EI spectra. However, in addition
to yielding the molecular mass more easily, such spectra may be more sensitive to small
changes in the structure: isomers yielding identical EI spectra often give rise to different
CI spectra.

For example, *cis*- and *trans*-1,4-cyclohexanediols yield essentially identical EI spectra.
In CI, the *trans* isomer yields an $(M + H - H_2O)^+$ peak that is relatively more intense than
that for the *cis* isomer [8], a phenomenon which allows one isomer to be distinguished from
the other. The difference is ascribed to the possibility of forming a hydrogen bond in the
case of the *cis* isomer and to the assisted fragmentation and cyclization in the case of the
trans isomer:

We shall label the even-electron cations using the symbol EE^+, standing for 'even-electron number', and the symbol $OE^{\bullet+}$ will label the radical cations, standing for 'odd-electron number'.

The production of a radical cation from an even-electron, or closed shell, ion is necessarily accompanied by that of a radical corresponding to the homolytic cleavage of a bond:

$$EE^+ \longrightarrow OE^{\bullet+} + R^{\bullet}$$

This process is usually highly endothermic and thus improbable. A statistical estimate based on many spectra shows that an even-electron ion yields even fragments in about 95 % of cases.

We thus have the following situations:

$$OE^{\bullet+} \begin{cases} \longrightarrow EE^{\oplus} + \dot{R} \\ \xrightarrow{\ r\ } OE^{\bullet+} + M \end{cases}$$

$$EE^{\oplus} \begin{cases} \xrightarrow{\ \times\ } OE^{\bullet+} + \dot{R} \\ \longrightarrow EE^{\oplus} + M \end{cases}$$

A cation with an odd number of electrons can fragment along two paths: either the production, through a cleavage i or α, of an EE^+ cation and a radical, or the production of a new radical cation ($OE^{\bullet+}$) and a molecule M after rearrangement. A cation with an even number of electrons (EE^+) usually can produce only a new EE^+. The same rules apply to negative ions. These two types of ions can thus be recognized by recalling that, *in the absence of nitrogen*, the following holds true:

$$OE^{\bullet+} \text{ or } OE^{\bullet-}: \text{even mass} \quad \text{and} \quad EE^+ \text{ or } EE^-: \text{odd mass}$$

7.4.3 Fragmentations Obeying the Parity Rule

The cleavage of the bond that is adjacent to the charged site, while observing a migration of the charge, is a common fragmentation process that occurs especially often when it allows the elimination of a small stable molecule. For example, protonated alcohols in soft ionization methods often lose water:

$$R - OH + H^+ \longrightarrow R - OH_2^+$$
$$R - OH_2^+ \longrightarrow R^+ + H_2O$$

McLafferty [9] proposed classifying the reactions of even-electron ions that obey the parity rule (an even ion yields an even ion + neutral fragment) as follows:

1 Cleavage of a bond with charge migration:

$\longrightarrow CH_3-CH_2^+ + H_2O$

2 Cleavage of a bond with cyclization and charge migration:

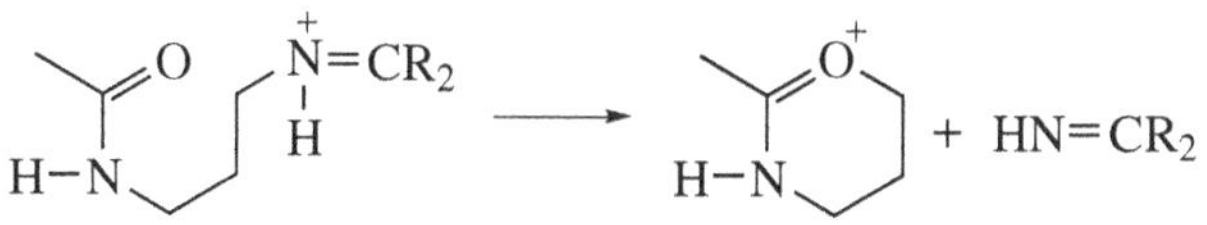

3 Cleavage of two bonds in a cyclic ion with charge retention:

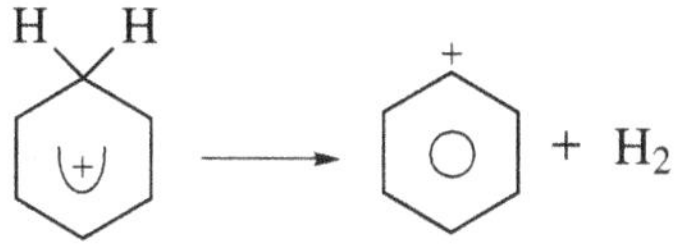

4 Cleavage of two bonds with rearrangement and charge retention:

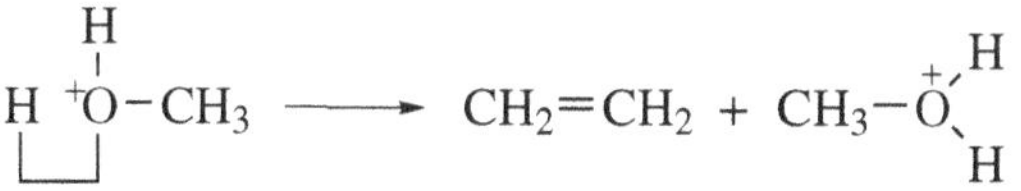

Type 1 and 2 reactions are the most frequent in the absence of collision activation. The first type occurs especially easily when the protonated site is less basic, as is shown by the following data concerning n-butyl-XR molecules [10] (PA = proton affinity). Of course, the ease with which these molecules separate from an organic cation or from a proton evolves in the same order:

$-XR$	NH_2	φ	SH	OH	I	Br	Cl
$[(M+H)-HXR)]^+/(M+H)^+$	0.04	0.08	0.15	19.5	240	>250	>250
PA (kcal mol^{-1})	207	183	175	164	145	141	140

This reaction is also influenced by the stability of the cation that is formed, as is shown by the comparison of the following data concerning various butylamines. A neutral molecule is lost even more easily when the cation product is stable. Thus, for example, in CI, a protonated tri-alkylamine loses ammonia more easily than a mono-alkylamine:

	n-butyl	s-butyl	t-butyl
$[(M+H)-NH_3]^+/(M+H)^+$	0.4	0.07	2.3

The type 2 reaction is in fact the same reaction that is favoured by the cyclization possibility if there is a heteroatom in a convenient position. The latter probably also provides anchimeric assistance to the expulsion of a neutral fragment. The example of cyclohexanediols provides a good illustration. As is shown by the following data, the comparison of ω-amino alcohols $H_2N(CH_2)_nOH$ with various linear chain lengths also offers an application of this principle:

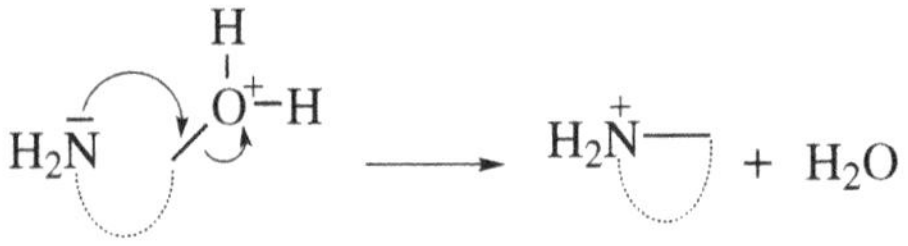

$$H_2\overset{\cdot\cdot}{N} \qquad \overset{H}{\underset{}{\overset{|}{\underset{}{\overset{+}{O}}}}}-H \longrightarrow H_2\overset{+}{N} + H_2O$$

n		2	3	4	5	6
$[(M+H) - H_2O]^+/(M+H)^+$		0.42	0.21	0.20	0.15	0.07

Moreover, the $(M+H-NH_3)^+$ ion is detected only for $n = 4$ or 5. In the case of $n = 5$:

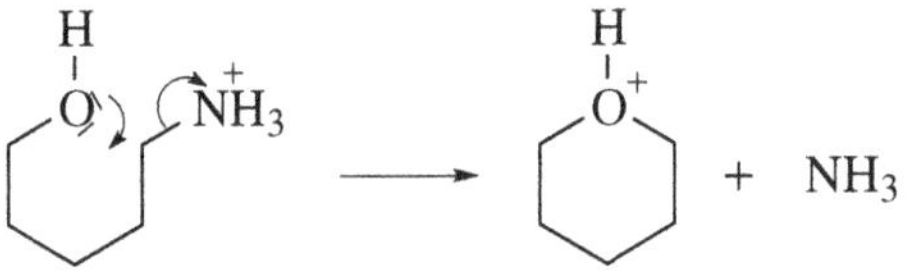

$$\overset{H}{\underset{}{\overset{|}{O}}} \quad \overset{+}{N}H_3 \longrightarrow \overset{H}{\underset{}{\overset{|}{\overset{+}{O}}}} + NH_3$$

Concerning type 3 and 4 reactions, Cooks and co-workers [11] noted that they are observed especially when the rearrangement is a four-centre one. The following examples illustrate this:

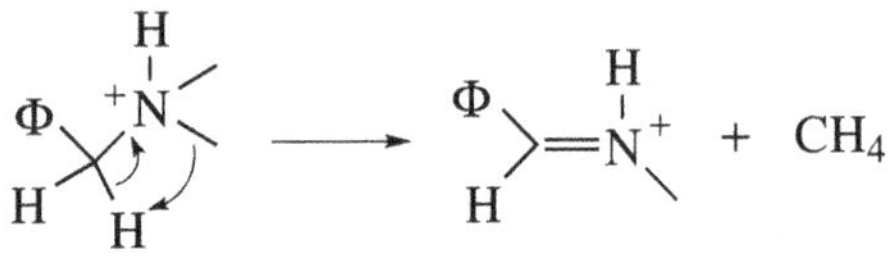

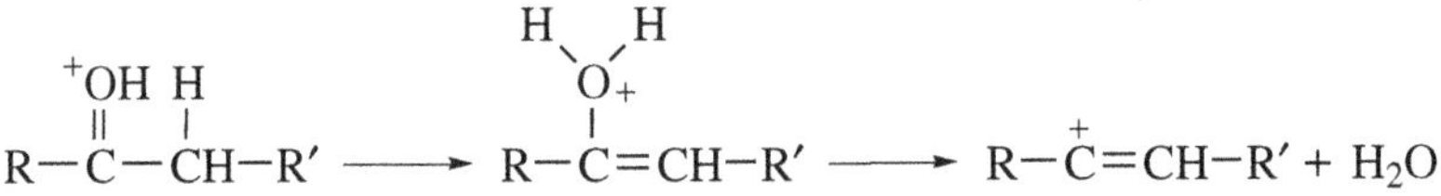

$$\Phi \overset{+}{N} \longrightarrow \Phi \overset{H}{\underset{H}{\diagdown}}=\overset{+}{N} + CH_4$$

$$\underset{R-\overset{||}{\underset{}{C}}-\overset{|}{\underset{}{C}}H-R'}{\overset{+OH \; H}{}} \longrightarrow \underset{R-\overset{|}{\underset{}{C}}=CH-R'}{\overset{H \diagdown O \diagup H}{\overset{+}{}}} \longrightarrow R-\overset{+}{C}=CH-R' + H_2O$$

Note also that in CI a ketone can lose water, which never occurs in EI. Moreover, esters often yield the ion corresponding to the protonated acid:

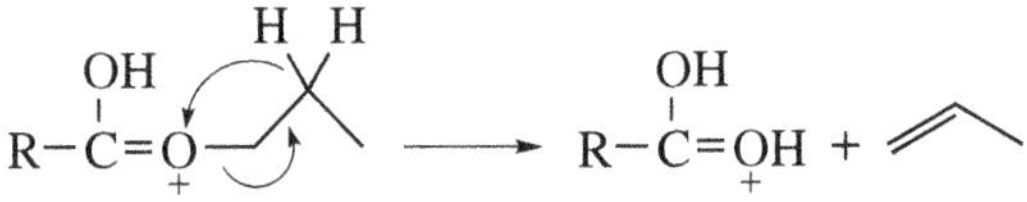

$$\underset{R-\overset{|}{\underset{+}{C}}=O}{\overset{OH}{}} \longrightarrow \underset{R-\overset{|}{\underset{+}{C}}=OH}{\overset{OH}{}} + \diagup\!\!\diagdown$$

The presence of two heteroatoms complicates the mechanism. In fact, propyl acetate yields protonated acetic acid as a base peak, whereas methylbutyrate yields an $(RCOOH + H)^+$ peak that is hardly detectable: the proton is derived from the alcohol. Experiments on propyl acetate labelled with deuterium at various positions on the propyl chain yield the following percentages concerning the origin of the proton during the rearrangement. This distribution seems fairly statistical:

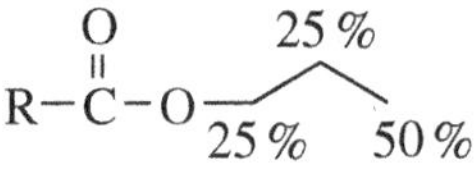

$$\underset{R-\overset{||}{\underset{25\%}{C}}-O}{\overset{O \quad 25\%}{}}\diagup\!\!\diagdown\,_{50\%}$$

The reaction may seem complicated but its analytical usefulness is very high. For example, some triglycerides yield, in CI, the three protonated acids and the pseudomolecular peak $(M + H)^+$.

Iminium ions with short chains fragment by forming an ion–neutral complex [12]. The slow step of the process is the 1,2 proton transfer leading to a secondary carbenium:

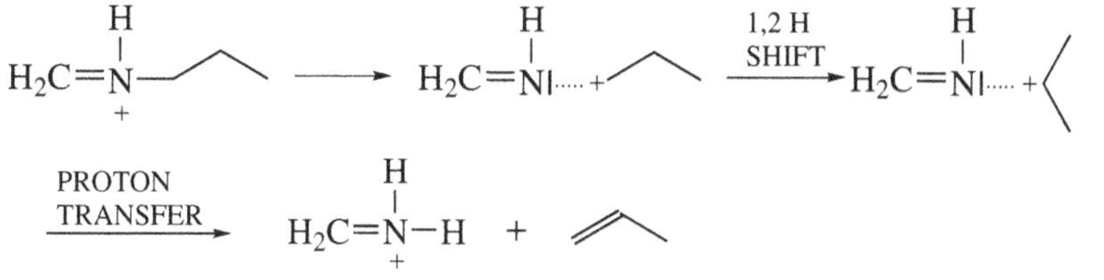

If the alkyl chain is longer, a McLafferty rearrangement, now occurring on an even-electron ion, becomes the dominant process. Labelling experiments show that the hydrogen originates from the γ position. The rearrangement thus results exclusively from a 1,5 shift, occurring through a six-atom ring:

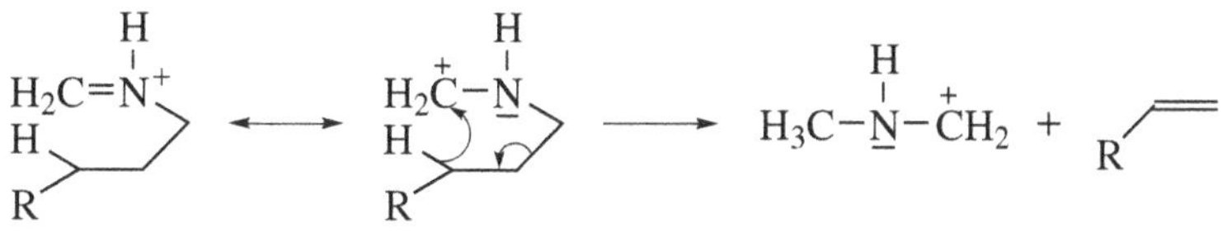

A similar mechanism could occur in other onium cases (oxonium, sulfonium, etc.).

7.4.4 Fragmentations Not Obeying the Parity Rule

The reactions considered up to now were limited to those of the McLafferty classification with regard to the parity rule. The reactions of even-electron ions (EE) that do not obey the parity rule are much rarer and more difficult to predict. They are often observed in ions with extended π systems, but they often imply complex rearrangements, as is shown in the case of tropylium:

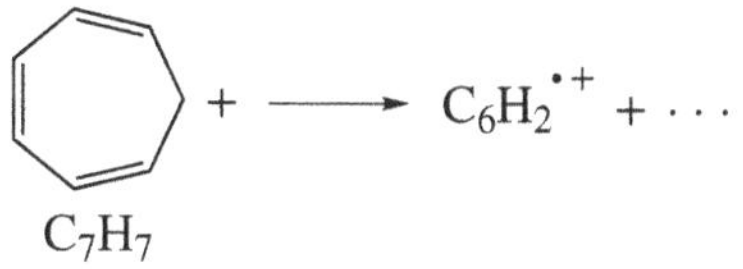

Note, however, that the main fragmentation path of the tropylium ion is the loss of acetylene.

It seems that the stability of the radical can play an important role. Thus nitroso compounds easily yields $\cdot\bar{N}{=}\dot{O}$, a stable radical.

7.5 Fragmentation Reactions of Negative Ions

The use of anions in mass spectrometric analysis was developed much later than the study of positive ions: the production of negative ions using the classical EI method is smaller than the production of positive ions by several orders of magnitude and commercial instruments were dedicated to the detection of positive ions. However, the situation has now completely changed, owing both to the discovery of the conversion dynode, allowing detection of the negative ions with high sensitivity, and to the development of many ionization techniques that produce negative ions efficiently.

The ions most often observed are even-electron $(M-H)^-$ anions. They are efficiently produced in the source only from compounds containing acidic functions. This allows some selectivity for their detection in mixtures.

In general, even-electron negative ions contain less energy than the positive ions, as explained in Chapter 1, and thus produce less fragments. This explains why the most important studies and applications of negative ion fragmentations make use of tandem mass spectrometry techniques. For some compounds, the sensitivity in the negative ion mode may be much higher than in the positive ion mode.

7.5.1 Fragmentation Mechanisms of Even-Electron Anions (EE^-)

The fragmentation of even-electron anions has been reviewed by Bowie [13, 14]. Observed reaction pathways include homolytic bond cleavage, the loss of one or several H• radicals being a common process. Alkyl radical loss is also observed [15, 16]:

$$H• \text{ loss: } (CH_2COCH_3)^- \longrightarrow •CH_2COCH_2^- + H•$$
$$\text{alkyl radical loss: } Ph^-CHOCH_3 \longrightarrow PhCHO^{•-} + CH_3•$$

Reactions through initial formation of an anion–neutral complex are often observed. They are followed by a displacement of the anion, a deprotonation, and an elimination process, such as [17]

$$^\ominus CH_2COCOCH_3 \longrightarrow [CH_3CO^\ominus(CH_2CO)] \longrightarrow CH_3-C{\overset{\ominus}{\underset{\diagdown O}{}}} + CH_2CO$$

Complexes of the hydride ion with neutrals are invoked to explain some observed reactions, especially the loss of hydrogen molecules from alcoholates and enolates:

$$CH_3-CH_2-O^\ominus \longrightarrow [H^\ominus(CH_3-CHO)] \longrightarrow [H_2(CH_2=CHO^\ominus)] \longrightarrow H_2 + CH_2=CHO^\ominus$$

Various rearrangements often result from internal nucleophilic condensation, intramolecular nucleophilic substitution, formation of an ion–molecule complex, and so on.

The fragmentation of ethylene glycol diacetate, shown in Figure 7.12, gives examples of many of these fragmentation mechanisms [18]. An almost identical spectrum is observed from the fragmentation of the negative pseudomolecular ion of ethylene glycol monoacetoacetate. This demonstrates the reversible rearrangement, step **a** in Figure 7.13.

From the diacetate ester (Figure 7.13), an ion–molecule complex consisting of the neutral ketene and the complementary alcoholate is formed. Either this complex dissociates to yield the ethylene glycol acetate anion A, which further fragments to yield the acetate anion E, or a proton transfer from the ketene to the alcoholate occurs in the complex, which after dissociation yields the ynolate ion G.

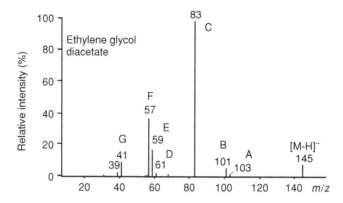

Figure 7.12
MS/MS product ion spectrum of the $[M-H]^-$ anion of ethylene glycol diacetate. Almost the same spectrum is observed from ethylene glycol monoacetoacetate. See Figure 7.13 for the explanation of the observed masses. Reproduced (modified) from Stroobant V., Rozenberg R., Bouabsa el M., and de Hoffmann E., J. Am. Soc. Mass Spectrom., 6, 498, 1995.

From the ethylene glycol monoacetoacetate, the acetoacetate anion is produced. The latter loses carbon dioxide to yield the acetone enolate ion F. Alternatively, an internal proton rearrangement and fragmentation yield an ion–molecule complex containing the ethylene glycol anion and acetylketene. This complex can dissociate to yield the free ethylene glycol anion D, or the acetylketene is ionized to produce the corresponding anion C.

7.5.2 Fragmentation Mechanisms of Radical Anions (OE$^{\bullet-}$)

Radical anions (OE$^{\bullet-}$) undergo single cleavage reactions. Bowie [19] showed that they occur α to the charged site or α to an atom that is conjugated with the charged site. They also give rise to rearrangement reactions.

Radical anions are not often observed. Fullerene [20] has no hydrogen atom in it, and thus ionization can only occur through electron capture or anion attachment. Figure 7.14 displays its spectrum obtained under negative ion desorption chemical ionization conditions with a $CH_4–N_2O$ mixture as the ionizing gas [21].

7.6 Charge Remote Fragmentation

Many studies have shown [22–24] that gas-phase ion fragmentation can occur at sites that are physically removed from the location of the charge. This is often called charge remote fragmentation (CRF) or sometimes remote (charge) site fragmentation. There is little if any involvement of the charge in the reaction. The charge must be stable in its location and not be able to move to the reaction site. Generally, the energy required to cause CRF is often considerable. CRF has been observed for both anions (sulfates, sulfonates, fatty acids, etc.) and cations (long-chain amines and phosphonium). For example, palmitic acid yields, with FAB, the anion $C_{16}H_{31}O_2^-$ which loses C_nH_{2n+2} starting from the alkyl end, as shown in

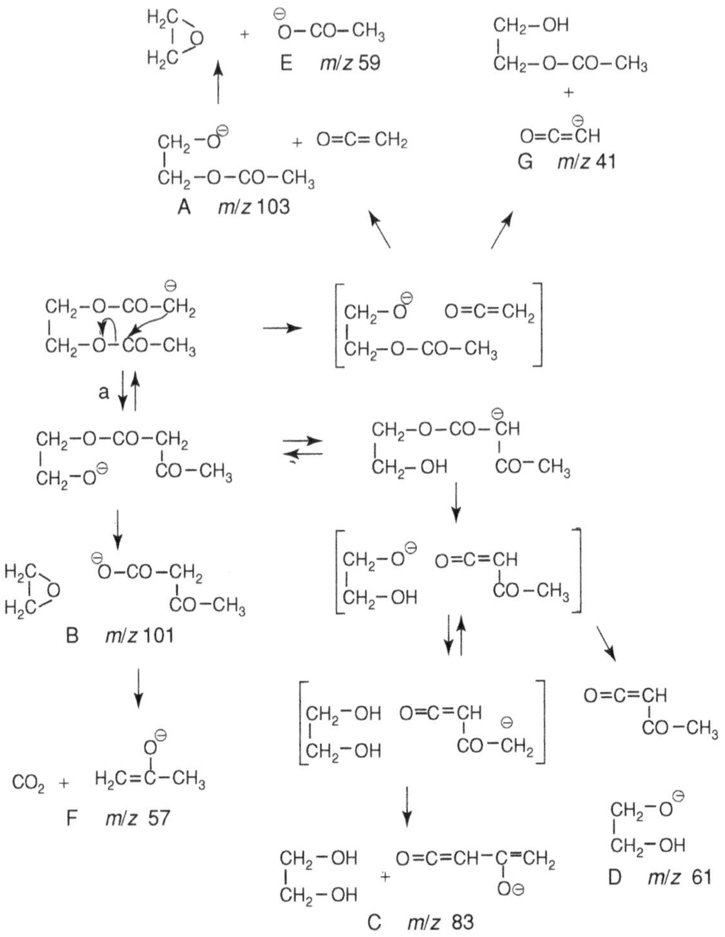

Figure 7.13
Fragmentation scheme of ethylene glycol diacetate and ethylene glycol monoacetoacetate anions. The spectrum is displayed in Figure 7.12. These two anions interconvert by the reversible pathway **a**, as demonstrated by the identity of the spectra of these two compounds. Other steps have been proved to be reversible by labelling experiments. Square brackets indicate ion–molecule complexes.

Figure 7.15. Other examples of CRF are discussed in Chapter 8. Charge remote fragmentations have been reviewed in [25, 26].

7.7 Spectrum Interpretation

We have seen a series of methods that allow us to work our way from a spectrum back to the molecular structure: information on the elemental composition, the molecular mass, the number of rings and unsaturations, the relationships between the structure and the

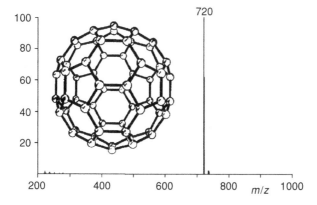

Figure 7.14
Negative ion DCI spectrum of fullerene, discovered by Kroto [20], using a mixture of CH_4 and N_2O as the ionizing gas. Small peaks are observed at 734 and 736 Th. Reproduced (modified) from Richter H., Dereux A., Gilles J.M., Guillaume C., Tiry P.A., Lucas A.A., Rozenberg R., Spote A., de Hoffmann E. and Girard C.H., Ber. Bunsenges. Phys. Chem., 98, 1329, 1994.

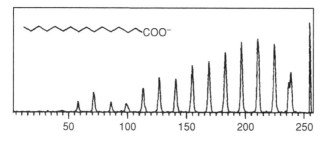

Figure 7.15
High-energy tandem mass spectrum of palmitate $(M-H)^-$ ion in the negative ion mode. Reproduced (modified) from Jensen N.J., Tomer K.B. and Gross M.L., J. Am. Chem. Soc., 107, 1863, 1985, with permission.

fragmentations, and so on. In this section we will study a few complementary aspects that are useful in the interpretation of spectra.

We saw that numerous ionization techniques exist that yield radical cations or radical anions, protonated or deprotonated molecules, and various adducts. These ions yield fragments with an even number of electrons (closed shell) or with an odd number of electrons (open shell). Even though the radical cations derived from electron ionization sources retain a privileged status in common mass spectrometry, the other ionization methods become increasingly common. Electron ionization is not possible for many categories of molecules. Therefore, we will not limit the discussion to radical cations.

Let us recall that the four main types of ions are closed-shell or open-shell, positive or negative ions.

7.7.1 Typical Ions

Some ions are typical of given structures. Note first that compounds containing hydrocarbon chains give rise to a series of ions distant from each other by 14 Da (–CH$_2$–). The mass where they appear depends on the group(s) that is (are) linked to them. Entirely saturated hydrocarbon ions appear at masses 15, 29, 43, 57, 71, 85, 99 Da, and so on.

Molecules with a benzene nucleus often yield a phenylium ion at m/z 77, accompanied by a fragment corresponding to acetylene loss at m/z 51. If an alkyl chain is bonded to the benzene nucleus, ions are observed at m/z 91, which are a mixture of benzylium and tropylium structures, which produce a fragment observed at m/z 65 by losing acetylene.

Trimethylsilyl derivatives are commonly used in GC/MS. When the molecule contains a hydroxyl, it fragments and yields (CH$_3$)$_3$Si$^+$, at m/z 73, and (CH$_3$)$_2$Si$^+$–OH, at m/z 75. When the molecule contains more than one (CH$_3$)$_3$SiO– group, an ion with mass 147 Da is systematically observed in the spectrum, even though both groups are remote from one another. This ion has the structure shown below. It is derived from the fragmentation of a complex between the ion (CH$_3$)$_3$Si$^+$ and the neutral remainder of the molecule:

$$CH_3-\underset{\underset{CH_3}{|}}{\overset{\overset{CH_3}{|}}{Si}}-O-\overset{\oplus}{Si}\overset{CH_3}{\underset{CH_3}{\diagdown}}$$

$$m/z = 147$$

7.7.2 Presence of the Molecular Ion

In EI the molecular ion is only weakly observed in the case of linear saturated hydrocarbons. The presence of branching usually entails the disappearance of this peak. However, an unsaturation, and especially an aromatic ring, makes the molecular ion peak more intense. The presence of electronegative saturated heteroatoms (oxygen, fluorine) normally prevents the observation of the molecular ion. In fact, its intensity depends on the groups that are present. Thus an aliphatic ester yields a weak or absent molecular ion, whereas an aromatic one usually yields an intense molecular ion peak.

Using soft ionization techniques in the positive ion mode, $(M-H)^+$ is most often observed in the case of saturated compounds and $(M+H)^+$ in the case of unsaturated compounds or heteroatom-containing compounds. However, halogen compounds or compounds containing an sp^3 oxygen often prevent the observation of the molecular ion. Thus, for example, it is very difficult to form the molecular ion or the pseudomolecular ion of acetals or orthoesters.

7.7.3 Typical Neutrals

The loss of a hydrogen atom is especially observed in EI starting from an aldehyde function, in the case of oxygen-containing or nitrogen-containing heterocycles, or when many hydrogen atoms are bonded to carbons α to nitrogen or oxygen atoms.

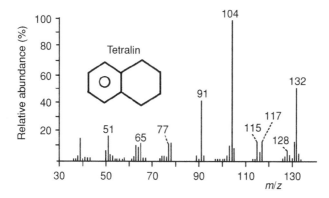

Figure 7.16
EI spectrum of tetralin. The complete aromatization through the loss of four hydrogen atoms is indicated by the presence of the peak at m/z 128. The most abundant ion results from the loss of ethylene. As in the case of most cyclanics, an intense ion is observed at $(M - 15)$, derived from the loss, following a rearrangement, of a methyl group. The typical ions benzylium (91) and phenylium (77) lose 26 Da, acetylene, and produce fragments at m/z 65 and 51, respectively.

The loss of a hydrogen molecule is observed starting from any ionic species (positive or negative, open- or closed-shell) every time it brings about an increased conjugation or aromaticity. It is especially common in the case of cyclic compounds. For example, the spectrum of tetralin (Figure 7.16) shows an intense ion corresponding to the loss of four hydrogen atoms.

The loss of 15 Da is typical for the elimination of a methyl group. Note that saturated rings often yield an intense loss of 15 Da. For example, the EI spectrum of cyclohexane, which has a molecular mass of 84 Da, shows a fragment with m/z 69 and a relative intensity of 25 %. This ion is also present in the EI spectrum of tetralin (Figure 7.16).

The 16 Da loss corresponds to a methane loss observed, as for the hydrogen loss, when it produces a conjugation or aromaticity gain, especially in cyclic compounds. Thus steroids or bile salts commonly lose methane from an angular methyl group. This 16 Da loss is also observed in N-oxides and sulfoxides, and then results from an oxygen atom loss.

The fragmentation resulting from water elimination is indicated by an 18 Da loss. It is common when a hydroxyl group is present, and for various types of ions. Molecules containing a carbonyl group do not commonly lose water except from their conjugated acids. When studying the spectrum, remember the possible 18 Da loss corresponding to $(H_2 + CH_4)$, mostly in the cases of non-aromatic rings.

Losses of 19 or 20 Da are typical of the presence of fluorine, whose atomic mass is 19 Da. Always beware, however, of possible consecutive losses of, for example, water and hydrogen that also lead to a 20 Da loss.

Some other neutral losses are typical of peculiar structures and sometimes have great analytical usefulness. Thus, methyl esters of fatty acids lose the three carbon atoms next to

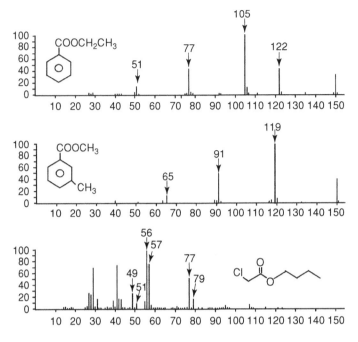

Figure 7.17
Spectra of three compounds having the same mass (150 Da) illustrating a few interpretative rules.

the carboxymethyl and a hydrogen atom through a specific rearrangement. This allows one to determine whether or not these carbon atoms carry substituents [27]. Thus, for example, the methyl ester of 2,4-dimethylhexadecanoic acid, with a molecular mass of 298 Da, yields a characteristic fragment at m/z 227, derived from the loss of 70 Da corresponding to carbon atoms 2–4 and to the two methyl groups they carry, and an additional hydrogen atom:

7.7.4 A Few Examples of the Interpretation of Mass Spectra

Figure 7.17 shows the spectra of three substances each with a mass of 150 Da, but with different structures.

The spectrum of ethyl benzoate (top) contains an ion of mass 122 Da. In the absence of nitrogen, it indicates an open-shell ion, or odd-electron ion. This is an application of the parity rules. It can thus only be derived from a rearrangement:

150   122

The charge-induced cleavage leads to the phenylium ion:

150   77

The latter loses acetylene in a retro-Diels–Alder reaction:

51

Another α cleavage initiated by the radical site leads to the benzoylium ion:

150   105

Compare this spectrum with the middle one in Figure 7.17, of a methyl ester carrying a methyl group on the benzene nucleus. Replacing the ethyl group by a methyl group makes the McLafferty rearrangement impossible: it is no longer possible to form an intermediate six-atom ring. The two α fragmentations remain that yield ions analogous to the previous case, but displaced by 14 Da ($105 \rightarrow 119$ and $77 \rightarrow 91$) because of the presence of the methyl group on the benzene ring. This is also observed after the retrocyclization, the ion at m/z 51 being shifted to 65 Th.

Note that in both cases the charge is always carried by the fragments that contain the aromatic ring: the latter have in this case the lowest ionization energy.

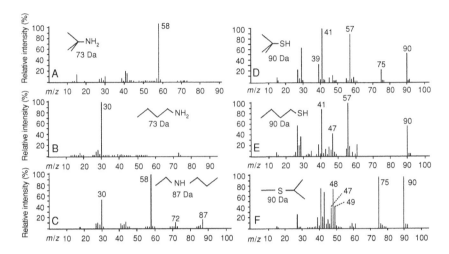

Figure 7.18
Comparison of spectra of amines, thiols and a thioether. The amine spectra display only very weak molecular ions. Thiols or thioethers have molecular ions that are relatively abundant, followed by the typical isotopic ^{34}S (4 %) peak.

The bottom spectrum in Figure 7.17 was obtained from another 150 Da compound that is very different. The molecular ion is not detected. The presence of an aromatic group favours the observation of the molecular ion, hence this compound probably does not contain such a group. The first intense fragment appears as a doublet at m/z 77 and 79, an isotopic cluster that is indicative of chlorine. It is formed by an α cleavage with respect to the carbonyl:

$$m/z\ 77;79$$

The m/z 57 ion, which does not contain a chlorine atom, since the ion at m/z 59 is almost absent, is derived from the cleavage of the bond adjacent to the sp^3 oxygen. The fragment at m/z 56 is an even-mass one: in the absence of nitrogen, this means that a radical cation is formed by a rearrangement:

$$m/z\ 56$$

Once again, the possibility of having a six-atom intermediate favours this McLafferty rearrangement which yields the most intense ion in the spectrum.

Finally, the ion at m/z 49 is accompanied by a peak at m/z 51 which indicates the presence of a chlorine atom. This means a cleavage α to the chlorine, yielding the $CH_2=Cl^+$ ion.

Other examples are shown in Figure 7.18. t-Butylamine and n-butylamine yield only one intense fragment, the ion corresponding to the radical-site-initiated fragmentation. These fragmentations give rise to the loss of a methyl radical in the first case, which gives m/z 58, and of a propyl radical in the second case, at m/z 30:

$$\text{(CH}_3)_3\text{C}-\overset{\bullet+}{\text{NH}}_2 \longrightarrow (\text{CH}_3)_2\text{C}=\overset{+}{\text{NH}}_2 \qquad m/z\ 58$$

$$\text{CH}_3\text{CH}_2\text{CH}_2\text{CH}_2-\overset{\bullet+}{\text{NH}}_2 \longrightarrow \text{H}_2\text{C}=\overset{+}{\text{NH}}_2 \qquad m/z\ 30$$

In contrast to the behaviour observed with amines, the spectra of *t*-butanethiol and of *n*-butanethiol are dominated by the loss of the HS$^{\bullet}$ radical formed by the cleavage of the adjacent bond. The ion of m/z 33, HS$^+$, is either absent or very weak. Thus the HS group is the one that removes the electron:

$$\text{(CH}_3)_3\text{C}-\overset{\bullet+}{\text{SH}} \longrightarrow \text{(CH}_3)_3\text{C}^+ + \ ^{\bullet}\text{SH} \qquad m/z\ 57$$

$$\text{CH}_3\text{CH}_2\text{CH}_2\text{CH}_2-\overset{\bullet+}{\text{SH}} \longrightarrow \text{C}_4\text{H}_9^+ + \ ^{\bullet}\text{SH} \qquad m/z\ 57$$

The hydrocarbon ion of m/z 57 fragments also, in the case of both *t*-butyl and *n*-butyl, yielding successively propenium and cyclopropenium ions:

$$\overset{+}{\text{C}_4\text{H}_9} \ (m/z\ 57) \xrightarrow{-\text{CH}_4} \ \text{CH}_2=\text{CH}-\overset{+}{\text{CH}}_2 \ (m/z\ 41) \xrightarrow{-\text{H}_2} \ \triangle^+ \ (m/z\ 39)$$

These compounds yield the fragment corresponding to the α cleavage as the medium-intensity fragment, appearing at m/z 75 and 47, respectively:

$$\text{(CH}_3)_3\text{C}-\overset{\bullet+}{\text{SH}} \longrightarrow (\text{CH}_3)_2\text{C}=\overset{+}{\text{SH}} \qquad m/z\ 75$$

$$\text{CH}_3\text{CH}_2\text{CH}_2\text{CH}_2-\overset{\bullet+}{\text{SH}} \longrightarrow \text{H}_2\text{C}=\overset{+}{\text{SH}} \qquad m/z\ 47$$

The secondary amine in spectrum C in Figure 7.18 shows two fragments corresponding to the cleavage:

$$\text{(CH}_3\text{CH}_2\text{CH}_2)-\overset{\bullet+}{\text{NH}}-(\text{CH}_2\text{CH}_2\text{CH}_3) \ (87\ \text{Da})$$

$$\xrightarrow{-\ ^{\bullet}\text{CH}_3} \ \text{H}_2\text{C}=\overset{+}{\text{NH}}-\text{CH}_2\text{CH}_2\text{CH}_3 \qquad m/z\ 72,\ \text{minor}$$

$$\xrightarrow{-\ ^{\bullet}\text{C}_2\text{H}_5} \ \text{CH}_3\text{CH}_2\text{CH}_2-\overset{+}{\text{NH}}=\text{CH}_2 \qquad m/z\ 58,\ \text{major}$$

Once again, the alkyl loss containing more atoms is favoured. The ion at m/z 58 loses ethylene and produces the m/z 30 fragment:

$$\text{CH}_3\text{CH}_2\text{CH}_2-\overset{+}{\text{NH}}=\text{CH}_2 \xrightarrow{-\text{C}_2\text{H}_4} \overset{+}{\text{NH}}_2=\text{CH}_2 \qquad m/z\ 30$$

A closed-shell ion fragments to yield another closed-shell ion, through the loss of a whole molecule, in this case ethylene. In such ions with an even number of electrons, a hydrogen atom transfer no longer requires a six-atom ring.

Methyl isopropyl thioether (spectrum F, Figure 7.18) yields a 75 Th ion fragment, corresponding to a methyl loss. This fragmentation can occur through either of two paths: a methyl loss by the isopropyl through an α cleavage, or loss of a methyl through the cleavage of the adjacent bond. The cleavage of the other bond adjacent to the sulfur is responsible for the formation of the CH_3S^+ ion at m/z 47. Ions of m/z 48 and 49 can only have the possible elemental compositions $CH_3SH^{\bullet+}$ and $CH_3SH_2^+$, respectively. The ion at m/z 48, with an even mass, results from a rearrangement and that at m/z 49 results from two successive rearrangements:

$$m/z\ 48$$

$$m/z\ 49$$

Note, however, that the first rearrangement, which is of the McLafferty type, occurs through a four-atom ring. In solution, such rearrangements most often imply a proton donor and a proton acceptor produced by the solvent. In the vapour phase at low pressures, only intramolecular processes can be considered.

References

1. Dunbar, R.C. (1992) *Mass Spectrom. Rev.*, **11**, 309.
2. Rosenstock, H.M., Wallenstein, M.B., Warharftig, A. and Eyring, H. (1952) *Proc. Natl. Acad. Sci. USA*, **38**, 667.
3. Marcus, R.A. (1952) *J. Chem. Phys.*, **20**, 359.
4. Longevialle, P. (1981) *Principes de la Spectrométrie de Masse des Substances Organiques*, Masson, Paris.
5. Gilbert, R.G. and Smith, S.C. (1990) *Theory of Unimolecular and Recombination Reactions*, Blackwell Scientific, Oxford.
6. McLafferty, F.W. (1980) *Interpretation of Mass Spectra*, 3rd edn, University Science Books, Mill Valley, CA.
7. McLafferty, F.W. and Turecek, F. (1993) *Interpretation of Mass Spectra*, 4th edn, University Science Books, Mill Valley, CA.
8. Winkler, F.J. and McLafferty, F.W. (1974) *Tetrahedron*, **30**, 29.
9. McLafferty, F.W. (1980) *Org. Mass Spectrom.*, **15**, 114.
10. Audier, H.E., Milliet, A., Perret, C., Tabet, J.C. and Varenne, P. (1978) *Org. Mass Spectrom.*, **13**, 315.
11. Sigsby, M.L., Day, R.J. and Cooks, R.G. (1979) *Org. Mass Spectrom.*, **14**, 273 and **14**, 556.
12. Veith, H.J. and Gross, J.H. (1991) *Org. Mass Spectrom.*, **26**, 1097 and **28**, 867 (1993)
13. Bowie, J.H. (1990) *Mass Spectrom. Rev.*, **9**, 349.
14. Bowie, J.H. (1994) *Experimental Mass Spectrometry*, Plenum Press, New York, pp. 1–38.
15. Foster, R.F., Tumas, W. and Brauman, J.I. (1983) *J. Chem. Phys.*, **79**, 4644.

16. Tumas, W., Foster, R.F. and Brauman, J.I. (1988) *J. Am. Chem. Soc.*, **110**, 2714.
17. O'Hair, R.A.J., Bowie, J.H. and Currie, G.J. (1988) *Aust. J. Chem.*, **41**, 57.
18. Stroobant, V., Rozenberg, R., Bouabsa el, M., Deffense, E. and de Hoffmann, E. (1995) *J. Am. Soc. Mass Spectrom.*, **6**, 498.
19. Bowie, J.H. (1984) *Mass Spectrom. Rev.*, **3**, 161.
20. Kroto, H. (1982) *Angew. Chem.*, **104**, 113.
21. Richter, H., Dereux, A., Gilles, J.M., Guillaume, C., Tiry, P.A., Lucas, A.A., Rozenberg, R., Spote, A., de Hoffmann, E. and Girard, C.H. (1994) *Ber. Bunsenges. Phys. Chem.*, **98**, 1329.
22. Jensen, N.J., Tomer, K.B. and Gross, M.L. (1985) J. Am. Chem. Soc., **107**, 1863.
23. Graul, S.T. and Squires, R.R. (1988) *J. Am. Chem. Soc.*, **110**, 607.
24. Tomer, K.B., Jensen, N.J. and Gross, M.L. (1986) *Anal. Chem.*, **58**, 2429.
25. Adams, J. (1990) *Mass Spectrom. Rev.*, **9**, 141.
26. Gross, M.L. (1992) *Int. J. Mass Spectrom. Ion Processes*, **118**, 137–65.
27. Odham, G. and Stenhagen, E. (1972) *Biochemical Applications of Mass Spectrometry*, John Wiley & Sons, Inc., New York, pp. 211–28.

8

Analysis of Biomolecules

8.1 Biomolecules and Mass Spectrometry

The determination of molecular weight is among the first measurements used to characterize biopolymers. Up to the end of the 1970s, the only techniques that provided this information were electrophoretic, chromatographic or ultracentrifugation methods. The results were not very precise (10–100 % relative error on average) because they depended also on characteristics other than the molecular weight, such as the conformation, the Stokes radius and the hydrophobicity. Thus the only possibility of knowing the exact molecular weight of a macromolecule remained its calculation based on its chemical structure.

At that time the mass spectrometric ionization techniques of electron ionization (EI) [1] and chemical ionization (CI) [2] required the analyte molecules to be present in the gas phase and were thus suitable only for volatile compounds or for samples subjected to derivatization to make them volatile. Moreover, the field desorption (FD) ionization method [3], which allows the ionization of non-volatile molecules with masses up to 5000 Da, was a delicate technique that required an experienced operator [4]. This limited considerably the field of application of mass spectrometry of large non-volatile biological molecules that are often thermolabile.

The development of desorption ionization methods based on the emission of pre-existing ions from a liquid or a solid surface, such as plasma desorption (PD) [5,6], fast atom bombardment (FAB) [7] or laser desorption (LD) [8], allowed a first breakthrough for mass spectrometry in the field of biomolecular analysis. Since then, the problem has no longer been the production of ions but rather that of analysing such high-mass singly charged ions, which are technically difficult to detect with good sensitivity and difficult to analyse with good resolution.

At the beginning of the 1990s, two new ionization methods, electrospray ionization (ESI) [9] and matrix-assisted laser desorption/ionization (MALDI) coupled to time-of-flight (TOF) analysers [10] that avoided such inconveniences, were developed and continue to revolutionize the role of mass spectrometry in biological research. These methods allow the high-precision analysis of biomolecules of very high molecular weight.

Today, mass spectrometry has become one of the most widely used analytical techniques in the life sciences [11,12]. The mass spectrometric analysis of different classes of biomolecules is reviewed in this chapter: peptides, proteins, nucleic acids, oligosaccharides and lipids. Several applications are detailed for each class. Metabolomics, which is the 'omics' science of metabolism, will also be examined at the end of the chapter.

Mass Spectrometry: Principles and Applications, Third Edition Edmond de Hoffmann and Vincent Stroobant
© Copyright 2007, John Wiley & Sons Ltd

Table 8.1   Post-translational modifications and corresponding average
mass variations.

Post-translational modification	Mass difference (Da)
Methylation	14.03
Propylation	42.08
Sulfation	80.06
Phosphorylation	79.98
Glycosylations by:	
Deoxyhexoses (Fuc)	146.14
Hexosamines (GlcN, GalN)	161.16
Hexoses (Glc, Gal, Man)	162.14
N-Acetylhexosamines (GlcNAc, GalNAc)	203.19
Pentoses (Xyl, Ara)	132.12
Sialic acid (NeuNAc)	291.26
Reduction of a disulfide bridge	2.02
Carbamidomethylation	57.03
Carboxymethylation	58.04
Cysteinylation	119.14
Ethylpyridylation	105.12
Acetylation	42.04
Formylation	28.01
Biotinylation	226.29
Farnesylation	204.36
Myristoylation	210.36
Pyridoxal phosphate Schiff condensation	231.14
Stearoylation	266.47
Palmitoylation	238.41
Lipoylation	188.30
Carboxylation of Asp or Glu	44.01
Deamidation of Asn or Gln	0.98
Hydroxylation	16.00
Methionine oxidation	16.00
Proteolysis of a peptide bond	18.02
Deamination from Gln to pyroglutamic	−17.03

8.2 Proteins and Peptides

Proteins and peptides are linear polymers made up of combinations of the 20 most common amino acids linked with each other by peptide bonds. Moreover, the protein produced by the ribosome may undergo covalent modifications, called post-translational modifications, after its incorporation of amino acids. Over 200 such modifications have been detected already [13,14], the most important being glycosylation, the formation of disulfide bridges, phosphorylation, sulfation, hydroxylation, carboxylation and acetylation of the N-terminal acid [15]. The most frequent are listed in Table 8.1 and a more comprehensive database of mass changes due to post-translational modifications of peptides and proteins is available on the Internet [16].

Mass spectrometry not only allows the precise determination of the molecular mass of peptides and proteins but also the determination of their sequences, especially when used with tandem mass techniques. Indeed, fragmentation of peptides and proteins gives sequence information that can be used for protein identification, *de novo* sequencing, and identification and localization of post-translational or other covalent modifications [17–24].

8.2.1 ESI and MALDI

The ionization methods that are used most often to study the peptides and proteins through mass spectrometry are ESI and MALDI. All of these techniques are characterized by the formation of stable ions (because they have only a low excess energy) and the absence of fragments.

ESI produces multiply charged ions, which allow the detection of large molecules with conventional mass spectrometers such as quadrupole, ion trap and magnetic instruments. The nanoelectrospray is an important development of this ionization method. With injection flows in the range of 20 m/min^{-1}, long-lasting signals can be obtained from minute quantities of sample, allowing numerous MS/MS experiments to be performed for structure elucidation. Recall that ESI sensitivity is not flow dependent up to the nanolitre per minute range. MALDI is a pulsed source, and is well suited for use with TOF mass spectrometers, or with spectrometers that allow the storage of ions such as ion traps and FTICR. The most frequently used is the TOF instrument. Over the last decade, these MALDI-TOF spectrometers have made important progress in the introduction of delayed extraction of ions, orthogonal injection of ions and reflectron flight tubes. In consequence, the performance of TOF analysers, in terms of resolution and mass accuracy, has greatly improved. Although not as great as that achievable with FTICR or orbitrap analysers, the resolution can now reach 20 000 FWHM and with a good mass calibration protocol, low-ppm mass accuracy can be obtained. The more sophisticated FTICR instruments are also being used with increasing success. These instruments are now using reaxialization and other improvements. With a 6 T magnet, a resolution of 2×10^6 is now possible at masses in the 2000 Th range. The complexity and cost of these instruments mean they are only available in a very limited number of laboratories. Another instrument development is to combine in sequence different analysers such as Q-TOF, LIT-ICR or LIT-orbitrap instruments, in order to increase the performance and the versatility and to allow multiple experiments to be performed.

The detection limit of ESI and MALDI depends on several factors, such as the nature of the sample and its preparation and purity, the instrument used and the skill of the operator. For peptides and proteins, the detection limit is in practice somewhere between femtomoles and picomoles, even if attomole limits have been reported [25–27].

Because the resolution needed to separate different peaks in the isotopic cluster of a small peptide is lower than the resolution of most analysers, the molecular mass that is measured corresponds to that calculated using the predominant isotope of each element. This is not so in the case of proteins. Because the resolution required to resolve the isotopic cluster increases with the mass or with the charge carried by the ion and because the resolution of analysers is limited, the various peaks in the isotopic cluster combine and form a single peak that spreads over several masses (15 Da at 10 000 Da and 45 Da at 100 000 Da)

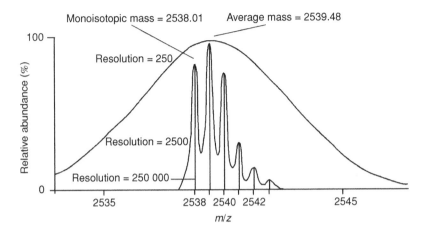

Figure 8.1
Mass spectrum of isotopic cluster of a singly protonated peptide ($C_{101}H_{145}N_{34}$ O_{44}) with monoisotopic mass $= 2538.015$ u and average mass $= 2539.483$ Da at a resolution of 250 (resolution typically obtained for linear TOF), 2500 (quadrupole) and 250 000 (FTICR). Redrawn (modified) from Suizdak G., Proc. Natl. Acad. Sci. USA, 91, 11290, 1994, with permission.

(Figure 8.1) [28]. Thus the molecular mass determined by these techniques corresponds to that calculated using the chemical mass of each element present in the protein. Both values are significantly different from one another. Indeed, the difference between the isotopic mass and the chemical mass of a peptide or of a protein is about 1 Da per 1500 Da. Hence, specifying the mass we are talking about is important. The choice is determined by the resolution of the analyser and the mass of the ion being analysed.

The mass measurement errors also depend on a large number of factors. A typical measurement error of 0.01 % can be obtained routinely. However, these errors can be higher than 0.1 % in the worst case (MALDI-LTOF) or to be lower than 0.0001 % in the best case (ESI-FTICR).

The characteristic absence of fragment ions allows the analysis of complex mixtures without any previous separation. However, the analysis of mixtures of proteins with close molecular masses is limited by the resolution of the analyser being used. For example, investigation of a mixture of a 10 000 Da protein and its oxidation product corresponding to the addition of one oxygen atom requires a resolution of at least 1000, whereas the resolution required to analyse the mixture is 10 000 if the protein has a mass of 100 000 Da. Another factor that plays a role in the possibility of analysing this type of mixture is the relative quantity of each component. A compound whose abundance is 10 % of the main component requires less resolution than the same mixture with the same compound having only a 1 % relative abundance.

Moreover, the analysis of complex mixtures is complicated by the phenomenon of competitive ionization [29]. This phenomenon, observed in MALDI and in ESI, is characterized by the suppression of molecular species ions of some peptides when they are in the presence of other peptides in the mixture. The signal corresponding to these peptides may disappear completely and thus these peptides may not be detected even though they yield an easily detectable signal when analysed individually.

Competitive ionization may be avoided by varying the pH conditions or the matrix, through chemical derivatization of the peptides contained in this mixture, or through the partial fractionation of the mixture through reversed-phase liquid chromatography so that each fraction contains peptides of a similar hydrophobicity.

For both ESI and MALDI, the concentration of the sample and the complexity of the contaminants play important roles in both the sensitivity and the mass accuracy. Biological samples most often are diluted solutions of peptides or proteins containing a great number of contaminants. These two problems, dilution and contaminants, are not easy to handle, especially when the total amount of sample is low, such as picomoles.

The contaminants are of different origin and include buffers, non-volatile salts, detergents and many compounds of unknown origin. ESI can tolerate only low quantities of contaminant ions. These ions can reduce the abundance of the ions from the compound of interest and can even totally suppress them. They also very often result in the formation of adduct ions, further reducing the sensitivity by distribution of the ion current over several species. Furthermore, they may complicate the determination of the molecular mass, or reduce the accuracy of the molecular mass if some adducts are not separated from the ions of the protonated molecule.

Generally, MALDI is more tolerant than ESI to many contaminants. This can in part be due in part to some separation occurring during the crystallization of the sample with the matrix [30]. Whatever the ionization method, the quality of the mass spectrum is higher if the contamination is reduced.

The second problem is the generally low concentration of the compound of interest in the biological samples. The volumes needed for the analysis are very low, in the microlitre range for both MALDI and ESI [31], and only part of it is actually consumed during the analysis. But the concentration has a marked influence on the observed spectra.

As a rule, a separation method should be used for both purification and concentration of the sample. The classic method for peptides and proteins is a reverse-phase liquid chromatography preparation of the sample, followed by a concentration step (often lyophilization) of the fraction of interest. During those steps performed on very small quantities of sample, loss on the sample can occur if care is not taken to avoid it. Lyophilization, for instance, can lead to the loss of the sample absorbed on the walls of the vial. The use of separation methods on-line with the mass spectrometer often are preferred. Micro- or nano-HPLC [32,33] and capillary electrophoresis [34], both coupled mainly to electrospray ionization/mass spectrometry (ESI-MS), are used more and more.

8.2.2 Structure and Sequence Determination Using Fragmentation

8.2.2.1 Fragmentation of Peptides

In order to generate structural data by mass spectrometry, the molecule that is studied must undergo fragmentation of one or several bonds to match the m/z of the resulting fragments with the chemical structure. However, the various techniques we have considered so far imply the formation of stable ions that do not yield any fragment. This property is used to facilitate the determination of the molecular mass of peptides or proteins, even when they appear in a complex mixture. However, the same property leads to a lack of information concerning the structure. This drawback is overcome by transferring at least the extra energy required by fragmentation to the stable ions produced during the ionization. Although various techniques allow an energy transfer, the most common method remains collision-induced dissociation (CID). Consequently, tandem mass spectrometry (MS/MS) has become an essential technique for structural analysis of peptides and proteins.

This technique consists of selecting the ion to be fragmented using a first mass analyser and sending it into a collision cell, where it collides with uncharged gas atoms. Thus the kinetic energy is transformed partly into vibrational energy and the resulting fragments are analysed by a second analyser, hence the name CID tandem mass spectrometry or CID MS/MS. If the instrument resolution is sufficient, the first analyser can select only the isotopic peak containing the main isotopes, such as ^{12}C and ^{16}O, which allows a fragmentation spectrum free from complex isotopic clusters (especially at high masses) to be obtained.

The tandem mass spectra may be obtained using many different instruments, such as sector, reflectron TOF, ion trap, triple quadrupole, ICR or hybrid instruments. The main difference from a practical point of view is in the kinetic energy of the ions. In magnetic and TOF instruments the precursor ion kinetic energy is several kiloelectronvolts whereas in the other types of analysers, such as the quadrupole, ion trap or ICR, the ion kinetic energy never exceeds 100 eV. This difference influences the fragmentation process [35]. As will be discussed in more detail below, the high-energy tandem mass spectra present a broader range of fragmentation pathways, some of which are not observed at low energy. A greater number of fragment ions often carry more information but also increase the complexity of the spectra, thereby rendering their interpretation more difficult.

The fragmentation of peptides can be observed also by a technique named post-source decay (PSD) when reflectron TOF instruments are used. In this technique, which is not only used for peptide analysis, the ions of the molecular species produced by MALDI indeed contain enough energy to fragment but this metastable fragmentation occurs during the flight between the source and the detector. With an linear TOF spectrometer the fragments reach the detector together with the precursor ions. In contrast, they will have different flight times after passing through the reflectron and thus their masses can thus be determined. A chosen precursor can be selected by a gating system at the origin of the flight tube. The resolution for this selection is low. This is in general sufficient to select a peptide in a mixture, but not to select one isotopic peak. MALDI-TOF/TOF instruments remove several disadvantages of MALDI-TOF PSD instruments and allow high-energy CID and high-speed analysis. Indeed, in these instruments, ions of high velocity that are produced by MALDI are selected with an ion gate and subject to CID. The resulting fragments are further accelerated and analysed in the second reflectron TOF analyser.

The fragmentation of peptides can also be obtained by FTICR instruments. Besides the most commonly used activation method, namely CID, the activation can alternatively be performed without gas by infrared multiphoton dissociation (IRMPD) and electron capture dissociation (ECD). These methods fragment peptide ions in the ICR cell by emitting a laser beam or electron beam, respectively.

The MS/MS analysis of many peptides with known sequences permits identification of the various existing fragmentation processes [36]. The high- and low-energy fragmentation spectra of a peptide are presented in Figure 8.2.

From a practical point of view, the fragments may be classified in either of two categories: those derived from the cleavage of one or two bonds in the peptide chain and those that also undergo a cleavage of the amino acid lateral chain. The nomenclature suggested by

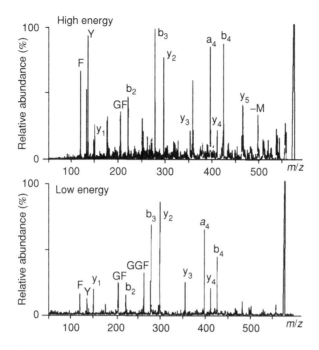

Figure 8.2
High- and low-energy fragmentation spectra of methionine–enkephaline (sequence YGGFM). The notation used is described in the text. Reproduced (modified) from Bean M.F., Carr S.A., Thorne G.C., Reilly M.H. and Gaskell S.J., Anal. Chem., 63, 1473, 1991, with permission.

Roepstorff and Fohlman [37] and later modified by Biemann [38] allows the labelling of the various fragments that are obtained.

The first fragments that were identified were produced by the cleavage of a bond in the main chain. The cleavage of a bond in that peptide chain can occur in either of three types of bonds, $C\alpha$–C, C–N or N–$C\alpha$, which yields six types of fragments that are respectively labelled a_n, b_n, c_n when the positive charge is kept by the N-terminal side and x_n, y_n, z_n when the positive charge is kept by the C-terminal side. The c_n and y_n fragments implicate the transfer of two extra hydrogen atoms, the first one responsible for the protonation and the second one originating from the other side of the peptide. The subscript n indicates the number of amino acids contained in the fragment. Figure 8.3 shows the various types of fragments produced through the cleavage of a bond in the peptide chain.

The mass difference between consecutive ions within a series allows the identity of the consecutive amino acids to be determined (see Table 8.2) and thus deduction of the peptide sequence. Indeed, the 20 common amino acid residues have distinctive elemental compositions and consequently distinctive masses. There is one exception with Leu and Ile, which are isomers. However, a low-accuracy measurement may be incapable of discriminating between Gln and Lys, which differ by 0.036 u. In addition, there are combinations of amino acid residues that yield the same nominal mass or even the same elemental composition.

Table 8.2   Mass increments of the various amino acids.

Amino acid	Code (3 letters)	Code (1 letter)	Monoisotopic mass	Chemical mass
Glycine	Gly	G	57.021 47	57.052
Alanine	Ala	A	71.037 12	71.079
Serine	Ser	S	87.032 03	87.078
Proline	Pro	P	97.052 77	97.117
Valine	Val	V	99.068 42	99.133
Threonine	Thr	T	101.047 68	101.105
Cysteine	Cys	C	103.009 19	103.144
Isoleucine	Ile	I	113.084 07	113.160
Leucine	Leu	L	113.084 07	113.160
Asparagine	Asn	N	114.042 93	114.104
Aspartate	Asp	D	115.026 95	115.089
Glutamine	Gln	Q	128.058 58	128.131
Lysine	Lys	K	128.094 97	128.174
Glutamate	Glu	E	129.042 60	129.116
Methionine	Met	M	131.040 49	131.198
Histidine	His	H	137.058 91	137.142
Phenylalanine	Phe	F	147.068 42	147.177
Arginine	Arg	R	156.101 12	156.188
Tyrosine	Tyr	Y	163.063 33	163.17
Tryptophan	Trp	W	186.079 32	186.213

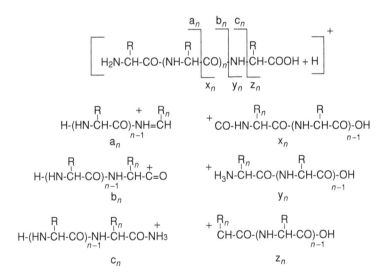

Figure 8.3
Main fragmentation paths of peptides in CID tandem mass spectrometry.

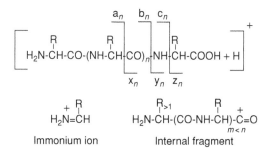

Figure 8.4
Fragments derived from a double cleavage of the peptide main chain.

Normally the spectra show several incomplete series of ions that produce redundant data and make the spectrum very complex and difficult to interpret. The peptide sequence can be deduced even if several incomplete series of ions are present, provided that there are other series of ions present and as long as there are overlaps in the deduced sequence information.

The types of fragment ions observed in an MS/MS spectrum depend on many factors including the amino acid composition, the peptide sequence, the amount of internal energy transferred, the ion activation method used, and so on. There is a marked difference between the fragmentations observed at high and low energy. At high energy, all the fragmentations described in Figure 8.3 can be generated in principle. However, all those fragments are not observed in the spectra, because the fragmentation can be influenced by the nature of the amino acids present in the sequence, as will be shown below. Unlike low-energy CID, ions do not readily lose water or ammonia. At low energy, the observed fragments are mostly the b_n and y_n ions. These fragments then lose small molecules such as water or ammonia from the functional groups on the side chains of the amino acids.

Two other types of fragments found in most spectra result from cleavage of at least two internal bonds in the peptide chain. The first type is called an internal fragment because these fragments have lost the initial N- and C-terminal sides [39]. They are represented by a series of simple letters corresponding to the fragment sequence. Fortunately, this type of ion is often only weakly abundant and, because they rarely contain more than three or four amino acid residues, they appear in the spectrum among the low masses. These peaks confirm the sequence but are often more of a nuisance than a help. Peptides containing a proline are an exception to this as the proline imino group is included in a five-atom ring and thus has a higher proton affinity than the other amide bonds in the peptide. Hence protonation and cleavage of the proline amide bond are a favoured to yield an internal fragment, extending from the proline in the direction of the C-terminal side. The various fragments requiring cleavage of two bonds are shown in Figure 8.4.

The second type of fragment that results from multiple cleavage of the peptide chain appears among the low masses in the spectrum. These are the immonium ions of the amino acids, labelled by a letter corresponding to the parent amino acid code. Even though these fragments are rarely observed for all of the peptide amino acids, those that appear yield information concerning the amino acid composition of the sample, especially when other diagnostic ions in the low-mass region are taken into account [40]. A list of immonium ions commonly found in spectra is given in Table 8.3 [41,42].

Table 8.3 Masses of the low-mass ions characteristic of natural amino acids, most often immonium ions.

Amino acid	Characteristic mass
Proline (P)	70
Valine (V)	72
Leucine (L)	86
Isoleucine (I)	86
Methionine (M)	104
Histidine (H)	110
Phenylalanine (F)	120
Tyrosine (Y)	136
Tryptophan (W)	159

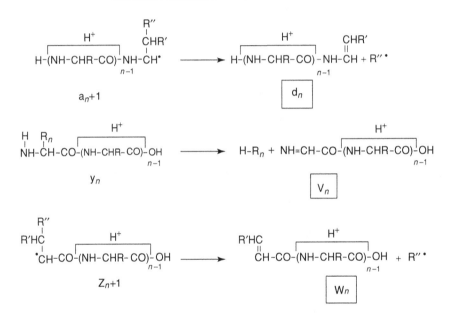

Figure 8.5
Fragmentation paths yielding the ions characteristic of amino acid lateral chains.

In addition to the ions described earlier, three new types of fragments that require cleavage of the peptide chain and the amino acid lateral chain were highlighted only in the high-energy spectra. These fragments are useful for distinguishing the isomers Leu and Ile. Figure 8.5 shows the mechanisms and structures of the corresponding fragments [43–46].

Two types of these fragments result from cleavage of the bond between the β and γ carbon atoms of the side chain of the amino acids. These fragments are symbolized as d_n or w_n respectively according to whether the positive charge is retained by the N-terminal or the C-terminal fragment. They are useful to distinguish the isomers Leu and Ile. Amino acids carrying an aromatic group attached to the β carbon atom (His, Phe, Tyr, Trp) either do not display these two fragments or have very low abundances.

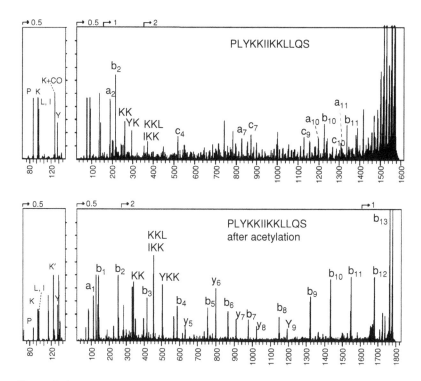

Figure 8.6
Influence of the presence and the position of the charge on the resulting fragments. Reproduced (modified) from Biemann K., Meth. Enzymol., 193, 455, 1990, with permission.

The last type of these fragments results from the complete loss of the side chain and is symbolized as v_n. This fragment containing the C-terminal moiety is intense for amino acids that do not easily yield w_n type fragments. No equivalent containing the N-terminal moiety has ever been observed.

The types of fragment ions observed in an MS/MS spectrum are influenced by charge position and charge delocalization. Protonation occurs mainly at the more basic sites. In peptides, the terminal amino group is basic. If more protons are added they will be located first at other basic amino acids, if present, and then on the amide groups. These last will be more statistically distributed along the chain.

It has been shown that charged ions from small peptides that do not contain basic amino acids display comparable abundances for all the y_n and b_n ions. When the chain becomes longer, b_n ions become favoured [47].

However, the presence and the position of a basic amino acid within the peptide influence the fragmentation process. The presence of basic amino acids such as Arg, Lys, His or Pro at the C-terminal amino acid induces mainly the formation of ions containing the C-terminal side (y_n, v_n and w_n), whereas the presence of these amino acids on the N-terminal side favours the formation of ions containing the N-terminal side (a_n, b_n and d_n). However, the absence of a basic site within the peptide is characterized by a distribution of ions containing one of the two sides (y_n, b_n), as shown in Figure 8.6 [48].

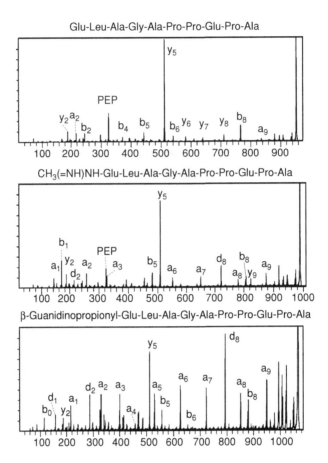

Figure 8.7
Influence of the charge delocalization on the type of fragmentation obtained. Reproduced from Martin S.A., Johnson
R.S., Costello C.E. and Biemann K., in 'The Analysis of Peptides and Protiens by Mass Spectrometry' edited by McNeal
C.J., Wiley & Sons, New York, 1988, pp. 135–150, with
permission.

Similarly, increasing the localization of the positive charge on the peptide increases the
a_n and d_n fragments that end up outnumbering the b_n and y_n fragments if the charge is
carried on the N-terminal side or increases the v_n and w_n fragments if the charge is carried
on the C-terminal side, as shown in Figure 8.7 [49].

The rule describing the relationship between the position and the charge localization on
the one hand and the fragmentation process on the other is summarized in Figure 8.8.

The discussion up to now has only been concerned on the CID spectra of singly charged
precursor ions, but the development of ESI allows the study of the dissociation of multiply
charged ions. As a rule, the fragmentation spectra of multiply protonated peptides, at
high as well as at low energy, display fragments analogous to the one observed for the
monocharged species [50,51]. Multiply charged ions require lower acceleration voltages

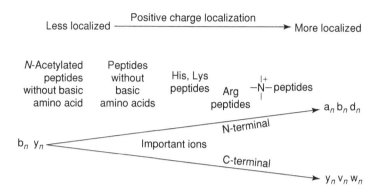

Figure 8.8
Charge and fragmentation with respect to the nature of the peptide.

before CID because the kinetic energy is proportional to neV, where n is the number of charges, e is the electron charge and V is the acceleration voltage. At first glance they appear to fragment easier. It is best, when possible, to acquire spectra of different charge states because some fragments can be more abundant for some charge states.

The dissociation of multiply charged ions is interesting but can lead to more complicated spectra. Firstly, the presence of protons on different protonation sites induces fragmentations from a variety of starting points. Secondly, the ions obtained can have different charge states and the charge state has to be determined. Because two adjacent peaks of the isotopic cluster are separed by 1 Da, if the number of charges is n, they will appear as being separated by $1/n$ Th. If the resolution of the instrument is sufficient, the charge state can be determined by this way. Tryptic peptides have two basic sites, one at each terminal, and thus easily yield doubly protonated ions: on the basic C-terminal amino acid and on the free amino group of the N-terminal. As a consequence, they mainly yield monocharged b_n and y_n ions, as shown by the example in Figure 8.9 [52,53].

Fragmentation of peptides can be obtained in MALDI during ionization as observed by ion source decay (ISD). In this case, the most abundant fragment ions observed in ISD are c_n and y_n ions. Fragmentation of peptides can be obtained also with an reflectron TOF instrument by the PSD technique [54,55]. An example of such a spectrum is displayed in Figure 8.10. The fragment ion types observed in MALDI-TOF PSD are close to those observed at low-energy CID [56]. With MALDI-TOF/TOF instruments where collision gas is used, the spectra are similar to those observed at high-energy CID. All ion series can be accompanied by ions resulting from losses of ammonia or water.

Fragmentation of peptides can also be observed with FTICR instruments. Infrared multiple photon dissociation (IRMPD) and electron capture dissociation (ECD) have been introduced as two alternative dissociation methods to the low-energy CID method. The IRMPD method produces many fragments that make the spectrum very complex and difficult to interpret. Some of the fragment types observed with IRMPD are b_n and y_n type ions or these ions that have lost ammonia or water. However, most of them are not these types of fragment ions.

ECD has recently been introduced as an alternative activation method to obtain fragmentation of multiply protonated peptides [57]. An example of an ECD fragmentation spectrum

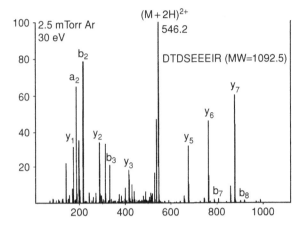

Figure 8.9
The CID ESI MS/MS spectrum of a doubly charged peptide. Reproduced from Finnigan MAT documentation, with permission.

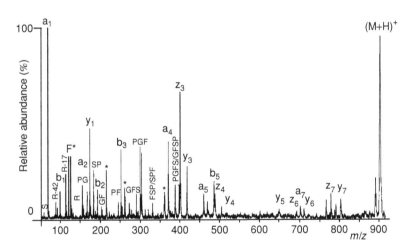

Figure 8.10
The PSD MALDI spectrum of Des-Arg[1] bradykinine, whose sequence is PPGF-SPFR. The starred peaks correspond to the loss of a 28 Da fragment. Reproduced (modified) from Rouse J.C., Yu W. and Martin S.A., J. Am. Mass Spectrom., 6, 822–835, 1996, with permission.

is displayed in Figure 8.11. As already mentioned in Chapter 4, CID and IRMPD induce ergonic processes whereby cleavages of the weakest bonds of the precursor ions are usually observed. For peptides, the backbone amide bond has the lowest energy barrier to fragment and predominantly b_n and y_n ions are formed. Furthermore, post-translational modifications often have lower energy barriers than those of backbone cleavage. This can lead to loss

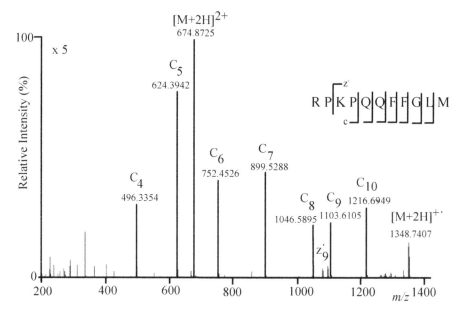

Figure 8.11
The ECD ESI MS/MS spectrum of a doubly charged ion of the substance P with FTICR MS. Reproduced (modified) from ThermoFinnigan documentation, with permission.

of information on the localization of post-translational modifications. In contrast to CID and IRMPD, ECD induces non-ergodic processes. So the fragmentation is independent of the bond dissociation energy, and thus the peptide backbone cleavage is essentially sequence independent. ECD induces more general backbone cleavage than other methods and thus provides more extensive sequence information that is ideal for *de novo* sequencing. Furthermore, the labile post-translational modifications are not dissociated by this non-ergodic dissociation upon fragmentation of the peptide backbone.

Therefore, ECD is a powerful tool for structural analysis of peptides and proteins that is complementary to the other ion activation methods. However, ECD is not compatible with instruments such as ion traps or QTOF. As a consequence, ECD analysis of peptides and proteins is typically performed on FTICR mass spectrometers.

To bring the power of ECD to ion trap analysers, a new ECD-like activation method has been developed [58]. This method, which is called electron transfer dissociation (ETD), uses gas-phase ion/ion chemistry to transfer an electron from singly charged aromatic anions to multiply charged ions. The mechanism of this method and the observed fragmentation pathways are analogous to those observed in ECD.

ECD and ETD produce a radical cation $[M + nH]^{(n-1)+\bullet}$ which can dissociate to produce mainly c_n and $z_n^\bullet$ fragment ions. The mechanism of ECD is detailed in Figure 8.12. The fragmentation by electron capture involves generation of an odd-electron hypervalent species $(RNH_3^\bullet)$ that dissociates to produce a hydrogen radical $(H^\bullet)$. Reaction of this hydrogen radical with carbonyl groups of the peptide backbone induces the formation of two series of complementary c_n and $z_n^\bullet$ fragment ions. It should be mentioned that the cyclic structure of proline does not allow formation of c_n and z_n fragments.

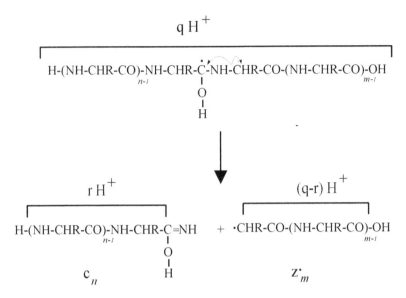

Figure 8.12
Fragmentation paths yielding the c and z ions after activation of a multiply protonated peptide by ECD.

In conclusion, because ECD and ETD induce fragmentation along the peptide backbone that is essentially sequence independent and preserves post-translational modifications, this method is an effective alternative dissociation method for peptides and proteins, especially for the analysis of post-translational modifications.

8.2.2.2 *Peptide and Protein* de novo *Sequencing*

Because mass spectrometry is entirely different from the other *de novo* sequencing techniques, it offers several advantages over the latter, such as the analysis of peptides within mixtures or of peptides that are blocked on the N-terminal side or carrying post-translational modifications, etc. Mass spectrometry is thus complementary to other existing methods. Mass spectrometry also is more sensitive and more rapid than these other methods. Acquisition of the fragmentation spectrum of a peptide requires only a few minutes. On the other hand, the interpretation of this spectrum is in many cases far from being simple. The complete sequence determination is generally difficult. Automatic algorithms have been developed for *de novo* sequencing, even if manual interpretation provides generally fewer errors.

Interpretation of the spectra is based on the mechanisms and the fragmentation pathways described above, as shown by the following example. A CID MS/MS fragment spectrum of a peptide with sequence Gly–Ile–Pro–Thr–Leu–Leu–Leu–Phe–Lys measured at high energy is shown in Figure 8.13. This spectrum contains the complete series of b_n ions, thus allowing one to deduce the peptide sequence from the N-terminal acid to the C-terminal acid, whereas the series of y_n ions allows identification of the sequence in the reverse direction. In fact, the mass difference of 97 Da between peak b_2 and b_3 indicates that the amino acid in position 3 corresponds to a proline (see Table 8.2). Similarly, the 147 Da difference between peaks y_1 and y_2 indicates that, the amino acid in the next-to-last position is a phenylalanine. The m/z values of ions w_3, w_4, w_5 and w_8 imply that the amino acid in

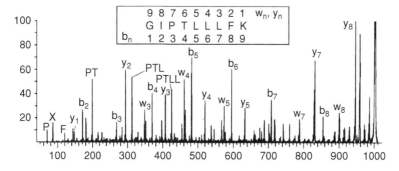

Figure 8.13
The FAB MS/MS high-collision energy trace of a mass 1001 peptide with sequence Gly–Ile–Pro–Thr–Leu–Leu–Leu–Phe–Lys. Reproduced (modified) from Biemann K., Biomed. Environ. Mass Spectrom., 16, 99, 1988, with permission.

positions 3, 4 and 5 starting from the C-terminal side are leucines, whereas the amino acid in position 8 is an isoleucine. The presence of a proline induces the formation of internal fragments labelled PT, PTL and PTLL that allow one to verify the deduced sequence. The peaks labelled P, F and X represent the immonium ions and indicate the presence of proline, phenylalanine and leucine and/or isoleucine.

Interpretation of the fragmentation spectra can be difficult and time consuming. It can be simplified if the peptide is modified in such a way that one type of fragmentation is favoured. This can be achieved by derivatizing the amino terminal group with N-succinimidyl-2-(3-pyridyl) acetate, which promotes b_n fragments [59]. Modification of the peptide also can make the distinction between C- and N-terminal fragments easier.

Derivatization also allows specific residues to be detected and located by comparing the spectra before and after derivatization. The derivatizations most often used for this purpose are: the formation of ethyl esters from Asp, Glu and C-terminal carboxylic acids; acetylation of the N-terminal amino group and Lys; and the Edman reaction for the N-terminal amino acid [60].

Distinction of the fragments containing the C- or the N-terminal can be achieved also by introducing an ^{18}O atom on the C-terminal of peptides obtained by tryptic digest. Indeed, if the digestion is done using a 50/50 mixture of $H_2^{16}O$ and $H_2^{18}O$, fragments containing the carboxyl terminal are identified easily by the fact that they are dedoubled [61]. They will appear as a doublet of peaks of about the same abundance separated by of $2/n$ Th, where n is the number of charges. The decision to use this method is to be taken before the digestion, whereas the other derivatization methods must be applied after digestion.

Sequence determination by mass spectrometry is based on the mass of the amino acids. However, Leu–Ile and Lys–Gln have the same nominal masses. Differentiating Leu from Ile may be achieved using d_n and w_n fragments if they are present. To favour this type of fragment, it is possible to derivatize the peptide to localize a positive charge on the N-terminal side in order to favour the formation of d_n fragments, or to localize a positive charge on the C-terminal side in order to favour the formation of w_n fragments [62].

The distinction between these two peptides can also be performed at low energy by fragmenting the immonium ions from these amino acids at m/z 86. However, this method can be applied only to peptides that contain either leucines or isoleucines [63]. Differentiation between Lys or Gln residue is performed by high-resolution instruments or by acetylation of the amine group using acetic anhydride [39].

Several algorithms have been developed to interpret tandem mass spectra of peptides. Two different approaches are followed. The first approach is the *de novo* spectral interpretation that involves automatically interpreting the spectra using the table of amino acid masses [64–67]. A second approach has been developed for the interpretation of fragmentation spectra. This approach searches the database to find the best sequence that matches the spectrum. However, this method is not able to interpret the spectra. For instance, with the algorithms named SEQUEST or MASCOT, a correlation is searched between the fragmentation spectrum and peptide sequences contained in a database. First, the algorithm looks for all the peptides in the database that have the same mass as the precursor ions. Then, a measure of the similarity between the predicted fragments from the sequence obtained from the database and the fragments present in the spectrum of the sample allows the most probable sequence to be proposed [68–71].

The use of ion trap, FTICR or orbitrap analysers, which allow MS^n spectra to be obtained, presents two major advantages compared with classical MS/MS techniques. The dissociation of a fragment ion can lead to the observation of diagnostic ions not present in the fragmentation spectrum of the molecular species. Furthermore, the different fragments obtained from the molecular species are identified not only on the basis of their mass but also of their fragments. Furthermore, FTICR or orbitrap analysers allow the analysis of fragment ions at high resolution.

De novo sequencing by MS/MS has been extended to larger peptides and to proteins. Two different approaches are used for this end. A traditional approach, which is referred to as bottom-up protein sequencing, is based on mass spectrometry fragmentation in the gas phase of peptides derived from protein digestion. An alternative approach has recently emerged and is referred to as top-down protein sequencing. This relatively new approach directly involves the fragmentation by MS/MS of intact proteins.

The strategy for bottom-up protein sequencing using mass spectrometry starts with the precise determination of the molecular mass of that protein using MALDI or ESI. This result allows one to verify the sequence that is determined ultimately and also to judge the homogeneity of the sample. The protein then is subjected to reduction and alkylation of the cysteine bridges. Determination of the molecular mass allows one to determine the number of cysteines present in the protein.

In order to generate two different peptide series, the protein is digested with trypsin, which specifically cleaves on the C-terminal side of Lys and Arg, and by protease V8, which specifically cleaves on the C-terminal side of Glu and Asp (see Table 8.4).

About 0.5–200 picomoles per digestion is sufficient for MS/MS analysis. After a reversed-phase HPLC fractionation, the molecular mass of each peptide is determined. Peptides with masses lower than 3000 Da that yield important signals are sequenced using MS/MS. Sequenced peptides then are assembled by matching the peptides in both series while considering the molecular masses of the large unsequenced peptides that contain several smaller sequenced peptides or the known sequence of a homologous protein.

Frequently a third enzymatic digestion is necessary to eliminate all the possible ambiguities. Using trypsin as a proteolysis enzyme allows one partially to distinguish Lys from Gln, two isobars. In fact, the trypsin specifically cleaves on the C-terminal side of Lys and Arg, so all the amino acids with mass 128 ending a peptide can be identified as Lys. However, the absence of a cleavage cannot be used as proof in identifying an amino acid

Table 8.4 Specific chemical or enzymatic cleavage of polypeptides. The cleavage is carboxyl side of X residue and amino side of Y residue.

Reagent	Cleavage site
Chemical cleavage	
Cyanogen bromide	Met-Y
Formic acid	Asp-Pro
Hydroxylamine	Asn-Gly
2-Nitro-5-thiocyanobenzoate	X-Cys
Phenyl isothiocyanate (Edman)	Terminal amino group of peptide
Enzymatic cleavage	
Trypsin	Arg-Y, Lys-Y
Chymotrypsin	Tyr-Y, Phe-Y, Trp-Y
Endoprotease V8	Glu-Y (Asp-Y)
Endoprotease Asp-N	X-Asp (X-Cys)

with mass 128 because Gln, as the tryptic cleavage may be incomplete. The only way to distinguish Lys from Gln within a peptide is to use a high-resolution instrument or acetylate the amine groups using acetic anhydride [39]. Figure 8.14 shows the strategy used in sequencing a protein using mass spectrometry.

The strategy for top-down protein sequencing using mass spectrometry is based directly on the analysis of intact protein ions by MS/MS [72]. Although analysis of intact proteins has been reported for all mass analysers, only MALDI/TOF, MALDI/TOF/TOF, ESI/QTOF, ESI/IT, ESI/FTICR and ESI/OT have been used for top-down analysis of proteins because they have the ability to generate sequence information from intact proteins in the gas phase. But in practice, the most successful technique is ESI/FTICR that provides sufficient resolution and mass accuracy to efficiently sequence large protein ions. The high resolution of this analyser allows the determination of fragment ion charge states. Furthermore, another advantage of the FTICR instrument is the development of several activation methods especially useful for fragmentation of intact protein ions. These different activation methods (that is collision-induced dissociation, infrared multiphoton dissociation and electron capture dissociation) give generally complementary data but electron capture dissociation provides more extensive sequence coverage and in some cases can cleave almost any peptide bond in proteins.

TOF-TOF instruments can also analyse small intact proteins but this instrument leads to low sequence coverage. Indeed, as observed for CID, proteins above 5000 Da produce sequential b_n and y_n fragment ions principally at the termini of the protein. Nevertheless, the most predominant fragment ions correspond to cleavage at the C-terminal to aspartic or glutamic acid residues and at the N-terminal to proline residues in the protein [73].

In comparison with the bottom-up approach, the top-down approach for *de novo* protein sequencing is faster and can be applied on proteins in mixture. However, for analysis of post-translational modifications, these two approaches are complementary and a combination of them is interesting. The bottom-up approach allows detection of low-stoichiometric modifications even if a high-stoichiometric modification is missed. Furthermore, enzymatic digestion leads to the loss of all connectivity information from the original protein species. On the other hand, the top-down approach allows observation of the global pattern of

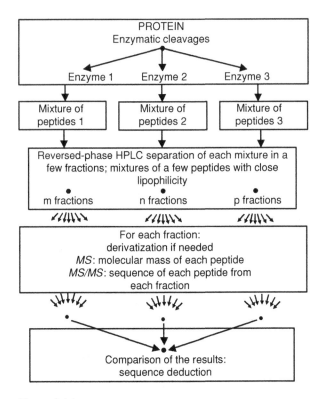

Figure 8.14
Strategy followed in bottom-up sequencing proteins using
mass spectrometry.

modification and the determination of modifications that are together on the same protein
species. However, low-stoichiometric modifications are missed.

8.2.3 Applications

The determination of a peptide or a protein molecular mass with high precision allows
many problems to be solved, such as protein identification, detection of mutations within
proteins, identification and localization of post-translational modifications, verification of
structure and purity of synthetic peptides and proteins produced by genetic engineering, and
even indirect sequencing. In opposition to direct sequencing based on gas-phase fragmentations, indirect sequencing is based on the generation of sequence-specific information from
methods other than gas-phase methods, e.g. solution-phase reactions. However, in spite of
these various possibilities, MS/MS is often used. Indeed, MS^n allows results based only on
the molecular mass determination to be confirmed and specified.

8.2.3.1 *Protein Identification*

Classically, protein identification requires total or partial determination of their sequences
by the method of Edman degradation. In some particular cases, protein identification can

also be based on their recognition by specific antibodies. Edman degradation limits this method to the isolated polypeptides that are not blocked on the N-terminal side. Moreover, this technique requires a long analysis time and a large amount of sample [74].

Without any doubt, mass spectrometry is now the most efficient way to identify proteins [75–78]. The method is based on comparison of the data obtained from the mass spectrometry with those predicted for all the proteins contained in a database. The efficiency of the method results from the development of mass spectrometry into a rapid and sensitive method to analyse peptides and proteins and also from the availability of larger and larger databases. In October 2006, these databases contained more than 2 400 000 non-redundant sequences. Furthermore, the data obtained from genomic sequences after translation in the six lecture frames also can be used. The databases based on expressed sequence tags (ESTs) are another usable source for search. They are composed of sequences based on cDNA fast sequencing. They are limited to short lengths, about 300 bases, and contain many errors but they correspond to coding sequences. Despite their defects, they are very useful for identification of proteins by mass spectrometry [79].

Accurate determination of the molecular mass of a protein is useful for its identification and the determination of its purity. Owing to the possibility of post-translational modifications that are not contained in the database, the correct mass often is not sufficient for identification of a protein but it does facilitate the search.

On the other hand, the peptide mixture obtained by cleavage of a protein by enzymes or specific reactants is characteristic of a protein, the post-translational modifications only affect a small number of the obtained peptides. Furthermore, fragmentation in the gas phase for sequencing by MS/MS is easier for peptides than for proteins and generally gives better sequence coverage. However, because the techniques are improved, the sequencing of intact proteins by MS/MS becomes more and more powerful [80–82]. In consequence, the bottom-up and the top-down approaches can be followed for identifying protein species.

The widely used strategy for protein identification is depicted in Figure 8.15. This strategy thus is to cleave the protein either by trypsin, V8 protease, Lys-C endoprotease or by a reactant such as BrCN. The mixture then is analysed by mass spectrometry to obtain the molecular masses of the largest possible number of peptides. The two ionization methods, MALDI and ESI, are used. MALDI is best to use if one wants to avoid chromatographic separation, because it yields very simple spectra, has a better sensitivity and is not so sensitive to the presence of contaminants. However, ESI can be coupled directly with HPLC or capillary electrophoresis (CE) if a separation is wanted. Furthermore, ESI is often used with mass spectrometers that allow MS/MS data to be easily obtained.

The profile of the masses of the peptides obtained by one of these methods is compared by means of a computer with all predicted peptide digests from a database of proteins to identify the best possible matches. This is termed peptide mass fingerprinting (PMF). A protein generally can be identified using the mass of four to six of its cleavage peptides having masses in the range 700–3000 Da and determined with an accuracy of 0.05 to 0.0005 %. This is improved further if the molecular mass of the protein also is provided [83–86].

The specificity and reliability can be improved by increasing the number of peptide masses used, by increasing the precision of the measured masses and by using different digestion enzymes with different specificities. The digest also can be reanalysed after one cycle of Edman degradation, after a derivatization reaction, or after deuteraytion, to identify the N-terminal amino acid, to obtain information on the amino acid composition or to find out the number of exchangeable hydrogen atoms.

Table 8.5 Different protein identification programs available on the Internet.

Name	URL
PeptideSearch	http://www.narrador.embl-heidelberg.de/GroupPages/Homepage.html
Profound	www.prowl.rockefeller.edu/
MASCOT	http://www.matrixscience.com/search_form_select.html
MS-FIT	www.prospector.ucsf.edu/

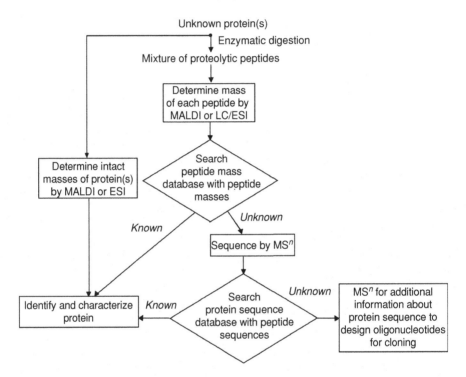

Figure 8.15
Strategy for the identification of proteins based on mass spectrometry only. This method allows protein identification from a 2D gel electrophoresis or from a mixture with certainty, quickly and at high sensitivity.

Various algorithms have been developed for the identification of proteins from the masses of the compounding peptides, and these are available through the Internet as shown in Table 8.5.

This method can be improved further by using peptide sequence information obtained by MS/MS. Indeed, the sequence or even part of the sequence of a peptide is more specific than its molecular mass. This approach, termed peptide sequence tag, allows a protein to be identified from a partial sequence and from mass differences between this sequence and the N-terminal and the C-terminal of the peptide resulting from cleavage of the protein [87,88].

There is a second approach widely used for protein identification that is closely related to PMF. This approach is also based on MS/MS but does not require the interpretation in terms of sequence of the observed fragments. The observed masses of fragment ions are compared with those expected for the various proteolytic peptides deduced from each protein contained in the database. As already described earlier, this method uses the SE-QUEST or MASCOT algorithms for this comparison. This is termed peptide fragmentation fingerprinting (PFF). The partial sequence information that is contained in a tandem mass spectrum of a peptide is more specific than the information based on the precise molecular mass of this peptide. Indeed, two peptides with the same amino acid contents but different sequences have the same molecular mass but different fragmentation patterns.

These methods of course need the protein be present in a database. If this is not so, the method of choice is to obtain some short sequences of part of the unknown protein. Oligonucleotide probes then are synthesized and used to identify the gene coding for this protein. The gene then is cloned and sequenced.

Mass spectrometry thus allows the identification of proteins with a high degree of confidence from a very small quantity of sample. The method is fast and can be applied even to mixtures of proteins. The great number of peptides obtained after cleavage of the mixture of proteins is separated by HPLC before being analysed by mass spectrometry. 2D HPLC, which combines cation exchange chromatography in the first dimension and cap or nano reverse-phase chromatography in the second dimension, is the best way to perform separation of complex peptide mixtures. The proteins of the mixture then can be identified. The possibility to be able to identify several mixed proteins is important, because the final product of many biological experiments, e.g. immunoprecipitation, is actually a mixture. Moreover, this capability can be very useful for the study of protein complexes and of protein–protein interactions [89–91].

Polyamide gel electrophoresis is certainly the method most often used for protein separation. This most powerful electrophoresis method combines a separation based on the isoelectric point in the first dimension and a size separation in the second dimension. Separation of several thousands proteins in one analytical operation is possible. The proteins to identify then are often presented to the mass spectrometrist as a spot or a band on a gel. The methods described before remain applicable but several dedicated modifications have been proposed. They differ by digestion step of the protein [92–94]. Either the proteins are digested directly in gel, or they are extracted first by electroelution or electroblotting.

Evolution of the separation technology and the databases, combined with the speed and sensitivity of the mass spectrometry analysis for the identification, render possible the analysis of all the proteins from a population of cells (proteome) [95,96]. This approach even allows variations in the expression of different proteins as the consequence of a stress to be studied.

8.2.3.2 Detection and Characterization of Mutations

Mass spectrometry allows the detection and characterization of mutations within proteins, whether they are natural or obtained through directed mutagenesis [97–100]. The approach generally involves three steps: molecular mass determination of the intact protein to detect the mutation, PMF of an enzymatic digest to identify mutated peptides, and finally, MS/MS to determine or confirm the position and the nature of the amino acid that has mutated.

A point mutation within a protein leads to the molecular mass variation of this protein. This variation is equal to the difference in mass between the wild type and mutant residues. Indeed,18 of the 20 natural amino acid residues have distinctive molecular masses. There is one exception with Leu and Ile, which are isomers. For the others, the mass differences among amino acid substitutions range from 0.0364 Da for Gln/Lys to 129.0578 Da for

Gly/Trp (see Table 8.6). Therefore, all substitutions between these amino acids lead to the molecular mass variation that can be theoretically detected by mass spectrometry.

The region of the protein containing the mutation can be determined by a proteolytic digestion of this protein followed by mass spectrometric analysis of the resulting peptides. Peptides containing the mutation present the same variation of molecular mass as the intact protein. On the other hand, mutation-free peptides have measured molecular masses in agreement with those calculated for the wild-type protein.

Thus a change in a peptide molecular mass indicates the position of the mutation whereas the difference between the native peptide molecular mass and that of the mutant peptide allows determination of the nature of the amino acid that has mutated as long as only one possibility exists. And in some cases, it is even possible to determine the position of the mutation in the peptide when the mutated residue is found only once in the peptide. Otherwise, MS/MS is necessary.

Table 8.6 lists the mass differences due to the substitutions of a given amino acid by another in the sequence of a peptide chain [101]. The accuracy of the molecular mass determination is critical for the success of the mutation characterization. The required accuracy depends not only on the determined molecular mass, but also on the detected substitution. Indeed, the characterization of Gln/Lys substitution needs less accurate mass measurement with a 1 kDa peptide than with a 40 kDa protein. In the same way, the characterization of Gly/Trp substitution is easier than the characterization of Gln/Lys substitution.

Hemoglobin is a protein for which a large number of mutants have been identified through mass spectrometry [102]. For example, a mutant of human hemoglobin (Hb) β chain, the Hb Miyazono mutant (β79Asp–Glu), was characterized using this strategy (Figure 8.16). MALDI-TOF is used to detect mutants of the β chain. Indeed, the MALDI mass spectrum of globins from this hemoglobin presents two peaks for the β chain, where the mutant chain is 14 Da heavier than the normal chain. After a tryptic digestion, the MALDI analysis of the peptides that are obtained shows two new peptides, T9m and T8+T9m. The peptide T9m is detected at m/z 1683.90, which is 14,01 Da heavier than that of normal peptide T9 (sequence 67–82 detected at m/z 1669.89). This difference suggests five different substitutions: G/A, S/T, V/I, V/L, N/Q or D/E. Because all of these residues (G, S, V, N and D) are present in the normal peptide T9 and some of them (G and D) are present in double, the nature and position of the mutation cannot be uniquely determined by the determination of a mass shift. In this case, MS/MS is used to determine the structure of the mutated peptide and allows one to conclude on the mutation of Asp 79 in Glu 79.

The problem is slightly different in the case of mutants obtained by directed mutagenesis. In that case, the expected modification is known and it is thus not necessary to verify the whole sequence. Comparing the molecular mass calculated from the mutant known sequence with that precisely measured using MALDI or ESI is sufficient. However, in the case when the mass difference between the substituted amino acids is smaller than the error in the measured molecular mass of the protein, a cleavage is necessary to obtain peptides with smaller masses. A cleavage is also necessary when several mutations were introduced.

The top-down approach, which sequences intact proteins by gas-phase fragmentation in the mass spectrometer, can also be used to characterize mutant proteins [103]. The position of the mutated amino acid can be unambiguously determined in each case by the characterization of product ions resulting from fragmentation on either side of the mutation site.

Table 8.6 Monoisotopic mass differences (Da) observed between various amino acids. The residue in the column corresponds to the expected amino acid whereas the residue in the row corresponds to the mutant amino acid.

	Gly	Ala	Ser	Pro	Val	Thr	Cys	Leu/Ile	Asn	Asp	Gln	Lys	Glu	Met	His	Phe	Arg	Tyr	Trp
Gly	—	14.01565	30.01057	40.0313	42.04695	44.02622	45.98773	56.0626	57.02147	58.00548	71.03712	71.0735	72.02113	74.01903	80.03745	90.04695	99.07965	106.04187	129.05785
Ala	-14.01565	—	15.99492	26.01565	28.0313	30.01057	31.97208	42.04695	43.00582	43.98983	57.02147	57.05785	58.00548	60.00338	66.0218	76.0313	85.064	92.02622	115.0422
Ser	-30.01057	-15.99492	—	10.02073	12.03638	14.01565	15.97716	26.05203	27.0109	27.99491	41.02655	41.06293	42.01056	44.00846	50.02688	60.03638	69.06908	76.0313	99.04728
Pro	-40.0313	-26.01565	-10.02073	—	2.01565	3.99492	5.95643	16.0313	16.99017	17.97418	31.00582	31.0422	31.98983	33.98773	40.00615	50.01565	59.04835	66.01057	89.02655
Val	-42.04695	-28.0313	-12.03638	-2.01565	—	1.97927	3.94078	14.01565	14.97452	15.95853	29.0109	29.02655	29.97418	31.97208	37.9905	48	57.0327	63.99492	87.0109
Thr	-44.02622	-30.01057	-14.01565	-3.99492	-1.97927	—	1.96151	12.03638	12.99525	13.97926	27.0109	27.04728	27.99418	29.99281	36.01123	46.02073	55.05343	62.01565	85.03163
Cys	-45.98773	-31.97208	-15.97716	-5.95643	-3.94078	-1.96151	—	10.07487	11.03374	12.01775	25.04939	25.08577	26.0334	28.0313	34.04972	44.05922	53.09192	60.05414	83.07012
Leu/Ile	-56.0626	-42.04695	-26.05203	-16.0313	-14.01565	-12.03638	-10.07487	—	0.95887	1.94288	14.97452	15.0109	15.95853	17.95643	23.97485	33.98435	43.01705	49.97927	72.99525
Asn	-57.02147	-43.00582	-27.0109	-16.99017	-14.97452	-12.99525	-11.03374	-0.95887	—	0.98401	14.01565	14.05203	14.99966	16.99756	23.01598	33.02548	42.05818	49.0204	72.03638
Asp	-58.00548	-43.98983	-27.99491	-17.97418	-15.95853	-13.97926	-12.01775	-1.94288	-0.98401	—	13.03164	13.06802	14.01565	16.01355	22.03197	32.04147	41.07417	48.03639	71.05237
Gln	-71.03712	-57.02147	-41.02655	-31.00582	-29.0109	-27.0109	-25.04939	-14.97452	-14.01565	-13.03164	—	0.03638	0.98401	2.98191	9.00033	19.00983	28.04253	35.00475	58.02073
Lys	-71.0735	-57.05785	-41.06293	-31.0422	-29.02655	-27.04728	-25.08577	-15.0109	-14.05203	-13.06802	-0.03638	—	0.94763	2.94553	8.96395	18.97345	28.00615	34.96837	57.98435
Glu	-72.02113	-58.00548	-42.01056	-31.98983	-29.97418	-27.99418	-26.0334	-15.95853	-14.99966	-14.01565	-0.98401	-0.94763	—	1.9979	8.01632	18.02582	27.05852	34.02074	57.03672
Met	-74.01903	-60.00338	-44.00846	-33.98773	-31.97208	-29.99281	-28.0313	-17.95643	-16.99756	-16.01355	-2.98191	-2.94553	-1.9979	—	6.01842	16.02792	25.06062	32.02284	55.03882
His	-80.03745	-66.0218	-50.02688	-40.00615	-37.9905	-36.01123	-34.04972	-23.97485	-23.01598	-22.03197	-9.00033	-8.96395	-8.01632	-6.01842	—	10.0095	19.0422	26.00442	49.0204
Phe	-90.04695	-76.0313	-60.03638	-50.01565	-48	-46.02073	-44.05922	-33.98435	-33.02548	-32.04147	-19.00983	-18.97345	-18.02582	-16.02792	-10.0095	—	9.0327	15.99492	39.0109
Arg	-99.07965	-85.064	-69.06908	-59.04835	-57.0327	-55.05343	-53.09192	-43.01705	-42.05818	-41.07417	-28.04253	-28.00615	-27.05852	-25.06062	-19.0422	-9.0327	—	6.96222	29.9782
Tyr	-106.04187	-92.02622	-76.0313	-66.01057	-63.99492	-62.01565	-60.05414	-49.97927	-49.0204	-48.03639	-35.00475	-34.96837	-34.02074	-32.02284	-26.00442	-15.99492	-6.96222	—	23.01598
Trp	-129.05785	-115.0422	-99.04728	-89.02655	-87.0109	-85.03163	-83.07012	-72.99525	-72.03638	-71.05237	-58.02073	-57.98435	-57.03672	-55.03882	-49.0204	-39.0109	-29.9782	-23.01598	—

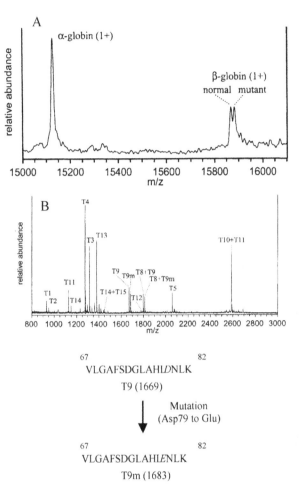

Figure 8.16
A: MALDI-TOF mass spectrum of human hemoglobin Hb Miyazono. Two β-globins (one normal and one mutant) are detected. B: MALDI spectrum of the tryptic cleavage of the β-globin from Hb Miyazono mutant of human hemoglobin. Mutated peptides T9m and T8+T9m are detected. By tandem mass spectrometry, the mutation is characterized as the substitution 79D-E. Reproduced (modified) from Wada Y., Journal of Chromatography B, 781, 291–301, 2002, with permission.

8.2.3.3 Verification of the Structure and Purity of Peptides and Proteins

Mass spectrometry also can play an important role in the verification of the structure and purity of synthetic peptides [104–106]. The development of automatic synthesizers has made the production of synthetic peptides increasingly easier. A large number of errors, however, may occur during or after the synthesis (see Table 8.7) the majority of which are

Table 8.7   Average mass differences for selected synthetic problems.

Mass difference (Da)	Problem	Affected residues
−32	Desulfurization	Cys
−18	Dehydratation, diketopiperazine	Asp, Glu, C-terminus, Gln
−17	Loss of NH_3 with cyclization	Asn, Gln
−2	Disulfur bond formation	Cys
+1	Deamidation	Asn, Gln
+14	Methylation	Asp, Glu, Lys, N-terminus
+16	Oxidation	Met, Trp
+28	Formylation	Ser, Thr, Tyr, Lys, N-terminus
+32	Di-oxidation	Met, Trp
+48	Tri-oxidation	Cys, Trp
+56	tBut (incomplete deprotection)	
+90	Anisylation	Asp, Glu
+99	Tetramethyluronium coupling agent/Amino acid > 1	N-terminus
+100	tBoc (incomplete deprotection)	
+106	Thioanisylation	Asp, Glu
+222	Fmoc (incomplete deprotection)	
+242	Trityl (incomplete deprotection)	

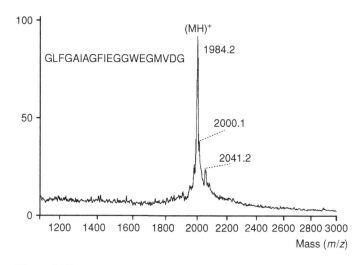

Figure 8.17
The MALDI spectrum of a synthetic peptide. Reproduced from
Finnigan MAT documentation, with permission.

detectable by mass spectrometry. The tandem mass spectrum allows one to confirm the
nature of the modification but also to determine its position within the peptide.

For example, Figure 8.17 shows the MALDI spectrum of a synthetic peptide with a
molecular mass of 1984.2 Da. Two compounds appear next to the sought peptide. The mass
difference between these two impurities and the synthetic peptide suggests that partial
oxidation of methionine and glycine addition occurred during the synthesis.

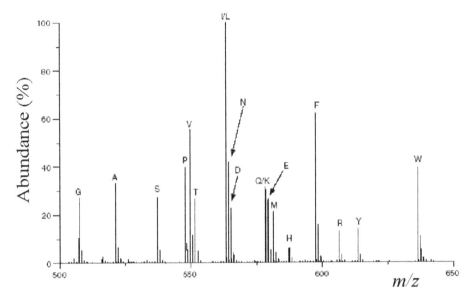

Figure 8.18
ESI spectrum of a peptide library TQTXT. The letters correspond to the amino acid in
position X. Isobaric amino acids (Q/K) and 13C-isobaric amino acids (L/N, N/D and Q/E)
cannot be distinguished at the low resolution used. The difference in abundance between
the peaks is due to the different response factors and suppression effects. For this reason,
quantification of peptide libraries based on a simple mass measurement is not always
possible. Reproduced (modified) from Schlosser G., Takats Z., Vekey G., Pocsfalvi G.,
Malorni A., Windberg E., Kiss A. and Hudecz F., Journal of Peptide Sciences, 9, 361–374,
2003, with permission.

Mass spectrometry also plays an important role in combinatorial chemistry [107,108]
and more particularly for structural analysis of peptide libraries [109,110]. These libraries
generally contain a large numbers of samples that are in many cases highly complex.
Indeed, peptide libraries based on solid-phase synthesis often contain millions of compounds. Control synthesis demands fast and reliable analytical techniques able to provide
information about sample compositions. As shown in Figure 8.18, with a small peptide
library [111] mass spectrometry is the method of choice for the control of synthesis. Mass
spectrometry can be used as a standalone technique. In this case, FTICR or OT instruments
with high resolution must be used for highly complex compound mixture libraries. But
generally, mass spectrometry is used to identify and to quantify peptides from the libraries
after separation, usually by liquid chromatography or CE. Furthermore, mass spectrometry
is not only limited to synthesis control, but also increasingly used for the identification of
active compounds from complex libraries.

Mass spectrometry also allows rapid verification of the fidelity and homogeneity of
proteins produced by genetic engineering [112–114]. In most cases, verifying the sequence
is not sufficient. It is necessary to check whether all the wanted covalent modifications are
present.

Figure 8.19 outlines the general strategy to characterize recombinant proteins by mass
spectrometry. If the accurately measured molecular mass of the produced protein is in
agreement with that calculated from the DNA sequence, then the deduced amino acid

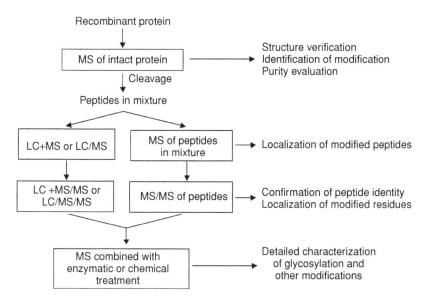

Figure 8.19
General strategy to characterize recombinant proteins.

sequence is correct, assignment of the N- and C-terminals is correct, no post-transitional modifications occur and the amino acids are not chemically modified. In contrast, if the measured molecular mass does not fit with the calculated one, then there is either a sequence error or a modification of the protein. In this case, the protein is cleaved in peptides, and the peptides analysed by mass spectrometry. Peptides whose mass or sequence do not correspond to the ones expected from the DNA sequence provide information on the errors or modifications that occurred. The top-down strategy, which fragments and analyses intact protein ions by MS/MS, is an alternative approach for the recombinant protein characterization [115].

For example, ESI/MS allowed the analysis of various proteins derived from the HIV virus and obtained by genetic engineering, such as the p18 protein [116]. Beyond the main peak series corresponding to the protein with a measured molecular mass of 14 590 Da (as opposed to a calculated molecular mass of 14 589 Da), the spectrum shown in Figure 8.20 contains two other series. The first minor series (8 %) indicated by the letter T and giving a measured mass of 12 651 Da corresponds to the C-terminal side of p18 cleaved at position 111–112, whereas the second series (3 %) indicated by the letter D with a measured mass of 29 175 Da corresponds to the dimer formed by the linkage of two cysteines of two different proteins.

8.2.3.4 *Peptide Ladder Sequencing*

An approach to peptide sequencing without using MS/MS has been described [117]. This two-step approach is called peptide ladder sequencing. First, a set of sequence-defining peptide fragments, each differing from the next by one amino acid, is generated by wet chemistry from a polypeptide chain that must be sequenced. Second, mass spectrometry is used to read out the complete fragment set in a single operation. MALDI-TOF is the instrument of choice for the determination of the molecular masses of these peptides

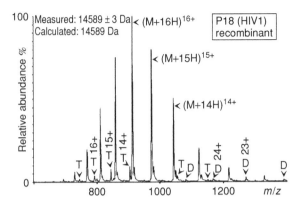

Figure 8.20
The ESI spectrum of the p18 recombinant protein obtained by genetic engineering. Reproduced (modified)
from Van Dorsselaer A., Bitsch F., Green B., Jarvis S.,
Lepage P., Bischoff R., Klobe H.V.J. and Roitsh C.,
Biomed. Environ. Mass Spectrom., 19, 692, 1990, with
permission.

because of the simplicity of its spectra and its high sensitivity. The mass spectrum contains
ions corresponding to each polypeptide species present. The mass differences between
consecutive peaks correspond to amino acid residues, and their order of occurrence in the
data set defines the sequence of amino acids in the original peptide chain. The sensitivity
of peptide ladder sequencing, picomole total amounts of peptide samples, is comparable
with that of existing Edman methods, with the potential for far greater sensitivity.

Several methods were proposed to obtain this set of peptide fragments. The first method
suggested is based on a rapid stepwise degradation using the Edman reagent phenyl isothiocyanate (PITC) in the presence of a small amount of terminating agent phenyl isocyanate
(PIC). A small proportion of peptide chain blocked at the amino terminus is generated in
each step. A predetermined number of cycles is performed without intermediate separation
or analysis. The following scheme illustrates the PITC reaction with the terminal amino
group of the peptide to liberate a cyclic derivative of the N-terminal amino acid corresponding to a phenylthiohydantoin amino acid and to leave an intact peptide shortened by one
amino acid. The principle of the method and an example are presented in Figure 8.21.

$$\text{Ph}-N=C=S + H_2N\underset{COOH}{\overset{O}{\underset{\|}{C}}}NH-R \longrightarrow \text{Ph}-N\underset{O}{\overset{S}{\diagup}}NH\text{(COOH)} + H_2N-R$$

R = Peptide chain

In another method, a volatile degradation reactant is used, the trifluoroethylisothio-
cyanate, to avoid any need for a complex cleaning method. Production of the complete set
of fragment peptides is performed by adding, at each cycle, the same amount of the original
protein. There is no need to add a terminating agent. This allows one to use, later on, a
derivatization reagent of the free-terminal amine to improve the detection sensitivity [118].

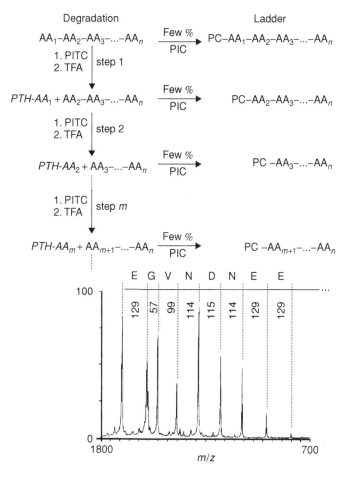

Figure 8.21
(Top) peptide ladder sequencing principle. Phenyl isothiocyanate (PITC) produces phenylthiohydantoin (PTH) of the terminal amino acid and a new peptide with one less amino acid. Phenyl isocyanate (PIC), in low quantity, produces N-terminal phenylcarbamate (PC) from a small fraction of each peptide. (Bottom) example of sequencing of [Glu^1]fibrinopeptide B. Reproduced (modified) from Chait B.T., Wang R., Beavis R.C. and Kent S.B.H., Science, 262, 89, 1993, with permission.

The two preceding methods allow the sequencing from the N-terminal side. Other methods have been developed to obtain a set of peptides characteristic of the C-terminal. To do this, enzymatic or chemical degradation methods have been used [119,120]. If the enzymatic degradation is used, the peptide sample is divided into several parts, each treated with an increasing amount of carboxypeptidase. This allows fragment peptides covering all the sequence to be generated.

The same approach has been applied to proteins. This protein sequencing technique, which is termed mass analysis of polypeptide ladders (MAP), allows rapid determination of protein sequences and characterization of protein modifications [121]. Indeed, proteins could be readily acid-hydrolysed after a brief exposure to microwave irradiation to produce predominately two series of peptide fragments, one containing the N-terminal side and the other containing the C-terminal side. MALDI-TOF analysis of the resulting hydrolysate gives a simple mass spectrum that allows the reading of the protein sequence.

8.2.3.5 Non-covalent Protein Complexes and Tridimensional Structural Information

Mass spectrometry and particularly ESI can be used for the study of the non-covalent interactions of proteins. Indeed, ESI is a sufficiently soft ionization method to allow, under appropriate circumstances, the observation of non-covalent complexes of proteins formed in solution and transferred in the gas phase [122–127].

The first interest is to obtain from the information in the gas phase results that are representative of the species present in the liquid phase. However, transposition of the results has to be done with many caution. The forces responsible for the molecular interactions in the liquid phase are not the same as those in the gas phase. Indeed, the main interactions acting in solution result from van der Waals forces, hydrophobic forces and hydrogen bonding. In the gas phase, electrostatic forces are predominant. Even if a great number demonstrate that complexes in the gas phase reflect the properties in the liquid phase, no generalization can be made. Each case is unique. Several controls have to be performed in order to confirm that the gas-phase observations are related to the liquid-phase behaviour [128,129].

Several experimental methods can be used to study non-covalent interactions. Among them, let us mention various spectroscopic methods and analytical ultracentrifugation. Each method presents advantages and inconveniences [130]. Nuclear magnetic resonance and X-ray diffraction allow information to be obtained on the structure of proteins in solution or in the solid state. However, these two methods require large quantities of high-purity material and are slow. The advantages of mass spectrometry are rapidity, sensitivity and specificity. Mass spectrometry allows rapid determination of the stoichiometry of the complex, the molecular mass is determined directly. Mass spectrometry also allows information to be obtained on the relative stability of the complex by studying the energetics of its dissociation.

Electrospray ionization has allowed the observation of a great number of non-covalent complexes: protein–protein, protein–metal ion, protein–drug and protein–nucleic acid. About one-third of the proteins exist as multimeric forms. Mass spectrometry allows the study of their quaternary structure. This has been done for alcohol dehydrogenase (ADH) from horse liver and from yeast. The ESI spectra are displayed in Figure 8.22. The horse liver ADH is observed to be dimeric whereas that of yeast is tetrameric [131].

These observations are in agreement with structural studies based on other techniques. The observed abundances of the various forms are dependent on the experimental conditions. An increase in the energy transferred to the ions in the ESI source leads to dissociation to monomers.

From a practical aspect, the ESI conditions that allow the survival of non-covalent complexes are not the standard ones. Normally, to obtain a maximum of sensitivity and to have a stable signal, the polypeptides are analysed in basic or acidic solution in the presence of organic solvents. However, non-covalent complexes are not stable under acidic or basic conditions, or in the presence of organic solvents.

Similarly, the ESI source parameters are important for observing non-covalent complexes, and the parameters that enable them to be observed are not the common ones. During

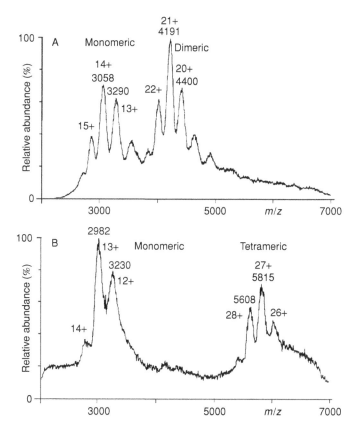

Figure 8.22
The ESI mass spectra of alcohol dehydrogenase under native form obtained from (A) horse liver and (B) yeast. Reproduced from Loo J.A., J. Mass Spectrom., 30, 180, 1995, with permission.

the ESI process, ions are desolvated in a first step by collisions occurring at the interface between the atmospheric pressure and the vacuum, and depend on the potential difference between the electrodes, nozzle and skimmer or capillary and skimmer. To observe fragile non-covalent complexes, this collision-activated desolvation must be reduced. It is interesting to note that most observed complexes carry a reduced number of charges. They thus appear at more elevated m/z values, and analysers with a sufficient mass range should be used.

MALDI has also been successfully used to study a number of intact non-covalent complexes such as protein quaternary structures [132–135]. However, MALDI has not yet contributed widely to the study of these complexes because it requires the crystallization of the sample with the matrix. Furthermore, the energy deposited to desorb the ions is not clearly known. In consequence, MALDI generally induces dissociation of the non-covalent interactions and leads to the formation of non-specific aggregates. Nevertheless, weak non-covalent interactions can survive during the MALDI process to allow the direct detection of intact complexes if specific methods, which have been developed to preserve these interactions, are followed [136].

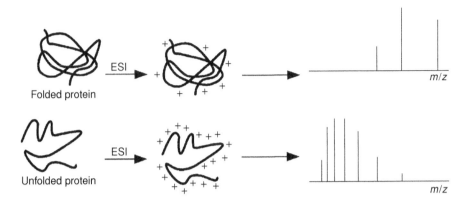

Figure 8.23
Characteristic charge distribution of native or denatured protein.

Mass spectrometry also provides information on the tridimensional structure of proteins. Often, the information from mass spectrometry complements those obtained by other techniques such as circular dichroism, nuclear magnetic resonance or fluorescence. In some circumstances, mass spectrometry, by its speed and sensitivity, allows information to be obtained that is impossible to obtain by other techniques.

Protein analysis by ESI/MS can occur on the native (folded) or on the denatured (unfolded) form [137,138]. Indeed, the conformation of the protein in solution has an important effect on the peak distribution observed in the ESI spectrum, as shown in Figure 8.23. The distribution of the charges on the denatured protein appears broader and with a larger number of charges, i.e. larger z value, than for the native one. This reflects a larger solvent-exposed surface area and better accessibility during the ionization process to the basic amino acids in the unfolded structure. Furthermore, under conditions that reduce the charge-transfer reactions in the gas phase, charge state distributions of protein ions depend only on the solvent-exposed surface in solution [139]. So, the surface areas of proteins and protein complexes can be estimated from the average charge in ESI mass spectra acquired under native conditions.

The dependence of the protein conformation on the charge state distribution of its ions produced by ESI can be applied to study protein conformation. This also allows the denaturation process to be followed over this time and sometimes possible intermediates to be detected (Figure 8.24) [140,141].

Thus, direct information on protein structure in space can be obtained from mass spectrometry. However, such information is most often obtained in an indirect way. This approach uses mass spectrometry to analyse mass changes produced by reactions that are sensitive to protein conformation For example, hydrogen/deuterium exchange (HDX) occurring over time in solution can be followed by mass spectrometry [142–145]. As the protein unfolds, more and more exchangeable sites are exposed to the solvent. Exchangeable hydrogens located inside the folded protein only exchange hydrogen for deuterium at a very low rate. The increase of protein mass over time is measured with high-resolution mass spectrometry. Thus, the role of mass spectrometry is limited to the determination of the number of deuterium atoms incorporated. However, these results can be interpreted in terms of global conformation and stability of the protein in solution. Information even can be obtained on selected regions of the protein. Indeed, the localization of the deuterium

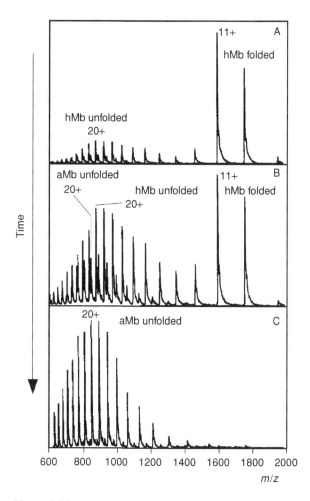

Figure 8.24
Results from the acidic denaturation of myoglobin as a function of the time. Three species are observed: native hMb (heme + myoglobin), denatured hMb (heme + denaturated hemoglobin) and denatured aMb (denaturated apomyoglobin). Reproduced (modified) from Konermann L., Rosell F.I., Mauk A.G. and Douglas D.J., Biochemistry, 34, 5554–5559, 1997, with permission.

incorporation can be determined in the sequence of the protein by MS/MS analysis of its peptic fragments that are produced after the labelling reaction. Compared with nuclear magnetic resonance (NMR), mass spectrometry is much more sensitive and can be used to measure hydrogen/deuterium exchange for proteins and protein complexes of large size and complexity.

A similar method is based on the observation of irreversible reactions with a functional group of the side chain of an amino acid [146–150]. This method allows the accessibility of the solvent to a selected amino acid to be measured. The role of mass spectrometry is

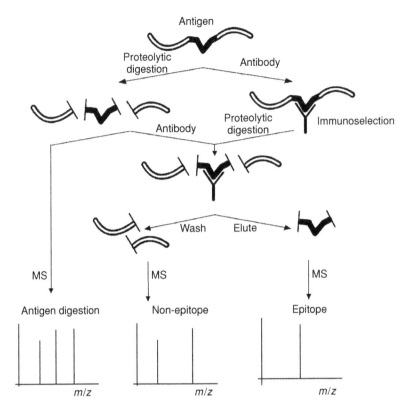

Figure 8.25
Two different methods for the identification by mass spectrometry of an
epitope recognized by a given antibody. The selection by affinity for the
antibody is performed either before or after the enzymatic cleavage.

to determine the number of modified amino acids and their position in the sequence of the
protein.

Another method is based on the proteolysis of the protein [151–154]. A possible cleavage
site will only be cleaved by a protease if it is accessible. Thus, sites located inside the protein
or surrounded by bulky groups will be resistant to proteolysis. In contrast sites located in
flexible areas or sites are not highly structured are easily cleaved. On the basis of the mass
of the obtained peptides, mass spectrometry allows the sites accessible for cleavage to be
located. This method allows also the determination of the region of a protein responsible for
the interaction with a ligand. Indeed, a variation in the conformation of a protein changes
the accessibility towards a protease. Similarly, the interaction of a protein with a ligand
results in the protection of some sites to the cleavage. The strategy in use is based on the
qualitative as well as quantitative comparison of the peptides obtained after proteolytic
cleavage of the protein in the presence or absence of the ligand.

A specific application of this last method is the identification of epitopes recognized by a
given antibody [155–157]. As illustrated in Figure 8.25, two possible approaches are used.
In the first one, the protein presenting the epitope is digested by a protease and the peptides
obtained that present the epitope are selected by the antibody. The mass spectrometry

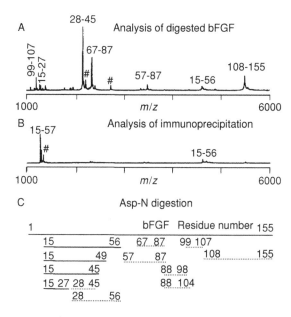

Figure 8.26
The MALDI mass spectrum obtained from the peptides resulting from the enzymatic digestion (Asp-N) of the bFGF protein: (A) spectrum before and (B) spectrum after immunoprecipitation with an antibody directed against the recombinant protein. The peaks marked # correspond to either Cu adducts or doubly charged ions. (C) Schematic representation of the observed peptides. A continuous line corresponds to a peptide that links to the antibody, and a dotted line to a peptide that does not link. Reproduced (modified) from Zhao Y.M., Muir T.W., Kent S.B.H., Tisher E., Scardina J.M. and Chait B.T., Proc. Natl. Acad. Sci. USA, 93, 4020–4024, 199, with permission.

analysis, in general MALDI, allows peptides to be identified that are bound to the antibody and thus present the epitope.

The alternative method consists of the proteolysis of the protein–antibody complex. Mass spectrometry allows those sites to be observed that are protected against enzymatic cleavage. Such sites are implied in the protein–antibody interaction. This method allows identification of nonlinear epitopes, whereas the first method is limited to linear epitopes.

Figure 8.26 illustrates the application of the first method to an antibody directed against the recombinant protein bFGF. Twelve peptides are identified in the MALDI spectrum of the mixture obtained after enzymatic cleavage non-bounded (Figure 8.26A). After immunoprecipitation with the antibody, the peptides are washed out, and only four peptides are detected as shown in Figure 8.26B. A comparison of the sequences of those peptides allows the epitope to be located somewhere in the residues 15–27, as explained in Figure 8.26C.

8.3 Oligonucleotides

Oligonucleotides, also named nucleic acids (DNA or RNA), are linear polymers of nucleotides. The nucleotides are composed of a heterocyclic base, a sugar and a phosphate group. There are five different bases. Cytosine (C), thymine (T) and uracil (U) are pyrimidine bases and adenine (A) and guanine are purine bases. Both DNA and RNA not only differ by their base composition (ACGT for DNA and ACGU for RNA) but also by the nature of the sugar part (2'-deoxy-D-ribose in DNA and D-ribose in RNA). The nucleosides and the deoxynucleosides are formed by binding the N-9 position of the purines or the N-1 position of the pyrimidines to the C-1' of ribose or deoxyribose, respectively.

The structures of the main nucleosides are presented in Figure 8.27. In the polymer chain, the nucleosides are linked together by the phosphate groups attached to the 3' hydroxyl of one nucleoside to the 5' hydroxyl of the adjacent nucleoside. Moreover, these oligonucleotides may undergo natural covalent modifications that are most often present in tRNA and in rRNA, or unnatural ones that result from reactions with exogenous compounds. An important aspect of the research in this field consists of characterizing these structural modifications. Mass spectrometry plays an important role in highlighting these modifications and in determining their structures and positions within the oligonucleotide.

Mass spectrometry not only allows precise determination of the molecular weight of oligonucleotides, but also in a direct or indirect way the determination of their sequences.

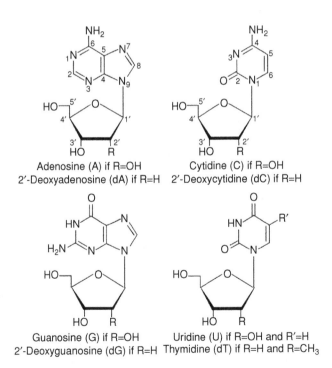

Figure 8.27
Structures of the main nucleosides making up the oligonucleotides.

Reviews on the mass spectrometry analysis of oligonucleotides have been published in the last few years. [158–161]

8.3.1 Mass Spectra of Oligonucleotides

For many years, mass spectrometry has had an important role in the analysis of nucleic acids. However, older work was limited to the analysis of nucleotides because the EI and CI methods did not allow oligonucleotide analysis. With the advent of FAB and PD, small oligonucleotides comprising up to about 10 bases have been analysed.

Because the nucleic acids are the most polar of the biopolymers, they have benefited much from the advent of ESI and MALDI because such methods allow the analysis of very small quantities of oligonucleotides, starting at some tens of femtomoles. However, the oligonucleotides are not so easy to analyse as the peptides and proteins. One reason for this is the formation of strong bonds between the phosphodiester group and the alkali metal atoms, mostly Na and K, leading to the formation of adducts containing several of these atoms. The oligonucleotides are best detected in the negative ion mode, and the observed ions have the general formula $(M - [n + m]H + mNa/K)^{n-}$. This distribution of the ions of the molecular species over several mass values leads to a reduction of the signal-to-noise ratio. The separation of the individual ions requires high resolution. Generally, they are not separated and result in one large peak. This reduces the accuracy of the mass determination. Furthermore, mixtures containing compounds with small mass differences will not be separated. This difficulty can be overcome, however, by adding ammonium ions to the sample. [162] These will replace the alkali metal ions by binding to the phosphodiester group, but in the gas phase ammonia is released, leaving the free oligonucleotide.

A second difficulty in the analysis of oligonucleotides is their easy fragmentation. As will be discussed in more detail, the oligonucleotides tend to lose their nucleic bases by a 1,2-elimination mechanism. Fragmentation of the nucleic chain at either the 5′ or 3′ side of the sugar having lost a base then can occur. This stability problem is more important for MALDI than for ESI analysis.

Oligonucleotides can be analysed in both positive and negative ion modes. However, negative ion mode give better sensitivity and resolution, especially in ESI.

Figure 8.28 displays the MALDI spectra of a DNA 40-mer in the negative ion mode and an RNA 195-mer in the positive mode. [163,164] The MALDI spectra are dominated by either the protonated $(M + H)^+$ or deprotonated $(M - H)^-$ ions of the molecular species. Multimers of general formula $(nM \pm H)^\pm$ are generally detected, but multiply charged ions are sometimes present, mainly for large oligonucleotides. As mentioned before, metal adducts also are often present.

The success of the analysis of oligonucleotides is strongly dependent on the choice of the matrix and the preparation of the sample. A major improvement has resulted from the introduction of new matrices specific for the oligonucleotides. An example is 3-hydroxypicolinic acid. [165] Several strategies for sample preparation have been adopted, with the principal goal of eliminating interference from the ubiquitous alkali metal ions: a combination of the use of cation exchange columns to replace the alkali metal ions by ammonium and the addition of some ammonium salt to the sample.

The response of nucleotides to MALDI depends also on their nature, their size and their composition in nucleic bases. Oligonucleotides containing only thymine give a much

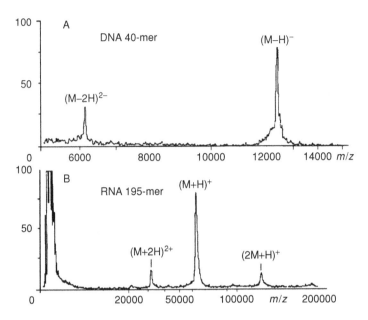

Figure 8.28
(A) The MALDI linear TOF mass spectrum of a DNA 40-mer in neg-
ative ion mode. (B) The MALDI reflection TOF mass spectrum of an
RNA 195-mer in the positive ion mode. In both cases, the spectra
were obtained using 3-HPA as matrix and a laser at 337 nm. Repro-
duced (modified) from Wu K.J., Shaler T. A. and Becker C.H., Anal.
Chem., 66, 1637, 1994 and Kirpekar F., Nordhoff E., Kristiansen
K., Roepstorff P., Lezius Z., Hahner S., Karas M. and Hillkamp F.,
Nucleic Acids Res., 22, 3688, 1994, with permission.

stronger signal than those composed of the other bases. Thymine is more difficult to protonate than the other bases. It is thus also more difficult to eliminate, which results in better
stability of the polymer chain. The RNA is easier to analyse than the DNA. This also appears
to result from the fact that 1,2-elimination of the base occurs only with deoxyribose; the
presence of the 2'-hydroxyl group in RNA hinders the elimination reaction.

When the size of the nucleotide increases it becomes more difficult both to ionize and to
detect. Furthermore, the formation of adducts is increased and the fragmentation reaction
also increases. These facts together explain why larger oligonucleotides are detected with a
lower signal-to-noise ratio and a lower resolution, or the signal may be totally suppressed.
Dependence on the size results in suppression of the largest nucleotides when equimolar
mixtures with small ones are analysed. [166]

The error on the measured molecular mass obtained by MALDI varies between 0.01
to 0.05 % for small oligonucleotides. This error increases to 0.5 % or even more for large
nucleotides. Indeed, at higher masses the signals from adducts or resulting from the loss
of a base are more abundant and cannot be resolved, any more, causing a broadening of
the peak. The error on the centroid determination increases. One could consider that the
maximum size that can be analysed reasonably by MALDI is around 50 bases for DNA.
For RNA more than 100 bases have been analysed with success. [164]

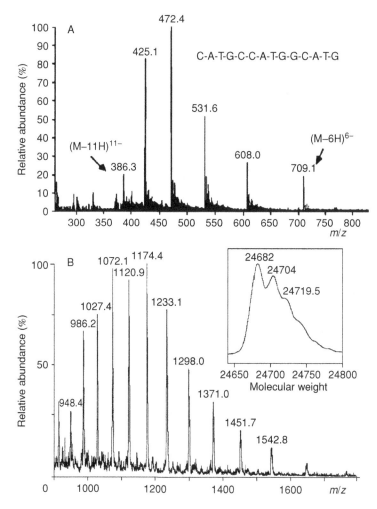

Figure 8.29
(A) Negative ion ESI mass spectrum of a 4260 Da synthetic oligonucleotide. (B) The ESI mass spectrum of tRNA[Val1] from *E. coli*; the calculated mass is 24 681 Da. Reproduced (modified) from Fenn J.B., Mann
M., Meng C.K., Wong S.F., Mass Spectrom., Rev., 9, 37, 1990 and Lim-
back P., Crain P.F. and Mc Closkey J.A., J. Am. Soc. Mass Spectrom., 6,
27, 1995, with permission.

The ESI mass spectra of a synthetic oligonucleotide CATGCCATGGCATG and tRNA
of more than 24 kDa are displayed in Figure 8.29. [167,168] The ESI mass spectra display a series of peaks of different charge states resulting from the loss of several protons
$(M - nH)^{n-}$ as well as metal adducts $(M - [n + m]H + mNa/K)^{n-}$ whose importance depends primarily on the sample preparation. The charge state is generally around one charge
per 1000 Da as a mean, depending on the analysed nucleotides and on the experimental
conditions: nature of the solution and source parameters.

As for MALDI, elimination of the metal adducts is critical to the success of the analysis.
One strategy consists of precipitating the oligonucleotide with an ethanol solution containing

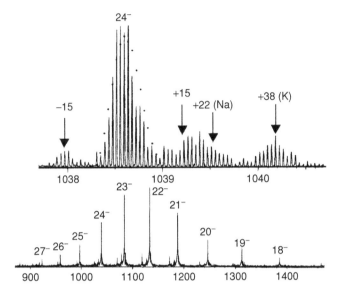

Figure 8.30
The ESI/FTMS trace of tRNA^{Phe} with a zoom of the isotopic cluster for the 24$^-$ anion. Peaks at -15 and $+15$ Da are due to minor impurities. Reproduced (modified) from Little D.P., Thannhauser T.W. and McLafferty F.W., Proc. Natl. Acad. Sci USA, 92, 2318, 1995, with permission.

ammonium acetate and then redisolving it in a dilute aqueous solution of ammonia. This solution also may contain divalent ion chelating agents, which allows one to observe up to a 50-mer devoid of sodium adducts; without precautions only up to one sodium per phosphate group can be observed. Another strategy is the use of a 20/80 water/acetonitrile solution containing imidazole, pyridine and acetic acid. This solution not only allows the sodium adducts to be reduced but also reduces the abundance of charge states, thereby increasing the sensitivity and simplifying the spectra.

Molecular weights determined by ESI are generally more precise that those obtained by MALDI. Errors as low as 0.01 % are often obtained after appropriate sample preparation. Oligonucleotides containing more than 100 bases have been analysed successfully by ESI. [168]

Electrospray ionization with the high resolving power of Fourier transform mass spectrometry (FTMS) makes possible the detection of adducts and subpicomole impurities that confuse lower resolution measurements, as shown in Figure 8.30. This method achieves accurate determination of molecular weights (<0.002 %) and permits the verification of 50–100-mer DNA and RNA sequences. [169,170]

8.3.2 Applications of Mass Spectrometry to Oligonucleotides

The simplest and also most common application of mass spectrometry is the accurate determination of molecular weight to verify the presence of an oligonucleotide of a

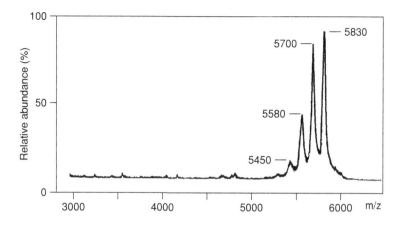

Figure 8.31
The MALDI/TOF mass spectrum of an oligonucleotide that has undergone
a partial depurination reaction. Reproduced (modified) from Van Ausdall
D.V. and Marshall W.S., Anal. Biochem., 256, 220–228, 1998, with permission.

given sequence that is wanted, predicted or wished. This particularly applies for synthetic
oligonucleotides or oligonucleotides produced by techniques of molecular biology, such as
polymerase chain reaction (PCR).

The use of oligonucleotides produced with automatic synthesizers becomes more and
more frequent. However, synthetic oligonucleotides can be considerably contaminated or,
worse, their sequence may be different from expected due to the depurination reaction,
non-removed protecting groups, deletions, etc. The association of mass spectrometers with
automatic synthesizers has greatly enhanced the degree of confidence for the oligonucleotides produced. [171,172] Indeed, the precise determination of molecular weight of the
oligonucleotides produced not only allows confirmation of the expected sequence but, in
cases of defects, allows identification of the modification that has occurred or the nature of
the contaminants.

An important advantage of mass spectrometry over other techniques such as electrophoresis or chromatography is its high speed. For instance, the association of MALDI/TOF with
automatic preparation of the samples, together with automated acquisition of the spectra,
allows the analysis of more than 100 oligonucleotides in 90 min. [172]

Another important advantage of mass spectrometry is its capacity to verify the incorporation of modified nucleotides. Figure 8.31 displays the mass spectrum of a synthetic
oligonucleotide whose theoretical mass is 5838 Da. In addition to the desired nucleotide,
other compounds are present. The mass difference between the peaks (111–151 Da) corresponds to the masses of the nucleic bases. This suggests that depurination reactions have
occurred during the synthesis.

Determination of the size or molecular weight of nucleic acids is crucial in molecular
biology. This importance is increased by the introduction of the PCR, which allows an
individual to be identified (e.g. a pathogen bacteria) or the genetic links between individuals
and allows the prediction or diagnosis of diseases.

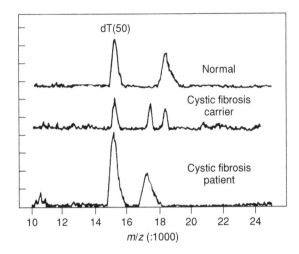

Figure 8.32
The MALDI/TOF mass spectrum of an oligonucleotide produced by PCR amplification of a part of the gene responsible for cystic fibrosis. The top spectrum is obtained from a normal individual, whereas the central and bottom spectra are obtained, respectively, from a healthy heterozygote carrier and an ill homozygote. Reproduced (modified) from Chang L., Tang K., Shell M., Ringelberg C., Matteson K.J., Allman S.L. and Chen C.H., Rapid Comm. Mass Spectrom., 9, 772, 1995, with permission.

Indeed, the PCR allows amplification to a detectable level of a determined region of DNA from whatever the source. Other applications in medicine, agriculture or the environmental sciences also are based on the identification of nucleic acids.

The classical methods for analysing the nucleic acids are based on electrophoresis, either on gels or in columns. In this approach, the nucleic acids are separated according to their molecular weights and detected by markers that are colored, fluorescent or radioactive. Important drawbacks are the length of the analysis time, the low resolution achieved and the need to mark the molecules. Owing to its capacity to assign quickly and accurately the molecular weight of the nucleic acids, mass spectrometry is an interesting alternative.

Up to now, exploitation of these capabilities has been limited. Only PCR products or digest obtained with restriction enzymes have been analysed in this way. [173,174] Some examples of detection of a mutations by mass spectrometry have been presented.

For instance, MALDI analysis allows the detection of a mutation often present in patients with cystic fibrosis (see Figure 8.32). [175] This mutation corresponds to the deletion of three base pairs, which leads to the loss of a phenylalanine residue in the protein. Using appropriate primers for the PCR reaction, the normal gene presents a 59-base fragment whereas the mutated one presents a 56-base fragment.

Determination of the molecular weight of an oligonucleotide allows its composition in to be assigned nucleic bases. [176]

If the oligonucleotide is small (about 5-mer) and the mass resolution at least 0.01 %, the base composition can be assigned unambiguously. Improving the mass resolution combined with other information such as the number of bases, possible limitations to the composition, etc. will allow the determination of the composition in bases to be extended for larger oligonucleotides, or at least limit the number of possible candidate oligonucleotides. For instance, a measure of the molecular weight with a precision of 0.01 % of a 14-mer nucleotide allows the complete composition to be assigned in bases if it is supposed that it contains only one guanine residue.

Determination of the molecular weight of oligonucleotides also will allow assignment of their sequence in an indirect way, as compared with the direct way based on MS/MS. One approach allowing the sequencing of oligonucleotides is known as 'DNA ladder sequencing'. [177,178] In a first step, a series of nucleic fragments, differing from each other by one nucleotide, is produced in solution from the unknown oligonucleotide. This is performed by enzymatic digestion by exonucleases such as SV phosphodiesterase or CP phosphodiesterase, which allow the nucleotides to be removed sequentially from 3'-OH or 5'-OH, respectively, To obtain fragments that cover the whole sequence, a small volume of the solution is set apart at different times during the digestion, or alternatively the solution is divided into parts and each one is treated with an increasing concentration of enzyme.

All these collections of fragments then are analysed by mass spectrometry. The sequence is assigned from the mass differences between two consecutive fragments: 289 Da for dC, 304 for dT, 313 for dA, 329 for dG, 329 for A, 305 for C, 345 for G and 306 Da for U. For DNA the smallest difference is 9 Da (difference between A and T), whereas for RNA it is only 1 Da (difference between U and C). The accuracy of the mass determination is thus important. This approach allows up to about 30 bases to be sequenced, as displayed in Figure 8.33. [179]

The series of nucleotidic fragments can be obtained also by chemical digestion. [180] This allows the degradation of oligonucleotides that are non-sensitive to nucleases. However, by chemical digestion three types of fragments can be produced. Indeed, any potential site can be cleaved in either of two directions: towards the 3' end or towards the 5' end. The redundancy of these fragments always introduces possible confusion, but sometimes can be helpful to obtain the full sequence. Furthermore, fragments resulting from the cleavage at two sites are useless for sequence determination and add to the confusion.

Another sequencing approach based on mass spectrometry makes use of the sequencing products from the Sanger method, but mass spectrometry replaces the standard electrophoretic method. [166,181]

In the Sanger method, an oligonucleotide complementary to that used for sequencing is produced using a DNA polymerase. However, in the solution of the four regular nucleoside triphosphates is placed a small quantity of the dideoxy analog of one of the four nucleoside triphosphates. The result is to stop the polymerization every time a dideoxy base is included, because the 3' hydroxyl end is lacking. This leads to a mixture of oligonucleotides of variable length, but always ending with the same dideoxy base.

Four similar mixtures can be obtained by repeating the operation with the four possible dideoxy bases. The use of MALDI mass spectrometry to analyse the sequencing products is particularly attractive in view of the shorter analysis time that it requires: several hours are need to perform an electrophoresis, whereas obtaining a mass spectrum needs only a

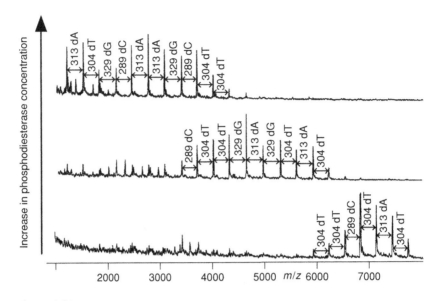

Figure 8.33
The MALDI/TOF mass spectrum of a 25-mer oligonucleotide after enzymatic digestion with a type I phosphodiesterase at three different concentrations. The different
peaks correspond to successive cleavages from the 3' side to the 5' one. Reproduced
(modified) from a PerSeptive biosystems documentation, with permission.

few minutes. The reading of the sequence from the four mass spectra corresponding to the
four dideoxy bases is completely similar to that performed after electrophoresis. The mass
resolution needed is not critical, as it is in ladder sequencing, because the mass differences
to observe correspond to one nucleotide, or about 300 Da. This approach is limited to
oligonucleotides of about 50 bases. Automatic DNA sequencers allow the routine analysis
of up to 500 bases. To be able to compete, both the mass limit and the sensitivity of
MALDI/TOF have to be increased by one to two orders of magnitude.

Electrospray ionization is a soft ionization technique allowing, under appropriate conditions, the transfer of non-covalent complexes from the liquid to the gas phase. The stoichiometry of the complex is easy to assign because it is its molecular weight that is actually
measured. Electrospray ionization has allowed the observation of non-covalent complexes
of oligonucleotides themselves (duplexes and tetraplexes), of drug–nucleotide and of nucleic acid–protein, [182–184] and thus is a useful method for the study of the hybridization
of native or chemically modified oligonucleotides, for the rapid detection of possible drug–
oligonucleotide interaction, etc. [185,186]

Mass spectrometry offers several advantages over classical methods for the characterization of non-covalent complexes: rapidity, simplicity and ability to work on mixtures. One
of the goals in this field is to transfer gas-phase data to the liquid phase. However, this
transposition has to be done with caution. Indeed, some of the complexes observed are non-
specific associations formed in the gas phase or aggregates resulting from the ESI process
itself. Also, some of the complexes present in the liquid phase might not be observed in the
mass spectrum.

8.3.3 Fragmentation of Oligonucleotides

The fragmentation of oligonucleotides in the gas phase can be a problem to be surmounted for the analysis of these molecules. However, this fragmentation can be used to determine the sequence of the oligonucleotide. The dissociation of oligonucleotides can occur in the source as the result of the energy excess that is imparted to the molecule during the desorption/ionization process. This is the case of FAB spectra of oligonucleotides. Indeed, the negative mode FAB mass spectra of oligonucleotides are characterized by the presence of two series of ions that are characteristic of the sequence: one containing the phosphate on the 5' side and the other containing the phosphate on the 3' side. Normally the intensity of the 5' terminal series is greater than that of the 3' series, and other ion series of masses 18 Da lower (water loss) or 80 Da higher (addition of HPO_3) are also present. A spectrum that is typical of a small oligonucleotide with sequence UGUU, [187] measured by negative mode FAB/MS, is shown in Figure 8.34.

Fragmentation of nucleotides can be observed also in MALDI spectra [188,189]. In principle, these fragmentations could allow the sequence to be deduced. However, coupling of MALDI to a TOF spectrometer and the variable kinetics of these fragmentations result, in the absence of specific experimental conditions, in broadening of the peak of the molecular species and a loss of resolution and sensitivity.

Only fragmentations that occur as fast as the desorption can be observed in the mass spectrum at their right m/z values in a linear TOF instrument. Dissociation that occurs later on can be observed only if the extraction from the source is delayed or if a reflectron analyser is used.

The coupling of the MALDI to a trapping analyser, either ion trap or FTICR, allows all those fragments to be detected. [190] The formation of these fragments depends on the matrix, the power of the laser beam and the sequence of the oligonucleotide submitted to the analysis.

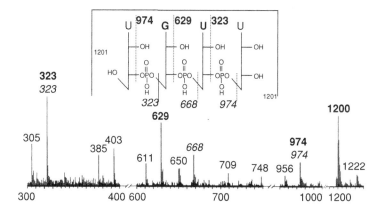

Figure 8.34
Negative mode FAB spectrum of an oligonucleotide with sequence UGUU. Reproduced (modified) from Grotjahn L., in 'Mass Spectrometry in Biomedical Research' edited by Gaskell S.J., Wiley, New York, 1986, pp. 215–234, with permission.

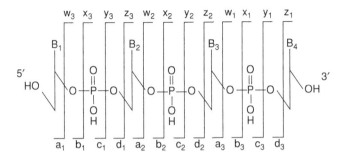

Figure 8.35
Nomenclature for the dissociation products from oligonucleotides.

Electrospray ionization produces stable ions of the molecular species, except if the
conditions in the source are adjusted to induce dissociation by collision in the nozzle–
skimmer region. [191]

Dissociation of the nucleotides can be obtained also by adding internal energy to the
stable ions. Combining this induced dissociation with MS/MS allows fragment ions to be
obtained only from the selected precursor. This allows the analysis of oligonucleotides from
a mixture and also suppresses the low-mass ions originating from the matrix when FAB
or MALDI are used. With ESI, the spectra are simplified because only one charge state
is selected. An increasing number of mass spectrometry analyses of oligonucleotides use
MS/MS. Results obtained by coupling an ESI or MALDI source to an FTICR analyser or
by coupling an ESI source to a triple quadrupole instrument have been reported. [190,192]
However, the large majority of reported results are obtained by coupling ESI to an ion trap
mass spectrometer, which allows CID/MS^n to be performed. [193–195]

A systematic nomenclature has been developed by McLuckey to describe the fragments
observed when oligonucleotides are studied by mass spectrometry. This is illustrated in
Figure 8.35. [194] Cleavage of the four possible types of phosphodiester bond yields eight
types of fragments, symbolized a_n, b_n, c_n and d_n if they contain the 5'-OH group and w_n,
x_n, y_n and z_n for those containing the 3'-OH group. The subscript n represents the number
of residues contained in the fragment, and thus indicates the position of the cleavage.
The additional loss of a base is indicated by parentheses with, if possible, the identity of
the base. For example, the fragment a_3-B_3(A) indicates a cleavage of the bond between the
ribose carbon atom and the oxygen atom of the phosphodiester group at position 3, with
the additional loss of an adenosine base at this same position.

This nomenclature is similar to that proposed for peptides. Note, however, that d_n and
w_n fragments are not similar to those symbolized with the same letters for peptides. It
is also interesting to mention that another nomenclature for the fragments observed in
MALDI spectra has been proposed by Viarr and completed by Nordhoff. [188,196] This
nomenclature, however, has the drawback that it cannot be applied systematically to all
possible fragments. It characterizes three kinds of fragments, denoted X-, X*- and Y-,
corresponding respectively to d_n, a_n-B_n and w_n.

As a rule, the fragmentation schemes of oligonucleotides are similar for the various types
of ion sources or analysers used. The differences observed can be explained either by the
charge state or the energy content.

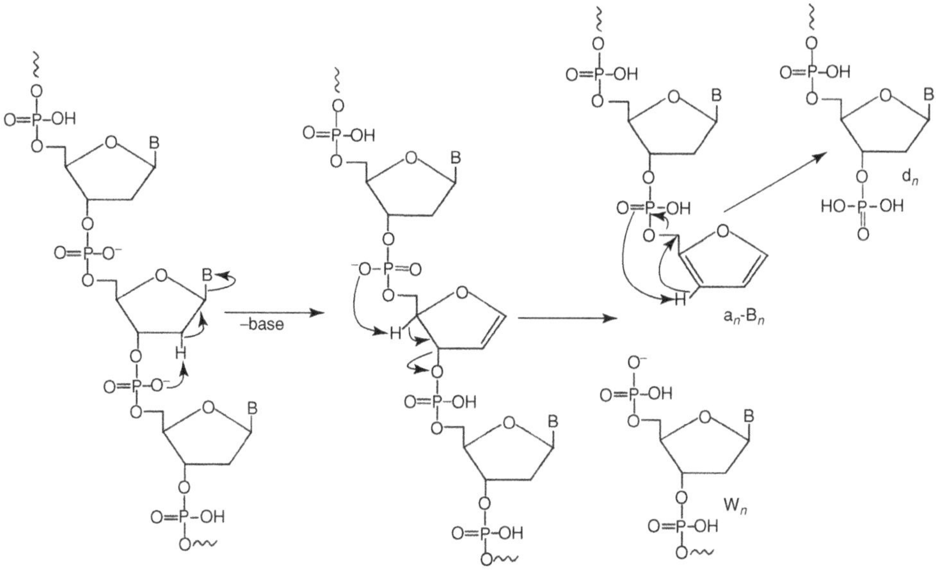

Figure 8.36
Fragmentation pathways of oligonucleotides that necessitate a charge proximate to the cleaved site.

One of the most frequently observed fragmentations corresponds to the loss of one of the nucleic bases as a neutral or an anion. The tendency to produce the anion rather than the neutral increases for higher charge states and longer nucleotides chains.

Other often-observed fragments correspond to the cleavage of the nucleic chain to yield three series of ions of types a_n, a_n-B_n and d_n characteristic of the sequence from 5' to 3', and one series of ions of type w_n characteristic of the inverse sequence, from 3' to 5'. Of course, the abundances of these ions depend on experimental factors, such as the ionization method. The MALDI spectra display abundant d_n and w_n ions whereas ESI spectra are generally dominated by a_n-B_n and w_n ions.

Other types of fragmentations resulting, for instance, from a rearrangement or a double cleavage are also observed. The abundance of double cleavage ions increases at higher internal energy, for higher charge states and for less stable oligonucleotide ions.

Several fragmentation pathways have been proposed to explain the formation of complementary ions of types a_n-B_n and w_n. [188,197–199] The first one includes two steps with the loss of the nucleic base as neutral or anion via a 1,2-elimination, as shown in Figure 8.36. This elimination may be favoured by an intramolecular base catalysis by the negatively charged oxygen atom from the 3' phosphate group. Then, this intermediate can yield the a_n-B_n and w_n fragments by cleavage of the 3'C—O bond of the phosphodiester group. The mechanism again is a 1,2-elimination, as shown in Figure 8.36. The reaction is favoured by the formation of the resonance-stabilized furan cycle. It can be catalyzed by the negative charge of the 5' or 3' phosphate, removing the 4' hydrogen atom of the ribose. The type d_n ions are produced by the mechanism depicted in Figure 8.36 from type a_n-B_n ions if they contain sufficient excess of internal energy. [200]

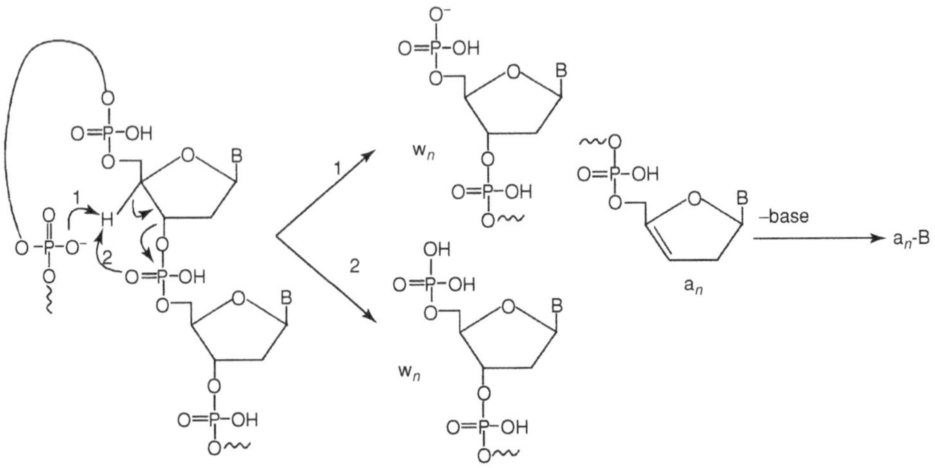

Figure 8.37
Fragmentation pathways of oligonucleotides that do not necessitate either the prior loss of a nucleic base or a charge located near the cleavage site.

A second fragmentation pathway has been proposed for type a_n-B_n and w_n ions. This pathway does not require the loss of a nucleic base or the proximity of the charge to the cleavage site. It thus allows observation of these ions to be explained in the absence of adjacent charge. The first step is the direct cleavage of the $3'C$—O bond of the phosphodiester group to produce the a_n and w_n fragments. The a_n ions then dissociate by the loss of a nucleic base to yield ions of type a_n-B_n.

Two mechanisms have been suggested to explain this fragmentation pathway. The first is a 1,2-elimination involving the transfer of the $4'$ hydrogen atom to a remote-charged phosphodiester, as represented in Figure 8.37, reaction 1. In the second mechanism, this $4'$ hydrogen atom is transferred to the oxygen atom of a non-charged $5'$ phosphate group through a six-atom cyclic transition state, as shown in reaction 2 of Figure 8.37.

An algorithm has been developed for the interpretation of tandem mass spectra of oligonucleotides. [201] This greatly facilitates the interpretation of mass spectra of oligonucleotides and is based on the general fragmentation scheme of these molecules.

On the basis of the recognition of specific fragments, it first builds a sequence for both the $5'$ and $3'$ sides. Then, it adjusts by cross-checking the two sequences. Then, on the basis of the measured mass and the constraints of the nucleic base composition, it rejects incorrect candidate sequences.

The determination of the sequence, based on the gas-phase fragmentation of oligonucleotides and subsequent identification of the fragments, is potentially very interesting. It is the fastest way to sequence oligonucleotides. However, the fragmentation spectra are often very complicated and may be difficult to interpret. They need a sufficient resolution to separate all the fragments produced and the mass determination has to be sufficiently accurate to assign unambiguously the observed fragments.

These requirements explain why the mass spectrometry method is limited to about 20-mer to obtain structural information and to 10-mer for the complete sequencing.

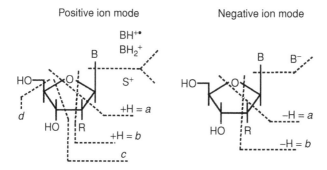

Figure 8.38
General fragmentation diagram of nucleosides.

8.3.4 Characterization of Modified Oligonucleotides

Oligonucleotides can undergo covalent modifications. [202] These can be natural ones such as those present in tRNA and rRNA but they can also result from reactions with exogenous substances and are the markers indicating possible degradations of the cells. They also can be modified chemically to create new drugs. Almost all these modifications are accompanied by variations in mass, and mass spectrometry is thus useful to identify their nature and position in the sequence. [203–205]

A general fragmentation diagram was drawn up based on the MS and MS/MS of various nucleosides measured in positive and negative modes by different ionization techniques (EI, [206] CI, [207] FAB, [208] ESI [209]), as shown in Figure 8.38. The most abundant fragmentation results from the cleavage of the bond between the nucleobase and the sugar.

Knowing this fragmentation diagram and using MS/MS, we can detect and identify the structures of the modified nucleosides directly in the nucleoside mixture obtained by enzymatic digestion or hydrolysis, without any previous separation. In fact, the analysis of the enzymatic digestion or the hydrolysis of a modified oligonucleotide contains the molecular species ion of the modified nucleoside in addition to the molecular species ions of the natural nucleosides. The MS/MS analysis of this ion of the modified nucleoside allows one to deduce its structure from the fragments obtained.

The recognition of a molecular species ion in a spectrum may be difficult. Hence, a scan of neutral losses with mass 132 Da allows one to detect selectively the protonated molecular ions of the various nucleosides contained in the mixture, because the production of an intense BH_2^+ ion from MH^+ by the loss of a sugar molecule (132 Da for a ribose) is an important characteristic of positive ion spectra. A modified nucleoside, N^6-isopentenyladenosine, was identified in tRNA^{Tyr} of *E.coli* using this method [210] (Figure 8.39).

Other methods (described in Figure 8.40) based on mass spectrometry, such as gas chromatography/mass spectrometry (GC/MS) [211] or high-performance liquid chromatography/mass spectrometry (HPLC/MS), [212–215] also allow one to detect and identify the modified constituents of oligonucleotides. These methods are good for small sample quantities of around 1 μg, and allow the detection of a modified nucleotide within a mixture of 10^7 nucleotides.

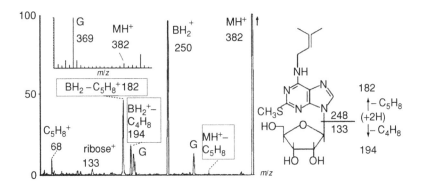

Figure 8.39
Collision-induced dissociation FAB/MS/MS trace of N^6-isopentenyladenosine (25 ng) contained in an unpurified hydrolysate of *E.coli* tRNA^{Tyr}. Reproduced (modified) from Nelson C.C. and McCloskey J.A., Adv. Mass. Spectrom., 11A, 260, 1989, with permission.

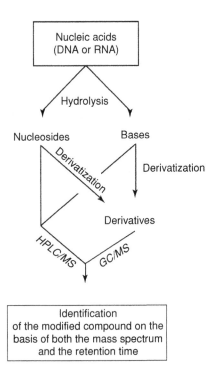

Figure 8.40
Strategy followed in order to detect and identify the modified constituents of oligonucleotides by HPLC/MS or GC/MS techniques.

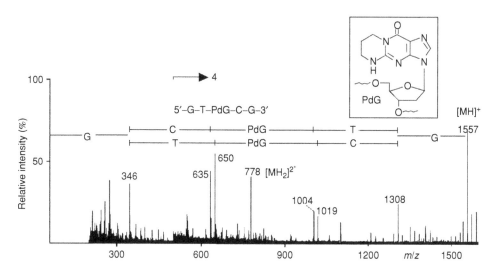

Figure 8.41
Collision-induced dissociation FAB/MS/MS spectrum of an oligonucleotide containing a modified base. Reproduced (modified) from Iden C.R. and Rieger R.A., Biomed. Environ. Mass Spectrom., 18, 617, 1989, with permission.

Localization of a modified nucleotide also is done using mass spectrometry. Several strategies are proposed. A first step is to determine the nature of the modified nucleotide as described above. Then, oligonucleotides of smaller sizes (less than 10 bases) are obtained from the starting oligonucleotide using an enzyme such as RNAse T1. These different oligonucleotides then are separated and the fraction that contains the modification is analysed by mass spectrometry. The molecular weight determination of the oligonucleotide, which presents this modification, allows its composition to be determined in nucleic bases. On the basis of this composition and sequence of gene, the modification can be localized. [216]

Another method to localize a modified nucleotide is based on sequencing of the oligonucleotide containing the modified nucleotide, using MS if it is previously purified or using MS/MS if it still lies in a mixture. The position and the molecular weight of the modified nucleotide can be determined from the fragment ions, as is shown in Figure 8.41. [217]

8.4 Oligosaccharides

Oligosaccharides are molecules formed by the association of several monosaccharides linked to each other through glycosidic bonds. The monosaccharides that are most often present in oligosaccharides are given in Table 8.8.

The determination of the complete structure of oligosaccharides is more difficult than for proteins or oligonucleotides because it requires the determination of additional specificities as a consequence of the isomeric nature of monosaccharides and their capacity to form linear or branched oligosaccharides. Knowing the structure of an oligosaccharide requires not only the determination of its monosaccharide sequence and its branching pattern, but also the isomer position and the anomeric configuration of each of its glycosidic bonds. Moreover, the monosaccharides in their cyclic forms can have pyran/furan isomers. The determination

Table 8.8 Monosaccharides present in oligosaccharides (the molecular weight of unmodified oligosaccharides may be calculated by summing the mass increments of the various residues and adding the mass of one water molecule).

Monosaccharide	Examples	Formula	Monoisotopic mass	Chemical mass
Pentose	Arabinose (Ara) Ribose (Rib) Xylose (Xyl)	$C_5H_8O_4$	132.04	132.12
Deoxyhexose	Fucose (Fuc) Rhamnose (Rha)	$C_6H_{10}O_4$	146.06	146.14
Hexose	Glucose (Glc) Galactose (Gal) Mannose (Man)	$C_6H_{10}O_5$	162.06	162.14
N-Acetyl-aminohexose	N-Acetylglucosamine (GlcNAc) N-acetylgalactosamine (GalNAc)	$C_8H_{13}NO_5$	203.08	203.19
Sialic acid	N-Acetylneuraminate (NeuAc)	$C_{11}H_{17}NO_8$	291.09	291.26

of all of this structural information can be obtained by mass spectrometry, even if generally it requires the use of various techniques. The mass spectrometry of oligosaccharides has been reviewed. [218–220]

8.4.1 Mass Spectra of Oligosaccharides

The use of mass spectrometry to characterize oligosaccharides is not new. Indeed, GC/MS has been used for several decades to identify monosaccharides or very small oligosaccharides. The use of GC/MS needs the prior derivatization of these molecules by methylation, acetylation or trimethylsilylation.

Today (2000), GC/MS is still the method of choice for the determination of the monosaccharide composition of oligosaccharides and is widely used. The analysis starts with the hydrolysis or alcoholysis of the oligosaccharide to monosaccharides, which after derivatization (mostly by trimethylsilylation) are analysed by GC/MS. [221] Monosaccharide identification is based on comparison of retention time and fragmentation pattern with references.

Gas chromatography/mass spectrometry also is used largely in methylation analysis, which allows the position of the glycosidic bonds to be determined on a residue in an oligosaccharide. [222] In its most used version, permethylation of the oligosaccharide is performed prior to hydrolysis. Then, the partially methylated monosaccharides produced are reduced to the corresponding alditols and peracetylated, yielding partially methylated alditol acetates (PMAA). Analysis of these molecules by GC/MS allows, on the basis of the straightforward interpretation of the fragmentation, the position of the former glycosidic bonds to be determined.

Mass spectrometry analysis of oligosaccharides without hydrolysis has started with the FAB ionization technique and has developed with ESI and MALDI. Fast atom bombardment usually generates a weak signal, whereas ESI is not as efficient for native oligosaccharides as MALDI. Indeed, native oligosaccharides do not contain either acidic or basic groups, and thus are not easily ionized in ESI. The response for native oligosaccharides is much weaker than for peptides or proteins [223]. However, phosphorylated, sulfated or sialic-acid-containing oligosaccharides are well ionized by ESI in negative ion mode.

In general, oligosaccharides are analysed by ESI after derivatization, either by methylation or acetylation. Reductive amination of the aldehyde or ketone group also is used. [224] The derivatized products are more lipophylic, allowing the use of more volatile organic solvents, which improves the production of ions by ESI. Reductive amination together favours the formation of positive ions by the introduction of a basic site or a permanent charge. Permethylation is preferred for derivatization because the resulting mass increase is lower and thus larger oligosaccharides can be analysed. Derivatization not only improves the signal but also can bring more structural information, especially with MS/MS, as will be discussed below.

Whatever the ionization method used, the mass spectra of oligosaccharides analysed by ESI, MALDI or FAB display intense ions of the molecular species resulting from protonation $(M + H)^+$ or cationization by an alkali metal ion $(M + alkali metal)^+$ in the positive ion mode or from deprotonation $(M - H)^-$ in the negative ion mode. In ESI, multiply charged ions also are produced.

Alkali metal adducts often are observed, and even in negative ion mode $(M + Na - 2H)^-$ are often abundant for monocharged ions. If they can be avoided, not only does the signal increase because it is not divided any more over several species but also better MS/MS spectra can be obtained. To eliminate these adducts, glassware should not be used during sample work-up. Then, addition of acid or, better still, ammonium acetate allows their interference to be reduced further. However, often alkali metal salts are added at low concentrations to suppress the protonated species. This is easier to achieve, but fragmentation of these adducts yields less sequence information than protonation.

The FAB spectra present source fragmentation, but with ESI few fragments are observed, as shown in Figure 8.42. A MALDI source produces ions of the molecular species containing more energy, but with linear flight tubes the fragments are not observed because they arrive at the detector at the same flight time as their precursor. These fragments can be observed, however, either by delaying the extraction from the source or by the use of reflector instruments. They are observed even better if the MALDI source is coupled to a spectrometer using ion storage, such as ion trap or FTMS instruments.

Owing to the limited number of monosaccharides present in oligosaccharides, measurement of the molecular weight allows the composition of monosaccharides such as pentoses, hexoses, etc. to be deduced but it is not able to distinguish between different isobaric monosaccharides. Thus hexoses, for example, cannot be differentiated on this basis into glucose, mannose, etc.

The molecular weight of the oligosaccharide will, however, allow structures to be proposed if the number of possible structures is limited. For example, oligosaccharides linked to asparagine (termed an N-linkage) from animal glycoproteins have compositions limited to a few different monosaccharides. They all have a common pentasaccharidic core made up of three mannose residues and two N-acetylglucosamines. Taking this information into account, the structures of the different oligosaccharides observed in Figure 8.42 have been elucidated. For instance, the oligosaccharide observed at m/z 1810 as a sodium adduct can have only the composition $(hexose)_5(N\text{-acetylhexose})_4(deoxyhexose)_1$ and its most probable structure is displayed in Figure 8.42. It is clear that these structures, based only on molecular weight, have to be confirmed by cross-checking with other information.

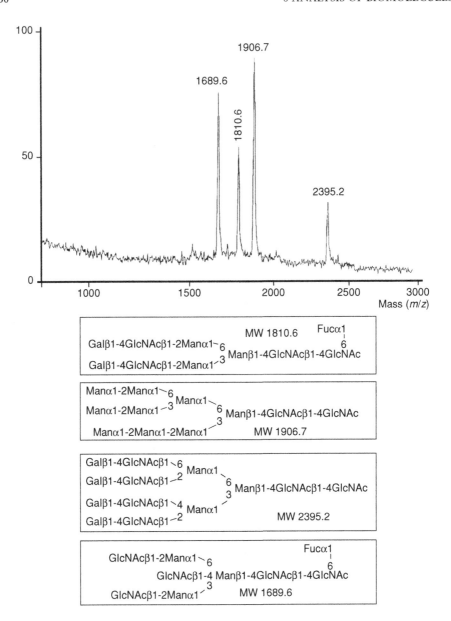

Figure 8.42
The MALDI spectrum of a mixture of four oligosaccharides derived from the cleavage of a glycoprotein. Reproduced from Finnigan MAT documentation, with permission.

8.4.2 Fragmentation of Oligosaccharides

Because the soft ionization methods used for oligosaccharides produce few fragments, collision-induced dissociation (CID) or post-source decay (PSD) must be used for structural study. These two techniques have been applied to deprotonated, protonated or alkaliated

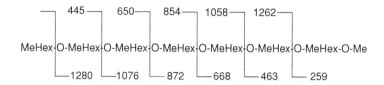

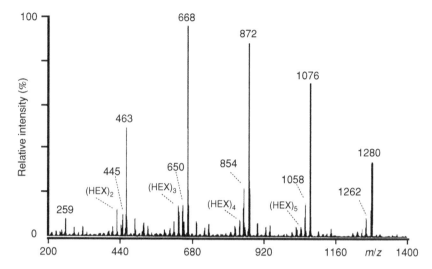

Figure 8.43
Collision-induced dissociation ESI/MS/MS trace of doubly charged sodium adduct of methylated maltoheptulose ($m/z = 760.7$). Reproduced (modified) from Reinhold V.N., Reinhold B.B. and Costello C.E., Anal. Chem., 67, 1772, 1995, with permission.

molecular ions from native or derivatized oligosaccharides using FAB, [225–230] ESI [231–234] or MALDI. [235–237]

The fragmentation of oligosaccharides is strongly influenced by a large number of factors, such as the ionization method, the analyser, the nature of the derivatization, the nature of the molecular species, etc. However, five series of fragment ions can be observed. As shown in the spectrum presented in Figure 8.43, the first two series of ions, generally abundant, come from the cleavage of one glycosidic bond. Thus, they contain the reducing or the non-reducing end. Both of these fragments, by undergoing cleavage of a second glycosidic bond, lead to a third series of ions called internal fragments. They do not contain the two initial termini any more (cleavage at both ends). The two last series of ions, generally less abundant, are due to a double cleavage across the glycosidic ring and contain either the reducing side or the non-reducing side.

A nomenclature suggested by Domon and Costello [238] was developed to characterize the various fragments obtained by mass spectrometry whatever the method used to produce them. The fragments retaining the charge on the non-reducing side are called A, B or C, and those retaining the charge on the reducing side are called X, Y and Z, depending on whether they cut the ring or the glycosidic bond (see Figure 8.44). The subscript corresponds to the

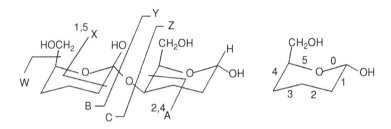

Figure 8.44
Nomenclature suggested by Domon and Costello.

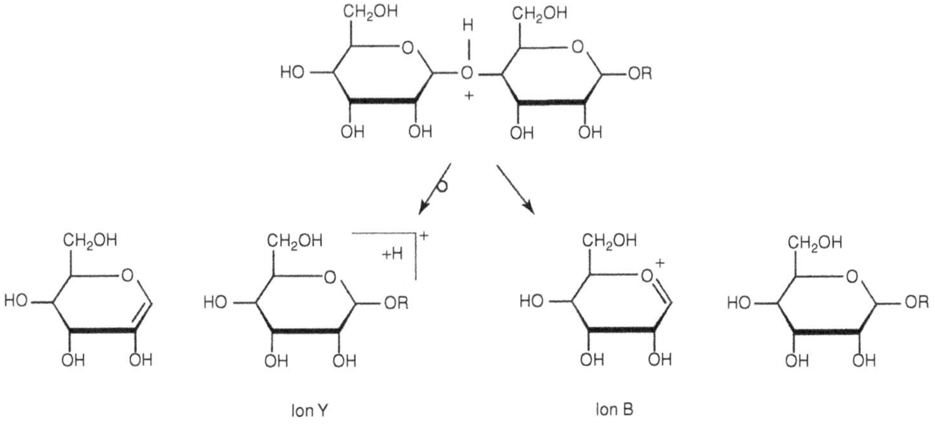

Ion Y   Ion B

Figure 8.45
Mechanism for the formation of B and Y ions in positive mode.

number of the ruptured glycosidic bond whereas the superscript at the left of the A and X fragments corresponds to bonds that were broken in order to observe these fragments, the bonds being numbered as is indicated in the figure. The letter α, β, etc., which may be attached to the subscript number, indicates the branch involved in the cleavage, provided, of course, that the oligosaccharide is branched.

The fragments that are observed most often in positive mode spectra correspond to cleavage of the glycosidic bond, with oxygen atom retention on the reductive part, by the mechanism shown in Figure 8.45. This yields B and Y ions.

For the fragments derived from a double cleavage across the ring (A and X ions), their formation is favoured under high-energy collisions from the sodium adducts. [239] These fragments are produced through a charge remote fragmentation (CRF) mechanism shown in Figure 8.46. This mechanism implies two decomposition paths yielding $^{1,5}X$, $^{1,3}A$ and $^{3,5}A$ ions in one case and $^{0,2}X$, $^{2,4}A$ and $^{0,4}A$ ions in the other case. It is also interesting to observe that the derivatized oligosaccharide carrying a preformed charge due to prior reductive amination of the oligosaccharide leads exclusively to the fragment ions that contain the reducing end. [240,241]

In general, MS/MS allows one to determine the sequence and the branching pattern of oligosaccharides. The isomer position of each of their glycosidic bonds also can be

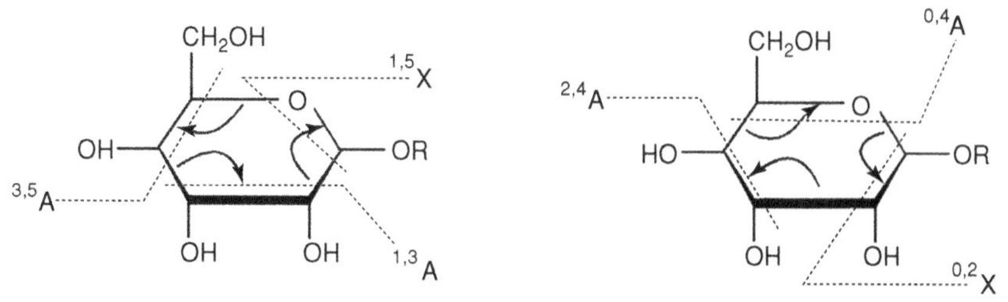

Figure 8.46
Formation mechanism of A and X ions in positive mode.

determined. On the other hand, the anomeric configuration of glycosidic bonds and the distinction of diastereoisomeric monosaccharides are seldom accessible by this technique.

The ions derived from cleavage of the glycosidic bond allow one to determine the sequence and the branching pattern of oligosaccharides. Indeed, the mass difference between fragment ions within the same series allows one to deduce the sequence of oligosaccharides. Such a determination is detailed in Figure 8.47A, which shows the spectrum of a peracetylated pentasaccharide, lacto-N-fucopentose (LNF-I). [242] This spectrum is dominated by the series of B oxonium ions. Ions with m/z 273, 561, 848, 1136 and 1424 correspond to fragments B_1 (Fuc-Ac 3), B_2 (Fuc-Hex-Ac 6), B_3 (Fuc-Hex-GlcNAc-Ac 8), B_4 (Fuc-Hex-GlcNAc-Hex-Ac 11) and B_5 (Fuc-Hex-GlcNAc-Hex-Hex-Ac 14). These fragments make it possible to ascribe a complete and unambiguous sequence to this oligosaccharide; however, they do not allow one to distinguish between the two diastereoisomers galactose and glucose (Hex = Gal or Glc).

The same principle may be used to determine the branching pattern. Figure 8.47B shows the FAB/MS/MS trace of a branched isomer of the pentasaccharide LNF-II. As opposed to the linear isomer, the branched isomer is characterized by two monosaccharidic B_1 ions with masses 273 Da ($B_1\beta$, Fuc-Ac 3) and 331 Da ($B_1\alpha$, Hex-Ac 4) and by the absence of disaccharidic B_2 ions at mass 561 Da. As in the linear structure, the ions with a higher mass sequence are also observed at m/z 848 (B_2, trisaccharidic: Hex-[Fuc]-GlcNac-Ac 8), 1136 (B_3, Hex-[Fuc]-GlcNac-Hex-Ac 11) and 1424 (B_4, Hex-[Fuc]-GlcNac-Hex-Hex-Ac 14) and allow one to establish the remaining sequence because the latter does not contain any further branching.

The ESI/MS/MS fragmentation spectrum of permethylated LNF-II is displayed in Figure 8.48. [243] This spectrum is somewhat more complicated than the one from the peracetylated derivative, but allows additional structural information to be obtained. As for the peracetylated derivative, this spectrum contains the ions characteristic of the sequence observed at m/z 638 (B_2, trisaccharidic: Hex-[Fuc]-GlcNAc-Me 8) and 842 (B_3, Hex-[Fuc]-GlcNAc-Hex-Me 11), which allow the branched structure and the sequence on the reducing side to be established.

However, this spectrum is different from the peracetylated one because it displays more secondary fragments. This process, due to the presence of HexNac, leads from the B_2 ion to the preferential loss of the substituent at the 3 position of the HexNac residue, in this case a tetramethylated hexose, to produce the E_2 fragment at m/z 402. From this last ion, the loss of the substituent at the 4 position, here a trimethylated deoxyhexose, yields the E_2' fragment at m/z 196. This spectrum, as with that of the peracetylated compound, allows

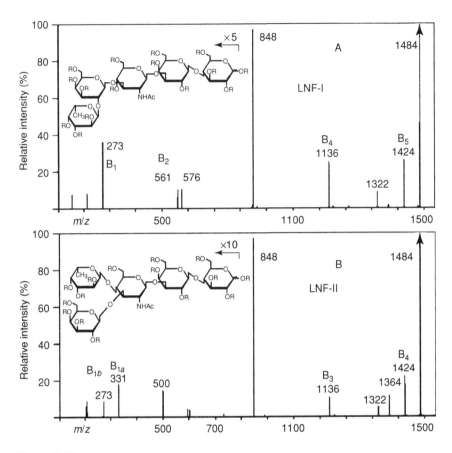

Figure 8.47
Collision-induced dissociation FAB/MS/MS traces of two peracetylated oligosaccha-rides: (A) LNF-I and (B) LNF-II, two branching isomers. Reproduced (modified) from Domon B., Muller D.R. and Richter W.J., Biomed. Environ. Mass Spectrom., 19, 390, 1990, with permission.

the sequence to be assigned but furthermore provides information that the 3-position of the HexNac bears a hexose and the 4 position a deoxyhexose.

The best results using this method are obtained with protonated molecular ions $(M + H)^+$ of derivatized or native oligosaccharides. However, the peracetylated or permethylated derivatives are superior not only in terms of response but also in their ability to identify the internal fragments derived from the double cleavage of glycosidic bonds. Compared with the fragments having the same number and the same type of residues but produced by the cleavage of only one glycosidic bond, the internal fragments present two free hydroxyls groups. Thus, these fragment ions have distinct masses and are easily differentiated. This is not the case of the native oligosaccharides and this can complicate, or even compromise, the determination of the sequence and lead to erroneous information.

The A and X ions derived from the cleavage of two bonds in the glycosidic ring and the W ions from the cleavage between carbons 5 and 6 allow a clear identification of the positional isomers of each glycosidic bond in the linear or branched oligosaccharides by

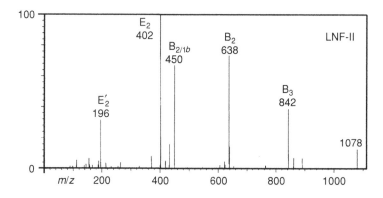

Figure 8.48
Collision-induced dissociation ESI/MS/MS spectrum of the LNF II
oligosaccharide. Reproduced (modified) from Viseux N., de Hoffmann
E. and Domon B., Anal. Chem., 69, 3193–3198, 1997, with permission.

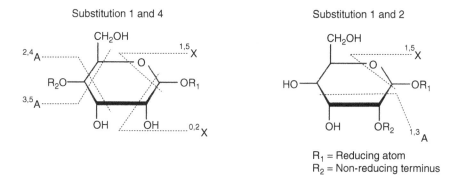

Figure 8.49
Diagnostic fragments for monosaccharides whose hydroxyl groups in either position
1 and 2 or 1 and 4 participate in glycosidic bonds.

their presence and/or their absence. For example, the cleavage of the ring with hydroxyls at
positions 1 and 4 involved in glycosidic bonds will allow only the formation of $^{1,5}X$, $^{0,2}X$,
$^{3,5}A$ and $^{2,4}A$ ions. On the other hand, as shown in Figure 8.49, the implication of hydroxyls
at positions 1 and 2 will allow only the formation of $^{1,5}X$ and $^{1,3}A$ ions. The diagnostic
fragments useful in determining the positional isomers of the glycosidic bonds are listed in
Table 8.9.

The example shown in Figure 8.50 includes the structure of the oligosaccharide, the
tandem mass spectrum of its $(M + Na)^+$ adduct and a summary table of the diagnostic
peaks.

The best results that are obtained by this method use the molecular species ions $(M +$
alkali metal$)^+$ of native or derivatized oligosaccharides measured at high energy.

The tandem mass spectrometric analysis of oligosaccharides thus allows one to obtain
valuable structural information. However, determination of the complete structures from

Table 8.9   Potentially diagnostic ions in determining the positional isomers of glycosidic bonds.

Positional isomer	Fragments						
	W_i	$^{0,2}X_i$	$^{1,5}X_i$	$^{0,4}A_i$	$^{1,3}A_i$	$^{2,4}A_i$	$^{3,5}A_i$
1 and 2	−	+	+	−	+	−	−
1 and 3	−	+	+	−	+	+	−
1 and 4	−	+	+	−	−	+	+
1 and 6	+	+	+	+	−	−	+

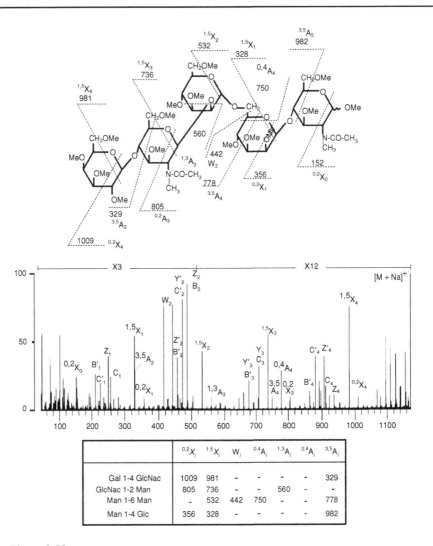

	$^{0,2}X_i$	$^{1,5}X_i$	W_i	$^{0,4}A_i$	$^{1,3}A_i$	$^{2,4}A_i$	$^{3,5}A_i$
Gal 1-4 GlcNac	1009	981	-	-	-	-	329
GlcNac 1-2 Man	805	736	-	-	560	-	-
Man 1-6 Man	-	532	442	750	-	-	778
Man 1-4 Glc	356	328	-	-	-	-	982

Figure 8.50

Collision-induced dissociation FAB MS/MS trace of the $(M + Na)^+$ of an oligosaccharide, measured at high energy. It allows one to determine the positional isomers of all its glycosidic bonds. Reproduced (modified) from Lemoine J., Fournet B., Despeyroux D., Jennings R., Rosenberg R. and de Hoffmann E., J., Am.Soc. Mass Spectrom., 4, 197, 1993, with permission.

such information alone is difficult. This task is still more difficult if residues contain highly labile groups such as sialic acid, N-acetylaminohexoses or fucose, because the spectra are dominated by fragments from these residues. The use of the MS^n capabilities of the ion trap or FTICR instruments is thus a great help. [244,245] There are three major advantages of MS^n over traditional MS/MS: dissociation of a fragment can lead to new fragments not observed from the ion of the molecular species; the various first-generation fragments observed are identified not only on the basis of their mass but also of their fragmentation; and filiation of the fragments can be established experimentally.

8.4.3 Degradation of Oligosaccharides Coupled with Mass Spectrometry

There are a variety of chemical or enzymatic degradation methods of oligosaccharides that have been used in conjunction with mass spectrometry. [246,247] Three of these methods are described in detail below.

The first method is based on the selective oxidation by chromium trioxide of the β-anomer of derivatized hexoses in order to yield a ketoester, as shown in Figure 8.51. Thus, this method allows the determination of the anomeric configuration of the various glycosidic bonds. [248] The oligosaccharide mass difference observed before and after oxidation allows one to determine the number of β-bonds that are present: an $N \times 14$ Da increment corresponds to the presence of N β-bonds. The oxidized bond positions may be detected using fragmentations of the glycosidic bonds. Finally, a soft methanolysis at room temperature may counter a possible low intensity of the ions derived from cleavage of the glycosidic bond, because it cleaves preferentially the ester bonds formed at the oxidized β-anomers.

The second method relies on enzymatic degradation of oligosaccharides by endoglucosidases. These enzymes catalyze selectively the hydrolysis of glucosidc bonds of monosaccharides located at the non-reducing end. The specificity of the different endoglucosidase concerns the nature of the monosaccharides at the reducing end or the configuration of the anomeric carbon atom involved in the glycosidic bond. It can also be sensitive to the nature of the monosaccharides at the penultimate monosaccharide. As an example,

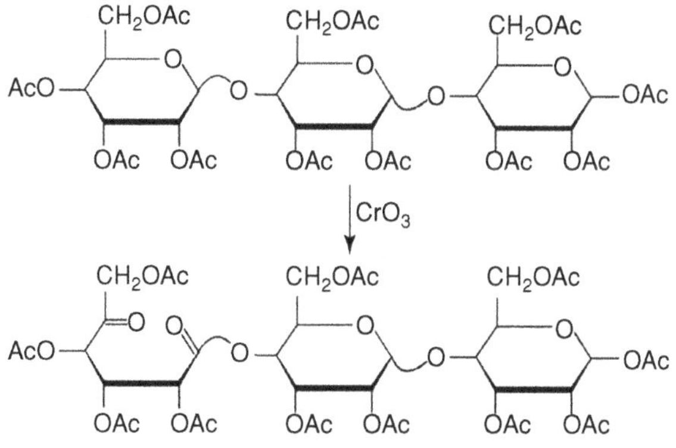

Figure 8.51
Chromium oxide action on the α- and β-anomers of the glycosidic bonds.

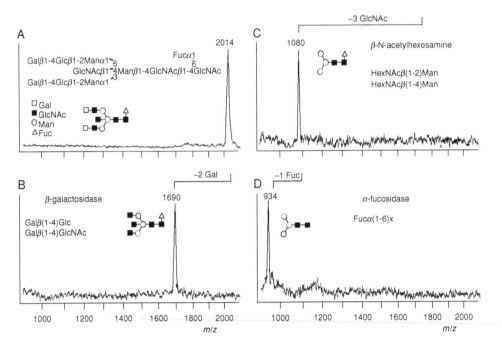

Figure 8.52
The MALDI spectrum from successive enzymatic digests of a native oligosaccharide. The structural information obtained from the specificity of the exoglucosidases used is indicated by italics in the complete structure in A. Reproduced (modified) from Kuster B., Naven T.J.P. and Harvey D.J., J. Mass Spectrom., 31, 1131–1140, 1996, with permission.

the β-galactosidase of *Streptococcus pneumoniae* cleaves the bonds Gal(β1-4)Glc or Gal(β1-4)GlcNAc whereas the α-L-fucosidase from bovine epididyme cleaves only the Fuc(α1–6)x bond.

The full structural determination of the structure of an oligosaccharide by this method includes complete sequencing by successive enzymatic digests using specific exoglycosidases and analysis of the products of each hydrolysis by mass spectrometry to observe the effectiveness of the hydrolysis by the decrease of molecular weight. [249]

A combination of the specific enzymes with the sensitivity, speed and accuracy of mass spectrometry allows complete sequence determination, including the anomeric configuration. According to the specificity of the enzymes, positional isomers of the glycosidic bonds can be recognized.

An application of this method is displayed in Figure 8.52. [250] The sequential treatment of the studied oligosaccharide with three specific exoglycosidases allows two Gal, three GlcNac and one Fuc residue to be removed successively, as shown by the detected ions of the molecular species.

In the case of an oligosaccharide of totally unknown structure, this method can be long and costly due to the need to use many different enzymes. However, it has the advantage that it can be applied to complex mixtures of oligosaccharides. [251]

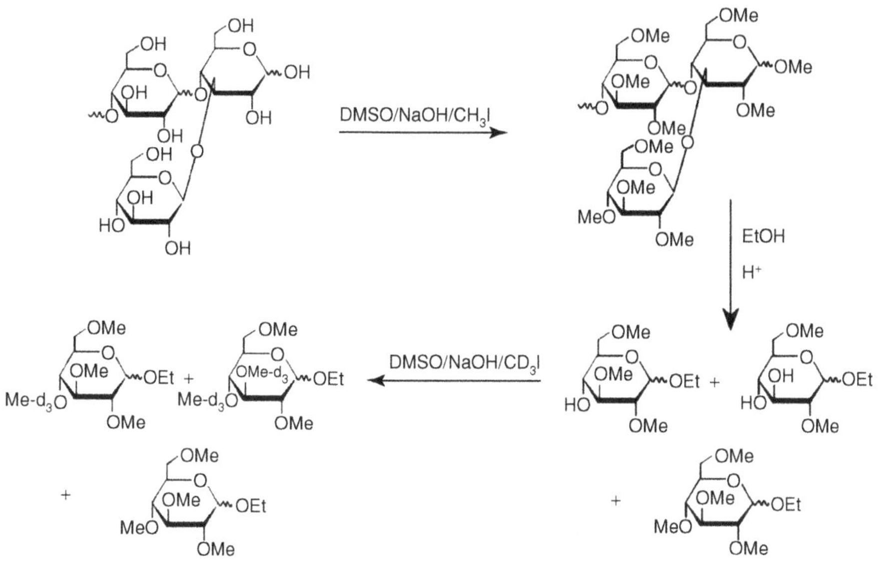

Figure 8.53
Sequence of reactions allowing permethylated ethylglycosides to be selectively deuteromethylated.

The last method described here for the degradation of oligosaccharides is a modification of the methylation analysis described in section 8.4.2. [252]

It is based on a derivatization process that starts with ethanolysis of the permethylated oligosaccharide (Figure 8.53). The partly methylated ethylglycosides are deuteromethylated on the hydroxyl groups made free by the ethanolysis step. The fingerprint of the glycosidic positions is held by deuteromethylation of these positions. The ethyl group specific to the 1 position will allow alcohol elimination from this position to be distinguished from the other alcohol eliminations (loss of methanol or deuterated methanol). The different monosaccharides derivatives then are analysed by chemical ionization GC/MS/MS.

Gas chromatographic analysis allows the determination of the monosaccharides present by comparing the retention times with those of reference compounds. It is noteworthy to mention that the chromatographic profile is quite simple, because the deuterated compounds are not separated from the non-deuterated, even if a small difference in retention time can be evidenced by comparison of specific ion chromatograms. Each monosaccharide can give a maximum of four peaks, corresponding to the α/β and the pyran/furan isomers.

Analysis by positive ion chemical ionization of the permethylated ethylglycosides leads to spontaneous fragmentation of the molecular species. As shown in Figure 8.54, this fragmentation produces oxonium ions (resulting from the loss of ethanol) at m/z 219 (for non-deuterated hexoses). The oxonium ions then can lose a methanol molecule to yield a fragment 32 mass units lower (m/z 187 for the non-deuterated hexoses).

The masses of the oxonium ions will allow the number of deuteromethyl groups present to be determined and so yield information on the position of the monosaccharide in the chain. Indeed, the oxonium ion will appear at m/z 219 (d_0) for a hexose in a non-reducing

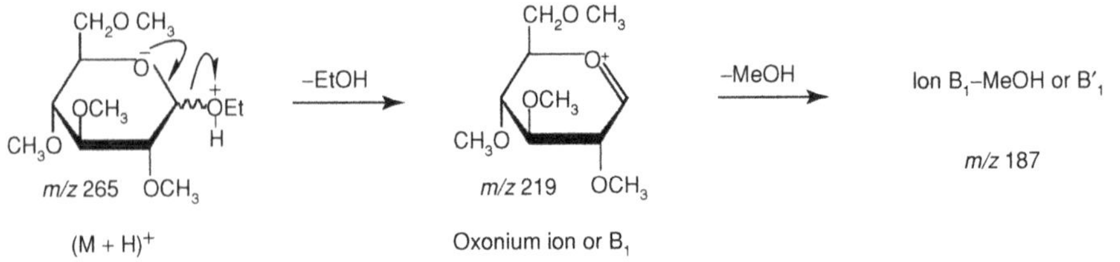

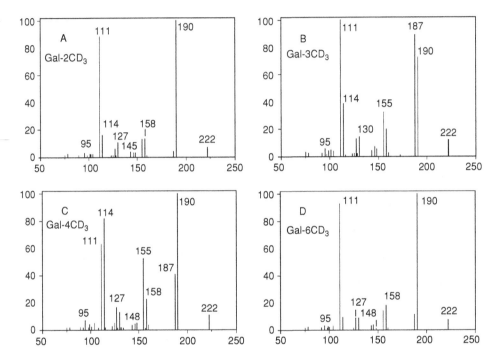

Figure 8.54
Fragmentation scheme of ethylglycosides produced in a CI source.

Figure 8.55
The CI/MS/MS mass spectra of oxoniums ions from galactose: (A) Gal-2CD$_3$; (B) Gal-3CD$_3$; (C) Gal-4CD$_3$; (D) Gal-6CD$_3$. Copyright © E. de Hoffmann.

terminal position, at m/z 222 (d$_3$) for a hexose in an internal position and at m/z 225 (d$_6$) or 228 (d$_9$) for a hexose at a branching point. This same information can be obtained from the oxonium ions of N-acetylhexoses at m/z 260 (d$_0$), 263 (d$_3$) and 266 (d$_6$) or for a deoxyhexose at m/z 189 (d$_0$), 192 (d$_3$) and 195 (d$_6$), respectively.

Subsequent analysis of the fragments of the oxonium ions by CID/MS/MS at low energy allows the positions of the deuterated methyl groups to be assigned corresponding to the branching positions. This is illustrated in Figure 8.55, displaying the product ion spectra obtained by MS/MS of the oxonium ions of Gal derivatives deuteromethylated on the 2, 3, 4 or 6 position, respectively. It is possible to differentiate the various positions of the CD$_3$.

This differentiation is based mainly on the ratios of the intensities of the two couples of ions at m/z 187 and 190 and m/z 111 and 114, respectively. The monosaccharides substituted on the 2 and 6 positions display the same spectrum. They can, however, be distinguished clearly by the fragmentation spectrum of the (oxonium-MeOH) ions. A systematic study has shown that the MS/MS spectra are predictable and largely independent of the nature of the monosaccharide.

This method alone does not allow the determination of the structure of oligosaccharides. If, however, it is used in conjunction with an MS/MS analysis of the whole permethylated oligosaccharides, the complete structure can be deduced.

Figure 8.56 describes the application of this method to the LNF-II oligosaccharide. The mixture of derivatized monosaccharides obtained after permethylation, ethanolysis and deuteropermethylation is analysed by chemical ionization GC/MS/MS. The GC chromatogram obtained is displayed along with the ion chromatograms of different oxonium ions. The multiplicity of the chromatographic traces results from the presence of the α/β and pyran/furan isomers for each glycoside.

From the retention times, compared with those of standards, the presence of Fuc, Glc, GlcNac and two Gal residues is clearly established. The masses detected for these different monosaccharides allow the positions of these residues to be assigned. The Fuc and one of the Gal residues are located at the non-reducing ends, based on the mass of their oxonium ions at m/z 189 and 219, respectively. Indeed, they are not deuterated. The second Gal residue and the Glc residue are positioned inside the oligosaccharide, because they include one deuteromethyl group as shown by the masses of the oxonium ions at m/z 222. Finally, the oxonium ion of GlcNac appears at m/z 266, corresponding to the incorporation of six deuterium atoms. This GlcNac is thus located at a branching point.

The next step is to assign the positions of the deuteromethyl groups and thus of the glycosidic bonds. This is deduced from MS/MS fragmentation spectra of the corresponding oxonium ions. The spectra obtained are displayed in Figure 8.56 and lead to the following conclusions: Glc is deuterated at the 4 position, Gal at the 3 position and GlcNac at the 3 and 4 positions.

This information together with that from the MS/MS fragmentation spectrum of the whole permethylated oligosaccharide (see Figure 8.48), allows the complete structure of LNF-II to be deducted, except for the anomeric configuration of the glycosidic bonds.

8.5 Lipids

Lipids are made up of many classes of very different molecules that all show solubility properties in organic solvents. Mass spectrometry plays a key role in the biochemistry of lipids. Indeed, mass spectrometry allows not only the detection and determination of the structure of these molecules but also their quantification. For practical reasons, only the fatty acids, acylglycerols and bile acids are discussed here, although other types of lipids such as phospholipids, [253–256] steroids, [257–259] prostaglandins, [260] ceramides, [261,262] sphingolipids [263,264] and leukotrienes [265,266] have been analysed successfully by mass spectrometry. Moreover, the described methods will be limited to those that are based only on mass spectrometry, even if the majority of these methods generally are coupled directly or indirectly with separation techniques such as GC or HPLC. A book on the mass spectrometry of lipids was published in 1993. [267]

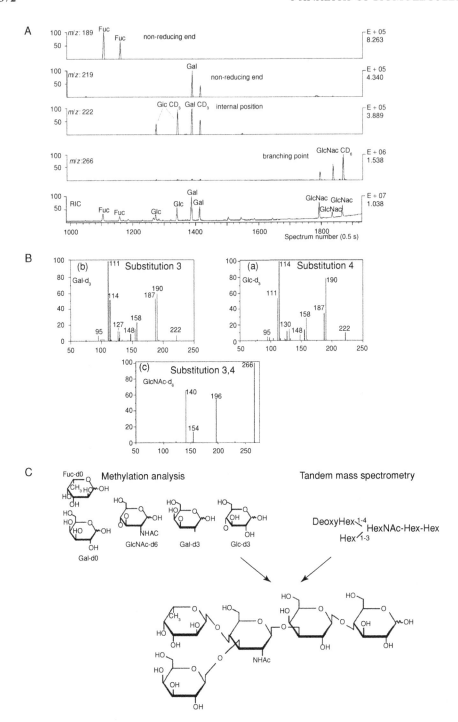

Figure 8.56
(A) The GC chromatogram of a mixture of derivatized monosaccharides obtained from the LNF-II oligosaccharide. (B) The MS/MS product ion spectra of the oxonium ions: (a) at *m/z* 222, monosubstituted Glc; (b) at *m/z* 222, monosubstituted Gal; (c) at *m/z* 266, disubstituted GlcNAc. (C) Deduction of the structure of LNF-II. Copyright © E. de Hoffmann.

8.5.1 Fatty Acids

Fatty acids are a class of molecules that are made up of a long hydrocarbon chain, of varying length and varying degrees of unsaturation, terminated by a carboxylic group. Some organisms such as bacteria, sponges and some plants can produce modified fatty acids: branchings; introduction of a cyclopropane, cyclopropene or epoxy ring; hydroxylations or alkoxylations; or even unusual unsaturations. These fatty acids are interesting not only for their particular biological activities but also for their rarity, which allows one to classify and identify the organisms that produce them much faster than by the classical characterization techniques.

The classical analysis method consists of extracting the lipids from the biological material and then characterizing the previously purified fatty acids by comparing their retention times with those of reference fatty acids, by degradative analyses and by using a series of spectroscopic methods. Numerous chromatographic techniques, including thin-layer chromatography (TLC), HPLC and GC, are used in the purification.

Because fatty acids derived from natural sources are present in a mixture, an ideal analysis method for these molecules should be applicable to mixtures without requiring a prior separation or derivatization. Mass spectrometry is an excellent tool for determining the structure of fatty acids present in a mixture. It is possible to determine not only the molecular weight and thus the elemental composition but also, in most cases, the nature and position of the branching and the other substituents on the carbon chain. [268,269] Furthermore, such an analysis requires low quantities ranging from 10 pg to 100 ng of total lipid, depending upon the analysed sample, the ionization method used and the configuration of the spectrometer. [270,271]

A general analysis method for fatty acids is based on the high-energy tandem mass spectrometry of carboxylate anions. Indeed, FAB, desorption chemical ionization (DCI), ESI or atmospheric pressure chemical ionization (APCI) spectra measured in the negative mode are characterized by the single presence of molecular ion species and their isotopic clusters and the absence of fragments. [272–274] This characteristic is used to facilitate the determination of the molecular weights of fatty acids even when they are present in a complex mixture, but it does not provide information concerning structure. This inconvenience may be overcome by transferring the extra energy necessary for fragmentation to the stable ions produced during ionization by using MS/MS.

The high-energy tandem mass spectrometric analysis of carboxylate ions from fatty acids yields a series of homologous fragments all separated by 14 Th. These fragments correspond to the formal loss of alkane: CH_4, C_2H_6, ..., C_nH_{2n+2}. They are produced by a highly specific 1,4-elimination mechanism (charge remote fragmentation) of a hydrogen molecule, as shown in Figure 8.57. [275] This mechanism has been confirmed by neutral fragment reionization, demonstrating that the lost neutral is indeed an alkene. [276] However, other mechanisms compatible with the experimental data, implying the homolytic rupture of C-C or C-H bonds, also were proposed. [277,278]

This loss of C_nH_{2n+2} starts at the terminal alkyl and progresses along the hydrocarbon chain. In the case of saturated fatty acids, the abundance of these fragments yields a characteristic spectrum such as that shown in Figure 8.58. [279,280]

The presence of unsaturation or substituents disrupts or disturbs the characteristics of the spectrum. The nature of the disruption allows one to distinguish the type of structural modification whereas the localization on the chain can be determined from the point where the perturbation occurs.

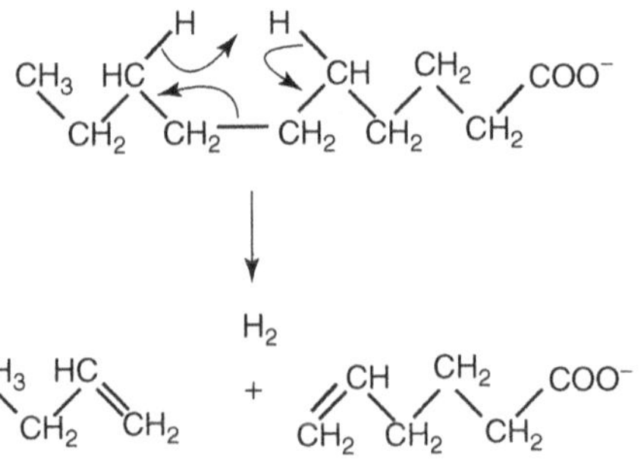

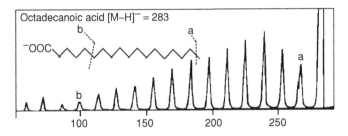

Figure 8.57
The 1,4-elimination mechanism of H_2 occurring along the fatty acid chain in order to yield fragments corresponding to the loss of an alkene.

Figure 8.58
Collision-induced dissociation FAB/MS/MS trace of octadecanoic acid acquired in the negative ion mode at high energy. Reproduced (modified) from Jensen N.J., Tomer K.B. and Gross M.L., Anal. Chem., 57, 2018, 1985, with permission.

The presence of branching on the hydrocarbon chain is indicated by the suppression of fragmentation at the branching point and by accentuation of fragmentation of the bond adjacent to the branching on the alkyl side. Thus the spectra of a branched fatty acid are characterized by a two-methylene 'hole', i.e. by two peaks in the series separated by 28 Th. The CID/MS/MS spectra of two branched fatty acids, presented in Figure 8.59, show how easy it is to localize the branching. [281]

The presence of an unsaturation within a fatty acid is indicated and its position is established by the absence of fragments derived from cleavages of this unsaturated bond and the adjacent ones. This corresponds, on the spectrum, to a four-carbon atom 'hole', i.e. by two peaks in the series separated by 54 Th (Figure 8.60). [282] This absence of ions results from the fact that cleaving a vinylic bond or a double bond is not energetically favoured. The localization of double bonds in unsaturated fatty acids is made more difficult and even impossible as the number of unsaturations increases because the process of losing alkanes is hidden by the loss of 45 Da (·COOH).

Other modifications of the hydrocarbon chain of fatty acids, such as hydroxylation or introduction of a cyclopropane, cyclopropene or epoxy ring, could be identified and located

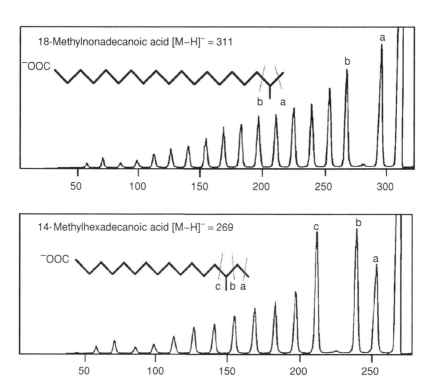

Figure 8.59
Collision-induced dissociation FAB/MS/MS traces of 18-methylnonadecanoic acid and 14-methylhexadecanoic acid, acquired in the negative ion mode at high energy. Reproduced (modified) from Jensen N.J. and Gross M.L., Lipids, 21, 362, 1986, with permission.

by this method. [283] High-energy MS/MS applied to $(M + Li)^+$ or $(M - H + 2Li)^+$ corresponding to fatty acids cationized by Li^+ (preferred over other alkali metals because they bind more strongly to carboxylic groups) yields results that are comparable with those obtained in negative ion mode. The only important difference lies with polyunsaturated fatty acids, which yield ions in the positive mode that are interpreted more easily.

An alternative method for the analysis of fatty acids by mass spectrometry is based on the preparation of picolinic esters. [284,285] These derivatives allow the GC separation of fatty acids and their EI spectra display abundant diagnostic fragment ions. These fragments correspond to C—C bond cleavages all along the hydrocarbon chain. The proposed mechanism starts with the production by EI of a radical-cation located at the nitrogen atom. This radical-cation rearranges to yield a distonic radical-cation by abstraction of a hydrogen atom from the chain. The C—C bonds then are broken by a reaction initiated at the radical site to yield an alkene and an alkyl radical. The presence of a modification along the hydrocarbon chain will cause an alteration of the mass spectrum. Figure 8.61 shows the mass spectra of two ramified fatty acid picolinic esters as examples. This method based on GC/MS does not require high-energy MS/MS.

Low-energy CID of the molecular ions of fatty acid methyl esters obtained by EI (70 eV) also has been studied. [286] Such fatty acids methyl esters decompose in the tandem

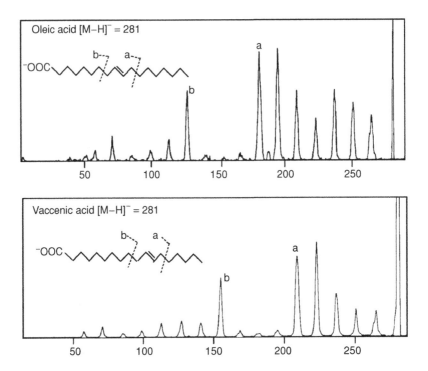

Figure 8.60
Collision-induced dissociation FAB/MS/MS traces of oleic acid and vaccenic
acid, measured in the negative mode at high energy. Reproduced (modified)
from Tomer K.B., Crow F.W. and Gross M.L., J. Am. Chem. Soc., 105, 5487,
1983, with permission.

quadrupole mass spectrometer to yield a regular homologous series of carbomethoxy ions.
Decomposition of the molecular ions of several methyl-branched fatty acid methyl esters, such as phytanic acid shown in Figure 8.62, reveals enhanced radical site cleavage at
the alkyl branching positions. This technique of low-energy CID of molecular ions generated by EI provides a sensitive, powerful and simple approach to the determination of
methyl- or alkyl-branched saturated fatty acids without the need for special derivatization or
high-energy MS/MS.

8.5.2 Acylglycerols

Acylglycerols are the fatty acid esters of glycerol. Because every alcohol function of glycerol
can be esterified, the molecules can contain one, two, or three fatty acids, termed mono-, di-
or triacylglycerols, respectively. Characterization of an acylglycerol molecule requires not
only the identification of its component fatty acids but also their positions in the glycerol
molecule (positional isomer).

The general procedure allowing the characterization of an acylglycerol consists of hy-
drolyzing the acylglycerol that has been purified previously by TLC or HPLC in order

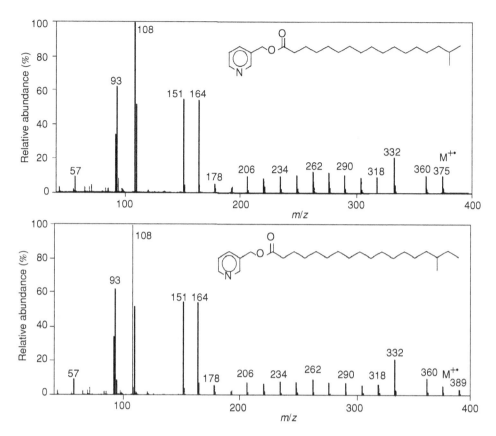

Figure 8.61
Electron ionization (25 eV) mass spectra of picolinic esters from *iso*-octadenoic acid and *ante*-isononadecanoic acid. The fragment ion at *m/z* 151 corresponds to a McLafferty rearrangement product. Reproduced (modified) from Harvey D.J., Biomed. Mass Spectrom., 9, 33, 1982, with permission.

to analyse the fatty acids obtained by the methods described earlier in this chapter. For determination of the positions of the various component fatty acids in an acylglycerol, no simple method exists that does not use mass spectrometry, in spite of the progress in acyl-glycerol analysis in the last few years using high-resolution GC or reverse-polarity HPLC. The most frequently used method consists of treating the purified acylglycerol with a lipase that specifically cleaves the central position ester bond and then characterizing the fatty acid that is liberated. Obviously the complete identification of acylglycerols using classical methods is long and difficult and requires a large quantity of sample. [287]

At present, mass spectrometry allows the analysis of an acylglycerol by identifying its various component fatty acids and their positions in the glycerol without necessitating prior chromatographic separation. The quantity of sample that is used during a mass spectrometric analysis is about 1 picomole. [288]

Various ionization techniques, such as EI, [289] CI, [290] DCI, [291] PD, [292] FAB, [293] APCI [294] and ESI, [295] have been used successfully. Overall, the acylglycerol

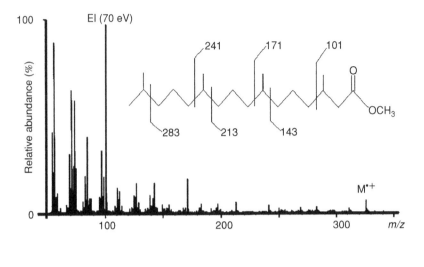

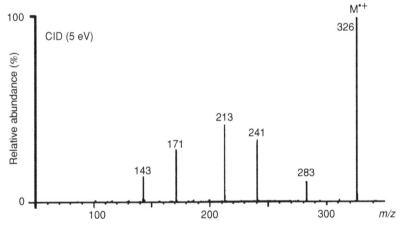

Figure 8.62
The EI/MS/MS spectrum of methyl 3,7,11,15-tetramethylhenadecanoate
(methyl phytanate) and the collision-induced dissociation EI/MS/MS spectrum
of the molecular ion (m/z 326) at low energy (5 eV). Reproduced (modified) from
Zirrolli J.A. and Murphy R.C., J. Am. Soc. Mass Spectrom., 4, 223, 1993, with
permission.

spectra obtained in the positive mode contain as predominant ions, in addition to the molecular ion $M^{\cdot+}$ in EI or the molecular species ions $(M + H)^+$, $(M + Na)^+$ or $(M + NH_4)^+$
in the case of other ionization techniques, the ions issued from the loss of every acyloxy
group present in the molecule, labelled $(M - R_nCOO)^+$, and the corresponding acylium
ions, labelled $(R_nCO)^+$. The labelled ions $(M - R_nCOO)^+$ have been referred to as
'diglyceride-type ions'. In the negative mode the predominant ions are the molecular
species ion $(M - H)^-$ and the ion corresponding to the fatty acids that are present, labelled
$(R_nCOO)^-$, as shown in Figure 8.63. The relative importance of these various ions depends,

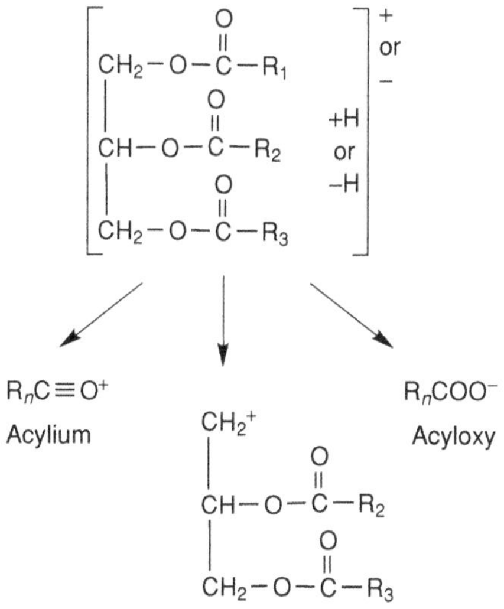

Figure 8.63
General fragmentation diagram of acylglycerols.

of course, on the ionization technique used. For example, in the positive mode, the intensity of the molecular species ion is weak in EI and FAB and stronger in CI and DCI, and is the only ion present in ESI.

Results obtained by CID/MS/MS product ion spectra of triacylglycerols in positive ion mode appear to be quite independent of the type of selected precursor as well as the ionization mode used to obtain the precursor ion. [293,295,296] These fragmentation spectra contain acylium ions of the fatty acids and, more abundantly, diacylglycerol-type ions.

Useful qualitative and quantitative information could be obtained from mass spectrometry. Starting from the measured mass of each acylglycerol, the elemental composition and the number of unsaturations can be determined. A chromatographic separation is no longer necessary because this method based on mass spectrometry allows one to determine the molecular weights of a mixture of various components. Hence an analysis using this method requires minimal preparation and achieves important savings in time. However, more information is necessary to determine the fatty acid composition of an acylglycerol. A triacylglycerol containing three C_{16} fatty acids has the same molecular weight as a tria- cylglycerol containing C_{14}, C_{16} and C_{18} fatty acids. This extra information is supplied by the fragments derived from cleavage of the fatty acid chains, which allow one to determine the molecular mass of each fatty acid linked to the glycerol. This fatty acid composition can be determined in the presence of other acylglycerols only if MS/MS is used, because all the acylglycerols in the mixture undergo fragmentation and yield ions that may overlap one another, thereby preventing correct interpretation of the spectrum.

It should be noted that only the elemental composition of each fatty acid is obtained, without other structural information on double bond position and isomerism, branching, etc. However, high-energy CID of the anions of the fatty acids allows this information to

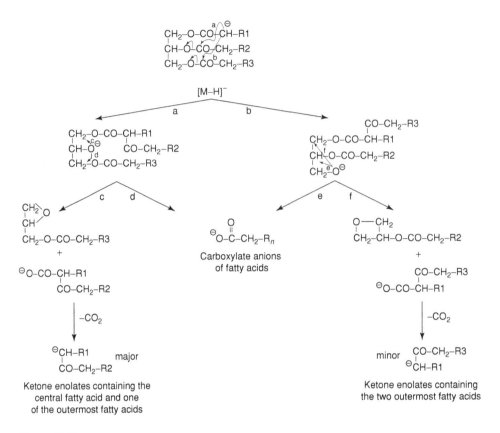

Figure 8.64
Claisen condensation mechanism responsible for the formation of ketones, allowing the determination of positional isomers.

be obtained. [297] For mono- and diacylglycerols, structural information can be obtained after derivatization with nicotinic acid. [298]

An unambiguous, rapid and sensitive method based on MS/MS of the deprotonated molecular ions of acylglycerols obtained in negative mode by DCI was developed in order to determine the position of each fatty acid linked to the glycerol molecule, i.e. to determine the positional isomers. [299]

The low-energy tandem mass spectra of the deprotonated molecular ions of acylglycerol contain a type of ion whose formal mass-based composition corresponds to a ketone obtained by the combination of two fatty acid chains with a carbonyl group, minus a proton. The ketone contains mainly the chains of the central fatty acid combined with one of the two external fatty acids, even if the ketone containing the two external fatty acids is present with a much weaker intensity. The formation of these ions may be explained by an internal Claisen condensation followed by a fragmentation induced by a nucleophile substitution and then by a decarboxylation, as shown in Figure 8.64.

This mechanism of formation explains the sensitivity of these ions for the positional isomers. Indeed, expulsion of the ketone containing the hydrocarbon chains of the central

Table 8.10 Masses observed in the various fragment spectra of the three deprotonated molecular ions present in the DCI spectrum of natural cocoa butter (P = palmitic acid, O = oleic acid, S = stearic acid).

$(M - H)^-$	$R_n COO^-$	$R_n COR'_n$	Deduced structure
831	255 (P), 281 (O)	475 (PO), *449*	POP
859	255 (P), 281 (O), 283 (S)	475 (PO), 503 (OS), *477*	POS
887	281 (O), 283 (S)	503 (SO), *505*	SOS

Numbers in italic indicate ions of low abundance.

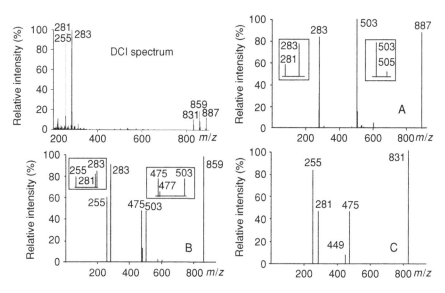

Figure 8.65
Desorption chemical ionization (DCI) mass spectrum of natural cocoa butter and collision-induced dissociation DCI/MS/MS traces of deprotonated molecular ions of 887 Th (A), 859 Th (B) and 831 Th (C). Reproduced (modified) from Stroobant V., Rozenberg R., Bouabsa E.M., Deffense E. and de Hoffmann E., J. Am. Soc. Mass Spectrom., 6, 498–506, 1995, with permission.

fatty acid combined with one of the outermost fatty acids (Figure 8.64, pathway c) requires the formation of a neutral epoxide, which is faster than the neutral oxetane formation necessary for expulsion of the ketone resulting from the condensation between the two outermost fatty acids (Figure 8.64, pathway f). This initial reaction thus occurs preferentially.

This method can be applied directly to acylglycerols present in a mixture, as is illustrated by the mass spectrometric analysis of natural cocoa butter. [299] This analysis (Figure 8.65) allowed the determination of the complete structure of the predominant acylglycerols in the cocoa (Table 8.10).

Table 8.11   Structure and nomenclature of various bile acids.

X = OH free bile acid
= NHCH$_2$CO$_2$H glycinoconjugated
= NHCH$_2$CH$_2$SO$_3$H tauroconjugated

Trivial name	Chemical name	R^1	R^2	R^3	R^4
Cholanic acid	5β-Cholan-24-oic acid	H	H	H	H
Lithocholic acid	5β-Cholan-24-oic-3α-ol acid	αOH	H	H	H
Hyodeoxycholic acid	5β-Cholan-24-oic-3α,6α-diol acid	αOH	αOH	H	H
Murocholic acid	5β-Cholan-24-oic-3α,6α-diol acid	αOH	βOH	H	H
Chenodeoxycholic acid	5β-Cholan-24-oic-3α,7α-diol acid	αOH	H	αOH	H
Ursodeoxycholic acid	5β-Cholan-24-oic-3α,7α-diol acid	αOH	H	βOH	H
Deoxycholic acid	5β-Cholan-24-oic-3α,12α-diol acid	αOH	H	H	αOH
Cholic acid	5β-Cholan-24-oic-3α,7α,12α-triol acid	αOH	H	α OH	αOH
Hyocholic acid	5β-Cholan-24-oic-3α,6α,7α-triol acid	αOH	αOH	αOH	H
Dehydrocholic acid	5β-Cholan-24-oic-3,7,12-trione acid	O	H	O	O

8.5.3 Bile Acids

Bile acids are a family of molecules derived from cholesterol. They are characterized by a 5β steroid ring made up of four fused cycles bearing a side chain attached to the C-17 carbon atom of the cycle, terminated by a carboxylic group. They differ from each other by the number and position of hydroxyl or keto groups and by the presence of unsaturations in the steroid cycle. Furthermore, these bile acids can exist as free carboxylic acids or as amide conjugates of the carboxylic groups with glycine (NH$_2$CH$_2$CO$_2$H) or taurine (NH$_2$CH$_2$CH$_2$SO$_3$H). Table 8.11 lists the structure of the bile acids encountered most often in the literature.

Mass spectrometry has become an indispensable method for the analysis of bile acids by virtue of its power to identify, assign structure and quantify free or conjugated bile acids, either pure or in mixtures. It is useful not only to study the metabolism of bile acids but also for the detection and diagnosis of metabolic diseases. Indeed, numerous metabolic diseases resulting from an alteration of the conversion of cholesterol to bile acids have been described, including peroxisomal disorders resulting in a block of β-oxidation of the lateral chain and other enzyme deficiencies interfering with the biochemistry of the side chain or the steroid nucleus.

Before the advent of soft ionization techniques, the analysis of bile acids was long and tedious and needed large sample quantities. First, the bile acids had to be extracted from the biological fluid and separated by lipophilic ion exchange chromatography into four classes:

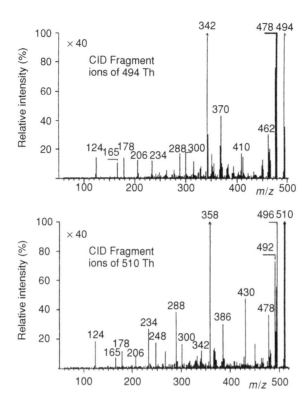

Figure 8.66
Collision-induced dissociation FAB/MS/MS traces of the
taurine conjugates of 7α-hydroxy-3-oxochol-4-en-24-oic
acid (*top*) and 7α, 12α-dihydro-3-oxochol-4-en-24-oic
acid (*bottom*). Reproduced (modified) from Libert R.,
Hermans D., Draye J.P., Van Hoof F., Sokal E. and de
Hoffmann E., Clin. Chem., 37, 2102–2110, 1991, with
permission.

unconjugated, tauro- or glycinoconjugated and sulfated. Next, each fraction separately was
hydrolyzed, extracted and derivatized. The bile acids then were identified and quantitated
in four GC/MS runs.

The emergence of soft ionization techniques such as FAB, thermospray (TSP), APCI
and ESI, combined with MS/MS, will considerably simplify these analysis. On a much
smaller sample quantity, bile acids can be analysed without prior separation or derivatization. Soft ionization techniques are well suited for such polar, non-volatile thermolabile
compounds.

The FAB, TSP, APCI and ESI mass spectra in the positive ionization mode display
the protonated molecular ion, generally accompanied by other adducts of the molecular
species and fragments resulting from the loss of one or several water molecules originating
from the ring hydroxyl groups. [300–302] As an acidic function is always present in these
compounds, they also yield intense ions in the negative ion mode. [303–305] The spectra

contain mainly deprotonated molecular ions. The presence of some adducts and a weak fragmentation sometimes also are observed. The relative importance of these ions depends on the sample preparation, the ionization method used and the experimental conditions. Generally, ESI spectra are simpler than FAB or TSP spectra.

The spectra thus contain few structural information and the use of MS/MS is very useful. The analysis of free or conjugated bile acids by high-energy MS/MS allows the observation of ions resulting from charge remote fragmentation (CRF) of the steroid cycle and the side chain. [306–308] Stereoisomers yield very similar spectra that do not allow differentiation between them in mixtures. But CRF provides information on the nature and position of substituents in the steroid ring. The position of double bonds also can be deduced by the presence of specific fragments or by the absence of some fragmentations.

The low-energy CID of bile acids is characterized by the presence of charge-induced fragmentations. [309] These fragmentations are often of little analytical interest because they result mostly in the loss of small molecules, such as water, formic acid, carbon dioxide, etc. However, the fragmentation of taurine conjugates in negative ion mode at low energy displays both charge-induced fragmentations CRF, yielding structural information on the ring substitution. [310]

This experimental observation is illustrated in Figure 8.66 by the tandem mass spectra of two bile acids of the Δ^4-3-oxo class: the taurine conjugates of 7α-hydroxy-3-oxochol-4-en-24-oic acid and 7α, 12α-dihydro-3-oxochol-4-en-24-oic acid. [311] Similar spectra were obtained for the Δ^4-3-hydroxy derivatives. The general fragmentation diagram deduced from these spectra is shown in Figure 8.67.

Knowing the fragmentation diagram of the bile acids, i.e. the structures of the fragments produced, allows one to determine the complete structure of the molecule. For example, the masses of fragments A and B indicate the presence or the absence of a hydroxyl group at position C-12.

The usefulness of low-energy MS/MS has been shown in the diagnosis of metabolic diseases by analysis of underivatized bile acids, conjugated or not, in complex biological samples such as urine or serum. A series of neutral loss and precursor ion scans was

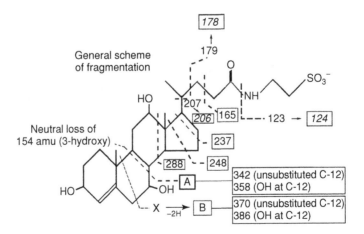

Figure 8.67
General fragmentation diagram of the Δ^4-cholenoic taurine conjugates [310].

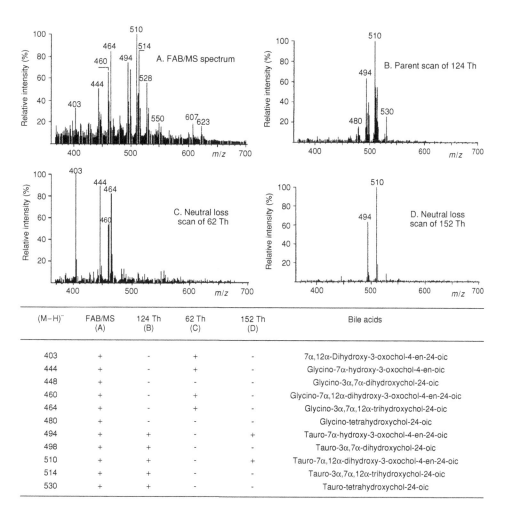

$(M-H)^-$	FAB/MS (A)	124 Th (B)	62 Th (C)	152 Th (D)	Bile acids
403	+	-	+	-	7α,12α-Dihydroxy-3-oxochol-4-en-24-oic
444	+	-	+	-	Glycino-7α-hydroxy-3-oxochol-4-en-24-oic
448	+	-	-	-	Glycino-3α,7α-dihydroxychol-24-oic
460	+	-	+	-	Glycino-7α,12α-dihydroxy-3-oxochol-4-en-24-oic
464	+	-	+	-	Glycino-3α,7α,12α-trihydroxychol-24-oic
480	+	-	-	-	Glycino-tetrahydroxychol-24-oic
494	+	+	-	+	Tauro-7α-hydroxy-3-oxochol-4-en-24-oic
498	+	+	-	-	Tauro-3α,7α-dihydroxychol-24-oic
510	+	+	-	+	Tauro-7α,12α-dihydroxy-3-oxochol-4-en-24-oic
514	+	+	-	-	Tauro-3α,7α,12α-trihydroxychol-24-oic
530	+	+	-	-	Tauro-tetrahydroxychol-24-oic

Figure 8.68
(A) The FAB spectrum of a mixture of 11 bile acids from a patient's urine, measured in negative mode. (B) Spectrum of the 124 Th precursors, allowing the selective detection of taurine conjugates. (C) Spectrum of 62 Da neutral losses. (D) Spectrum of 152 Da neutral losses. Reproduced (modified) from Libert R., Hermans D., Draye J.P., Van Hoof F., Sokal E., and de Hoffmann E., Clin. Chem., 37, 2102–2110, 111.

developed in order to allow the selective detection of molecular ion species of some bile acid classes present within a complex mixture. These scans mainly include: the scan of the precursor ions of 124 Th (anions of taurine), to detect all the taurine conjugates; the scan of neutral losses of 152 and 154 Da (mass of fragment A), to characterize the Δ^4-3-oxo and Δ^4-3-hydroxy taurine conjugates; and the scan of the neutral loss of 62 Da (corresponding to the loss of $CO_2 + H_2O$), to detect the free hydroxylated C-12 bile acids and their glycine conjugates. This method is rapid because it does not require prior chromatographic separation, degradation or derivatization.

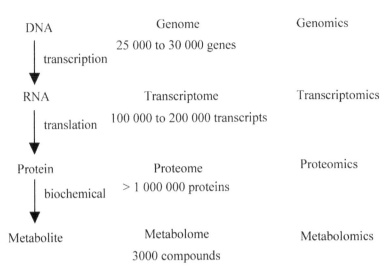

Figure 8.69
Relationship of metabolomics to the fields of genomics, transcriptomics and proteomics in 'systems biology'. Metabolome has been combined with genome, transcriptome and proteome to enable greater understanding of a biological system. The aim of these 'omics' sciences is the identification of all genes and their products (transcripts, proteins and metabolites) present in a specific biological sample.

The example in Figure 8.68 shows the analysis of a mixture containing 11 bile acids. [311] The FAB spectrum containing the molecular ion species of all of the bile acids present in the mixture (more intense than the background noise) is not sufficient to determine the composition of the mixture. Only the various specific scans allow the determination of the nature of all the bile acids present in the mixture, without however determining their stereochemistry.

8.6 Metabolomics

Metabolomics is the 'omics' science of metabolism. This term has been defined in analogy with genomics, transcriptomics and proteomics. As shown in Figure 8.69, genomics looks at the genome that is the entire collection of the genes of an organism. Similarly, transcriptomics looks at the transcriptome that is the complete set of mRNA transcripts and proteomics looks at the proteome that is the expressed set of proteins that are encoded by the genome. In the same manner, metabolome can be defined as the complete set of small molecules (non-polymeric compounds with a molecular weight lower than about 1000 Da) that are involved in general metabolic reactions and that are biosynthesized by a cell, tissue or organism [312]. Higher mass compounds are often polymers and their systematic identifications are termed according to the class of compounds, such as the 'glycome' for example. Metabolomics is the study that covers the identification and quantification of the metabolome under a given set of conditions. It also studies the dynamic changes in the metabolome.

The metabolome is a part of 'system biology', the umbrella that covers all the compounds produced in a biological system, from an organelle to a whole organism. Thus, system biology includes the genome, transcriptome, proteome and metabolome. Like transcriptome and proteome, metabolome is characterized by dynamic changes because it is sensitive to genetic or environmental changes. It is just one part of system biology though it may be the most useful because it is the most direct observation of the status of cellular physiology. Indeed, while the genome is what might be expressed and the proteome is what is actually expressed, the metabolome is what is done and represents the current status of a biological system. The increase in a transcript level does not always correlate with the increase in its corresponding protein level. And the increase in a protein level does not always correlate with the increase of its corresponding metabolite concentration because, once synthesized, a protein is not necessarily enzymatically active or its activity might be regulated.

In the whole biosphere, it is estimated that there are about 100 000 compounds with a molecular weight lower than about 1000 Da produced in living species. In a human, the metabolome is estimated to be no higher than about 3000 compounds. In comparison, the number of genes is 25 000 to 30 000, the number of transcripts is 100 000 to 200 000 and it is estimated that there are up to 1 000 000 different proteins, including post-translational modifications. Because the metabolome is composed of a limited number of compounds, metabolomics provides data that are less complex and more quantitative than genomics, transcriptomics and proteomics.

8.6.1 Mass Spectrometry in Metabolomics

The determination of the genome and the transcriptome did not rely much on mass spectrometry, other techniques allowing the gene sequence to be reached more efficiently. For the proteome, at the contrary, mass spectrometry has a central role. No other technique can compete with its efficiency to determine protein sequence quickly on very low sample quantities. But knowledge of the genome, the transcriptome and the proteome does not reveal the phenotype of a living system, as it is difficult or impossible to establish a direct link between the protein and its enzymatic activity, and thus the produced metabolites. There is almost no way to predict the substrates and metabolites of unknown enzymes whose total sequence is known. And it often appears that even for known enzymes the known activities are only partial. To understand the cell, it is thus an essential task to link the expression of proteins to the produced metabolites.

The elucidation of the metabolome is particularly challenging owing to the diverse chemical nature of the small molecules. Indeed, genome, transcriptome and proteome, which are biopolymers composed of a low number of building blocks, are very similar chemically. On the contrary, metabolome is based on small molecules that are very diverse in their structures and properties. For this reason, they are not easier to characterize than polymeric compounds. To get a complete metabolome is still an important challenge. First, one has to rely on efficient sample handling and separation techniques, loss of compounds at these steps being the rule. Then, identification and quantification of these compounds are also challenging. Mass spectrometry and nuclear magnetic resonance (NMR) spectroscopy are the most widely used techniques for metabolome analysis, although other techniques including electrochemical array detection have been used. NMR can measure simultaneously all kinds of small molecule from the metabolome. Furthermore, the sample can be recovered for further analyses. However, NMR also has significant limitations of sensitivity. Therefore, it is only applied to high-concentration compounds and cannot be used to analyse the most interesting compounds in lower concentrations.

In comparison with NMR, mass spectrometry is more sensitive and, thus, can be used for compounds of lower concentration. While it is easily possible to measure picomoles of compounds, detection limits at the attomole levels can be reached. Mass spectrometry also has the ability to identify compounds through elucidation of their chemical structure by MS/MS and determination of their exact masses. This is true at least for compounds below 500 Da, the limit at which very high-resolution mass spectrometry can unambiguously determine the elemental composition. In 2005, this could only be done by FTICR. Orbitrap appears to be a good alternative, with a more limited mass range but a better signal-to-noise ratio. Furthermore, mass spectrometry allows relative concentration determinations to be made between samples with a dynamic range of about 10 000. Absolute quantification is also possible but needs reference compounds to be used. It should be mentioned that if mass spectrometry is an important technique for metabolome analysis, another key tool is specific software to manipulate, summarize and analyse the complex multivariant data obtained.

Mass spectrometry in metabolome analysis and its applications were reviewed in 2005 [313,314]. It can be used as a standalone technique. In this case, the sample is infused directly into the mass spectrometer with no prior separation. For instance, Marshall has shown that it is possible to observe over 10 000 compounds in one sample by high-resolution FTICR [315]. Indeed FTMS with its high resolution offers a peak capacity exceeding that of HPLC by a factor of over 200. But generally, mass spectrometry is used to identify and to quantify compounds of the metabolome after separation, usually by chromatography. Other separation methods such as CE have also been applied. The initial mass spectrometric technique in the metabolome analysis is GC/MS. It has been widely used because it is a powerful method that presents very high chromatographic resolution. However, some large and polar compounds cannot be analysed by GC because this separation method is limited to molecules that were either volatile or could be made volatile by derivatization. Owing to the limitations of GC/MS, LC/MS has become a more popular technique for metabolome analysis. Indeed, a much wider range of analytes can potentially be analysed by LC/MS. In spite of a lower chromatographic resolution, several thousand compounds are also observable in one HPLC/MS run. Thus, in principle the metabolome should be accessible through a limited number of mass spectrometry analyses. The most used ionization methods in this field are atmospheric pressure ionizations, mainly ESI and APCI.

8.6.2 Applications

Metabolomics has emerged as a useful technique in diverse fields such as clinical diagnostics, drug discovery and plant biochemistry. While the elucidation of all the small molecules of a cell, tissue or organism is an important goal, 'differential metabolomics' can yield very interesting results through comparison between the metabolome of a reference and the same sample after modification, be it by a drug, a genetic modification, a disease, a stress or any other action.

It is known that metabolomics is now widely used in the pharmaceutical industry, as the difference between the metablome of an organism without and with the influence of a drug is of paramount interest. This could give information not only on the drug metabolites, but

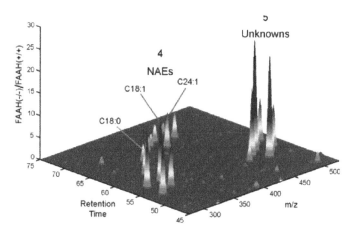

Figure 8.70
Ratio of the ion intensities FAAH(−/−)/FAAH(+/+) observed at the different masses during elution of extracts from the brain and spinal cord of knockout mice (FAAH(−/−)) and normal mice (FAAH(+/+)). This plot reveals two regions where compounds are in much higher concentration in knockout mice. The first region corresponds to a known class of substrates for FAAH that are related to ethanolamine amides (NAE). This class of substrates is much larger than previously described because several unrecognized compounds of this class, such as the C24:1 compound, were also found. Very low-abundance compounds of this class, such as anandamide (C20:4), were also detected. The second region is particularly interesting and corresponds to an unknown class of substrates for FAAH. Reproduced from Saghatelian A., Trauger S.A., Want E.J., Hawkins E.G., Siuzdak G. and Cravatt B.F., Biochemistry, 43, 14332–14339, 2004, with permission.

principally on the drug's action on biological pathways. Metabolomics can be important at every stage of the drug discovery process. By accurately measuring the metabolome changes between healthy and diseased patients, biomarker compounds of various diseases can be identified and become diagnostic for these diseases. By mapping these changes to known metabolic pathways, enzymes that are responsible for these changes can be deduced and thus the enzyme that is critical to a disease can be identified. To validate this enzyme as a good candidate for drug discovery, it can be inactivated by an inhibitor or a genetic modification. The induced effect on the metabolome can be compared with the disease. If the metabolomes are similar, then the inactivated enzyme is a good candidate. Unfortunately, there are few detailed descriptions of complete metabolome analysis in the literature. The paper detailed in the next section is an exception.

As an example of the application of metabolomics, we will refer to the substrates of the enzyme fatty acyl amide hydrolase (FAAH) that regulates several brain lipids that have interesting pharmacological properties including effects on the control of pain. It is well known that some fatty acyl amides of ethanolamine are substrates of FAAH, such as the ethanolamide of arachidonic acid (anandamide), which is an endogenous ligand of cannabinoid receptors.

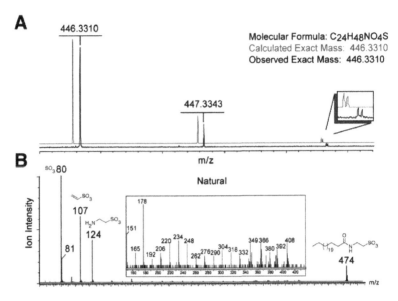

Figure 8.71
Identification and structural characterization of an unknown class of sub-
strates for FAAH. (A) By high-resolution mass spectrometry, the high-
accuracy mass measurements of a compound of this class gives an exact
mass of 446.3310 that corresponds to a molecular formula of $C_{24}H_{48}NO_4S$.
(B) By MS/MS analysis, the structure of this compound is assigned as the
C24:0 fatty acyl amide of taurine (NAT). Reproduced from Saghatelian
A., Trauger S.A., Want E.J., Hawkins E.G., Siuzdak G. and Cravatt B.F.,
Biochemistry, 43, 14332–14339, 2004, with permission.

But are there other endogenous substrates for this enzyme? A method based on separation
techniques and mass spectrometry has been developed to identify endogenous substrates
of enzymes by analysis of metabolomes from wild-type and enzyme-inactivated organisms
[316]. Indeed, the accumulation of metabolites resulting from the inactivation of the enzyme
would be considered candidate endogenous substrates for this enzyme. This method based
on comparative metabolomics is called discovery metabolites profiling (DMP).

Knockout mice lacking the gene to synthesize FAAH would accumulate all the possible
endogenous substrates of this enzyme. Thus, a comparison of the metabolomes of normal
and knockout mice would allow the identification of all substrates of FAAH, present mainly
in the brain and spinal cord. After exhaustive extraction from these tissues, the mixture
was first analysed by HPLC/ESI/MS in both the positive and negative ion mode. A detailed
comparison of the mass spectra obtained from normal and knockout mice is presented in
Figure 8.70. The results are displayed in a 3D map plot, which consists of ion intensity
ratios in the knockout and normal mice samples plotted over a mass range and retention
times. Dividing the ion intensities observed for the knockout mice by those observed at the
same mass and retention time for normal mice yields a quantitative measure of variations
in the metabolome induced by the inactivation of FAAH. Figure 8.70 reveals two regions
where compounds eluted in much higher concentrations in knockout mice. The first region

corresponds to a known class of substrates (fatty acyl amides of ethanolamine, NAE), the second one to unknown substrates.

Identification and characterization of these unknown compounds is possible through accurate mass measurements and MS/MS. Accumulated HPLC fractions of unknowns were then first analysed by high-resolution mass spectrometry using FTMS in negative ion mode. High-resolution instruments are capable of measuring the mass of these unknowns with sufficient accuracy to allow the determination of their elemental compositions. For instance, as shown in Figure 8.71, one of these unknown compounds has an exact mass of 446.3310 u, corresponding to an elemental composition of $C_{24}H_{48}NO_4S$.

These same fractions were then analysed by MS/MS using a Q-TOF in the negative ion mode. As displayed in Figure 8.71 for one of these unknowns, two series of peaks are observed. The first series, which dominates the spectra, corresponds to low-mass ions (m/z 80, 107 and 124) suggestive of the presence of taurine conjugate. The second series corresponds to lower intensity ions distant from each other by 14 u that are characteristic of compounds containing hydrocarbon chains.

These experiments show altogether that the unknowns are amides of taurine from several fatty acids. Therefore, this new class of endogenous substrates of FAAH corresponds to fatty acyl amides of taurine (NAT) with very long-chain fatty acids. The structure of this new class of substrates of FAAH is confirmed by chemical synthesis and comparison of retention times and mass spectra obtained for these synthetic products with those of the natural products.

References

1. Bleakney, W. (1929) *Phys. Rev.*, **34**, 157.
2. Munson, M.S.B. and Field, F.H. (1966) *J. Am. Chem. Soc.*, **88**, 2621.
3. Beckey, H.D. (1963) *Z. Anal. Chem.*, **197**, 80.
4. Morris, H.R. (1980) *Soft Ionization Biological Mass Spectrometry*, Heyden, London.
5. Thorgersen, D.F., Skowronsky, R.P. and Macfarlane, R.D. (1974) *Biochem. Biophys. Res. Commun.*, **60**, 616.
6. Harkansson, P., Kamenski, I., Sundquist, B. *et al.* (1982) *J. Am. Chem. Soc.*, **104**, 2498.
7. Barber, M., Bordoli, R.S., Sedwick, R.D. and Tyler, A.N. (1981) *J. Chem. Soc. Chem. Commun.*, **7**, 325.
8. Cotter, R.J. (1984) *Anal. Chem.*, **56**, 485A.
9. Fenn, J.B., Mann, M., Meng, C.K. and Whitehouse, C.M. (1989) *Science*, **246**, 64.
10. Karas, M. and Hillenkamp, F. (1988) *Anal. Chem.*, **60**, 2299.
11. Siuzdak, G. (2003) *The Expanding Role of Mass Spectrometry in Biotechnology*, MCC Press, San Diego, CA.
12. Burlingame, A.L. (ed.) (2005) *Methods in Enzymology, Biological Mass Spectrometry*, vol. **204**, Academic Press, New York.
13. Krihna, R. and Wold, F. (1997) *Protein Structure – A Practical Approach*, 2nd edn (ed. T.E. Creighton), Oxford University Press, New York, pp. 391–402.
14. Walsh, C.T. (2006) *Posttranslational Modifications of Proteins: Expanding Nature's Inventory*, Roberts and Co. Publishers, Greenwood Village, CO.
15. http://www.ebi.ac.uk/RESID/ (21 March 2007).
16. http://www.abrf.org/index.cfm/dm.home (21 March 2007).
17. Aebersold, R. and Mann, M. (2003) Mass spectrometry-based proteomics. *Nature*, **422**, 198–207.

18. Standing, K.G. (2003) Peptide and protein de novo sequencing by mass spectrometry. *Curr. Opin. Struct. Biol.*, **13**, 595–601.
19. Sechi, S. and Oda, Y. (2003) Quantitative proteomics using mass spectrometry. *Curr. Opin. Chem. Biol.*, **7**, 70–7.
20. Jensen, O.N. (2004) Modification-specific proteomics: characterization of post-translational modifications by mass spectrometry. *Curr. Opin. Chem. Biol.*, **8**, 33–41.
21. Meng, F.Y., Forbes, A.J., Miller, L.M. and Kelleher, N.L. (2005) Detection and localization of protein modifications by high resolution tandem mass spectrometry. *Mass Spectrom. Rev.*, **24**, 126–34.
22. Burlingame, A.L. (ed.) (2005) *Methods in Enzymology, Mass Spectrometry: Modified Proteins and Glycoconjugates*, vol. **405**, Academic Press, New York.
23. Medzihradszky, K.F. (2005) *Methods in Enzymology, Biological Mass Spectrometry*, vol. **204**, (ed. A.L. Burlingame), Academic Press, New York, pp. 209–44.
24. Hernandez, P., Muller, M. and Appel, R.D. (2006) Automated protein identification by tandem mass spectrometry: issues and strategies. *Mass Spectrom. Rev.*, **25**, 235–54.
25. Vorm, O., Roepstorff, P. and Mann, M. (1994) *Anal. Chem.*, **66**, 3281.
26. Valaskovic, G.A., Kelleher, N.L. and McLafferty, F.W. (1996) Attomole protein characterization by capillary electrophoresis mass spectrometry. *Science*, **273** (5279), 1199–1202.
27. Martin, S.E., Shabanowitz, J., Hunt *et al.* (2000) Subfemtomole MS and MS/MS peptide sequence analysis using nano-HPLC micro-ESI Fourier transform ion cyclotron resonance mass spectrometry. *Anal. Chem.*, **72**, 4266–74.
28. Suizdak, G. (1994) *Proc. Natl. Acad. Sci. USA*, **91**, 11290.
29. Cohen, S.L. and Chait, B.T. (1996) Influence of matrix solution conditions on the MALDI-MS analysis of peptides and proteins. *Anal. Chem.*, **68** (1), 31–7.
30. Beavis, R.C. and Chait, B.T. (1990) *Anal. Chem.*, **62**, 1836.
31. Wilm, M., Shevchenko, A., Houthaeve, T. *et al.* (1996) Femtomole sequencing of proteins from polyacrylamide gels by nano-electrospray mass spectrometry. *Nature*, **379** (6564), 466–9.
32. Davis, M.T., Stahl, D.C. and Lee, T.D. (1994) *J. Am. Soc. Mass Spectrom.*, **5**, 571.
33. Emmet, M.R. and Caprioli, R.M. (1994) *J. Am. Soc. Mass Spectrom.*, **5**, 605.
34. Wahl, J.H., Goodlett, D.R., Udseth, H.R. and Smith, R.D. (1993) *Electrophoresis*, **14**, 448.
35. Bean, M.F., Carr, S.A., Thorne, G.C. *et al.* (1991) *Anal. Chem.*, **63**, 1473.
36. Papayannopoulos, I.A. (1995) *Mass Spectrom. Rev.*, **14**, 49.
37. Roepstorff, P. and Fohlman, J. (1984) *Biomed. Mass Spectrom.*, **11**, 601 and **12**, 631 (1985)
38. Biemann, K. (1988) *Biomed. Environ. Mass Spectrom.*, **16**, 99.
39. Hunt, D.F., Yates, J.R., Shabanowitz, J. *et al.* (1986) *Proc. Natl. Acad. Sci. USA*, **83**, 6233.
40. Johnson, R.S. and Biemann, K. (1989) *Biomed. Environ. Mass Spectrom.*, **18**, 945.
41. Heerma, W. and Kulik, W. (1988) *Biom. Environ. Mass Spectrom.*, **16**, 155.
42. Falik, A.M., Hines, W.M., Medzihradsky, K.F. *et al.* (1993) *J. Am. Soc. Mass Spectrom.*, **4**, 882.
43. Johnson, R.S., Martin, S.A., Biemann, K. *et al.* (1987) *Anal. Chem.*, **59**, 2621.
44. Tomer, K.B., Crow, F.W. and Gross, M.L. (1983) *J. Am. Chem. Soc.*, **105**, 5487.
45. Johnson, R.S., Martin, S.A. and Biemann, K. (1988) *Int. J. Mass Spectrom. Ion Processes*, **86**, 137.
46. Alexander, A.J. and Boyd, R.K. (1989) *Int. J. Mass Spectrom. Ion Processes*, **90**, 211.
47. vanDongen, W.D., Ruijters, H.F.M., Luinge, H.J. *et al.* (1996) Statistical analysis of mass spectral data obtained from singly protonated peptides under high-energy collision-induced dissociation conditions. *J. Mass Spectrom.*, **31** (10), 1156–62.
48. Biemann, K. (1990) *Methods Enzymol.*, **193**, 455.

49. Martin, S.A., Johnson, R.S., Costello, C.E. and Biemann K. (1988) *The Analysis of Peptides and Proteins by Mass Spectrometry* (ed. C.J. McNeal), John Wiley & Sons, Inc., New York, pp. 135–50.

50. Tang, X.J., Thibault, P. and Boyd, R.K. (1993) *Anal. Chem.*, **65**, 2824.

51. Downard, K.M. and Biemann, K. (1994) *J. Am. Soc. Mass Spectrom.*, **5**, 966.

52. Covey, T.R., Huang, E.C. and Henion, J.D. (1991) *Anal. Chem.*, **63**, 1193.

53. Watkins, P.F.E., Jardine, I. and Zhou, J.X.G. (1991) *Biochem. Soc. Trans.*, **19**, 957.

54. Kaufmann, R., Spengler, B. and Lutzenkirchen, F. (1993) *Rapid Commun. Mass Spectrom.*, **7**, 902.

55. Huberty, M.C., Vath, J.E., Yu, W. and Marti, S.A. (1993) *Anal. Chem.*, **65**, 2791.

56. Rouse, J.C., Yu, W. and Martin, S.A. (1996) A comparison of the peptide fragmentation obtained from a reflector matrix-assisted laser desorption-ionisation time-of-flight mass and a tandem 4-sector mass spectrometer. *J. Am. Mass Spectrom.*, **6** (9), 822–35.

57. Zubarev, R.A., Kelleher, N.L. and McLafferty, F.W. (1998) Electron capture dissociation of multiply charged protein cations. A nonergodic process. *J. Am. Chem. Soc.*, **120**, 3265–6.

58. Syka, J.E.P., Coon, J.J., Schroeder, M.J. *et al.* (2004) Peptide and protein sequence analysis by electron transfer dissociation mass spectrometry. *Proc. Natl. Acad Sci. USA*, **101**, 9528–33.

59. Cardenas, M.S., Van Der Heeft, E. and de Jong, A.P.J.M. (1997) On-line derivatization of peptides for improved sequence analysis by micro-column liquid chromatography coupled with electrospray ionization tandem mass spectrometry. *Rapid Commun. Mass Spectrom.*, **11** (12), 1271–8.

60. Hendrickson, R.C., Skipper, J.C., Shabanowitz, J. *et al.* (1997) *Immunology Methods Manual* (ed. I. Lefkovits), Academic Press, San Diego, CA, vol. **2**, pp. 605–38.

61. Shevchenko, A., Chernushevich, I., Ens, W. *et al.* (1997) Rapid 'de novo' peptide sequencing by a combination of nanoelectrospray, isotopic labeling and a quadrupole/time-of-flight mass spectrometer. *Rapid Commun. Mass Spectrom.*, **11** (9), 1015–24.

62. Vath, J.E. and Biemann, K. (1990) *Int. J. Mass Spectrom. Ion Processes*, **100**, 287.

63. Hulst, A.G. and Kientz, C.E. (1996) Differentiation between the isomeric amino acids leucine and isoleucine using low-energy collision-induced dissociation tandem mass spectrometry. *J. Mass Spectrom.*, **31** (10), 1188–90.

64. Hines, W.M., Faliek, A.M., Burlingame, A.L. and Gibson, B.W. (1992) *J. Am. Mass Spectrom.*, **3**, 326.

65. Taylor, J.A. and Johnson, R.S. (2001) Implementation and uses of automated de novo peptide sequencing by tandem mass spectrometry. *Anal. Chem.*, **73**, 2594–604.

66. Fernandez-De-Cossio, J., Gonzalez, J., Satomi, Y. *et al.* (2001) Automated interpretation of low-energy collision induced dissociation spectra by SeqMS, a software aid for de novo sequencing by tandem mass spectrometry. *Electrophoresis*, **21**, 1694–9.

67. Dancik, V., Addona, T.A., Clauser, K.R. *et al.* (1999) De novo peptide sequencing via tandem mass spectrometry. *J. Comput. Biol.*, **6**, 327–42.

68. Eng, J.K., McCormack, A.L. and Yates, J.R. (1994) *J. Am. Soc. Mass Spectrom.*, **5**, 976.

69. Yates, J.R., Eng, J.K. and McCormack, A.L. (1995) *Anal. Chem.*, **67**, 1426.

70. Yates, J.R., Eng, J.K., Clauser, K.R. and Burlingame, A.L. (1996) Search of sequence databases with uninterpreted high-energy collision-induced dissociation spectra of peptides. *J. Am. Soc. Mass Spectrom.*, **7** (11) 1089–98.

71. Liska, A.J. and Shevchenko, A. (2003) Combining mass spectrometry with database interrogation strategies in proteomics. *Trends Anal. Chem.*, **22**, 291–8.

72. Kelleher, N.L. (2004) Top-down proteomics. *Anal. Chem.*, **76**, 197A–203A.

73. Liu, Z. and Schey, K.L. (2005) Optimization of a MADLI TOF-TOF mass spectrometer for intact protein analysis. *J. Am. Soc. Mass Spectrom.*, **16**, 482–90.

74. Gevaert, K. and Vandekerckhove, J. (2000) Protein identification methods in proteomics. *Electrophoresis*, **21**, 1145–54.
75. Jungblut, P. and Thiede, B. (1997) Protein identification from 2-DE gels by MALDI mass spectrometry. *Mass Spectrom. Rev.*, **16** (3), 145–62.
76. Lin, D., Tabb, D.L. and Yates, J.R. (2003) Large-scale protein identification using mass spectrometry. *Biochim. Biophys. Acta – Proteins Proteomics*, **1646**, 1–10.
77. Wysocki, V.H., Resing, K.A., Zhang, Q.F. and Cheng, G.L. (2005) Mass spectrometry of peptides and proteins. *Methods*, **35**, 211–22.
78. Kolker, E., Higdon, R. and Hogan, J.M. (2006) Protein identification and expression analysis using mass spectrometry. *Trends Microbiol.*, **14**, 229–35.
79. Deissler, H., Wilm, M., Genc, B. *et al.* (1997) Rapid protein sequencing by tandem mass spectrometry and cDNA cloning of p20-CGGBP: a novel protein that binds to the unstable triplet repeat 5′-d(CGG)(n)-3′ in the human FMR1 gene. *J. Biol. Chem.*, **272** (27), 16761–8.
80. Nemeth-Cawley, J.F., Tangarone, B.S. and Rouse, J.C. (2003) Top down characterization is a complementary technique to peptide sequencing for identifying protein species in complex mixtures. *J. Proteome Res.*, **2**, 495–505.
81. Bogdanov, B. and Smith, R.D. (2005) Proteomics by FTICR mass spectrometry: top down and bottom up. *Mass Specrom. Rev.*, **24**, 168–200.
82. Han, X., Jin, M., Breuker, K. and McLafferty, F.W. (2006) Extending top-down mass spectrometry to proteins with masses greater than 200 kilodaltons. *Science*, **314**, 106–12.
83. Henzel, W.J., Billeci, T.M., Stults, J.T. *et al.* (1993) *Proc. Natl. Acad. Sci USA*, **90**, 5011.
84. Mann, M., Hojrup, P. and Roepstorff, P. (1993) *Biol. Mass Spectrom.*, **22**, 338.
85. Pappin, D., Hojrup, P. and Bleasby, A.J. (1993) *Curr. Biol.*, **3**, 327.
86. Yates, J.R., Speicher, S., Griffin, P.R. and Hunkapiller, T. (1993) *Anal. Biochem.*, **214**, 397.
87. Mann, M. and Wilm, M. (1994) *Anal. Chem.*, **66**, 4390.
88. Mann, M. (1996) A shortcut to interesting human genes: peptide sequence tags, expressed-sequence tags and computers. *Trends Biochem. Sci.*, **21** (12), 494–5.
89. McCormack, A.L., Schieltz, D.M., Goode, B. *et al.* (1997) Direct analysis and identification of proteins in mixtures by LC/MS/MS and database searching at the low-femtomole level. *Anal. Chem.*, **69** (4), 767–76.
90. Warnock, D.E., Fahy, E. and Taylor, S.W. (2004) Identification of protein associations in organelles, using mass spectrometry-based proteomics. *Mass Spectrom. Rev.*, **23**, 259–80.
91. Ethier, M., Lambert, J.P., Vasilescu, J. and Figeys, D. (2006) Analysis of protein interaction networks using mass spectrometry compatible techniques. *Anal. Chim. Acta*, **564**, 10–18.
92. Shevchenko, A., Wilm, M., Vorm, O. and Mann, M. (1996) Mass spectrometric sequencing of proteins from silver stained polyacrylamide gels. *Anal. Chem.*, **68** (5), 850–8.
93. Eckerskorn, C. and Grimm, R. (1996) Enhanced in situ gel digestion of electrophoretically separated proteins with automated peptide elution onto mini reversed-phase columns. *Electrophoresis*, **17** (5), 899–906.
94. Wilm, M., Shevchenko, A., Houthaeve, T. *et al.* (1996) Femtomole sequencing of proteins from polyacrylamide gels by nano-electrospray mass spectrometry. *Nature*, **379** (6564), 466–9.
95. Shevchenko, A., Jensen, O.N., Podtelejniikov, A.V. *et al.* (1996) Linking genome and proteome by mass spectrometry: large-scale identification of yeast proteins from two dimensional gels. *Proc. Natl. Acad. Sci. USA*, **93** (25), 14440–5.
96. Aebersold, R. and Mann, M. (2003) Mass spectrometry-based proteomics. *Nature*, **422** 198–207.

97. Lewis, J.K., Krone, J.R. and Nelson, R.W. (1998) Mass spectrometric methods for evaluating point mutations. *BioTechniques*, **24** (1), 102.

98. Wada, Y. (2002) *The Protein Protocols Handbook*, 2nd edn, Humana Press, Totowa, NJ, pp. 681–92.

99. Tanaka, K., Takenaka, S., Tsuyama, S. and Wada, Y. (2006) Determination of unique amino acid substitutions in protein variants by peptide mass mapping with FT-ICR MS. *J. Am. Soc. Mass Spectrom.*, **17**, 508–13.

100. Shimizu, A., Nakanishi, T., and Miyazaki, A. (2006) Detection and characterization of variant and modified structures of proteins in blood and tissues by mass spectrometry. *Mass Spectrom. Rev.*, **25**, 686–712.

101. http://www.expasy.org/tools/findmod/aa_subst_average.html and http://www.expasy.org/tools/findmod/aa_subst_monoisotopic.html (21 March 2007).

102. Wada, Y. (2002) Advanced analytical methods for hemoglobin variants. *J. Chromatogr. B*, **781**, 291–301.

103. Scherperel, G., Yan, H.G., Wang, Y. and Reid, G.E. (2006) Top-down characterization of site-directed mutagenesis products of Staphylococcus aureus dihydroneopterin aldolase by multistage tandem mass spectrometry in a linear quadrupole ion trap. *Analyst*, **131**, 291–302.

104. Papayannopoulos, I.A. and Biemann, K. (1987) *Peptide Res.*, **5**, 83.

105. Smart, S.S., Mason, T.J., Bennell, P.S. *et al.* (1996) High-throughput purity estimation and characterisation of synthetic peptides by electrospray mass spectrometry. *Int. J. Peptide Protein Res.*, **47**, 47–55.

106. Arttamangkul, S., Arbogast, B., Barofsky, D. and Aldrich, J.V. (1997) Characterization of synthetic peptide byproducts from cyclization reactions using on-line HPLC-ion spray and tandem mass spectrometry. *Lett. Peptide Sci.*, **3**, 357–70.

107. Enjalbal, C., Martinez, J. and Aubagnac, J.L. (2000) Mass spectrometry in combinatorial chemistry. *Mass Spectrom. Rev.*, **19**, 139–61.

108. Kassel, D.B. (2001) Combinatorial chemistry and mass spectrometry in the 21st century drug discovery laboratory. *Chem. Rev.*, **101**, 255–67.

109. Metzger, J.W., Wiesmüller, K.-H., Kienle, S. *et al.* (1996) *Combinatorial Peptide and Non-peptide Libraries* (ed. G. Jung), Verlag Chemie, Weinheim, pp. 247–86.

110. Sussmuth, R.D. and Jung, G. (1999) Impact of mass spectrometry on combinatorial chemistry. *J. Chromatogr. B*, **725**, 49–65.

111. Schlosser, G., Takats, Z., Vekey, G. *et al.* (2003) Mass spectrometric analysis of combinatorial peptide libraries derived from the tandem repeat unit of MUC2 mucin. *J. Peptide Sci.*, **9**, 361–74.

112. Andersen, J.S. (1995) *Biochem. Soc. Trans.*, **23**, 917.

113. Flensburg, J. and Belew, M. (2003) Characterization of recombinant human serum albumin using matrix-assisted laser desorption ionization time-of-flight mass spectrometry. *J. Chromatogr. A*, **1009**, 111–17.

114. Hoffman, R.C., Jennings, L.L., Tsigelny, I. *et al.* (2004) Structural characterization of recombinant soluble rat neuroligin 1: mapping of secondary structure and glycosylation by mass spectrometry. *Biochemistry*, **43**, 1496–1506.

115. Wang, L.T., Amphlett, G., Lambert, J.M. *et al.* (2005) Structural characterization of a recombinant monoclonal antibody by electrospray time-of-flight mass spectrometry. *Pharm. Res.*, **22**, 1338–49.

116. Van Dorsselaer, A., Bitsch, F., Green, B. *et al.* (1990) *Biomed. Environ. Mass Spectrom.*, **19**, 692.

117. Chait, B.T., Wang, R., Beavis, R.C. and Kent, S.B.H. (1993) *Science*, **262**, 89.

118. Bartlet-Jones, M., Jeffery, W.A., Hansen, H.F. and Pappin, D.J.C. (1994) *Rapid Commun. Mass Spectrom.*, **8**, 737.

119. Thiede, B., Salnikow, J. and WittmannLiebold, B. (1997) C-terminal ladder sequencing by an approach combining chemical degradation with analysis by matrix-assisted-laser-desorption ionization mass spectrometry. *Eur. J. Biochem.*, **244** (3), 750–4.
120. Patterson, D.H., Tarr, G.E., Renier, F.E. and Martin, S.A. (1995) *Anal. Chem.*, **67**, 3971.
121. Zhong, H.Y., Zhang, Y., Wen, Z.H. and Li, L. (2004) Protein sequencing by mass analysis of polypeptide ladders after controlled protein hydrolysis. *Nat. Biotechnol.*, **22**, 1291–6.
122. Gamen, B., Li, Y.T. and Henion, J.D. (1991) *J. Am. Chem. Soc.*, **113**, 7818.
123. Katta, V. and Chait, B.T. (1991) *J. Am. Chem. Soc.*, **113**, 8534.
124. Loo, J.A. (1997) Studying noncovalent protein complexes by electrospray ionization mass spectrometry. *Mass Spectrom. Rev.*, **16** (1), 1–23.
125. Loo, J.A. (2000) Electrospray ionization mass spectrometry: a technology for studying noncovalent macromolecular complexes. *Int. J. Mass Spectrom.*, **200**, 175–86.
126. Kriwacki, R., Reisdorph, N. and Siuzdak, G. (2004) Protein structure characterization with mass spectrometry. *Spectroscopy*, **18**, 34–47.
127. Heck, A.J.R. and Van Den Heuvel, R.H.H. (2004) Investigation of intact protein complexes by mass spectrometry. *Mass Spectrom. Rev.*, **23**, 368–89.
128. Daniel, J.M., Friess, S.D., Rajagopalan, S. *et al.* (2002) Quantitative determination of noncovalent binding interactions using soft ionization mass spectrometry. *Int. J. Mass Spectrom.*, **216**, 1–27.
129. Hofstadler, S.A. and Sannes-Lowery, K.A. (2006) Applications of ESI-MS in drug discovery: interrogation of noncovalent complexes. *Nat. Rev. Drug Discovery*, **5**, 585–98.
130. Cai, X.M. and Dass, C. (2003) Conformational analysis of proteins and peptides. *Curr. Org. Chem.*, **7**, 1841–54.
131. Loo, J.A. (1995) *J. Mass Spectrom.*, **30**, 180.
132. Fan, X. and Beavis, R.C. (1993) *Org. Mass Spectrom.*, **28**, 1424.
133. Glocker, M.O., Bauer, S.H.J., Kast, J. *et al.* (1996) Characterization of specific noncovalent protein complexes by UV matrix-assisted laser desorption ionization mass spectrometry. *J. Mass Spectrom.*, **31** (11), 1221–7.
134. Cohen, L.R.H., Strupat, K. and Hillenkamp, F. (1997) Analysis of quaternary protein ensembles by matrix assisted laser desorption/ionization mass spectrometry. *J. Am. Soc. Mass Spectrom.*, **8**, 1046–52.
135. Bolbach, G. (2005) Matrix-assisted laser desorption/ionization analysis of non-covalent complexes: fundamentals and applications. *Curr. Pharm. Des.*, **11**, 2535–57.
136. Zehl, M. and Allmaier, G. (2005) Instrumental parameters in the MALDI-TOF mass spectrometric analysis of quaternary protein structures. *Anal. Chem.*, **77**, 103–10.
137. Loo, J.A., Edmonds, C.G., Udseth, H.R. and Smith, R.D. (1990) *Anal. Chem.*, **62**, 693.
138. Chowdhury, S.K., Katta, V. and Chait, B.T. (1990) *J. Am. Chem. Soc.*, **112**, 9012.
139. Kaltashov, I.A. and Mohimen, A. (2005) Estimates of protein surface areas in solution by electrospray ionization mass spectrometry. *Anal. Chem.*, **77**, 5370–9.
140. Konermann, L., Rosell, F.I., Mauk, A.G. and Douglas, D.J. (1997) Cytochrome c folding kinetics studied by time-resolved electrospray ionization mass spectrometry. *Biochemistry*, **34** (18), 5554–9.
141. Grandori, R. (2003) Electrospray-ionization mass spectrometry for protein conformational studies. *Curr. Org. Chem.*, **7**, 1589–603.
142. Katta, V. and Chait, B.T. (1991) *Rapid Commun. Mass Spectrom.*, **5**, 214.
143. Thevenon-Emeric, G., Kozlowski, J., Zhang, Z. and Smith, D.L. (1992) *Anal. Chem.*, **64**, 2456.
144. Lanman, J., Lam, T.T., Emmett, M.R. *et al.* (2004) Key interactions in HIV-1 maturation identified by hydrogen-deuterium exchange. *Nat. Struct. Mol. Biol.*, **11**, 676–7.
145. Wales, T.E. and Engen, J.R. (2006) Hydrogen exchange mass spectrometry for the analysis of protein dynamics. *Mass Spectrom. Rev.*, **25**, 158–70.

146. Suckeau, D., Mak, M. and Przybylski, M. (1992) *Proc. Natl. Acad. Sci. USA*, **89**, 5630.
147. Glocker, M.O., Borchers, C., Fielder, W. *et al.* (1994) *Bioconjugate Chem.*, **5**, 583.
148. Maleknia, S.D. and Downard, K. (2001) Radical approaches to probe protein structure, folding, and interactions by mass spectrometry. *Mass Spectrom. Rev.*, **20**, 388–401.
149. Meng, F.Y., Forbes, A.J., Miller, L.M. and Kelleher, N.L. (2005) Detection and localization of protein modifications by high resolution tandem mass spectrometry. *Mass Spectrom. Rev.*, **24**, 126–34.
150. Sinz, A. (2006) Chemical cross-linking and mass spectrometry to map three-dimensional protein structures and protein-protein interactions. *Mass Spectrom. Rev.*, **25**, 663–82.
151. Massotte, D., Yamamoto, M., Scianimanico, S. *et al.* (1993) *Biochemistry*, **32**, 13787.
152. Cohen, S.L., Ferré-D'Amaré, A.R., Burley, S.K. and Chait, B.T. (1995) *Protein Sci.*, **4**, 1088.
153. Shields, S.J. and Oyeyemi, O. (2003) Mass spectrometry and non-covalent protein-ligand complexes: confirmation of binding sites and changes in tertiary structure. *J. Am. Soc. Mass Spectrom.*, **14**, 460–70.
154. Stroh, J.G., Loulakis, P., Lanzetti, A.J. and Xie, J.L. (2005) LC-mass spectrometry analysis of N- and C-terminal boundary sequences of polypeptide fragments by limited proteolysis. *J. Am. Soc. Mass Spectrom.*, **16**, 38–45.
155. Suckau, D., Kohl, J., Shneider, K. *et al.* (1990) *Proc. Natl. Acad. Sci. USA*, **87**, 9848.
156. Zhao, Y.M., Muir, T.W., Kent, S.B.H. *et al.* (1996) Mapping protein-protein interactions by affinity-directed mass spectrometry. *Proc. Natl. Acad. Sci. USA*, **93** (9), 4020–4.
157. Macht, M., Marquardt, A., Deininger, S.O. *et al.* (2004) 'Affinity-proteomics': direct protein identification from biological material using mass spectrometric epitope mapping. *Anal. Bioanal. Chem.*, **378**, 1102–11.
158. Murray, K.K. (1996) DNA sequencing by mass spectrometry. *J. Mass Spectrom.*, **31** (11), 1203–15.
159. Limbach, P.A. (1996) Indirect mass spectrometric methods for characterizing and sequencing oligonucleotides. *Mass Spectrom. Rev.*, **15** (5), 297–336.
160. Nordhoff, E., Kirpekar, F. and Roepstorff, P. (1996) Mass spectrometry of nucleic acids. *Mass Spectrom. Rev.*, **15** (2), 67–138.
161. Crain, P.F. and McCloskey, J.A. (1998) Applications of mass spectrometry to the characterization of oligonucleotides and nucleic acids. *Curr. Opin. Biotechnol.*, **9** (1), 25–34.
162. Stults, J.T. and Masters, J.C. (1991) *Rapid Commun. Mass Spectrom.*, **5**, 359.
163. Wu, K.J., Shaler, T.A. and Becker, C.H. (1994) *Anal. Chem.*, **66**, 1637.
164. Kirpekar, F., Nordhoff, E., Kristiansen, K. *et al.* (1994) *Nucleic Acids Res.*, **22**, 3688.
165. Wu, K.J., Stedding, A. and Becker, C.H. (1993) *Rapid Commun. Mass Spectrom.*, **7**, 142.
166. Fitzgerald, M.C., Zhu, L. and Smith, L.M. (1993) *Rapid Commun. Mass Spectrom.*, **7**, 895.
167. Fenn, J.B., Mann, M., Meng, C.K. and Wong, S.F. (1990) *Mass Spectrom. Rev.*, **9**, 37.
168. Limbach, P., Crain, P.F. and Mc Closkey, J.A. (1995) *J. Am. Soc. Mass Spectrom.*, **6**, 27.
169. Little, D.P., Chorus, R.A., Speir, J.P. *et al.* (1994) *J. Am. Chem. Soc.*, **116**, 4893.
170. Little, D.P., Thannhauser, T.W. and McLafferty, F.W. (1995) *Proc. Natl. Acad. Sci. USA*, **92**, 2318.
171. Ball, R.W. and Packman, L.C. (1998) Matrix-assisted laser desorption ionization time-of-flight mass spectrometry as a rapid quality control method in oligonucleotide synthesis. *Anal. Biochem.*, **246** (2), 185–94.
172. Van Ausdall, D.V. and Marshall, W.S. (1998) Automated high-throughput mass spectrometric analysis of synthetic oligonucleotides. *Anal. Biochem.*, **256** (2), 220–8.
173. Lui, Y.H., Bai, J., Lubman, D.M. and Venta, P.J. (1995) *Anal. Chem.*, **67**, 3482.
174. Doktycz, M.J., Hurst, G.B., Goudarzi, S.H. *et al.* (1995) *Anal. Biochem.*, **230**, 205.

175. Chang, L., Tang, K., Shell, M. *et al.* (1995) *Rapid Commun. Mass Spectrom.*, **9**, 772.
176. Pomerantz, S.C., Kowalak, J.A. and McCloskey, J.A. (1993) *J. Am. Soc. Mass Spectrom.*, **4**, 204.
177. Pieles, U., Zürcher, W., Shär, M. and Moser, H.E. (1993) *Nucleic Acids Res.*, **21**, 3191.
178. Limbach, P.A., McCloskey, J.A. and Crain, P.F. (1994) *Nucleic Acids Symp. Ser.*, **31**, 127.
179. Smirnov, I.P., Roskey, M.T., Juhasz, P. *et al.* (1996) Sequencing oligonucleotides by exonuclease digestion and delayed extraction matrix-assisted laser desorption ionization time-of-flight mass spectrometry. *Anal. Biochem.*, **238** (1), 19–25.
180. Keoug, T., Baker, T.R., Dobson, R.L.M. *et al.* (1993) *Rapid Commun. Mass Spectrom.*, **7**, 195.
181. Shaler, T.A., Tan, Y., Wickham, J.N. *et al.* (1995) *Rapid Commun. Mass Spectrom.*, **9**, 942.
182. Goodlett, D.R., Camp II, D.G., Hardin, C.C. *et al.* (1993) *Biol. Mass Spectrom.*, **22**, 181.
183. Pocsfalvi, G., Dilanda, G., Ferranti, P. *et al.* (1997) Observation of non-covalent interactions between beauvericin and oligonucleotides using electrospray ionization mass spectrometry. *Rapid Commun. Mass Spectrom.*, **11** (3), 265–72.
184. Cheng, X.H., Morin, P.E., Harms, A.C. *et al.* (1996) Mass spectrometric characterization of sequence-specific complexes of DNA and transcription factor PU.1 DNA binding domain. *Anal. Biochem.*, **239** (1), 35–40.
185. Triolo, A., Arcamone, F.M., Raffaeli, A. and Salvadori, P. (1997) Non-covalent complexes between DNA-binding drugs and double-stranded deoxyoligonucleotides: a study by ionspray mass spectrometry. *J. Mass Spectrom.*, **32** (11), 1186–94.
186. Bayer, E., Bauer, T., Schmeer, K. *et al.* (1994) *Anal. Chem.*, **66**, 3858.
187. Grotjahn, L. 1986 *Mass Spectrometry in Biomedical Research* (ed. S.J. Gaskell), John Wiley & Sons, Inc., New York, pp. 215–34.
188. Nordhoff, E., Karas, M., Cramer, R. *et al.* (1995) *J. Mass Spectrom.*, **30**, 99.
189. Juhasz, P., Roskey, M.T., Smirnov, I.P. *et al.* (1996) Applications of delayed extraction matrix-assisted laser desorption ionization time-of-flight mass spectrometry to oligonucleotide analysis. *Anal. Chem.*, **68** (6), 941–6.
190. Stemmler, E.A., Buchanan, M.V., Hurst, G.B. and Hettich, R.L. (1995) *Anal. Chem.*, **67**, 2924.
191. Little, D.P., Chorush, R.A., Spier, J.P. *et al.* (1994) *J. Am. Chem. Soc.*, **116**, 4893.
192. Boschenok, J. and Sheil, M.M. (1996) Electrospray tandem mass spectrometry of nucleotides. *Rapid Commun. Mass Spectrom.*, **10** (1), 144–9.
193. Little, D.P., Aaserud, D.J., Valaskoviic, G.A. and McLafferty, F.W. (1996) Sequence information from 42–108-mer DNAs (complete for a 50-mer) by tandem mass spectrometry. *J. Am. Chem. Soc.*, **118** (39), 9352–9.
194. McLuckey, S.A., Van Berkel, G.J. and Glish, G.L. (1992) *J. Am. Soc. Mass Spectrom.*, **3**, 60.
195. McLuckey, S.A. and Goudarzi, H.S. (1993) *J. Am. Chem. Soc.*, **115**, 12085.
196. Viari, A., Ballini, J.P., Vigny, P. *et al.* (1987) *Biomed. Environ. Mass Spectrom.*, **16**, 225.
197. Barry, J.P., Vouros, P., Schepdael, A.V. and Law, S.J. (1995) *J. Mass Spectrom.*, **30**, 993.
198. McLuckey, S.A. and Goudarzi, H.S. (1993) *J. Am. Chem. Soc.*, **115**, 12085.
199. Bartlett, M.G., McCloskey, J.A., Manalili, S. and Griiffey, R.H. (1996) The effect of backbone charge on the collision-induced dissociation of oligonucleotides. *J. Mass Spectrom.*, **31** (11), 1277–83.
200. Hettich, R.L. and Stemmler, E.A. (1996) Investigation of oligonucleotide fragmentation with matrix-assisted laser desorption ionization Fourier-transform mass spectrometry and sustained off-resonance irradiation. *Rapid Commun. Mass Spectrom.*, **10** (3), 321–7.

201. Ni, J.S., Pomerantz, S.C., Rozenski, J. *et al.* (1996) Interpretation of oligonucleotide mass spectra for determination of sequence using electrospray ionization and tandem mass spectrometry. *Anal. Chem.*, **68** (13), 1989–99.

202. Rozenski, J., Crain, P.F. and McCloskey, J.A. (1999) The RNA modification database: 1999 update. *Nucleic Acids Res.*, **27** (1), 196–7.

203. Chiarelli, M.P. and Lay, O.J. (1992) *Mass Spectrom. Rev.*, **11**, 447.

204. McCloskey, J.A. and Crain, P.F. (1992) *Int. J. Mass Spectrom. Ion Processes*, **118/119**, 593.

205. Crain, P.F. and McCloskey, J.A. (1994) *Biological Mass Spectrometry* (eds T. Matsuo, R.M. Caprioli and M.L. Gross), John Wiley & Sons, Inc., New York, pp. 509–37.

206. Biemann, K. and McCloskey, J.A. (1962) *J. Am. Chem. Soc.*, **84**, 2005.

207. Wilson, M.S. and McCloskey, J.A. (1975) *J. Am. Chem. Soc.*, **97**, 3436.

208. Crow, F.W., Tomer, K.B., Gross, M.L. *et al.* (1984) *Anal. Biochem.*, **139**, 243.

209. Chaudhary, A.K., Nokubo, M., Oglesby, T.D. *et al.* (1995) *J. Mass Spectrom.*, **30**, 1157.

210. Nelson, C.C. and McCloskey, J.A. (1989) *Adv. Mass Spectrom.*, **11A**, 260.

211. Dizdaroglu, M. (1990) *Methods Enzymol.*, **193**, 842.

212. Crain, P.F., Hashizume, T., Nelson, C.C. *et al.* (1990) *Biological Mass Spectrometry* (eds A.L. Burlingame and J.A. McCloskey), Elsevier, Amsterdam, pp. 509–25.

213. Pomerantz, S.C. and McCloskey, J.A. (1990) *Methods Enzymol.*, **193**, 796.

214. Apruzzese, W.A. and Vouros, P. (1998) Analysis of DNA adducts by capillary methods coupled to mass spectrometry: a perspective. *J. Chromatogr.*, **794** (1–2), 97–108.

215. Esmans, E.L., Broes, D., Hoes, I. *et al.* (1998) Liquid chromatography mass spectrometry in nucleoside, nucleotide and modified nucleotide characterization. *J. Chromatogr.*, **794** (1–2), 109–27.

216. Kowalak, J.A., Pomerantz, S.C., Crain, P.F. and McCloskey, J.A. (1993) *Nucleic Acids Res.*, **21**, 4577.

217. Iden, C.R. and Rieger, R.A. (1989) *Biomed. Environ. Mass Spectrom.*, **18**, 617.

218. Reinhold, V.N., Reinhold, B.B. and Chan, S. (1994) *Biological Mass Spectrometry Present and Future* (eds T. Matsuo, R.M. Caprioli, M.L. Gross and Y. Seyama), John Wiley & Sons, Inc., New York, pp. 403–62.

219. Reinhold, V.N., Reinhold, B.B. and Costello, C.E. (1995) *Anal. Chem.*, **67**, 1772.

220. Harvey, D.J., Naven, T.J.P. and Kuster, B. (1996) Identification of oligosaccharides by matrix-assisted laser desorption ionization and electrospray MS. *Biochem. Soc. Trans.*, **24** (3), 905–12.

221. Merkle, R.K. and Poppe, I. (1994) *Methods in Enzymology*, vol. **230** (eds J.K. Lennarz and G.W. Hart), Academic Press, New York, pp. 1–15.

222. Hellerqvist, C.G. (1990) *Methods in Enzymology*, vol. **193** (ed. J.A. McCloskey), Academic Press, New York, pp. 554–73.

223. Bahr, U., Pfenninger, A., Karas, M. and Stahl, B. (1997) High sensitivity analysis of neutral underivatized oligosaccharides by nanoelectrospray mass spectrometry. *Anal. Chem.*, **69** (22), 4530–5.

224. Dell, A. (1990) *Methods in Enzymology*, vol. **193** (ed. J.A. McCloskey), Academic Press, New York, pp. 647–60.

225. Gillece-Castro, B.L. and Burlingame, A.L. (1990) *Biological Mass Spectrometry* (eds A.L. Burlingame and J.A. McCloskey), Elsevier, Amsterdam, pp. 411–36.

226. Garozzo, D., Giuffrida, M., Impallomeni, G. *et al.* (1990) *Anal. Chem.*, **62**, 279.

227. Carr, S.A., Reinhold, V.N., Green, B.N. and Hass, J.R. (1985) *Biomed. Mass Spectrom.*, **12**, 288.

228. Bosso, C., Heyraud, A. and Patron, L. (1991) *Org. Mass Spectrom.*, **26**, 321.

229. Domon, B., Muller, D.R. and Richter, W.J. (1989) *Org. Mass Spectrom.*, **24**, 357.

230. Orlando, R., Bush, A. and Fenselau, C. (1990) *Biomed. Environ. Mass Spectrom.*, **19**, 747.

231. Duffin, K.L., Welply, J.K., Huang, E. and Henion, J.D. (1992) *Anal. Chem.*, **64**, 1440.
232. Phillips, N.J., Apicella, M.A., Griffiss, J.M. and Gibson, B.W. (1993) *Biochemistry*, **32**, 2003.
233. Linsley, K., Chan, S.-Y., Chan, S. *et al.* (1994) *Anal. Biochem.*, **219**, 207.
234. Reinhold, B.B., Chan, S.-Y., Chan *et al.* (1995) *Org. Mass Spectrom.*, **29**, 736.
235. Spengler, B., Kirch, D., Kaufmann, R. and Lemoine, J. (1994) *Org. Mass Spectrom.*, **29**, 782.
236. Perreault, H. and Costello, C.E. (1994) *Org. Mass Spectrom.*, **29**, 720.
237. Cancilla, M.T., Penn, S.G., Caroll, J.A. and Lebrilla, C.B. (1996) Coordination of alkali metals to oligosaccharides dictates fragmentation behavior in matrix assisted laser desorption ionization Fourier transform mass spectrometry. *J. Am. Chem. Soc.*, **118** (28), 6736–45.
238. Domon, B. and Costello, C.E. (1988) *Glycoconj. J.*, **5**, 397.
239. Lemoine, J., Fournet, B., Despeyroux, D. *et al.* (1993) *J. Am. Soc. Mass Spectrom.*, **4**, 197.
240. Domon, B., Muller, D.R. and Richter, W.J. (1994) *Org. Mass Spectrom.*, **29**, 713.
241. Lemoine, J., Chirat, F. and Domon, B. (1996) Structural analysis of derivatized oligosaccharides using post-source decay matrix-assisted laser desorption ionization mass spectrometry. *J. Mass Spectrom.*, **31** (8), 908–12.
242. Domon, B., Muller, D.R. and Richter, W.J. (1990) *Biomed. Environ. Mass Spectrom.*, **19**, 390.
243. Viseux, N., de Hoffmann, E. and Domon, B. (1997) Structural analysis of permethylated oligosaccharides by electrospray tandem mass spectrometry. *Anal. Chem.*, **69** (16), 3193–8.
244. Solouki, T., Reinhold, B.B., Costello, C.E. *et al.* (1998) Electrospray ionization and matrix-assisted laser desorption/ionization Fourier transform ion cyclotron resonance mass spectrometry of permethylated oligosaccharides. *Anal. Chem.*, **70** (5), 857–64.
245. Sheeley, D.M. and Reinhold, V.N. (1998) Structural characterization of carbohydrate sequence, linkage, and branching in a quadrupole ion trap mass spectrometer: neutral oligosaccharides and N-linked glycans. *Anal. Chem.*, **70** (14), 3053–9.
246. Angel, A.S. and Nilsson, B. (1990) *Methods in Enzymology*, vol. **193** (ed. J.A. McCloskey), Academic Press, New York, pp. 721–30.
247. Cancilla, M.T., Penn, S.G. and Lebrilla, C.B. (1998) Alkaline degradation of oligosaccharides coupled with matrix-assisted laser desorption/ionization Fourier transform mass spectrometry: a method for sequencing oligosaccharides. *Anal. Chem.*, **70** (4), 663–72.
248. Khoo, K.H. and Dell, A. (1990) *Glycobiology*, **1**, 83.
249. Sutton, C.W., O'Neil, J.A. and Cottrell, J.S. (1994) *Anal. Biochem.*, **218**, 34.
250. Kuster, B., Naven, T.J.P. and Harvey, D.J. (1996) Rapid approach for sequencing neutral oligosaccharides by exoglycosidase digestion and matrix-assisted laser desorption ionization time-of-flight mass spectrometry. *J. Mass Spectrom.*, **31** (10), 1131–40.
251. Harvey, D.J., Rudd, P.M., Bateman, R.H. *et al.* (1994) *Org. Mass Spectrom.*, **29**, 753.
252. Viseux, N., de Hoffmann, E. and Domon, B. (1996) An integrated methodology for structural analysis of glycans based on methylation and tandem mass. *Glycobiology*, **6** (7), 121.
253. Jensen, N.J. and Gross, M.L. (1988) *Mass Spectrom. Rev.*, **7**, 41.
254. Gage, D.A., Huang, Z.H. and Sweeley, C.C. (1994) *Mass Spectrometry Clinical and Biomedical Applications* (ed. M.D. Desiderato), Plenum Press, New York, pp. 53–87.
255. Kim, H.Y., Wang, T.C.L. and Ma, Y.C. (1994) *Anal. Chem.*, **66**, 3977.
256. Murphy, R.C. and Harrson, K.A. (1994) *Mass Spectrom. Rev.*, **13**, 57.
257. Tomer, K.B. and Gross, M.L. (1989) *Biomed. Environ. Mass Spectrom.*, **15**, 89.

258. Tabet, J.C. (1994) *Applications of Mass Spectrometry to Organic Stereochemistry* (eds J.S. Splitter and F. Turecek), VCH, New York, pp. 543–91.

259. Bean, K.A. and Henion, J.B. (1997) Direct determination of anabolic steroid conjugates in human urine by combined high-performance liquid chromatography and tandem mass spectrometry. *J. Chromatogr. B*, **690** (1–2), 65–75.

260. Zirrolli, J.A., Davoli, E., Bettazzoli, L. *et al.* (1990) *J. Am. Soc. Mass Spectrom.*, **1**, 325.

261. Ann, Q. and Adams, J. (1993) *Anal. Chem.*, **65**, 7.

262. Gu, M., Kerwin, J.L. and Watts, J.D. (1997) Ceramide profiling of complex lipid mixtures by electrospray ionization mass spectrometry. *Anal. Biochem.*, **244** (2), 347–56.

263. Adams, J. and Ann, Q. (1993) *Mass Spectrom. Rev.*, **12**, 51.

264. Mano, N., Oda, Y. and Yamada, K. (1997) Simultaneous quantitative determination method for sphingolipid metabolites by liquid chromatography/ionspray ionization tandem mass spectrometry. *Anal. Biochem.*, **244** (2), 291–300.

265. Wheelan, P., Zirrolli, J.A. and Murphy, R.C. (1996) Electrospray ionization and low energy tandem mass spectrometry of polyhydroxy unsaturated fatty acids. *J. Am. Soc. Mass Spectrom.*, **7** (2), 140–9.

266. Murphy, R.C. (1995) *J. Mass Spectrom.*, **30**, 5.

267. Murphy, R.C. (1993) *Mass Spectrometry of Lipids, Handbook of Lipid Research*, vol. **7**, Plenum Press, New York.

268. Jensen, N.J. and Gross, M.L. (1987) *Mass Spectrom. Rev.*, **6**, 497.

269. Kuksis, A. and Myher, J.J. (1995) *J. Chromatogr.*, **671**, 35.

270. Tomer, K.B., Jensen, N.J. and Gross, M.L. (1986) *Anal. Chem.*, **58**, 2429.

271. Kerwin, J.L. and Torvik, J.J. (1996) Identification of monohydroxy fatty acids by electrospray mass spectrometry and tandem mass spectrometry. *Anal. Biochem.*, **237** (1), 56–64.

272. Jensen, N.J., Tomer, K.B. and Gross, M.L. (1985) *J. Am. Chem. Soc.*, **105**, 5487.

273. Bambagiotti, M.A., Coran, S.A., Vincieri, F.F. *et al.* (1986) *Org. Mass Spectrom.*, **21**, 485.

274. Kerwin, J.L., Wiens, A.M. and Ericsson, L.H. (1996) Identification of fatty acids by electrospray mass spectrometry and tandem mass spectrometry. *J. Mass Spectrom.*, **31**, 184.

275. Adams, J. (1990) *Mass Spectrom. Rev.*, **9**, 141.

276. Cordero, M.M. and Wesdemiotis, C. (1994) *Anal. Chem.*, **66**, 861.

277. Nizigiyimana, L., Van Den Heuvel, H. and Claeys, M. (1997) Comparison of low- and high-energy collision-induced dissociation tandem mass spectrometry in the analysis of ricinoleic and ricinelaidic acid. *J. Mass Spectrom.*, **32** (3), 277–86.

278. Cheng, C.F., Pittenauer, E. and Gross, M.L. (1998) Charge-remote fragmentations are energy-dependent processes. *J. Am. Soc. Mass Spectrom.*, **9** (8), 840–4.

279. Jensen, N.J., Tomer, K.B. and Gross, M.L. (1985) *J. Am. Chem. Soc.*, **107**, 1863.

280. Jensen, N.J., Tomer, K.B. and Gross, M.L. (1985) *Anal. Chem.*, **57**, 2018.

281. Jensen, N.J. and Gross, M.L. (1986) *Lipids*, **21**, 362.

282. Tomer, K.B., Crow, F.W. and Gross, M.L. (1983) *J. Am. Chem. Soc.*, **105**, 5487.

283. Tomer, K.B., Jensen, N.J. and Gross, M.L. (1987) *Anal. Chem.*, **59**, 1576.

284. Harvey, D.J. (1982) *Biomed. Mass Spectrom.*, **9**, 33.

285. Dobson, G. and Christie, W.W. (1996) Structural analysis of fatty acids by mass spectrometry of picolinyl esters and dimethyloxazoline derivatives. *TrAC Trends Anal. Chem.*, **15** (3), 130–7.

286. Zirrolli, J.A. and Murphy, R.C. (1993) *J. Am. Soc. Mass Spectrom.*, **4**, 223.

287. Ruizgutierrez, V. and Barron, L.J.R. (1995) *J. Chromatogr.*, **671**, 113.

288. Laakso, P. (1996) Analysis of triacylglycerols – approaching the molecular composition of natural mixtures. *Food Rev. Int.*, **12** (2), 199–250.

289. Hites, R.A. (1970) *Anal. Chem.*, **42**, 1736.
290. Murata, T. and Takahashi, S. (1977) *Anal. Chem.*, **49**, 728.
291. Schulte, E., Hohn, M. and Rapp, U. (1981) *Fresenius Z. Anal. Chem.*, **307**, 115.
292. Showell, J.S., Fales, H.M. and Sokoloski, E.A. (1989) *Org. Mass Spectrom.*, **24**, 632.
293. Evans, C., Traldi, P., Bambagiotti-Alberti, M. *et al.* (1991) *Biol. Mass Spectrom.*, **20**, 351.
294. Neff, W.E. and Byrdwell, W.C. (1995) *J. Am. Oil Chem. Soc.*, **72**, 5487.
295. Duffin, K.L., Henion, J.D. and Shieh, J.J. (1991) *Anal. Chem.*, **63**, 1781.
296. Anderson, M.A., Collier, L., Dilliplane, R. and Ayorinde, F.O. (1993) *J. Am. Oil Chem. Soc.*, **70**, 905.
297. Bambagiotti, M.A., Coran, S.A., Vincieri, F.F. *et al.* (1986) *Org. Mass Spectrom.*, **21**, 485.
298. Keusgen, M., Curtis, J.M. and Ayer, S.W. (1996) The use of nicotinates and sulfoquinovosyl monoacylglycerols in the analysis of monounsaturated n-3 fatty acids by mass spectrometry. *Lipids*, **31** (2), 231–8.
299. Stroobant, V., Rozenberg, R., Bouabsa, E.M. *et al.* (1995) Fragmentation of conjugate bases of esters derived from multifunctional alcohols including triacylglycerols. *J. Am. Soc. Mass Spectrom.*, **6**, 498–506.
300. Ballatore, A.M., Beckner, C.F., Caprioli, R.M. *et al.* (1983) *Steroids*, **41**, 197.
301. Ito, Y., Takeuchi, T., Ishii, D. *et al.* (1986) *J. Chromatogr.*, **358**, 201.
302. Roda, A., Giacchini, A.M. and Baraldini, M. (1995) *J. Chromatogr.*, **665**, 281.
303. Evans, J.E., Ghosh, A., Evans, B.A. and Natowicz, M.R. (1993) *Biol. Mass Spectrom.*, **22**, 331.
304. Scalia, S. and Games, D.E. (1992) *Org. Mass Spectrom.*, **27**, 1266.
305. Warrack, B.M. and Didonato, G.C. (1993) *Biol. Mass Spectrom.*, **22**, 101.
306. Lier, J.G., Kingston, E.E. and Beynon, J.H. (1985) *Biomed. Mass Spectrom.*, **12**, 95.
307. Tomer, K.B., Jensen, N.J. and Gross, M.L. (1986) *Biomed. Mass Spectrom.*, **13**, 265.
308. Griffiths, W.J., Egestad, B. and Sjovall, J. (1993) *Rapid Commun. Mass Spectrom.*, **7**, 235.
309. Eckers, C., New, A.P., East, P.B. and Haskins, N.J. (1990) *Rapid Commun. Mass Spectrom.*, **4**, 449.
310. Stroobant, V., Libert, R., Van Hoof, F. and de Hoffmann, E. (1995) Fast-atom bombardment mass spectrometry and low energy collision-induced tandem mass spectrometry of tauro conjugated bile acid anions. *J. Am. Soc. Mass Spectrom.*, **6**, 588–96.
311. Libert, R., Hermans, D., Draye, J.P. *et al.* (1991) Bile acids and conjugates identified in metabolic disorders by fast atom bombardment and tandem mass spectrometry. *Clin. Chem.*, **37**, 2102–10.
312. Beecher, C. (2003) *Metabolic Profiling: Its Role in Biomarker Discovery and Gene Function Analysis* (eds G.G. Harrigan and R. Goodacre), Kluwer Academic, Dordrecht, pp. 345–52.
313. Villas-Boas, S.G., Mas, S., Akesson, M. *et al.* (2005) Mass spectrometry in metabolome analysis. *Mass Spectrom. Rev.*, **24**, 613–46.
314. Brown, S.C., Kruppa, G. and Dasseux, J.-L. (2005) Metabolomics applications of FT-ICR mass spectrometry. *Mass Spectrom. Rev.*, **24**, 223–31.
315. Hughey, C.A., Rodgers, R.P. and Marshall, A.G. (2002) Resolution of 11,000 compositionally distinct components in a single electrospray ionization Fourier transform ion cyclotron resonance mass spectrum of crude oil. *Anal. Chem.*, **74**, 4145–9.
316. Saghatelian, A., Trauger, S.A., Want, E.J. *et al.* (2004) Assignment of endogenous substrates to enzymes by global metabolite profiling. *Biochemistry*, **43**, 14332–9.

9

Exercises

Questions

9.1. Calculate the isotopic distribution of:

   (a) $CH_2 = CHCl$
   (b) CH_2BrCl

9.2. The molecular ion of a product occurs at 59.9670 Th. Identify the product and calculate the relative abundances of the ions with a nominal mass of 62 Th or with the exact mass of 61.9620 Th.

9.3. Which resolution is required to distinguish the molecular ions of:

   (1)   CO and C_2H_4
   (2) $C_{10}H_{21}CHO$ and $C_{12}H_{26}$
   (3)   The monoisotopic peaks of Question 9.1?

9.4. Spectrum A in Figure 9.1 is the EI mass spectrum of S_8. Spectra B and C are the MS/MS fragmentation spectra of S_8 for the 256 Th and 258 Th precursors, respectively. Explain the observed isotope ratios.

9.5. The electrospray negative ion spectrum of an oligonucleotide displays the following ions: 2903.8 (40 %), 2580.9 (67 %), 2323.7 (100 %), 2111.4 (71 %) and 1935.7 (38 %). What is the molecular weight of this oligonucleotide derivative?

9.6. The specific interaction of Cu ions with a 26-residue peptide present at the surface of a human protein has been studied by mass spectrometry [1]. The positive ion electrospray spectra A and B (Figure 9.2) have been obtained, respectively, before and after addition of Cu(II).

   What is the molecular weight of this peptide?
   What is the maximum number of Cu atoms included in the peptide?
   Which is the oxidation stage of the Cu atoms?

9.7. How can you distinguish between the spectrum of n-butyraldehyde and that of *iso*-butyraldehyde?

9.8. The spectra displayed in Figure 9.3 originate from isomeric ketones analysed by electron ionization. They comprise only linear saturated chains. Give the developed formula of each of these ketones. Justify your answer by interpreting at least three fragments, one of them resulting from a rearrangement.

Mass Spectrometry: Principles and Applications, Third Edition Edmond de Hoffmann and Vincent Stroobant
© Copyright 2007, John Wiley & Sons Ltd

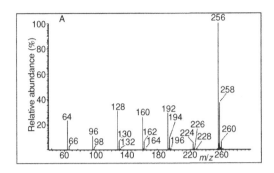

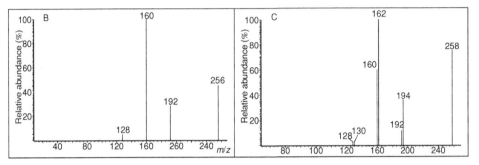

Figure 9.1
The EI mass spectrum (A) and MS/MS spectra (B, C) of S_8 for the 256 Th (B) and 258 Th (C) precursors.

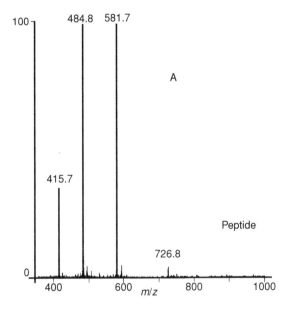

Figure 9.2
Positive ion electrospray spectra of a 26-residue peptide: (A) free peptide ; (B) in the presence of Cu(II) ions.

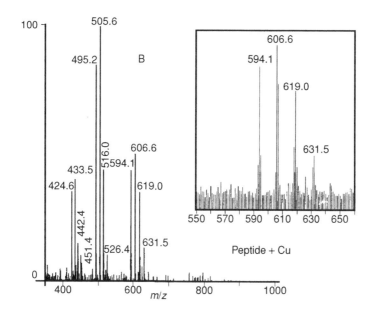

Figure 9.2
(*continued*)

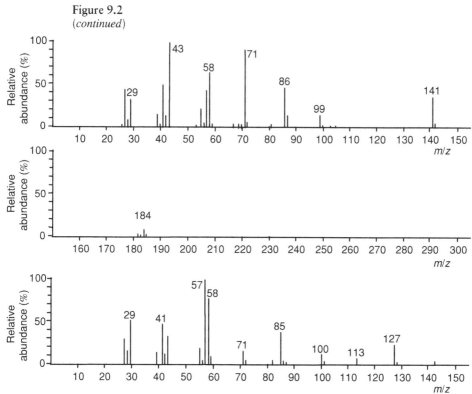

Figure 9.3
Spectra of three isomass ketones of molecular weight 184 Da.

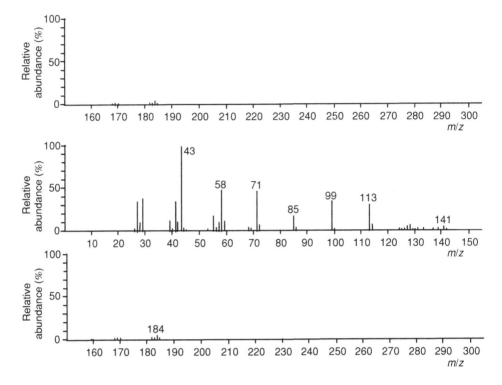

Figure 9.3
(*continued*)

9.9. A compound has a molecular peak at 115 Th. Where is the peak corresponding to the metastable loss of CH_3 in the spectrum obtained with a magnetic instrument using an EB configuration?

9.10. Explain the following: 'MIKE spectroscopy' and 'B/E linked scan'.

9.11. Interpret the main peaks in the following spectra:

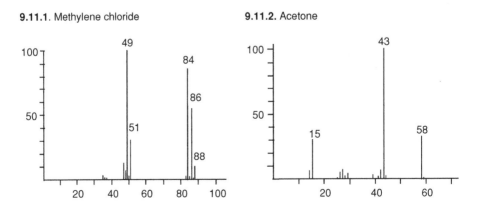

9.11.1. Methylene chloride

9.11.2. Acetone

9.11.3. Ethylamine

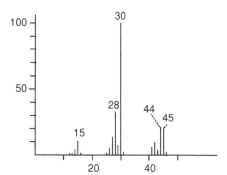

9.11.4. 1-Butene

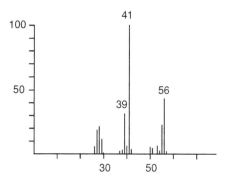

9.11.5. Ethylbenzene

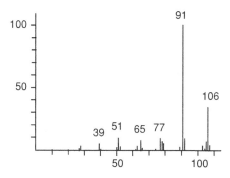

9.11.6. Tetralin

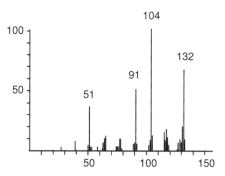

9.11.7. *n*-Butyraldehyde

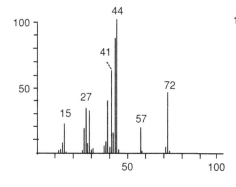

9.11.8. Cyclohexanol

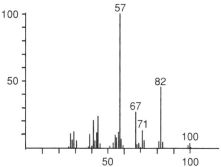

9.11.9. 2-Dodecanone

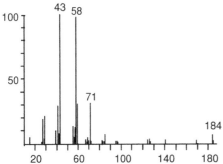

9.11.10. *n*-Dodecane

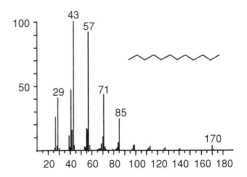

9.11.11. 1-Phenyl-*n*-hexane

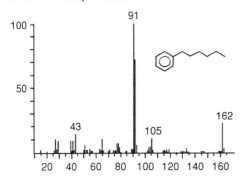

9.11.12. Cyclohexane

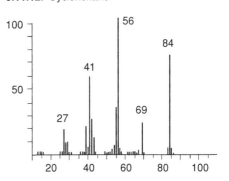

9.12. Identify the product and interpret the main peaks in the following:

9.12.1.

m/z	Relative abundance (%)
14	3.4
15	20
16	0.2
19	2
31	10
32	9.8
33	94
34	100
35	1.1

9.12.2.

m/z	Relative abundance (%)
12	3.2
13	4.4
14	4.6
16	1.5
28	32
29	100
30	88
31	1.2

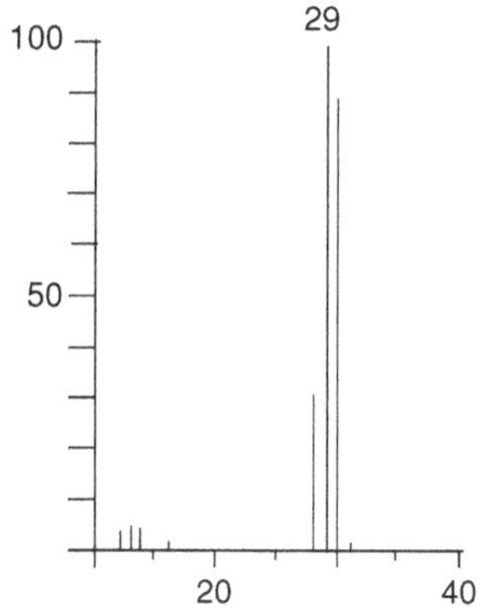

9.12.3.

m/z	Relative abundance (%)
15	15.3
27	37
28	32
29	44
39	12
41	27
42	12
43	100
44	3.2
57	2.3
58	12
59	0.5

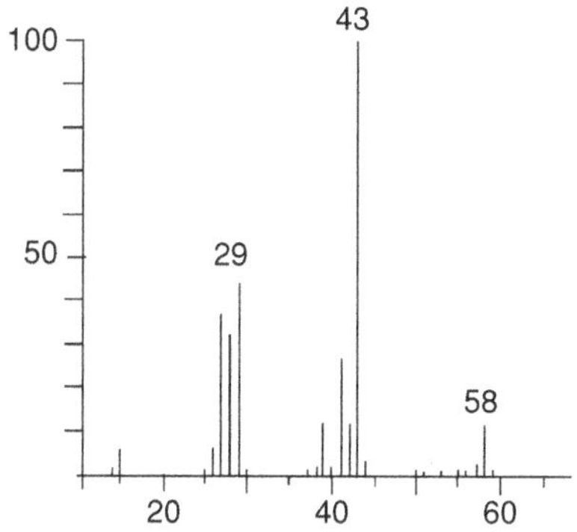

9.12.4.

m/z	Relative abundance (%)
26	38
27	74
28	12
29	4.4
44	14
45	32
55	74
56	2.5
57	0.2
71	4.2
72	100
73	3.5
74	0.5

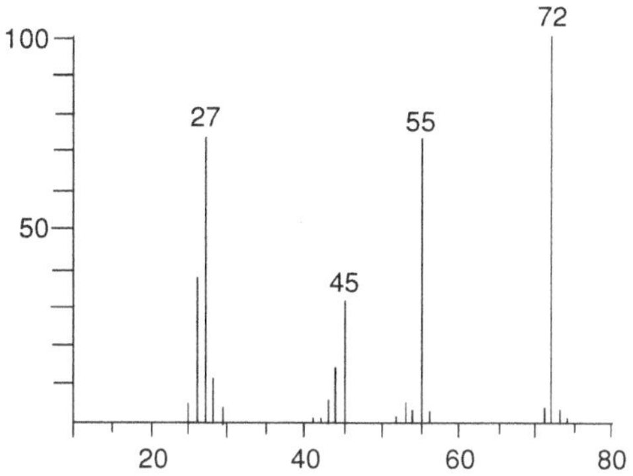

9.12.5.

m/z	Relative abundance (%)
14	5.1
19	8.3
33	36
52	100
53	0.5
71	30

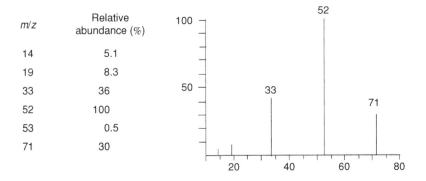

9.12.6.

m/z	Relative abundance (%)
19	0.2
31	1.7
35	3.1
37	1.2
50	6.4
69	100
70	1.2
85	18
86	0.2
87	5.8
104	0.7
106	0.2

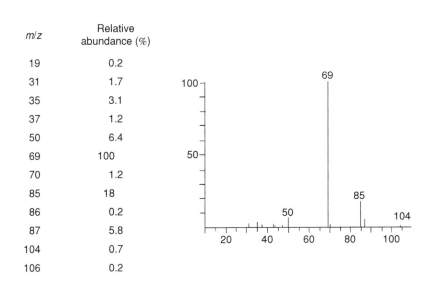

9.13. Identify the product that yields the following mass spectrum [2]:

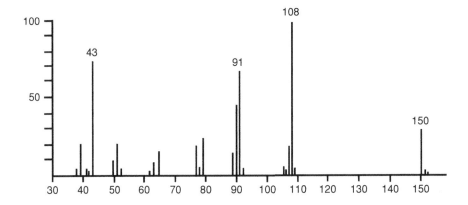

In IR, the main peaks appear at 3058, 2941, 1745, 1385, 1225, 1026, 749 and 697 cm^{-1}.
In ^{1}H-NMR:

δ	Intensity	Multiplicity
7.22	5	Singlet
5.00	2	Singlet
1.96	3	Singlet

9.14. Identify the product that yields the following mass spectrum [2]:

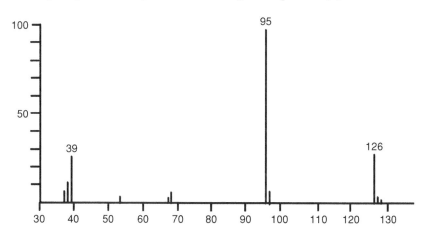

In IR, the main peaks appear at 3106, 2941, 1730, 1587, 1479, 1449, 1393, 1299, 1205, 1121 and 758 cm^{-1}.
In ^{1}H-NMR:

δ	Intensity	Multiplicity
7.51	1	Quadruplet
7.02	1	Quadruplet
6.45	1	Quadruplet
3.80	3	Singlet

In off-resonance decoupled ^{13}C-NMR:

δ	Multiplicity
160	Singlet
146	Doublet
144	Singlet
118	Doublet
112	Doublet
51	Quadruplet

9.15. Identify the product that yields the following mass spectrum [2]:

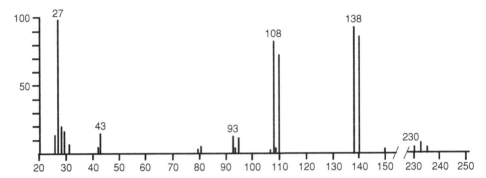

The IR spectrum shows the following main peaks: 2970, 2850, 1425, 1360, 1279, 1120 and 630 cm^{-1}.

In ^{1}H-NMR:

δ	Intensity	Multiplicity
3.40	1	Multiplet
3.88	1	Multiplet

9.16. Identify the product that yields the following mass spectrum [2]:

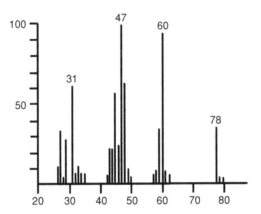

In IR, the main peaks appear at 3367, 3030, 2558, 1429, 1298, 1050 and 1020 cm^{-1}.

In off-resonance decoupled ^{13}C-NMR:

δ	Multiplicity
28	Triplet
64	Triplet

9.17. A 429 Da peptide gives the MS/MS product ion spectrum from its protonated molecular ion displayed in Figure 9.4. What is the sequence of this peptide?

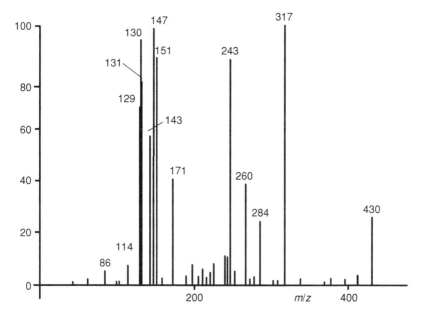

Figure 9.4
Fragment ion spectrum of a protonated peptide.

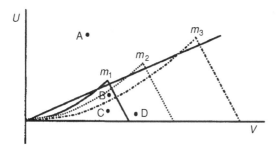

Figure 9.5
Stability diagram for three ions of mass m_1, m_2 and m_3 Th in U, V coordinates.

9.18. Consider Figure 9.5, representing the stability diagram for three ions of masses m_1, m_2 and m_3 Th in U, V coordinates. Explain which ions will be observed if the U, V values are adjusted successively to the values indicated by the points A–D in the figure.

9.19. A quadrupole with $r_0 = 1$ cm operating at an RF frequency of 0.5 MHz has a stability diagram with an apex at $a = 0.22$ and $q = 0.72$ for an ion of mass 2000 u. What are the corresponding U and V values?

9.20. The same quadrupole as in Question 9.19 operating in the 'RF-only' mode has values of $a = 0$ and $q = 0.91$. What is the minimum V_{rf} voltage allowing all the ions of less than 100 Th to be eliminated?

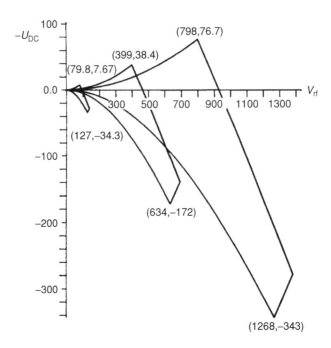

Figure 9.6
Stability area for ions simultaneously stable along r and z.

9.21. If a V_{rf} of 2 kV is applied to the same quadrupole as in Question 9.20, from which m/z value will the ions be unstable?

9.22. In an ideal ion trap, that is $r_0^2 = 2z_0^2$, with $r_0 = 1$ cm and $v = 1.1$ MHz, referring to the values for U and V given in Figure 9.6, calculate the a and q values at the apex of the stability diagram.

9.23. In an ideal ion trap, that is $r_0^2 = 2z_0^2$, with $r_0 = 1$ cm and $v = 1.6$ MHz, $U = 4$ V and $V = 400$ V for a nitrogen ion $N_2^{•+}$, calculate the corresponding a and q values.

9.24. In an ideal ion trap, that is $r_0^2 = 2z_0^2$, with $r_0 = 1$ cm and $v = 1.1$ MHz, $U = 4$ V and $V = 500$ V, what is the resonant frequency that should be applied to expel an ion of 100 Th?

9.25. A time-of-flight (TOF) analyser has the following characteristics: $V_s = 3$ kV, $V_R = 3130$ V, $D = 0.522$ m, $L_1 = 1$ m.

    (a) At which distance L_2 does it focalize ions having the same m/z $(m_p) = 578$ Th?
    (b) What is the focal distance for PSD ions with m/z $(m_f) = 560$ Th?
    (c) At which value should V_R be reduced in order to focalize the fragment ions at the same distance as the precursors?

Answers

9.1. Calculate the isotopic distribution of:

(a) $CH_2 = CHCl$

Answer:

	m/z nominal	m/z exact	Relative abundance (%)
$^{12}C_2H_3{}^{35}Cl$	62	61.992 32	74.11
$^{12}C^{13}CH_3{}^{35}Cl$	63	62.995 68	1.65
$^{12}C_2H_3{}^{37}Cl$	64	63.988 62	23.7
$^{13}C_2H_3{}^{35}Cl$	64	63.999 03	0.009
$^{13}C^{12}CH_3{}^{37}Cl$	65	64.991 98	0.53
$^{13}C_2H_3{}^{37}Cl$	66	65.995 33	0.003

(b) CH_2BrCl

Answer:

	m/z nominal	m/z exact	Relative abundance (%)
$^{12}CH_2{}^{79}Br^{35}Cl$	128	127.902 83	37.99
$^{13}CH_2{}^{79}Br^{35}Cl$	129	128.906 18	0.42
$^{12}CH_2{}^{79}Br^{37}Cl$	130	129.899 13	12.15
$^{12}CH_2{}^{81}Br^{35}Cl$	130	129.900 80	36.95
$^{13}CH_2{}^{79}Br^{37}Cl$	131	130.902 48	0.14
$^{13}CH_2{}^{81}Br^{35}Cl$	131	130.904 15	0.41
$^{12}CH_2{}^{81}Br^{37}Cl$	132	131.897 10	11.82
$^{13}CH_2{}^{81}Br^{37}Cl$	133	132.900 46	0.13

9.2. The molecular ion of a product occurs at 59.9670 Th. Identify the product and calculate the relative abundances of the ions with a nominal mass of 62 Th or with the exact mass of 61.9620 Th.

Answer:

It is COS. The ion at 61.9620 Th is due to $CO^{34}S$ and has a relative abundance of 4.44 %. The ion with a nominal mass of 62 Th results from different possible combinations of isotopes and has a relative abundance of 4.74 %. The combinations that contribute in a significant way to nominal mass 62 Th are $CO^{34}S$, $C^{18}OS$ and $^{13}CO^{33}S$.

9.3. Which resolution is required to distinguish the molecular ions of:

(1) CO and C_2H_4
(2) $C_{10}H_{21}CHO$ and $C_{12}H_{26}$
(3) The monoisotopic peaks of Question 9.1?

Answer:

(1) 770
(2) 4800
(3) 6200 (a) and 78 000 (b)

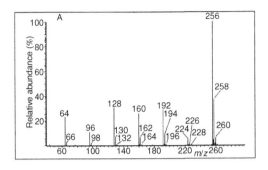

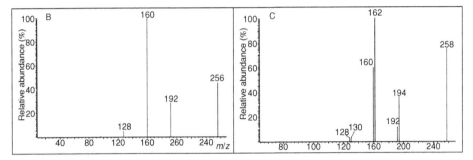

Figure 9.1

9.4. Spectrum A in Figure 9.1 is the EI mass spectrum of S_8. Spectra B and C are the MS/MS fragmentation spectra of S_8 for the 256 Th and 258 Th precursors, respectively. Explain the observed isotope ratios.

Answer:

The two main isotopes of sulfur are ^{32}S and ^{34}S. The latter is present with a natural abundance of 4.44 %, taking ^{32}S as 100 %. At m/z 256, the formula is S_8. The ion at 258 Th should have a relative intensity of $8 \times 0.0444 = 0.355$ or 35.5 %, and the ion at 260 Th corresponds to the probability of having two ^{34}S atoms together, thus $(8 \times 0.044)(7 \times 0.044) = 0.3552 \times 0.3108 = 0.1104$ or 11.04 %. The peak at 192 Th has the formula S_6. The 194 Th fragment will have an abundance of $6 \times 0.044 = 0.266$ or 26.6 % of the abundance at m/z 192. The 196 Th fragment will have an abundance corresponding to one out of six and one out of the remaining five being ^{34}S. This is thus $(6 \times 0.0444)(5 \times 0.044) = 0.2664 \times 0.222 = 0.059$ or 5.9 % of the abundance at m/z 192.

Spectrum B is the MS/MS product ion spectrum of 256 Th, thus isotopically pure $^{32}S_8$. Its fragments contain only ^{32}S and there are no isotope peaks.

Spectrum C is the MS/MS product ion spectrum of 258 Th, thus having the formula $^{34}S^{32}S_7$. One out of eight sulfur atoms is now ^{34}S. The fragment at m/z 160–162 has the formula S_5 and results from the loss of three sulfur atoms. The probability is thus 5/8 that the ^{34}S is saved, producing the m/z 162 fragment, and 3/8 that it is lost, which leads to the 160 Th fragment. The ratio of the relative abundances '160/162' is thus $(3/8)/(5/8) = 3/5$ or 60 %. At 128 and 130 Th, the formula is S_4 and thus the probability is equal for holding or losing the ^{34}S, leading to two peaks of equivalent abundance.

9.5. The electrospray negative ion spectrum of an oligonucleotide displays the following ions: 2903.8 (40%), 2580.9 (67%), 2323.7 (100%), 2111.4 (71100%) and 1935.7 (38100%). What is the molecular weight of this oligonucleotide derivative?

Answer:

Using the formula for the multiply charged negative ions, $z_1 = j(m_2 + m_p)/(m_2 - m_1)$, where m_p is the proton mass and for the extreme ions $j = 4$, one obtains $z_1 = 12$. The formula $M = z_i(m_i + m_p)$ yields $M = 23\,240.5, 23\,236.5, 23\,247.1, 23\,237.2$ and $23\,238.5$ Da for the five ions. The average value is $M = 23\,240 \pm 3$ Da.

9.6. The specific interaction of Cu ions with a 26-residue peptide present at the surface of a human protein has been studied by mass spectrometry [1]. The positive ion electrospray spectra obtained before and after addition of Cu(II) are displayed in Figure 9.2.

What is the molecular weight of this peptide?
What is the maximum number of Cu atoms included in the peptide?
Which is the oxidation stage of the Cu atoms?

Answer:

The molecular weight of the peptide is deduced from its electrospray spectrum (Figure 9.2A). If we suppose that the observed peaks in this spectrum correspond to different charge state of the peptide, than we can calculate the number of charges z_1 for the peak detected at m_1: $z_1 = j(m_2 - 1)/(m_2 - m_1)$ where j corresponds to the number of peaks +1 separating m_1 and m_2; z_1 has to be rounded to the nearest integer. The molecular weight is then $M = z_1(m_1 - 1)$.

For spectrum A, applying the formula with $m_1 = 484.8$, $m_2 = 581.7$ and $j = 1$, one obtains a charge state of +6 for the number of charges on the ions at m/z 484. From this charge state and the m/z value, one obtains 2902.8 Da for the molecular weight.

Table 9.1 gives the molecular weights deduced from every observed mass and their more accurate average values.

After addition of a Cu(II) salt to the solution, new peaks are observed as displayed in Figure 9.2B. Four new peaks appear at every charge state. The molecular weight determination of these peaks allows one to conclude that a maximum of four Cu atoms is incorporated. Table 9.2 summarizes the expected masses for different numbers of Cu atoms in various charge states and compares them with the experimental molecular weights. The copper is

Table 9.1   Molecular weight calculated from each observed mass

Number of charges	Detected m/z	Molecular weight
7+	415.7	2903.4
6+	484.8	2902.8
5+	581.7	2903.5
4+	726.8	2903.2
		Average = 2903.2

Table 9.2   Comparison of expected and observed molecular weight (MW) for different numbers of Cu atoms in various charge states

Number of Cu atoms	MW if Cu(0) [M + nCu]	MW if Cu(I) [M + nCu − nH]	MW if Cu(II) [M + nCu − 2nH]	Observed MW	Difference
$n = 0$	2903.02	2903.02	2903.02	2903.2	62.1
1	2966.52	2965.52	2964.52	2965.3	62.4
2	3030.02	3028.02	3026.02	3027.7	62.3
3	3093.52	3090.52	3087.52	3090.0	62.6
4	3157.02	3153.02	3149.02	3152.6	Average = 62.35

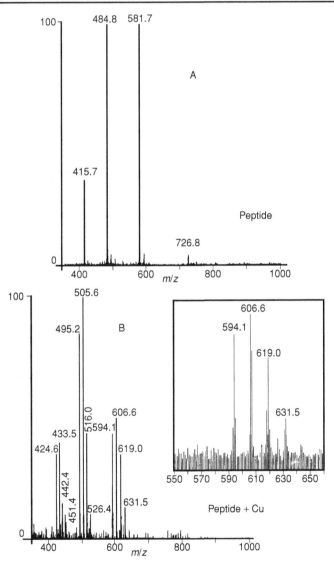

Figure 9.2

thus present as Cu(I). Indeed, for each Cu atom added the molecular weight increases by 62.35 Da, corresponding to the atomic weight of Cu (63.5) minus a proton. Because the observed charge state does not change with the addition of Cu atoms, it must be concluded that it is monocharged. In conclusion, the ions observed in spectrum B correspond to the formula $[M + xH + nCu - nH]^{x+}$, where x is the number of charges (varying from 4 to 7) and n is the number of Cu atoms (varying from 1 to 4).

9.7. How can you distinguish between the spectrum of *n*-butyraldehyde and that of *iso*-butyraldehyde?

Answer:

n-Butyraldehyde can undergo a McLafferty rearrangement yielding an ion at m/z 44 Th through rHi followed by αi, whereas *iso*-butyraldehyde cannot undergo such a rearrangement.

9.8. The spectra displayed in Figure 9.3 originate from isomeric ketones analysed by electron ionization. They comprise only linear saturated chains. Give the developed formula of each of these ketones. Justify your answer by interpreting at least three fragments, one of them resulting from a rearrangement.

Answer:

The molecular weight of these ketones is 184. The weight of the sum of the aliphatic chains is thus $184 - 28 = 156$, that is $C_{11}H_{24}$. Acylium ions are quite stable and α cleavage often occurs adjacent to the carbonyl group:

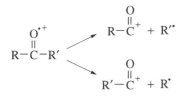

The first important observed fragments at the higher mass side appear at 141, 127 and 113 u for the three ketones, respectively. We can speculate that this corresponds to the higher mass acylium ion in each case.

For the top spectrum $184 - 141 = 43$ indicates that a C_3H_7 radical is lost. We then should expect the other acylium ion to result from the loss of C_8H_{17} and to appear at m/z 71. This is indeed an intense peak, as expected, more abundant than the 141 Th peak because the lost alkyl radical is larger. This first spectrum could correspond to a C_8H_{17}–CO–C_3H_7 ketone. We will now try to confirm this by looking at the rearrangement peaks. The McLafferty rearrangement is indeed often important for ketones. The three spectra shows a peak at m/z 58. The even mass indicates a rearrangement, but this one will not give information because it is the same for the three compounds.

In the first spectrum, another rearrangement peak appears at m/z 86. The following scheme shows the corresponding McLafferty rearrangement:

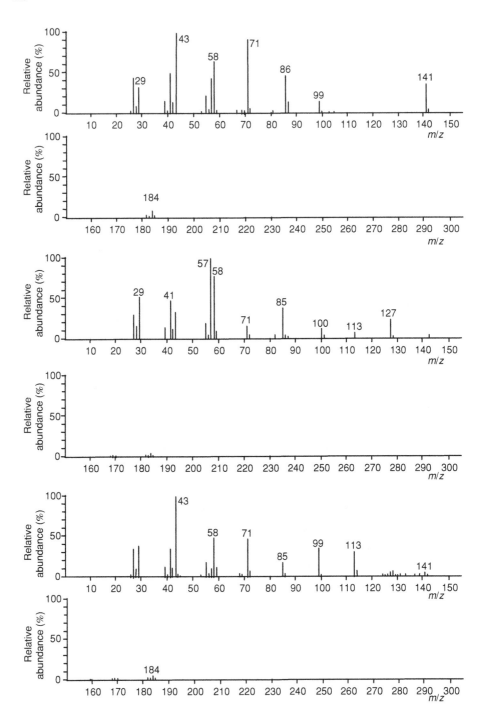

Figure 9.3

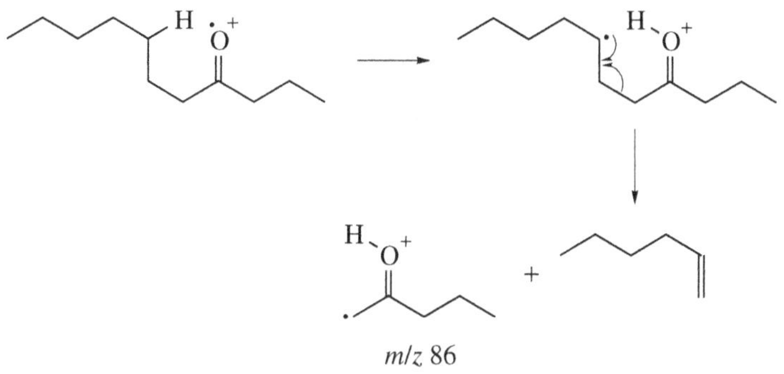

m/z 86

This confirms the proposed structure. The *m/z* 58, common to the three ketones, results from a second rearrangement and fragmentation of this *m/z* 86 ion:

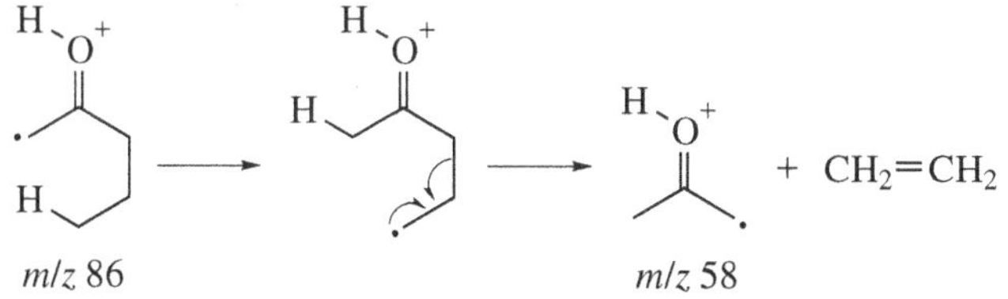

m/z 86   *m/z* 58

By the same reasoning, the two other ketones can be shown to be C_7H_{15}–CO–C_4H_9 and C_6H_{13}–CO–C_5H_{11}, respectively. For the latter, the McLafferty rearrangement gives a weak peak at *m/z* 114.

9.9. A compound has a molecular peak at 115 Th. Where is the peak corresponding to the metastable loss of CH_3 in the spectrum obtained with a magnetic instrument using an EB configuration?

Answer:

The metastable occurs at $100^2/115$ Th $= 86.9$ Th.

9.10. Explain the following: 'MIKE spectroscopy' and 'B/E linked scan'.

Answer:

MIKE stands for 'Mass-analysed Ion Kinetic Energy' and is a tandem mass spectrometry method that allows product ion spectra to be recorded with a magnetic instrument of 'inverse' or BE geometry. The parent ion m_p is selected with the magnetic analyser B and the metastable fragment ions m_f are analysed by scanning the electrical sector. The relevant equation is $E_p/E_f = m_p/m_f$. If kinetic energy E'_k is released during the fragmentation process, the kinetic energy of the fragments will be composed between $(E_{kf} + E'_k)$ and $(E_{kf} - E'_k)$. A broadening of the peak will result, which allows direct measurement of the released kinetic energy.

In *B/E* linked scan, the precursor ion is first focalized by the appropriate values of both and E_p. Then, *B* and *E* are scanned together in such a way that the ratio *B/E* remains constant. A fragment produced between the source and the analysers, either in the EB or BE configuration, will be focalized at values of B_f and E_f such that $B_p/E_p = B_f/E_f$. Because

$B_p/B_f = m_p/m_f$, the mass of the fragment can be determined. This method gives a better resolution than the MIKE method but does not allow measurement of the kinetic energy released.

9.11. Interpret the main peaks of the following spectra:

9.11.1. Methylene chloride

88: $^{37}Cl_2CH_2^{\cdot+}$
86: $^{37}Cl^{35}ClCH_2^{\cdot+}$
84: $^{35}Cl_2CH_2^{\cdot+}$
51: $^{37}ClCH_2^+$
49: $^{35}ClCH_2^+$

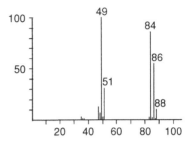

9.11.2. Acetone

58: $M^{\cdot+}$
43: CH_3CO^+
15: CH_3^+

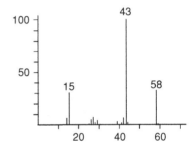

9.11.3. Ethylamine

45: $M^{\cdot+}$
44: $M^{\cdot+} - H$
30: $M^{\cdot+} - CH_3$
28: $CHNH^+$
15: CH_3^+

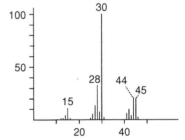

9.11.4. 1-Butene

56: $M^{\cdot+}$
55: $M^{\cdot+} - H$
41: $M^{\cdot+} - CH_3$
39: $C_3H_3^+$

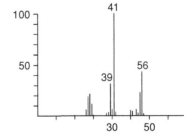

9.11.5. Ethylbenzene

106: M$^{\bullet+}$
  91: tropylium ion
  77: phenyl
  65: retro-Diels–Alder of 91
  51: retro-Diels–Alder of 77
  39: cyclopropenium

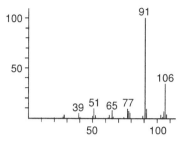

9.11.6: Tetralin

132: M$^{\bullet+}$
104: M$^{\bullet+}$ – C$_2$H$_4$ retro-Diels–Alder
  91: Tropylium ion
  51: C$_4$H$_3^+$

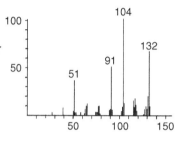

9.11.7. *n*-Butyraldehyde

  72: M$^{\bullet+}$
  57: M$^{\bullet+}$ – CH$_3^{\bullet}$
  44: C$_2$H$_4$O$^{\bullet+}$ McLafferty
  43: C$_3$H$_7^+$ σ cleavage
  41: C$_3$H$_5^+$
  39: C$_3$H$_3^+$
  29: C$_2$H$_5^+$
  15: CH$_3^+$

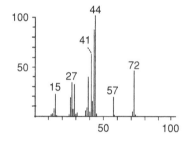

9.11.8. Cyclohexanol

100: M$^{\bullet+}$
  82: M$^{\bullet+}$ – H$_2$O
  71: M$^{\bullet+}$ – CHO$^{\bullet}$
  67: M$^{\bullet+}$ – H$_2$O – CH$_3^{\bullet}$
  57: M$^{\bullet+}$ – C$_2$H$_3$O$^{\bullet}$

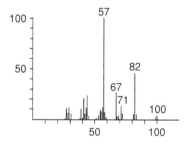

9.11.9. 2-Dodecanone

184: M$^{\bullet+}$
  85: C$_6$H$_{13}^+$ through σ cleavage
  71: C$_5$H$_{11}^+$
  58: C$_3$H$_6$O$^{\bullet+}$ McLafferty
  43: C$_3$H$_7^+$ through σ cleavage or
      C$_2$H$_3$O$^+$ through α cleavage

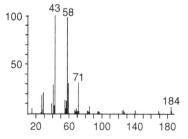

9.11.10. *n*-Dodecane

The main peaks are explained by σ
cleavage

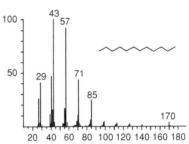

9.11.11. 1-Phenyl-*n*-hexane

162: M$^{\bullet+}$
105: M$^{\bullet+}$ – C$_4$H$_9$$^{\bullet}$
 91: tropylium ion
 43: C$_3$H$_7$$^{\bullet+}$

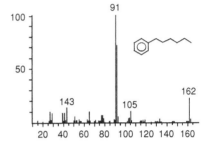

9.11.12: Cyclohexane

84: molecular ion.

69: ring opening,
hydrogen rearrangement,
then cleavage

56: α cleavage after ring
opening

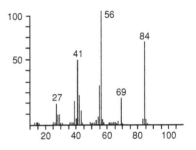

9.12. Identify the product and interpret the main peaks in the following:

9.12.1.

m/z	Relative abundance (%)	Interpretation
14	3.4	
15	20	CH$_3$$^+$
16	0.2	
19	2	F$^+$
31	10	CF$^+$
32	9.8	CHF$^{\bullet+}$
33	94	CH$_2$F$^+$
34	100	^{12}CH$_3$F$^{\bullet+}$
35	1.1	^{13}CH$_3$F$^{\bullet+}$

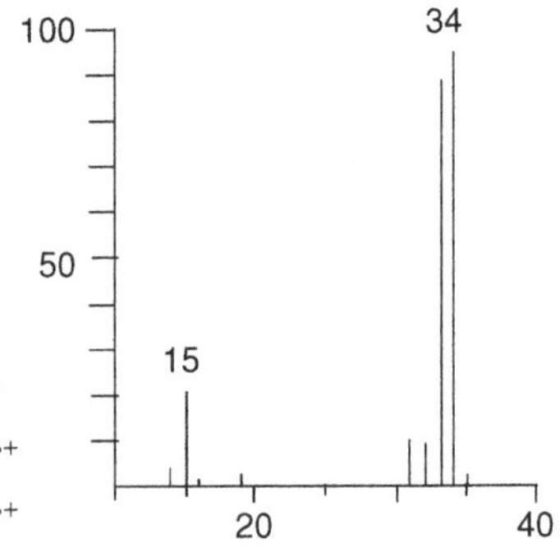

The intensity ratio of the peaks 35/34 indicates the presence of one C atom and the peak at 15 Th is ascribed to CH_3; if the molecular peak occurs at 34 Th, the $^\bullet CH_3$ loss leaves 19, which is due to fluorine; the formula that is proposed is thus CH_3F (methyl fluoride).

9.12.2.

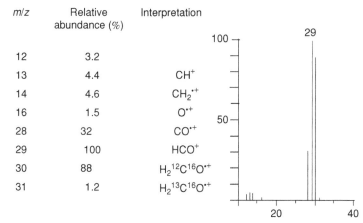

m/z	Relative abundance (%)	Interpretation
12	3.2	
13	4.4	CH^+
14	4.6	$CH_2^{\bullet +}$
16	1.5	$O^{\bullet +}$
28	32	$CO^{\bullet +}$
29	100	HCO^+
30	88	$H_2{}^{12}C^{16}O^{\bullet +}$
31	1.2	$H_2{}^{13}C^{16}O^{\bullet +}$

The intensity ratio of the peaks 31/30 indicates the presence of only one C; the remainder, 18, is due to H_2O. The structure that is suggested is thus $CH_2=O$ (formaldehyde).

9.12.3.

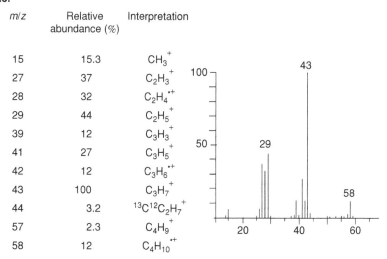

m/z	Relative abundance (%)	Interpretation
15	15.3	CH_3^+
27	37	$C_2H_3^+$
28	32	$C_2H_4^{\bullet +}$
29	44	$C_2H_5^+$
39	12	$C_3H_3^+$
41	27	$C_3H_5^+$
42	12	$C_3H_6^{\bullet +}$
43	100	$C_3H_7^+$
44	3.2	$^{13}C^{12}C_2H_7^+$
57	2.3	$C_4H_9^+$
58	12	$C_4H_{10}^{\bullet +}$

The intensity ratio of the peaks 59/58 indicates C_4; the intensity ratio 44/43 indicates C_3; if the molecular peak occurs at 58 Th, peak 43 indicates a $^\bullet CH_3$ loss; the crude formula C_4H_{10} is thus proposed. The peak at 29 Th due to $^+C_2H_5$ allows one to exclude isobutane; the structure that is proposed is thus n-butane. The main peaks that are observed are due to σ cleavages.

9.12.4.

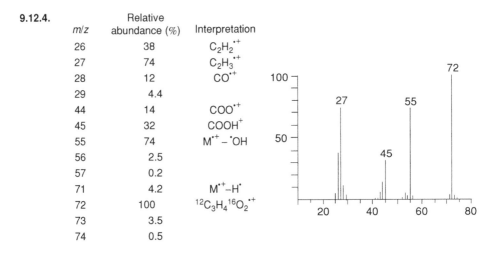

m/z	Relative abundance (%)	Interpretation
26	38	$C_2H_2^{\bullet+}$
27	74	$C_2H_3^{\bullet+}$
28	12	$CO^{\bullet+}$
29	4.4	
44	14	$COO^{\bullet+}$
45	32	$COOH^+$
55	74	$M^{\bullet+} - {}^{\bullet}OH$
56	2.5	
57	0.2	
71	4.2	$M^{\bullet+} - H^{\bullet}$
72	100	${}^{12}C_3H_4{}^{16}O_2{}^{\bullet+}$
73	3.5	
74	0.5	

The intensity ratio 73/72 points to the presence of C_3; the intensity ratio 74/72 points to the presence of O_2; the residual mass can be justified only by 4 H atoms, which leads to the crude formula $C_3H_4O_2$. If the molecular peak occurs at 72 Th, the peak at 55 Th can be due to an ${}^{\bullet}OH$ loss; the intensity ratios 57/56/55 confirm the existence of C_3 and O in this ion with a residual mass of 15 due to CH_3, which leads to the same crude formula. The number of unsaturations is $N = 2$. The peak at 45 Th due to $COOH^+$ leads us to suggest the following structure: $CH_2 = CH–COOH$ (acrylic acid).

9.12.5.

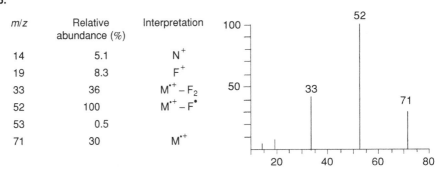

m/z	Relative abundance (%)	Interpretation
14	5.1	N^+
19	8.3	F^+
33	36	$M^{\bullet+} - F_2$
52	100	$M^{\bullet+} - F^{\bullet}$
53	0.5	
71	30	$M^{\bullet+}$

As the molecular ion has an odd mass, it must contain an odd number of nitrogen atoms. With one nitrogen, the residual mass 57 can be ascribed to F_3; the presence of one nitrogen is confirmed by the intensity ratio 53/52; the formula that is proposed, NF_3 (nitrogen trifluoride), is confirmed by the interpretation of the peaks that are observed.

9.12.6.

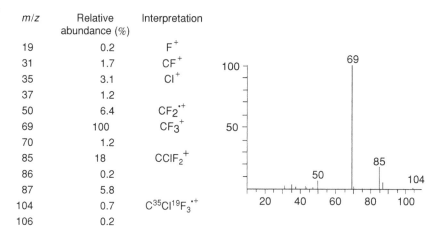

m/z	Relative abundance (%)	Interpretation
19	0.2	F^+
31	1.7	CF^+
35	3.1	Cl^+
37	1.2	
50	6.4	$CF_2^{\cdot+}$
69	100	CF_3^+
70	1.2	
85	18	$CClF_2^+$
86	0.2	
87	5.8	
104	0.7	$C^{35}Cl^{19}F_3^{\cdot+}$
106	0.2	

The intensity ratios 106/104 and 87/85 point to the presence of Cl with an $F^\bullet$ loss between the two; a 35 u loss leads to the 70/69 cluster whose intensity ratio points to the presence of one carbon atom. If the molecular peak occurs at 104 Th, the residual mass from CCl is 57, which can be explained by F_3. The formula we suggest is thus $CClF_3$ (chlorotrifluoromethane).

9.13. Identify the product that yields the following mass spectrum [2]:

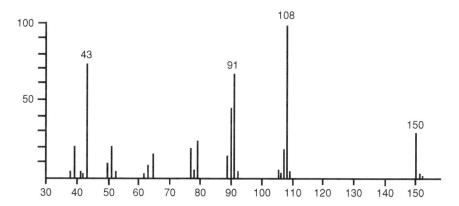

In IR, the main peaks appear at 3058, 2941, 1745, 1385, 1225, 1026, 749 and 697 cm^{-1}.
In ^{1}H-NMR:

δ	Intensity	Multiplicity
7.22	5	Singlet
5.00	2	Singlet
1.96	3	Singlet

Answer:

In the mass spectrum, the peak at m/z 150 can be the molecular peak (even number of nitrogen atoms). The isotopic cluster suggests $C_9H_{10}O_2$ as the most probable formula with five rings and/or double bonds.

In the IR spectrum, the peak at 1745 cm^{-1} corresponds to the unconjugated C=O stretch, the peak at 1225 cm^{-1} is the C–O–C stretch of an acetal, the peak at 1026 cm^{-1} can be the asymmetric stretch of COC, and the peaks at 749 and 697 cm^{-1} point to a singly substituted benzene.

All this leads us to suggest the formula (benzyl acetate):

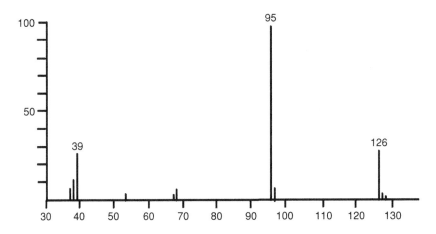

This is confirmed by the ^{1}H-NMR spectrum:

δ	Intensity	Assignment
7.22	5	Benzene protons
5.00	2	CH$_2$
1.96	3	CH$_3$

The main peaks in the mass spectrum are easily ascribed to the suggested formula:

150:	molecular peak
108:	COCH$_3$ loss with H transfer
91:	tropylium ion
77, 78:	benzene ions
43:	CH$_3$CO$^+$

9.14. Identify the product that yields the following mass spectrum [2]:

In IR, the main peaks appear at 3106, 2941, 1730, 1587, 1479, 1449, 1393, 1299, 1205, 1121 and 758 cm^{-1}.

In ^{1}H-NMR:

δ	Intensity	Multiplicity
7.51	1	Quadruplet
7.02	1	Quadruplet
6.45	1	Quadruplet
3.80	3	Singlet

In off-resonance decoupled ^{13}C-NMR:

δ	Multiplicity
160	Singlet
146	Doublet
144	Singlet
118	Doublet
112	Doublet
51	Quadruplet

Answer:

The mass spectrum shows a peak at m/z 126 which can be the molecular peak (even number of nitrogen atoms); the isotopic cluster corresponds to $C_6H_6O_3$ (the exact mass can be confirmed by high resolution) with four rings and two double bonds. The base peak at m/z 95 corresponds to M–OCH$_3$.

The IR spectrum shows a peak at 1730 cm^{-1} that can be ascribed to a conjugated C=O, peaks at 1205 and 1121 cm^{-1} that can be ascribed to an ester C–O stretch, the peak at 758 cm^{-1} indicates the presence of an aromatic ring, and the same is true for the peaks at 3106, 1587 and 1479 cm^{-1}.

The ^{1}H-NMR spectrum confirms the presence of six protons and the ^{13}C-NMR confirms the presence of six C.

All this leads us to suggest the formula (carboxymethylfuran):

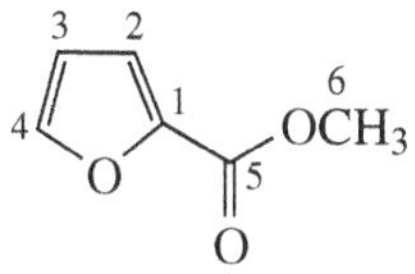

This structure is confirmed by the decoupled ^{13}C-NMR:

δ	Multiplicity	Number of H	Assignment
160	Singlet	0	C-5
146	Doublet	1	C-4
144	Singlet	0	C-1
118	Doublet	1	C-2
112	Doublet	1	C-3
51	Quadruplet	3	C-6

9.15. Identify the product that yields the following mass spectrum [2]:

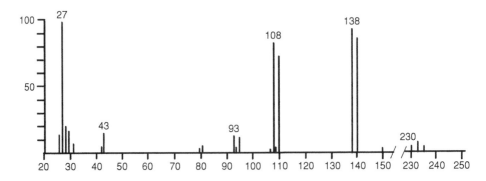

The IR spectrum shows the following main peaks: 2970, 2850, 1425, 1360, 1279, 1120 and 630 cm^{-1}.

In ^{1}H-NMR:

δ	Intensity	Multiplicity
3.40	1	Multiplet
3.88	1	Multiplet

Answer:

The mass spectrum with the isotopic cluster at m/z 230, 232, 234 indicates the presence of two bromine atoms; the remainder (72) can be ascribed to C_4H_8O. The symmetry of the ^{1}H-NMR spectrum indicates an AA'BB' symmetric structure, so we can suggest that the compound is di(bromoethyl) ether:

$$BrCH_2CH_2OCH_2CH_2Br$$

The IR spectrum confirms this structure:

2970 and 2850 cm^{-1}:	methylene stretch
1360 and 1279 cm^{-1}:	CH_2Br wagging
1120 cm^{-1}:	aliphatic ether
630 cm^{-1}:	CBr stretch

The main peaks in the mass spectrum are identified with this formula:

230:	molecular peak
138, 140:	CH_2Br loss and H transfer
108, 110:	OCH_2CH_2Br loss and H transfer
93, 95:	CH_2Br^+
27:	$C_2H_3^+$

9.16. Identify the product that yields the following mass spectrum [2]:

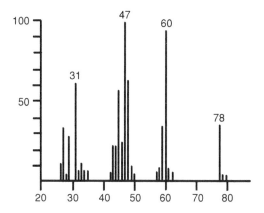

In IR, the main peaks appear at 3367, 3030, 2558, 1429, 1298, 1050 and 1020 cm^{-1}. In off-resonance decoupled ^{13}C-NMR:

δ	Multiplicity
28	Triplet
64	Triplet

Answer:

The molecular peak at m/z 78 indicates a molecule with an even number of nitrogen atoms. The intensity of the M + 2 peak indicates the presence of S; the remainder (46) could be due to C_2H_6O.

The crude formula that is suggested is thus C_2H_6OS, with no ring or double bond. The decoupled ^{13}C-NMR spectrum confirms the presence of two C; in the off-resonance spectrum, the triplets indicate that four out of the six hydrogens belong to neighbouring methylene groups. The structure that is proposed is thus 2-hydroxyethanethiol:

$$HOCH_2CH_2SH$$

The IR spectrum confirms such a structure:

3367 cm^{-1}:	OH stretch
2558 cm^{-1}:	SH stretch
1050 cm^{-1}:	CH_2OH primary alcohol

The main peaks in the mass spectrum are easily assigned:

78:	molecular peak
60:	H_2O loss
47:	CH_2SH^+
31:	CH_2OH^+

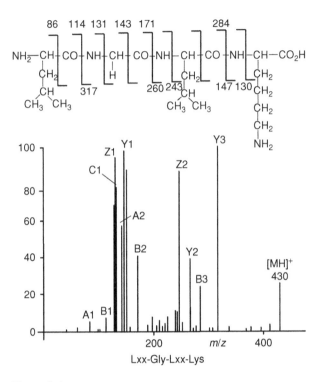

Figure 9.4
Fragment ion spectrum of a protonated peptide and (*top*)
the sequence of ions.

9.17. A 429 Da peptide gives the MS/MS product ion spectrum from its protonated molecular ion displayed in Figure 9.4. What is the sequence of this peptide?

Answer:

The difference between b_n and a_n fragments is 28 u. This applies to fragments 114/86 and 143/171. Thus $b_1 = 114$ and $b_2 = 171$ u. The N-terminal sequence is deduced: Lxx–Gly ..., with Lxx either Leu or Ile.

Similarly, y_n and z_n fragments differ by 17 Da, which applies to the couples of ions 130/147 and 243/260. Thus, $y_1 = 147$ and $y_2 = 260$. The C-terminal sequence is deduced: Lxx–Lys.

Combining these N-and C-terminal sequences, one gets the sequence Lxx–Gly–Lxx–Lys. The molecular weight is indeed 429 Da. This sequence is further confirmed by the observation of $y_3 = 317$ and $b_3 = 284$ u.

9.18. Consider Figure 9.5, representing the stability diagram for three ions of masses m_1, m_2 and m_3 Th in U, V coordinates. Explain which ions will be observed if the U, V values are adjusted successively to the values indicated by the points A–D in the figure.

Answer:

At A, none of these ions will be observed as the point is outside the stability curves. At B, both ions m_1 and m_2 will be observed because the point is inside their stability diagrams;

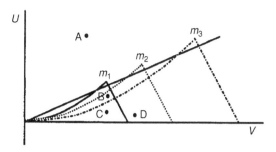

Figure 9.5

m_3 will not be observed. At C, the three ions will be observed because this point is within the three stability diagrams. Point D is inside the stability diagrams of m_2 and m_3 which will be observed; m_1 will not be detected.

9.19. A quadrupole with $r_0 = 1$ cm operating at an RF frequency of 0.5 MHz has a stability diagram with an apex at $a = 0.22$ and $q = 0.72$ for an ion of mass 2000 u. What are the corresponding U and V values?

Answer:

Applying the formula for a quadrupole,

$$a_u = \frac{8\,eU}{mr_0^2\omega^2} \quad \text{and} \quad q_u = \frac{4\,eV}{mr_0^2\omega^2}$$

with $e = 1.6 \times 10^{-19}$ C, $m = 2000 \times 1.66 \times 10^{-27} = 3.32 \times 10^{-24}$ kg, $r_0 = 0.01$ m and $\omega = 2\pi f = 2 \times 3.14 \times 500\,000 = 3.14 \times 10^6$ radians per second, one finds: $U = 563$ V and $V = 3.68$ kV.

9.20. The same quadrupole as in Question 9.19 operating in the 'RF-only' mode has values of $a = 0$ and $q = 0.91$. What is the minimum V_{rf} voltage allowing all the ions of less than 100 Th to be eliminated?

Answer:

$V_{min} = 233$ V.

9.21. If a V_{rf} of 2 kV is applied to the same quadrupole as in Question 9.20, from which m/z value will the ions be unstable?

Answer:

$m_{min} = 859$ Da.

9.22. In an ideal ion trap, that is $r_0^2 = 2z_0^2$, with $r_0 = 1$ cm and $v = 1.1$ MHz, referring to the values for U and V given in Figure 9.6, calculate the a and q values at the apex of the stability diagram.

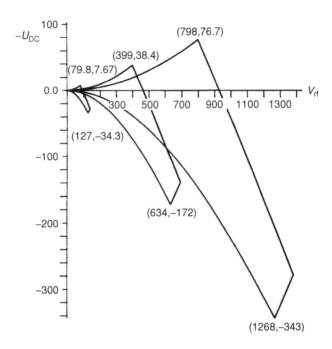

Figure 9.6

Answer:

Referring to the values for U and V given in Figure 9.6, one finds: $a_z = -0.67$ and $q_z = -1.24$ for the lower apex and $a_z = 0.15$ and $q_z = -0.78$ for the upper apex.

9.23. In an ideal ion trap, that is $r_0^2 = 2z_0^2$, with $r_0 = 1$ cm and $v = 1.6$ MHz, $U = 4$ V and $V = 400$ V for a nitrogen ion $N_2^{\bullet+}$, calculate the corresponding a and q values.

Answer:

Using the appropriate formula for an ideal ion trap,

$$a_u = a_z = -2a_r = \frac{-16zeU}{m\left(r_0^2 + 2z_0^2\right)\omega^2} \quad \text{and} \quad q_u = q_z = -2q_r = \frac{8zeV}{m\left(r_0^2 + 2z_0^2\right)\omega^2}$$

The results are:

$$a_z = -0.0109 \quad a_r = 0.0055$$
$$q_z = -0.55 \quad q_r = 0.27$$

9.24. In an ideal ion trap, that is $r_0^2 = 2z_0^2$, with $r_0 = 1$ cm and $v = 1.1$ MHz, $U = 4$ V and $V = 500$ V, what is the resonant frequency that should be applied to expel an ion of 100 Th?

Answer:

Applying the formula $\beta_z = (a_z + q_z^2/2)^{1/2}$ and $v_z = \beta_z v/2$, one finds $v = 153$ kHz.

9.25. A TOF analyser has the following characteristics: $V_s = 3$ kV, $V_R = 3130$ V, $D = 0.522$ m, $L_1 = 1$ m.

    (a) At which distance L_2 does it focalize ions having the same m/z $(m_p) = 578$ Th?

    (b) What is the focal distance for PSD ions with $m/z(m_f) = 560$ Th?

    (c) At which value should V_R be reduced in order to focalize the fragment ions at the same distance as the precursors?

Answer:

    (a) $x_p = V_s D/V_R = 0.5$ m; $L_{2p} = 4x - L_1 = 1$ m

    (b) $x_f = x_p m_f/m_p = 0.48$ m; $L_{2f} = 0.94$ m

    (c) $V_{rf} = V_{rp} \quad m_f/m_p = 3032$ V

References

1. Hutchens, T.W., Nelson, R.W., Allen, M.H., Li, C.M. and Yip T.T. (1992) *Biol. Mass Spectrom.*, **21**, 151.

2. Silverstein, R.M., Bassler, G.C. and Morril, T.C. (1991) *Spectrometric Identification of Organic Compounds*, 5th edn, John Wiley & Sons, Inc., New York.

Appendices

1 Nomenclature

Standard terms and definitions relating to mass spectrometry can be found in Chapter 12 of the IUPAC 'Orange Book' Compendium of Analytical Nomenclature [1] and in other sources [2–4].

A full list of old and new terms related to the field of mass spectrometry is given on the msterms.com web site [5]. The goal of this site is to provide information about and discussion of the IUPAC sponsored project to update and extend the standard terms and definitions for mass spectrometry. This site also contains a compilation of the latest draft terms, related discussion and Orange Book terms.

1.1 Units

1. The mass unit (u) is also called the dalton (Da). It is defined as one-twelfth of the mass of a ^{12}C atom:

$$1\,u = 1\,Da = 1.660\,540 \times 10^{-27}\,kg.$$

2. The thomson (Th) is the m/z unit:

$$1\,Th = 1\,u/e = 1.036\,426 \times 10^{-8}\,kgC^{-1}$$

3. The charge unit is the charge of the electron, absolute value:

$$e = 1.602\,177 \times 10^{-19}\,C$$

1.2 Definitions

1. Average mass or chemical mass: mass calculated using a weighted average of the natural isotopes for the atomic mass of each element. Example: average mass of $CH_3Br = (12.011\,15 + 3 \times 1.00797 + 79.904)\,u = 94.939\,06\,u$. This is the mass a chemist normally uses in stoichiometric calculations.

2. Nominal mass: mass calculated using the mass of the predominant isotope of each element rounded to the nearest integer value. Example: nominal mass of CH_3OH $= {}^{12}C^1H_3{}^{16}O^1H = (12 + 4 \times 1 + 16)\,u = 32\,u$.

3. Monoisotopic mass: mass calculated using the 'exact' mass of the predominant isotope of each element, which takes into account the mass defects. Example: $CH_3Br = {}^{12}C^1H_3{}^{79}Br = (12.0000 + 3 \times 1.007825 + 78.918336)\,u = 93.941\,011\,u$.

4. Mass defect: difference between the monoisotopic and nominal mass.

Mass Spectrometry: Principles and Applications, Third Edition Edmond de Hoffmann and Vincent Stroobant
© Copyright 2007, John Wiley & Sons Ltd

5. m/z: the ratio of the relative mass of an ion to its charge number. It is an abstract, unitless number that allows us to talk about, for example, $m/z = 60$. This is the definition we use in this book. If we define m/z as the mass-to-charge ratio, m is given in daltons and the charge is given in e, thus m/z is in thomsons. For example, an ion with mass $m = 200$ u and with charge $2e$ has $m/z = 100$ Th.

6. Relative mass (m): mass of an atom, a molecule or an ion divided by one-twelfth of the mass of the ^{12}C carbon atom. It is thus dimensionless. Thus m = (mass of a molecular or ionic species)/(mass of ^{12}C) $\times$ (1/12).

7. Mass number: sum of the protons and neutrons in an atom, molecule or ion.

8. Charge number: the total charge on an ion divided by the elementary charge (e).

9. Mass spectrum: plot of ion abundance versus mass-to-charge ratio normalized to most abundant ion in the spectrum.

10. Base peak: the most intense peak in the spectrum.

11. Isotopic peak: peak due to other isotopes of the same chemical formula but with a different isotopic composition.

12. Relative abundance: normalization in per cent relative to the base peak.

13. Relative intensity: ratio of the peak intensity to that of the base peak.

14. Percentage of the total intensity: abundance of an ion over the total abundance of all ions within a specified mass region.

1.3 Analysers

1. Mass spectrometer: an instrument that generates ions and measures their mass-to-charge ratio (m/z) and their relative abundances.

2. Mass spectrograph: an instrument that separates the ion beams according to their m/z ratio and records the abundance at every m/z ratio simultaneously, using a photographic plate or an array detector.

3. Field-free region: a section of a mass spectrometer in which there are no electric or magnetic fields.

4. Double-focusing spectrometer: an electromagnetic instrument that focuses the ions with a given m/z ratio even though their initial directions diverge and their kinetic energies are different.

5. Static field spectrometer: a spectrometer that separates the ion beams using fields that remain constant in time.

6. Dynamic field spectrometer: a spectrometer in which the separation of ion beams depends on fields that vary in time.

7. Electrostatic analyser: a kinetic energy focusing instrument that produces an electrostatic field that is perpendicular to the ions' displacement direction. Its effect is to bring all ions of the same kinetic energy to a single focus.

8. Magnetic analyser: an instrument that causes a direction focusing produced by a magnetic field that is perpendicular to the ions' displacement direction. Its effect is to bring to a common focus all the ions of a given momentum. This can be converted to m/z ratios if all the ions have the same kinetic energy.

9. Quadrupole analyser: a mass filter that produces a quadrupolar field with a DC component and an AC component such that only ions with a given m/z pass through.

10. Time-of-flight analyser: an instrument that measures the time of flight of the ions accelerated to known kinetic energies over a known fixed distance. As this time is a function of the mass, the mass can be calculated.

11. Reflectron (also known as an ion mirror): a device used in a time-of-flight mass spectrometer that retards and then reverses ion velocities in order to correct for the flight times of ions having the same mass but different kinetic energies

12. Cyclotron resonance analyser (also known as Penning ion trap): an instrument that confines ions by placing them in a static magnetic field. Inside the field, the ions are subject to the Lorentzian force that causes ions of a particular m/z to cycle at a specific frequency (cyclotron frequency).

13. Ion cyclotron-FTMS: cyclotron resonance instrument where the current induced by the ions in a plate is detected and submitted to a Fourier transform. This produces a mass spectrum with a very high-resolution capacity.

14. Ion trap analyser: a mass analyser in which ions are confined in space by means of one or a combination of electric fields, AC or DC, capable of storing ions within a selected range of m/z ratios and then to detect the ions according to their m/z values. There are different types of traps: Paul trap, linear traps, etc.

15. Orbitrap analyser: a mass analyser that store the ions in a quadro-logarithmic field. The ions are selectively detected according to their m/z values using their induced current. The treatment of this current by a Fourier transform yields their mass-to-charge ratios. It is an analyser with very high-resolution capacity.

16. Hybrid mass spectrometer: a mass spectrometer that combines analysers of different types to perform tandem mass spectrometry.

17. Tandem mass spectrometer: a mass spectrometer designed for MS/MS analysis.

1.4 Detection

1. Electron multiplier: an ion-to-electron detector that multiplies an electronic current by accelerating the electrons on the surface of an electrode. The collision yields a number of secondary electrons higher than the number of incident electrons. The secondary electrons are accelerated towards another electrode or parts of a continuous electrode to produce further secondary electrons, thereby continuing the process.

2. Electro-optical ion detector: an ion-to-photon detector that combines an ion and photon detection device. This type of detector operates by converting ions to electrons and then to photons. Ions strike a conversion dynode to produce electrons that in turn strike a phosphor and the resulting photons are detected by a photomultiplier.

3. Point collector: a detector in which the ion beam is focused onto a point and the individual ions arrive sequentially.

4. Focal plane collector: a detector for spatially disperse ion beams in which all ions simultaneously impinge on the detector plane.

5. Detection limit: the smallest sample quantity that yields a signal that can be distinguished from the background noise. See also sensitivity.

6. Resolution: the ratio $m/\delta m$ where m and $m + \delta m$ are the relative masses of the two ions that yield neighbouring peaks with a valley depth $x\%$ of the weakest peak's intensity. In the commercial description of mass spectrometers, $x = 50$ is normally used. However, 10% valley has been largely used in the past. Another definition entails using for δm the width of an isolated peak at $x\%$ of its maximum.

7. Sensitivity: the sensitivity of an instrument is the ratio of the ionic current change to the sample flux change in the source (in $C\,\mu g^{-1}$). The analytical sensitivity is the smallest quantity of compound yielding a definite signal-to-noise ratio, often $10{:}1$.

1.5 Ionization

1. Ionization: a process that yields an ion from an atom or from a neutral molecule.

2. Electron ionization: the ionization of a sample by a beam of electrons most often accelerated by a potential of about 70 eV.

3. Field ionization: the removal of an electron from the sample molecule by the application of an intense electric field.

4. Field desorption: the production of ions from a sample laid on a solid surface and submitted to an intense electric field.

5. Chemi-ionization: the production of an ion through the reaction of the sample molecule with another molecule that was excited beforehand:

$$M + A^* \longrightarrow MA^+ + e^-$$

6. Chemical ionization: the formation of a new ionized species when the sample molecule reacts with an ion.

7. Atmospheric pressure chemical ionization: a solution of the sample is pneumatically sprayed at atmospheric pressure and the solvent is ionized by a coronary discharge from a needle. The monocharged ions produced cause the chemical ionization of the analyte and the ions are accelerated towards the vacuum region.

8. Laser ionization: the production of ions by irradiation of the sample with a laser beam.

9. Photoionization: ionization using photons.

10. Thermal ionization: occurs when an atom or a molecule is ionized through an interaction with a heated solid surface or through its presence in a high-temperature gaseous environment.

11. Electron attachment: a resonance process whereby an electron is incorporated into an atomic or molecular orbital.

12. Ion pair formation: an ionization process where the only products are a positive ion fragment and a negative ion fragment.

13. Vertical ionization: a process in which an electron is removed from a molecule so quickly that a positive ion is produced without any change in position or momentum of the atoms. The ion that results is often in an excited state.

14. Adiabatic ionization: the electron is removed from the molecule in its fundamental state, and produces an ion also in its fundamental state. This require a change in position and/or momentum during ionization.

15. Dissociative ionization: a process in which a gas molecule decomposes in order to form products, one of which is an ion.

16. Electrospray: a solution of the sample in a solvent is pumped into the source through a capillary submitted to an electric field of several kilovolts per centimetre. This produces droplets and solvated ions in the gas phase. The solvated ions pass through a desolvation device under vacuum and are injected into the analyser. Electrospray produces multiply charged ions from large molecules.

17. Fast atom bombardment: the sample in solution in a non-volatile solvent such as glycerol is ionized under vacuum by 'bombardment' with accelerated neutrals such as argon.

1.6 Ion Types

1. Molecular ion: ion derived from the neutral molecule by loss or gain of an electron. For the physicist, a molecular ion is any ion containing more than one atom.

2. Ion of the molecular species: ion resulting from the ionization of a molecule by the addition or removal of a proton or an hydride, or by the formation of an adduct with another ion, such as Na^+, Cl^-, acetate, and so on. They allow the molecular mass to be deduced. In the past the terms 'pseudomolecular ion' or 'quasimolecular ion' have been used but should be avoided.

3. Precursor ion: ion that decomposes or changes its charge, yielding a product ion. Also named parent ion.

4. Fragment ion: ion derived from the decomposition of a precursor ion.

5. Metastable ion: ion that fragments spontaneously in a field-free region.

6. Adduct ion: ion formed from the interaction between two species, usually an ion and a molecule, containing all the atoms of the two species.

7. Radical ion (synonymous with odd-electron ion): ion containing an unpaired electron.

8. Cluster ion: ion formed by combination with two or more molecules either of the same species or of different species.

9. Distonic ion: radical ion in which the charge and radical sites are formally located on different atoms in the molecule.

10. Isotopic ion: any ion containing one or more of the less abundant, naturally occurring isotopes of the elements that make up its structure.

11. Multiply charged ion: ions containing several positive or negative charges.

12. Isobaric ion: ion of the same nominal mass (integral).

1.7 Ion–Molecule Reaction

1. Collision-activated or collision-induced dissociation: the rapidly moving projectile ion dissociates through interaction with a neutral immobile target atom or molecule.

2. Collision activation: the rapidly moving projectile ion is excited by interaction with an immobile neutral target atom or molecule.

3. Elastic collision: ion–neutral interaction in which the total kinetic energy of the collision partners remains unchanged.

4. Inelastic collision: ion–neutral interaction in which the total kinetic energy (and thus the internal energy) of the collision partners changes.

5. Charge stripping: ion–molecule reaction that increases the ion positive charge.

1.8 Fragmentation

1. Single electron transfer: half-arrow ⤴

2. Electron pair transfer: full arrow ⟶

3. Hydrogen rearrangement: $\xrightarrow{\text{rH}}$

2 Acronyms and Abbreviations

An extensive list that defines acronyms and abbreviations in the field of mass spectrometry was published in 2002 [6]. A single analytical technique or a type of instrument is abbreviated without hyphens or slashes. However, it is customary to use hyphens for a description of an instrument whereas an abbreviation that describes the method uses slashes. For example, LC-MS is an instrument where a liquid chromatograph is coupled with a mass spectrometer, while LC/MS is the method of liquid chromatography/mass spectrometry. Thus, one uses an LC-MS instrument to obtain a LC/MS spectrum.

ADC	Analogue-to-Digital Converter
AE	Appearance Energy
AED	Atomic Emission Detector
AMS	Accelerator Mass Spectrometry
AP	Appearance Potential
APCI	Atmospheric Pressure Chemical Ionization
API	Atmospheric Pressure Ionization
APPI	Atmospheric Pressure Photoionization
ARMS	Angle-Resolved Mass Spectrometry
CA	Collision Activation
CAD	Collision-Activated Dissociation
CAR	Collision-Activated Reaction
CE	Capillary Electrophoresis
CE/MS	Capillary Electrophoresis/Mass Spectrometry
CEM	Channel Electron Multiplier
CEMA	Channel Electron-Multiplier Array
CF-FAB	Continuous Flow Fast Atom Bombardment
CI	Chemical Ionization
CID	Collision-Induced Dissociation
CRF	Charge Remote Fragmentation
CS	Charge Stripping
CX	Charge Exchange
CZE	Capillary Zone Electrophoresis
DAC	Digital-to-Analogue Converter
DADI	Direct Analysis of Daughter Ions
DART	Direct Analysis in Real Time
DCI	Desorption Chemical Ionization
DE	Delayed Extraction
DEI	Desorption Electron Ionization
DESI	Desorption Electrospray Ionization
DI	Desorption Ionization
DIP	Direct-Insertion Probe
DLI	Direct Liquid Introduction
DLV	Direct Laser Vaporization
DTMS	Drift Tube Ion Mobility Spectrometry
EA	Electron Affinity
ECCI	Electron Capture Chemical Ionization
ECD	Electron Capture Dissociation
EE	Even Electron
EHD	Electron Hydrodynamic Desorption
EI	Electron Ionization (formerly Electron Impact)
EM	Electron Multiplier
ESA	Electrostatic Analyser
ESI	Electrospray, Electrospray Ionization
ETV	Electrothermal Vaporization
FA	Flowing Afterglow
FAB	Fast Atom Bombardment

FD	Field Desorption
FFR	Field-Free Region
FFT	Fast Fourier Transform
FI	Field Ionization
FIB	Fast Ion Bombardment
FID	Flame Ionization Detector
FTICR	Fourier Transform Ion Cyclotron Resonance
FTMS	Fourier Transform Mass Spectrometry
FVP	Flash Vacuum Pyrolysis
FWHM	Full Width at Half Maximum
GC	Gas Chromatograph
GC/MS	Gas Chromatography/Mass Spectrometry
GD	Glow Discharge
GD/MS	Glow Discharge/Mass Spectrometry
HPLC	High-Performance Liquid Chromatography
HPLC/MS	High-Performance Liquid Chromatography/Mass Spectrometry
HR	High Resolution
HRMS	High-Resolution Mass Spectrometry
IC	Ion Chromatograph
IC/MS	Ion Chromatography/Mass Spectrometry
ICP	Inductively Coupled Plasma
ICR	Ion Cyclotron Resonance
IDMS	Isotope Dilution Mass Spectrometry
IE	Ionization Energy (formerly Ionization Potential)
IKES	Ion Kinetic Energy Spectrometry
IMS	Ion Mobility Spectrometry
IP	Ionization Potential
IRMS	Isotope Ratio Mass Spectrometry
IRMPD	Infrared Multiple Photon Dissociation
ISD	In-Source Decay
ISP	Ionspray
IT	Ion Trap
ITD	Ion Trap Detector
ITMS	Ion Trap Mass Spectrometer
ITR	Integrating Transient Recorder
LC	Liquid Chromatograph
LC/MS	Liquid Chromatography/Mass Spectrometry
LD	Laser Desorption
LDLPMS	Laser Desorption Laser Photoionization Mass Spectrometry
LI	Laser Ionization
LIMS	Laser Ionization Mass Spectrometry
LIT	Linear Ion Trap
LMS	Laser Mass Spectrometry
LOD	Limit of Detection
LOQ	Limit of Quantitation
LIF	Laser-Induced Fluorescence
LPCI	Low-Pressure Chemical Ionization

LSIMS	Liquid Secondary Ion Mass Spectrometry
LTOF	Linear Time of Flight
MALD	Matrix-Assisted Laser Desorption
MALDI	Matrix-Assisted Laser Desorption/Ionization
MB	Molecular Beam
MBMS	Molecular Beam Mass Spectrometry
MCP	Multi-Channel Plate
MIKES	Mass-analysed Ion Kinetic Energy Spectrometry
MIMS	Membrane Introduction Mass Spectrometry
MPI	Multiphoton Ionization
MPD	Multiphoton Dissociation
MRM	Multiple Reactions Monitoring, also Metastable Reaction Monitoring
MS	Mass Spectrometry
MS/MS	Tandem Mass Spectrometry
MS^n	General designation of mass spectrometry to the nth degree
nano-ESI	Nanoelectrospray
NfR	Neutral Fragment Reionization
NICI	Negative Ion Chemical Ionization
NR	Neutralization Reionization
NRMS	Neutralization Reionization Mass Spectrometry
oa	Orthogonal Acceleration
oa-TOF	Orthogonal Acceleration Time of Flight
OE	Odd Electron
OT	Orbitrap
PA	Proton Affinity
PB	Particle Beam
PBM	Probability-Based Matching
PD	Plasma Desorption
PDMS	Plasma Desorption Mass Spectrometry
PI	Photoionization
PICI	Positive Ion Chemical Ionization
PID	Photo-Induced Dissociation
PSD	Post-Source Decay
PyMS	Pyrolysis Mass Spectrometry
q	Quadrupole device used in RF-only mode
Q	Quadrupole
QET	Quasi-Equilibrium Theory
QIT	Quadrupole Ion Trap
QUISTOR	Quadrupole Ion Store
R	Resolution
RA	Relative Abundance
RDA	Retro-Diels–Alder
RE	Recombination Energy
REC	Resonance Electron Capture
REMPI	Resonance-Enhanced Multiphoton Ionization
RF	Radio Frequency
RI	Relative Intensity

RIC	Reconstructed Ion Chromatogram
RIMS	Resonance Ionization Mass Spectrometry
RN	Resonance Neutralization
RP	Resolving Power
RR	Reaction Region
RTOF	Reflectron Time of Flight
SI	Surface Ionization
SID	Surface-Induced Dissociation
SIFT	Selected-Ion Flow Tube
SIM	Selected-Ion Monitoring
SIMS	Secondary Ion Mass Spectrometry
SIR	Selected-Ion Recording
SIS	Selected-Ion Storage
SNMS	Secondary Neutral Mass Spectrometry
SORI	Sustained Off-Resonance Irradiation
SRM	Selected Reaction Monitoring
SSMS	Spark Source Mass Spectrometry
SWIFT	Stored Waveform Inverse Fourier Transform
TCC	Time-Compressed Chromatography
TDC	Time-to-Digital Conversion
TI	Thermal Ionization
TIC	Total Ion Current
TID	Thermally Induced Dissociation
TIMS	Thermal Ionization Mass Spectrometry
TLC	Thin-Layer Chromatography
TLF	Time-Lag Focusing
TOF	Time of Flight
TSP	Thermospray

3 Fundamental Physical Constants [7, 8]

Quantity	Symbol	Value	Units
Speed of light in vacuum	c	299 792 458	m s^{-1}
Gravitation constant	G	6.672 59	10^{-11} m^3 kg^{-1} s^{-2}
Planck constant	h	6.626 075 5	10^{-34} J s
Elementary charge	e	1.602 177 33	10^{-19} C
Electron mass	m_e	9.109 389 7	10^{-31} kg
Proton mass	m_p	1.672 623 1	10^{-27} kg
Neutron mass	m_n	1.674 928 6	10^{-27} kg
Atomic mass unit, dalton	u, Da	1.660 5402	10^{-27} kg
Boltzmann constant	k	1.380 658	10^{-23} J K^{-1}
		1.602 177 33	10^{-19} J
Electronvolt	eV	23.060 54	kcal mol^{-1}
		96.485 3	kJ mol^{-1}
Avogadro constant	N_A	6.022 136 7	10^{23} mol^{-1}
Molar gas constant	R	8.314 510	J mol^{-1} K^{-1}

Prefix	Symbol	Factor	Prefix	Symbol	Factor
deca	da	10^1	deci	d	10^{-1}
hecto	h	10^2	centi	c	10^{-2}
kilo	k	10^3	milli	m	10^{-3}
mega	M	10^6	micro	μ	10^{-6}
giga	G	10^9	nano	n	10^{-9}
tera	T	10^{12}	pico	p	10^{-12}
peta	P	10^{15}	femto	f	10^{-15}
exa	E	10^{18}	atto	a	10^{-18}
zetta	Z	10^{21}	zepto	z	10^{-21}
yotta	Y	10^{24}	yocto	y	10^{-24}

4A Table of Isotopes in Ascending Mass Order [7]

Z	Symbol	Nominal mass	%	% rel.	Isotopic mass	Average mass
0	n	1	—		1.008 665	—
1	H	1	99.985	100	1.007 825	1.007 94
	D	2	0.015	0.015	2.014	
2	He	3	0.000 137	0.000 137	3.016 030	4.002 6
		4	≈100	100	4.00 260	
3	Li	6	7.5	8.010 8	6.015 121	6.941
		7	92.5	100	7.016 003	
4	Be	9	100	100	9.012 182	9.012 182
5	B	10	19.9	24.84	10.012 937	10.811
		11	80.1	100	11.009 305	
6	C	12	98.90	100	12.000 000	12.011
		13	1.10	1.112	13.003 355	
7	N	14	99.63	100	14.003 074	14.006 74
		15	0.37	0.37	15.000 108	
8	O	16	99.76	100	15.994 915	15.999 4
		17	0.04	0.04	16.999 133	
		18	0.20	0.20	17.999 160	
9	F	19	100	100	18.998 403	18.998 4
10	Ne	20	90.48	100	19.992 435	20.179 7
		21	0.27	0.298	20.993 843	
		22	9.25	10.22	21.991 264	
11	Na	23	100	100	22.989 768	22.989 8
12	Mg	24	78.99	100	23.985 042	24.305 0
		25	10.00	12.66	24.985 837	
		26	11.01	13.94	25.982 593	
13	Al	27	100	100	26.981 539	26.981 5
14	Si	28	92.21	100	27.976 927	28.085 5
		29	4.67	5.065	28.976 495	
		30	3.10	3.336	29.973 770	
15	P	31	100	100	30.973 762	30.973 8

(continued overleaf)

Z	Symbol	Nominal mass	%	% rel.	Isotopic mass	Average mass
16	S	32	95.03	100	31.972 070	32.066
		33	0.75	0.789	32.971 456	
		34	4.22	4.44	33.967 866	
		36	0.02	0.021	35.967 080	
17	Cl	35	75.77	100	34.968 852	35.453
		37	24.23	31.98	36.965 903	
18	Ar	36	0.337	0.338	35.967 545	39.948
		38	0.063	0.063 3	37.962 732	
		40	99.600	100	39.962 384	
19	K	39	93.2581	100	38.963 707	39.098 3
		40	0.0117	0.012 5	39.963 999	
		41	6.7302	7.22	40.961 825	
20	Ca	40	96.941	100	39.962 591	40.078
		42	0.647	0.667 42	41.958 618	
		43	0.135	0.139	42.958 766	
		44	2.086	2.152	43.955 480	
		46	0.004	0.004	45.953 689	
		48	0.187	0.193	47.952 533	
21	Sc	45	100	100	44.955 911	44.956
22	Ti	46	8.00	10.84	45.952 629	47.88
		47	7.3	9.892	46.951 764	
		48	73.8	100	47.947 947	
		49	5.51	7.466	48.947 871	
		50	5.4	7.317	49.944 792	
23	V	50	0.25	0.251	49.947 161	50.941 5
		51	99.75	100	50.943 962	
24	Cr	50	4.345	5.185	49.946 046	51.996 1
		52	83.79	100	51.940 509	
		53	9.50	11.34	52.940 651	
		54	2.365	2.82	53.938 882	
25	Mn	55	100	100	54.938 046	54.938 0
26	Fe	54	5.9	6.43	53.939 612	55.847
		56	91.72	100	55.934 939	
		57	2.1	2.29	56.935 396	
		58	0.28	0.305	57.933 277	
27	Co	59	100	100	58.933 198	58.933 2
28	Ni	58	68.27	100	57.935 346	58.693 4
		60	26.10	38.23	59.930 788	
		61	1.13	1.66	60.931 058	
		62	3.59	5.26	61.928 346	
		64	0.91	1.33	63.927 968	
29	Cu	63	69.17	100	62.929 598	63.546
		65	30.83	44.57	64.927 765	
30	Zn	64	48.6	100	63.929 145	65.39
		66	27.9	57.41	65.926 034	
		67	4.1	8.44	66.927 129	
		68	18.8	38.68	67.924 846	
		70	0.6	1.23	69.925 325	

Z	Symbol	Nominal mass	%	% rel.	Isotopic mass	Average mass
31	Ga	69	60.108	100	68.925 580	69.723
		71	39.892	66.37	70.924 700	
32	Ge	70	20.5	56.16	69.924 250	72.61
		72	27.4	75.07	71.922 079	
		73	7.8	21.37	72.923 463	
		74	36.5	100	73.921 177	
		76	7.8	21.37	75.921 401	
33	As	75	100	100	74.921 594	74.921 6
34	Se	74	0.9	1.80	73.922 475	78.96
		76	9.1	18.24	75.919 212	
		77	7.6	15.23	76.919 912	
		78	23.6	47.29	77.917 309	
		80	49.9	100	79.916 520	
		82	8.9	17.84	81.916 698	
35	Br	79	50.69	100	78.918 336	79.904
		81	49.31	97.28	80.916 289	
36	Kr	78	0.35	0.614	77.920 401	83.80
		80	2.25	3.947	79.916 380	
		82	11.6	20.35	81.913 482	
		83	11.5	20.175	82.914 135	
		84	57.0	100	83.911 507	
		86	17.3	30.35	85.910 610	
37	Rb	85	72.17	100	84.911 794	85.467 8
		87	27.83	38.562	86.909 187	
38	Sr	84	0.56	0.68	83.913 431	87.62
		86	9.86	11.94	85.909 267	
		87	7.00	8.5	86.908 884	
		88	82.58	100	87.905 619	
39	Y	89	100	100	88.905 849	88.906
40	Zr	90	51.45	100	89.904 703	91.224
		91	11.22	21.73	90.905 643	
		92	17.15	33.33	91.905 039	
		94	17.38	33.78	93.906 314	
		96	2.80	5.44	95.908 275	
41	Nb	93	100	100	92.906 377	92.906
42	Mo	92	14.84	61.50	91.906 808	95.94
		94	9.25	38.33	93.905 085	
		95	15.92	65.98	94.905 840	
		96	16.68	69.13	95.904 678	
		97	9.55	39.58	96.906 020	
		98	24.13	100	97.905 406	
		100	9.63	39.91	99.907 477	
43	Tc	98	100	100		
44	Ru	96	5.54	17.53	95.907 599	101.07
		98	1.86	5.89	97.905 267	
		99	12.7	40.19	98.905 939	
		100	12.6	38.87	99.904 219	
		101	17.1	54.11	100.905 582	

(continued overleaf)

Z	Symbol	Nominal mass	%	% rel.	Isotopic mass	Average mass
		102	31.6	100	101.904 348	
		104	18.6	58.86	103.905 424	
45	Rh	103	100	100	102.905 500	102.905
46	Pd	102	1.02	3.73	101.905 634	106.42
		104	11.14	40.76	103.904 029	
		105	22.33	81.71	104.905 079	
		106	27.33	100	105.903 478	
		108	26.46	96.82	107.903 895	
		110	11.72	42.88	109.905 167	
47	Ag	107	51.839	100	106.905 092	107.868
		109	48.161	94.90	108.904 757	
48	Cd	106	1.25	4.35	105.906 461	112.411
		108	0.89	3.10	107.904 176	
		110	12.49	43.47	109.903 005	
		111	12.80	44.55	110.904 182	
		112	24.13	83.99	111.902 758	
		113	12.22	42.53	112.904 400	
		114	28.73	100	113.903 357	
		116	7.49	26.07	115.904 754	
49	In	113	4.3	4.49	112.904 061	114.82
		115	95.7	100	114.903 880	
50	Sn	112	0.97	2.98	111.904 826	118.710
		114	0.65	1.99	113.902 784	
		115	0.36	1.10	114.903 348	
		116	14.53	43.58	115.901 747	
		117	7.68	23.57	116.902 956	
		118	24.22	73.32	117.901 609	
		119	8.58	26.33	118.903 310	
		120	32.59	100	119.902 200	
		122	4.63	14.21	121.903 440	
		124	5.79	17.77	123.905 274	
51	Sb	121	57.4	100	120.903 821	121.752
		123	42.6	74.22	122.904 216	
52	Te	120	0.095	0.28	119.904 048	127.60
		122	2.59	7.65	121.903 054	
		123	0.905	2.67	122.904 271	
		124	4.79	14.14	123.902 823	
		125	7.12	21.02	124.904 433	
		126	18.93	55.89	125.903 314	
		128	31.70	93.59	127.904 463	
		130	33.87	100	129.906 229	
53	I	127	100	100	126.904 476	126.904 5
54	Xe	124	0.10	0.37	123.905 894	131.29
		126	0.09	0.33	125.904 281	
		128	1.91	7.10	127.903 531	
		129	26.4	98.14	128.904 780	
		130	4.1	15.24	129.903 509	
		131	21.2	78.81	130.905 072	

Z	Symbol	Nominal mass	%	% rel.	Isotopic mass	Average mass
		132	26.9	100	131.904 144	
		134	10.4	38.866	133.905 395	
		136	8.9	33.09	135.907 214	
55	Cs	133	100	100	132.905 429	132.905
56	Ba	130	1.101	1.536	129.906 284	137.34
		132	0.097	0.135	131.905 045	
		134	2.42	3.77	133.904 493	
		135	6.59	9.2	134.905 671	
		136	7.81	10.9	135.904 559	
		137	11.32	15.8	136.905 815	
		138	71.66	100	137.905 235	
57	La	138	0.090	0.09	137.907 11	138.91
		139	99.91	100	138.906 347	
72	Hf	174	0.162	0.46	173.940 044	178.49
		176	5.206	14.83	175.941 406	
		177	18.606	53.01	176.943 217	
		178	27.297	77.77	177.943 696	
		179	13.629	38.83	178.945 812	
		180	35.100	100	179.946 545	
73	Ta	180	0.012	0.012	179.947 462	
		181	99.988	100	180.947 992	180.948
74	W	180	0.12	0.39	179.946 701	183.85
		182	26.3	85.67	181.948 202	
		183	14.28	46.51	182.950 220	
		184	30.7	100	183.950 928	
		186	28.6	93.16	185.954 357	
75	Re	185	37.40	59.74	184.952 951	186.207
		187	62.60	100	186.955 744	
76	Os	184	0.02	0.05	183.952 488	190.2
		186	1.58	3.85	185.953 830	
		187	1.6	3.90	186.955 741	
		188	13.3	32.44	187.955 860	
		189	16.1	39.27	188.958 137	
		190	26.4	64.39	189.958 436	
		192	41.0	100	191.961 467	
77	Ir	191	37.3	59.49	190.960 584	192.22
		193	62.7	100	192.962 917	
78	Pt	190	0.01	0.03	189.959 917	195.08
		192	0.79	2.34	191.961 019	
		194	32.9	97.34	193.962 655	
		195	33.8	100	194.964 766	
		196	25.3	74.85	195.964 926	
		198	7.2	21.30	197.967 869	
79	Au	197	100	100	196.966 543	196.967
80	Hg	196	0.15	0.50	195.965 807	200.59
		198	10.0	33.56	197.966 743	
		199	16.9	56.71	198.968 254	

(*continued overleaf*)

Z	Symbol	Nominal mass	%	% rel.	Isotopic mass	Average mass
		200	23.1	77.52	199.968 300	
		201	13.2	44.30	200.970 277	
		202	29.8	100	201.970 617	
		204	6.85	22.99	203.973 467	
81	Tl	203	29.524	41.89	202.972 320	204.383
		205	70.476	100	204.974 401	
82	Pb	204	1.4	2.67	203.973 020	207.2
		206	24.1	45.99	205.974 440	
		207	22.1	42.18	206.975 872	
		208	52.4	100	207.976 627	
83	Bi	209	100	100	208.980 374	208.980
90	Th	232	100	100	232.038 054	232.038
92	U	234	0.0055	0.005 5	234.040 946	238.03
		235	0.720	0.725	235.043 924	
		238	99.2745	100	238.050 784	

4B   Table of Isotopes in Alphabetical Order [7]

Z	Symbol	Nominal mass	%	% rel.	Isotopic mass	Average mass
47	Ag	107	51.839	100	106.905 092	107.868
		109	48.161	94.90	108.904 757	
13	Al	27	100	100	26.981 539	26.981 5
18	Ar	36	0.337	0.338	35.967 545	39.948
		38	0.063	0.0633	37.962 732	
		40	99.600	100	39.962 384	
33	As	75	100	100	74.921 594	74.921 6
79	Au	197	100	100	196.966 543	196.967
5	B	10	19.9	24.84	10.012 937	10.811
		11	80.1	100	11.009 305	
56	Ba	130	1.101	1.536	129.906 284	137.34
		132	0.097	0.135	131.905 045	
		134	2.42	3.77	133.904 493	
		135	6.59	9.2	134.905 671	
		136	7.81	10.9	135.904 559	
		137	11.32	15.8	136.905 815	
		138	71.66	100	137.905 235	
4	Be	9	100	100	9.012 182	9.012 182
83	Bi	209	100	100	208.980 374	208.980
35	Br	79	50.69	100	78.918 336	79.904
		81	49.31	97.28	80.916 289	
6	C	12	98.90	100	12.000 000	12.011
		13	1.10	1.112	13.003 355	

Z	Symbol	Nominal mass	%	% rel.	Isotopic mass	Average mass
20	Ca	40	96.941	100	39.962 591	40.078
		42	0.647	0.667 42	41.958 618	
		43	0.135	0.139	42.958 766	
		44	2.086	2.152	43.955 480	
		46	0.004	0.004	45.953 689	
		48	0.187	0.193	47.952 533	
48	Cd	106	1.25	4.35	105.906 461	112.411
		108	0.89	3.10	107.904 176	
		110	12.49	43.47	109.903 005	
		111	12.80	44.55	110.904 182	
		112	24.13	83.99	111.902 758	
		113	12.22	42.53	112.904 400	
		114	28.73	100	113.903 357	
		116	7.49	26.07	115.904 754	
17	Cl	35	75.77	100	34.968 852	35.453
		37	24.23	31.98	36.965 903	
27	Co	59	100	100	58.933 198	58.933 2
24	Cr	50	4.345	5.185	49.946 046	51.996 1
		52	83.79	100	51.940 509	
		53	9.50	11.34	52.940 651	
		54	2.365	2.82	53.938 882	
55	Cs	133	100	100	132.905 429	132.905
29	Cu	63	69.17	100	62.929 598	63.546
		65	30.83	44.57	64.927 765	
9	F	19	100	100	18.998 403	18.998 4
26	Fe	54	5.9	6.43	53.939 612	55.847
		56	91.72	100	55.934 939	
		57	2.1	2.29	56.935 396	
		58	0.28	0.305	57.933 277	
31	Ga	69	60.108	100	68.925 580	69.723
		71	39.892	66.37	70.924 700	
32	Ge	70	20.5	56.16	69.924 250	72.61
		72	27.4	75.07	71.922 079	
		73	7.8	21.37	72.923 463	
		74	36.5	100	73.921 177	
		76	7.8	21.37	75.921 401	
1	H	1	99.985	100	1.007 825	1.007 94
	D	2	0.015	0.015	2.014	
2	He	3	0.000 137	0.000 137	3.016 030	4.002 6
		4	≈100	100	4.002 60	
72	Hf	174	0.162	0.46	173.940 044	178.49
		176	5.206	14.83	175.941 406	
		177	18.606	53.01	176.943 217	
		178	27.297	77.77	177.943 696	
		179	13.629	38.83	178.945 812	
		180	35.100	100	179.946 545	

(continued overleaf)

Z	Symbol	Nominal mass	%	% rel.	Isotopic mass	Average mass
80	Hg	196	0.15	0.50	195.965 807	200.59
		198	10.0	33.56	197.966 743	
		199	16.9	56.71	198.968 254	
		200	23.1	77.52	199.968 300	
		201	13.2	44.30	200.970 277	
		202	29.8	100	201.970 617	
		204	6.85	22.99	203.973 467	
53	I	127	100	100	126.904 476	126.904 5
49	In	113	4.3	4.49	112.904 061	114.82
		115	95.7	100	114.903 880	
77	Ir	191	37.3	59.49	190.960 584	192.22
		193	62.7	100	192.962 917	
19	K	39	93.258 1	100	38.963 707	39.098 3
		40	0.011 7	0.012 5	39.963 999	
		41	6.730 2	7.22	40.961 825	
36	Kr	78	0.35	0.614	77.920 401	83.80
		80	2.25	3.947	79.916 380	
		82	11.6	20.35	81.913 482	
		83	11.5	20.175	82.914 135	
		84	57.0	100	83.911 507	
		86	17.3	30.35	85.910 610	
57	La	138	0.090	0.09	137.907 11	138.91
		139	99.91	100	138.906 347	
3	Li	6	7.5	8.010 8	6.015 121	6.941
		7	92.5	100	7.016 003	
12	Mg	24	78.99	100	23.985 042	24.305 0
		25	10.00	12.66	24.985 837	
		26	11.01	13.94	25.982 593	
25	Mn	55	100	100	54.938 046	54.938 0
42	Mo	92	14.84	61.50	91.906 808	95.94
		94	9.25	38.33	93.905 085	
		95	15.92	65.98	94.905 840	
		96	16.68	69.13	95.904 678	
		97	9.55	39.58	96.906 020	
		98	24.13	100	97.905 406	
		100	9.63	39.91	99.907 477	
7	N	14	99.63	100	14.003 074	14.006 74
		15	0.37	0.37	15.000 108	
11	Na	23	100	100	22.989 768	22.989 8
41	Nb	93	100	100	92.906 377	92.906
10	Ne	20	90.48	100	19.992 435	20.179 7
		21	0.27	0.298	20.993 843	
		22	9.25	10.22	21.991 264	
28	Ni	58	68.27	100	57.935 346	58.693 4
		60	26.10	38.23	59.930 788	
		61	1.13	1.66	60.931 058	
		62	3.59	5.26	61.928 346	
		64	0.91	1.33	63.927 968	

Z	Symbol	Nominal mass	%	% rel.	Isotopic mass	Average mass
8	O	16	99.76	100	15.994 915	15.999 4
		17	0.04	0.04	16.999 133	
		18	0.20	0.20	17.999 160	
76	Os	184	0.02	0.05	183.952 488	190.2
		186	1.58	3.85	185.953 830	
		187	1.6	3.90	186.955 741	
		188	13.3	32.44	187.955 860	
		189	16.1	39.27	188.958 137	
		190	26.4	64.39	189.958 436	
		192	41.0	100	191.961 467	
15	P	31	100	100	30.973 762	30.973 8
82	Pb	204	1.4	2.67	203.973 020	207.2
		206	24.1	45.99	205.974 440	
		207	22.1	42.18	206.975 872	
		208	52.4	100	207.976 627	
46	Pd	102	1.02	3.73	101.905 634	106.42
		104	11.14	40.76	103.904 029	
		105	22.33	81.71	104.905 079	
		106	27.33	100	105.903 478	
		108	26.46	96.82	107.903 895	
		110	11.72	42.88	109.905 167	
78	Pt	190	0.01	0.03	189.959 917	195.08
		192	0.79	2.34	191.961 019	
		194	32.9	97.34	193.962 655	
		195	33.8	100	194.964 766	
		196	25.3	74.85	195.964 926	
		198	7.2	21.30	197.967 869	
37	Rb	85	72.17	100	84.911 794	85.467 8
		87	27.83	38.562	86.909 187	
75	Re	185	37.40	59.74	184.952 951	186.207
		187	62.60	100	186.955 744	
45	Rh	103	100	100	102.905 500	102.905
44	Ru	96	5.54	17.53	95.907 599	101.07
		98	1.86	5.89	97.905 267	
		99	12.7	40.19	98.905 939	
		100	12.6	38.87	99.904 219	
		101	17.1	54.11	100.905 582	
		102	31.6	100	101.904 348	
		104	18.6	58.86	103.905 424	
16	S	32	95.03	100	31.972 070	32.066
		33	0.75	0.789	32.971 456	
		34	4.22	4.44	33.967 866	
		36	0.02	0.021	35.967 080	
51	Sb	121	57.4	100	120.903 821	121.752
		123	42.6	74.22	122.904 216	
21	Sc	45	100	100	44.955 911	44.956

(continued overleaf)

Z	Symbol	Nominal mass	%	% rel.	Isotopic mass	Average mass
34	Se	74	0.9	1.80	73.922 475	78.96
		76	9.1	18.24	75.919 212	
		77	7.6	15.23	76.919 912	
		78	23.6	47.29	77.917 309	
		80	49.9	100	79.916 520	
		82	8.9	17.84	81.916 698	
14	Si	28	92.21	100	27.976 927	28.085 5
		29	4.67	5.065	28.976 495	
		30	3.10	3.336	29.973 770	
50	Sn	112	0.97	2.98	111.904 826	118.710
		114	0.65	1.99	113.902 784	
		115	0.36	1.10	114.903 348	
		116	14.53	43.58	115.901 747	
		117	7.68	23.57	116.902 956	
		118	24.22	73.32	117.901 609	
		119	8.58	26.33	118.903 310	
		120	32.59	100	119.902 200	
		122	4.63	14.21	121.903 440	
		124	5.79	17.77	123.905 274	
38	Sr	84	0.56	0.68	83.913 431	87.62
		86	9.86	11.94	85.909 267	
		87	7.00	8.5	86.908 884	
		88	82.58	100	87.905 619	
73	Ta	180	0.012	0.012	179.947 462	180.948
		181	99.988	100	180.947 992	
43	Tc	98	100	100		
52	Te	120	0.095	0.28	119.904 048	127.60
		122	2.59	7.65	121.903 054	
		123	0.905	2.67	122.904 271	
		124	4.79	14.14	123.902 823	
		125	7.12	21.02	124.904 433	
		126	18.93	55.89	125.903 314	
		128	31.70	93.59	127.904 463	
		130	33.87	100	129.906 229	
90	Th	232	100	100	232.038 054	232.038
22	Ti	46	8.00	10.84	45.952 629	47.88
		47	7.3	9.892	46.951 764	
		48	73.8	100	47.947 947	
		49	5.51	7.466	48.947 871	
		50	5.4	7.317	49.944 792	
81	Tl	203	29.524	41.89	202.972 320	204.383
		205	70.476	100	204.974 401	
92	U	234	0.005 5	0.005 5	234.040 946	238.03
		235	0.720	0.725	235.043 924	
		238	99.274 5	100	238.050 784	
23	V	50	0.25	0.251	49.947 161	50.941 5
		51	99.75	100	50.943 962	

Z	Symbol	Nominal mass	%	% rel.	Isotopic mass	Average mass
74	W	180	0.12	0.39	179.946 701	183.85
		182	26.3	85.67	181.948 202	
		183	14.28	46.51	182.950 220	
		184	30.7	100	183.950 928	
		186	28.6	93.16	185.954 357	
54	Xe	124	0.10	0.37	123.905 894	131.29
		126	0.09	0.33	125.904 281	
		128	1.91	7.10	127.903 531	
		129	26.4	98.14	128.904 780	
		130	4.1	15.24	129.903 509	
		131	21.2	78.81	130.905 072	
		132	26.9	100	131.904 144	
		134	10.4	38.866	133.905 395	
		136	8.9	33.09	135.907 214	
39	Y	89	100	100	88.905 849	88.906
30	Zn	64	48.6	100	63.929 145	65.39
		66	27.9	57.41	65.926 034	
		67	4.1	8.44	66.927 129	
		68	18.8	38.68	67.924 846	
		70	0.6	1.23	69.925 325	
40	Zr	90	51.45	100	89.904 703	91.224
		91	11.22	21.73	90.905 643	
		92	17.15	33.33	91.905 039	
		94	17.38	33.78	93.906 314	
		96	2.80	5.44	95.908 275	

5 Isotopic Abundances (in %) for Various Elemental Compositions CHON (M = 100 %)

	M + 1	M + 2	Mass		M + 1	M + 2	Mass
12				**17**			
C	1.11	0.00	12.0000	HO	0.06	0.20	17.0027
13				H_3N	0.42	0.00	17.0266
CH	1.13	0.00	13.0078	**18**			
14				H_2O	0.07	0.20	18.0106
N	0.37	0.00	14.0031	**24**			
CH_2	1.14	0.00	14.0157	C_2	2.22	0.01	24.0000
15				**25**			
HN	0.39	0.00	15.0109	C_2H	2.24	0.01	25.0078
CH_3	1.16	0.00	15.0235	**26**			
16				CN	1.48	0.00	26.0031
O	0.04	0.20	15.9949	C_2H_2	2.25	0.01	26.0157
H_2N	0.40	0.00	16.0187				
CH_4	1.17	0.00	16.0313				

	M + 1	M + 2	Mass			M + 1	M + 2	Mass
27					**39**			
CHN	1.50	0.00	27.0109		C$_2$HN	2.61	0.02	39.0109
C$_2$H$_3$	2.27	0.01	27.0235		C$_3$H$_3$	3.38	0.04	39.0235
28					**40**			
N$_2$	0.74	0.00	28.0062		CN$_2$	1.85	0.01	40.0062
CO	1.15	0.20	27.9949		C$_2$O	2.26	0.21	39.9949
CH$_2$N	1.51	0.00	28.0187		C$_2$H$_2$N	2.62	0.02	40.0187
C$_2$H$_4$	2.28	0.01	28.0313		C$_3$H$_4$	3.39	0.04	40.0313
29					**41**			
HN$_2$	0.76	0.00	29.0140		CHN$_2$	1.87	0.01	41.0140
CHO	1.17	0.20	29.0027		C$_2$HO	2.28	0.21	41.0027
CH$_3$N	1.53	0.00	29.0266		C$_2$H$_3$N	2.64	0.02	41.0266
C$_2$H$_5$	2.30	0.01	29.0391		C$_3$H$_5$	3.41	0.04	41.0391
30					**42**			
NO	0.41	0.20	29.9980		N$_3$	1.11	0.00	42.0093
H$_2$N$_2$	0.77	0.00	30.0218		CNO	1.52	0.21	41.9980
CH$_2$O	1.18	0.20	30.0106		CH$_2$N$_2$	1.88	0.01	42.0218
CH$_4$N	1.54	0.01	30.0344		C$_2$H$_2$O	2.29	0.21	42.0106
C$_2$H$_6$	2.31	0.01	30.0470		C$_2$H$_4$N	2.65	0.02	42.0344
31					C$_3$H$_6$	3.42	0.04	42.0470
HNO	0.43	0.20	31.0058		**43**			
H$_3$N$_2$	0.79	0.00	31.0297		HN$_3$	1.13	0.00	43.0171
CH$_3$O	1.20	0.20	31.0184		CHNO	1.54	0.21	43.0058
CH$_5$N	1.56	0.01	31.0422		CH$_3$N$_2$	1.90	0.01	43.0297
32					C$_2$H$_3$O	2.31	0.21	43.0184
O$_2$	0.08	0.40	31.9898		C$_2$H$_5$N	2.67	0.02	43.0422
H$_2$NO	0.44	0.20	32.0136		C$_3$H$_7$	3.44	0.04	43.0548
H$_4$N$_2$	0.80	0.00	32.0375		**44**			
CH$_4$O	1.21	0.20	32.0262		N$_2$O	0.78	0.20	44.0011
33					H$_2$N$_3$	1.14	0.00	44.0249
HO$_2$	0.10	0.40	32.9976		CO$_2$	1.19	0.40	43.9898
H$_3$NO	0.46	0.20	33.0215		CH$_2$NO	1.55	0.21	44.0136
34					CH$_4$N$_2$	1.91	0.01	44.0375
H$_2$O$_2$	0.11	0.40	34.0054		C$_2$H$_4$O	2.32	0.21	44.0262
36					C$_2$H$_6$N	2.68	0.02	44.0501
C$_3$	3.33	0.04	36.0000		C$_3$H$_8$	3.45	0.04	44.0626
37					**45**			
C$_3$H	3.35	0.04	37.0078		HN$_2$O	0.80	0.20	45.0089
38					H$_3$N$_3$	1.16	0.00	45.0328
C$_2$N	2.59	0.02	38.0031		CHO$_2$	1.21	0.40	44.9976
C$_3$H$_2$	3.36	0.04	38.0157		CH$_3$NO	1.57	0.21	45.0215
					CH$_5$N$_2$	1.93	0.01	45.0453
					C$_2$H$_5$O	2.34	0.21	45.0340
					C$_2$H$_7$N	2.70	0.02	45.0579

	M + 1	M + 2	Mass		M + 1	M + 2	Mass
46				**54** (*cont.*)			
NO_2	0.45	0.40	45.9929	$C_2H_2N_2$	2.98	0.03	54.0218
H_2N_2O	0.81	0.20	46.0167	C_3H_2O	3.40	0.24	54.0106
H_4N_3	1.17	0.01	46.0406	C_3H_4N	3.76	0.05	54.0344
CH_2O_2	1.22	0.40	46.0054	C_4H_6	4.53	0.08	54.0470
CH_4NO	1.58	0.21	46.0293				
CH_6N_2	1.94	0.01	46.0532	**55**			
C_2H_6O	2.35	0.22	46.0419	CHN_3	2.24	0.02	55.0171
				C_2HNO	2.65	0.22	55.0058
47				$C_2H_3N_2$	3.01	0.03	55.0297
HNO_2	0.47	0.40	47.0007	C_3H_3O	3.42	0.24	55.0184
H_3N_2O	0.83	0.20	47.0248	C_3H_5N	3.78	0.05	55.0422
H_5N_3	1.19	0.01	47.0484	C_4H_7	4.55	0.08	55.0548
CH_3O_2	1.24	0.40	47.0133				
CH_5NO	1.60	0.21	47.0371	**56**			
				N_4	1.48	0.01	56.0124
48				CN_2O	1.89	0.21	56.0011
O_3	0.12	0.60	47.9847	CH_2N_3	2.25	0.02	56.0249
H_2NO_2	0.48	0.40	48.0085	C_2O_2	2.30	0.41	55.9898
H_4N_2O	0.84	0.20	48.0324	C_2H_2NO	2.66	0.22	56.0136
CH_4O_2	1.25	0.40	48.0211	$C_2H_4N_2$	3.02	0.03	56.0375
C_4	4.44	0.07	48.0000	C_3H_4O	3.43	0.24	56.0262
				C_3H_6N	3.79	0.05	56.0501
49				C_4H_8	4.56	0.08	56.0626
HO_3	0.14	0.60	48.9925				
H_3NO_2	0.50	0.40	49.0164	**57**			
C_4H	4.46	0.07	49.0078	HN_4	1.50	0.01	57.0202
				CHN_2O	1.91	0.21	57.0089
50				CH_3N_3	2.27	0.02	57.0328
H_2O_3	0.15	0.60	50.0003	C_2HO_2	2.32	0.41	56.9976
C_3N	3.70	0.05	50.0031	C_2H_3NO	2.68	0.22	57.0215
C_4H_2	4.47	0.07	50.0157	$C_2H_5N_2$	3.04	0.03	57.0453
				C_3H_5O	3.45	0.24	57.0340
51				C_3H_7N	3.81	0.05	57.0579
C_3HN	3.72	0.05	51.0109	C_4H_9	4.58	0.08	57.0705
C_4H_3	4.49	0.08	51.0235				
				58			
52				N_3O	1.15	0.20	58.0042
C_2N_2	2.96	0.03	52.0062	H_2N_4	1.51	0.01	58.0280
C_3O	3.37	0.24	51.9949	CNO_2	1.56	0.41	57.9929
C_3H_2N	3.73	0.05	52.0187	CH_2N_2O	1.92	0.21	58.0167
C_4H_4	4.50	0.08	52.0313	CH_4N_3	2.28	0.02	58.0406
				$C_2H_2O_2$	2.33	0.42	58.0054
53				C_2H_4NO	2.69	0.22	58.0293
C_2HN_2	2.98	0.03	53.0140	$C_2H_6N_2$	3.05	0.03	58.0532
C_3HO	3.39	0.24	53.0027	C_3H_6O	3.46	0.24	58.0419
C_3H_3N	3.75	0.05	53.0266	C_3H_8N	3.82	0.05	58.0657
C_4H_5	4.52	0.08	53.0391	C_4H_{10}	4.59	0.08	58.0783
54				**59**			
CN_3	2.22	0.02	54.0093	HN_3O	1.17	0.20	59.0120
C_2NO	2.63	0.22	53.9980				

	M + 1	M + 2	Mass		M + 1	M + 2	Mass
59 (*cont.*)				**63**			
H_3N_4	1.53	0.01	59.0359	HNO_3	0.51	0.60	62.9956
$CHNO_2$	1.58	0.41	59.0007	$H_3N_2O_2$	0.87	0.40	63.0195
CH_3N_2O	1.94	0.21	59.0246	H_5N_3O	1.23	0.21	63.0433
CH_5N_3	2.30	0.02	59.0484	CH_3O_3	1.28	0.60	63.0082
$C_2H_3O_2$	2.35	0.42	59.0133	CH_5NO_2	1.64	0.41	63.0320
C_2H_5NO	2.71	0.22	59.0371	C_4HN	4.83	0.09	63.0109
$C_2H_7N_2$	3.07	0.03	59.0610	C_5H_3	5.60	0.12	63.0235
C_3H_7O	3.48	0.24	59.0497				
C_3H_9N	3.84	0.05	59.0736	**64**			
				O_4	0.16	0.80	63.9796
60				H_2NO_3	0.52	0.60	64.0034
N_2O_2	0.82	0.40	59.9960	$H_4N_2O_2$	0.88	0.40	64.0273
H_2N_3O	1.18	0.20	60.0198	CH_4O_3	1.29	0.60	64.0160
H_4N_4	1.54	0.01	60.0437	C_3N_2	4.07	0.06	64.0062
CO_3	1.23	0.60	59.9847	C_4O	4.48	0.27	63.9949
CH_2NO_2	1.59	0.41	60.0085	C_4H_2N	4.84	0.09	64.0187
CH_4N_2O	1.95	0.21	60.0324	C_5H_4	5.61	0.12	64.0313
CH_6N_3	2.31	0.02	60.0563				
$C_2H_4O_2$	2.36	0.42	60.0211	**65**			
C_2H_6NO	2.72	0.22	60.0449	HO_4	0.18	0.80	64.9874
$C_2H_8N_2$	3.08	0.03	60.0688	H_3NO_3	0.54	0.60	65.0113
C_3H_8O	3.49	0.24	60.0575	C_3HN_2	4.09	0.06	65.0140
C_5	5.55	0.12	60.0000	C_4HO	4.50	0.27	65.0027
				C_4H_3N	4.86	0.09	65.0266
61				C_5H_5	5.63	0.12	65.0391
HN_2O_2	0.84	0.40	61.0038				
H_3N_3O	1.20	0.21	61.0277	**66**			
H_5N_4	1.56	0.01	61.0515	H_2O_4	0.19	0.80	65.9953
CHO_3	1.25	0.60	60.9925	C_2N_3	3.33	0.04	66.0093
CH_3NO_2	1.61	0.41	61.0164	C_3NO	3.74	0.25	65.9980
CH_5N_2O	1.97	0.21	61.0402	$C_3H_2N_2$	4.10	0.06	66.0218
CH_7N_3	2.33	0.02	61.0641	C_4H_2O	4.51	0.27	66.0106
$C_2H_5O_2$	2.38	0.42	61.0289	C_4H_4N	4.87	0.09	66.0344
C_2H_7NO	2.74	0.22	61.0528	C_5H_6	5.64	0.12	66.0470
C_5H	5.57	0.12	61.0078				
				67			
62				C_2HN_3	3.35	0.04	67.0171
NO_3	0.49	0.60	61.9878	C_3HNO	3.76	0.25	67.0058
$H_2N_2O_2$	0.85	0.40	62.0116	$C_3H_3N_2$	4.12	0.06	67.0297
H_4N_3O	1.21	0.42	62.0368	C_4H_3O	4.53	0.27	67.0184
H_6N_4	1.57	0.01	62.0594	C_4H_5N	4.89	0.09	67.0422
CH_2O_3	1.26	0.60	62.0003	C_5H_7	5.66	0.12	67.0548
CH_4NO_2	1.62	0.41	62.0242				
CH_6N_2O	1.98	0.21	62.0480	**68**			
$C_2H_6O_2$	2.39	0.42	62.0368	CN_4	2.59	0.02	68.0124
C_4N	4.81	0.09	62.0031	C_2N_2O	3.00	0.23	68.0011
C_5H_2	5.58	0.12	62.0157	$C_2H_2N_3$	3.36	0.04	68.0249
				C_3O_2	3.41	0.44	67.9898

	M + 1	M + 2	Mass		M + 1	M + 2	Mass
68 (*cont.*)				**72** (*cont.*)			
C_3H_2NO	3.77	0.25	68.0136	$C_2H_2NO_2$	2.70	0.42	72.0085
$C_3H_4N_2$	4.13	0.06	68.0375	$C_2H_4N_2O$	3.06	0.23	72.0324
C_4H_4O	4.54	0.28	68.0262	$C_2H_6N_3$	3.42	0.04	72.0563
C_4H_6N	4.90	0.09	68.0501	$C_3H_4O_2$	3.47	0.44	72.0211
C_5H_8	5.67	0.13	68.0626	C_3H_6NO	3.83	0.25	72.0449
				$C_3H_8N_2$	4.19	0.07	72.0688
69				C_4H_8O	4.60	0.28	72.0575
CHN_4	2.61	0.03	69.0202	$C_4H_{10}N$	4.96	0.09	72.0814
C_2HN_2O	3.02	0.23	69.0089	C_5H_{12}	5.73	0.13	72.0939
$C_2H_3N_3$	3.38	0.04	69.0328	C_6	6.66	0.18	72.0000
C_3HO_2	3.43	0.44	68.9976				
C_3H_3NO	3.79	0.25	69.0215	**73**			
C_3H_5N2	4.15	0.06	69.0453	HN_4O	1.54	0.21	73.0151
C_4H_5O	4.56	0.28	69.0340	CHN_2O_2	1.95	0.41	73.0038
C_4H_7N	4.92	0.09	69.0579	CH_3N_3O	2.31	0.22	73.0277
C_5H_9	5.69	0.13	69.0705	CH_5N_4	2.67	0.03	73.0515
				C_2HO_3	2.36	0.62	72.9925
70				$C_2H_3NO_2$	2.72	0.42	73.0164
CN_3O	2.26	0.22	70.0042	$C_2H_5N_2O$	3.08	0.23	73.0402
CH_2N_4	2.62	0.03	70.0280	$C_2H_7N_3$	3.44	0.04	73.0641
C_2NO_2	2.67	0.42	69.9929	$C_3H_5O_2$	3.49	0.44	73.0289
$C_2H_2N_2O$	3.03	0.23	70.0167	C_3H_7NO	3.85	0.25	73.0528
$C_2H_4N_3$	3.39	0.04	70.0406	$C_3H_9N_2$	4.21	0.07	73.0767
$C_3H_2O_2$	3.44	0.44	70.0054	C_4H_9O	4.62	0.28	73.0653
C_3H_4NO	3.80	0.25	70.0293	$C_4H_{11}N$	4.98	0.09	73.0892
C_3H_6N2	4.16	0.07	70.0532	C_6H	6.68	0.18	73.0078
C_4H_6O	4.57	0.28	70.0419				
C_4H_8N	4.93	0.09	70.0657	**74**			
C_5H_{10}	5.70	0.13	70.0783	N_3O_2	1.19	0.41	73.9991
				H_2N_4O	1.55	0.21	74.0229
71				CNO_3	1.60	0.61	73.9878
CHN_3O	2.28	0.22	71.0120	$CH_2N_2O_2$	1.96	0.41	74.0116
CH_3N_4	2.64	0.03	71.0359	CH_4N_3O	2.32	0.22	74.0355
C_2HNO_2	2.69	0.42	71.0007	CH_6N_4	2.68	0.03	74.0594
$C_2H_3N_2O$	3.05	0.23	71.0246	$C_2H_2O_3$	2.37	0.62	74.0003
$C_2H_5N_3$	3.41	0.04	71.0484	$C_2H_4NO_2$	2.73	0.42	74.0242
$C_3H_3O_2$	3.46	0.44	71.0133	$C_2H_6N_2O$	3.09	0.23	74.0480
C_3H_5NO	3.82	0.25	71.0371	$C_2H_8N_3$	3.45	0.05	74.0719
$C_3H_7N_2$	4.18	0.07	71.0610	$C_3H_6O_2$	3.50	0.44	74.0368
C_4H_7O	4.59	0.28	71.0497	C_3H_8NO	3.86	0.25	74.0606
C_4H_9N	4.95	0.10	71.0736	$C_3H_{10}N_2$	4.22	0.07	74.0845
C_5H_{11}	5.72	0.13	71.0861	$C_4H_{10}O$	4.63	0.28	74.0732
				C_5N	5.92	0.14	74.0031
72				C_6H_2	6.69	0.18	74.0157
N_4O	1.52	0.21	72.0073				
CN_2O_2	1.93	0.41	71.9960	**75**			
CH_2N_3O	2.29	0.22	72.0198	HN_3O_2	1.21	0.41	75.0069
CH_4N_4	2.65	0.03	72.0437	H_3N_4O	1.57	0.21	75.0308
C_2O_3	2.34	0.62	71.9847				

	M + 1	M + 2	Mass		M + 1	M + 2	Mass
75 (cont.)				**78** (cont.)			
$CHNO_3$	1.62	0.61	74.9956	$H_4N_3O_2$	1.25	0.41	78.0304
$CH_3N_2O_2$	1.98	0.41	75.0195	H_6N_4O	1.61	0.21	78.0542
CH_5N_3O	2.34	0.22	75.0433	CH_2O_4	1.30	0.80	77.9953
CH_7N_4	2.70	0.03	75.0672	CH_4NO_3	1.66	0.61	78.0191
$C_2H_3O_3$	2.39	0.62	75.0082	$CH_6N_2O_2$	2.02	0.41	78.0429
$C_2H_5NO_2$	2.75	0.43	75.0320	$C_2H_6O_3$	2.43	0.62	78.0317
$C_2H_7N_2O$	3.11	0.23	75.0559	C_3N_3	4.44	0.08	78.0093
$C_2H_9N_3$	3.47	0.05	75.0798	C_4NO	4.85	0.29	77.9980
$C_3H_7O_2$	3.52	0.44	75.0446	$C_4H_2N_2$	5.21	0.11	78.0218
C_3H_9NO	3.88	0.25	75.0684	C_5H_2O	5.62	0.32	78.0106
C_5HN	5.94	0.14	75.0109	C_5H_4N	5.98	0.14	78.0344
C_6H_3	6.71	0.18	75.0235	C_6H_6	6.75	0.19	78.0470
76				**79**			
N_2O_3	0.86	0.60	75.9909	HNO_4	0.55	0.80	78.9905
$H_2N_3O_2$	1.22	0.41	76.0147	$H_3N_2O_3$	0.91	0.60	79.0144
H_4N_4O	1.58	0.21	76.0386	$H_5N_3O_2$	1.27	0.41	79.0382
CO_4	1.27	0.80	75.9796	CH_3O_4	1.32	0.80	79.0031
CH_2NO_3	1.63	0.61	76.0034	CH_5NO_3	1.68	0.61	79.0269
$CH_4N_2O_2$	1.99	0.41	76.0273	C_3HN_3	4.46	0.08	79.0171
CH_6N_3O	2.35	0.22	76.0511	C_4HNO	4.87	0.29	79.0058
CH_8N_4	2.71	0.03	76.0750	$C_4H_3N_2$	5.23	0.11	79.0297
$C_2H_4O_3$	2.40	0.62	76.0160	C_5H_3O	5.64	0.32	79.0184
$C_2H_6NO_2$	2.76	0.43	76.0399	C_5H_5N	6.00	0.14	79.0422
$C_2H_8N_2O$	3.12	0.24	76.0637	C_6H_7	6.77	0.19	79.0548
$C_3H_8O_2$	3.53	0.44	76.0524	**80**			
C_4N_2	5.18	0.10	76.0062	H_2NO_4	0.56	0.80	79.9983
C_5O	5.59	0.32	75.9949	$H_4N_2O_3$	0.92	0.60	80.0222
C_5H_2N	5.95	0.14	76.0187	CH_4O_4	1.33	0.80	80.0109
C_6H_4	6.72	0.19	76.0313	C_2N_4	3.70	0.05	80.0124
77				C_3N_2O	4.11	0.26	80.0011
HN_2O_3	0.88	0.60	76.9987	$C_3H_2N_3$	4.47	0.08	80.0249
$H_3N_3O_2$	1.24	0.41	77.0226	C_4O_2	4.52	0.47	79.9898
H_5N_4O	1.60	0.21	77.0464	C_4O_2	4.52	0.47	79.9898
CHO_4	1.29	0.80	76.9874	C_4H_2NO	4.88	0.29	80.0136
CH_3NO_3	1.65	0.61	77.0113	$C_4H_4N_2$	5.24	0.11	80.0375
$CH_5N_2O_2$	2.01	0.41	77.0351	C_5H_4O	5.65	0.32	80.0262
CH_7N_3O	2.37	0.22	77.0590	C_5H_6N	6.01	0.14	80.0501
$C_2H_5O_3$	2.42	0.62	77.0238	C_6H_8	6.78	0.19	80.0626
$C_2H_7NO_2$	2.78	0.43	77.0477	**81**			
C_4HN_2	5.20	0.11	77.0140	H_3NO_4	0.58	0.80	81.0062
C_5HO	5.61	0.32	77.0027	C_2HN_4	3.72	0.05	81.0202
C_5H_3N	5.97	0.15	77.0266	C_3HN_2O	4.13	0.26	81.0089
C_6H_5	6.74	0.19	77.0391	$C_3H_3N_3$	4.49	0.08	81.0328
78				C_4HO_2	4.54	0.48	80.9976
NO_4	0.53	0.80	77.9827	C_4H_3NO	4.90	0.29	81.0215
$H_2N_2O_3$	0.89	0.60	78.0065	$C_4H_5N_2$	5.26	0.11	81.0453

	M + 1	M + 2	Mass		M + 1	M + 2	Mass
81 (*cont.*)				**85** (*cont.*)			
C_5H_5O	5.67	0.32	81.0340	$C_2HN_2O_2$	3.06	0.43	85.0038
C_5H_7N	6.03	0.14	81.0579	$C_2H_3N_3O$	3.42	0.24	85.0277
C_6H_9	6.80	0.19	81.0705	$C_2H_5N_4$	3.78	0.06	85.0515
				C_3HO_3	3.47	0.64	84.9925
82				$C_3H_3NO_2$	3.83	0.45	85.0164
C_2N_3O	3.37	0.24	82.0042	$C_3H_5N_2O$	4.19	0.27	85.0402
$C_2H_2N_4$	3.73	0.05	82.0280	$C_3H_7N_3$	4.55	0.08	85.0641
C_3NO_2	3.78	0.45	81.9929	$C_4H_5O_2$	4.60	0.48	85.0289
$C_3H_2N_2O$	4.14	0.26	82.0167	C_4H_7NO	4.96	0.29	85.0528
$C_3H_4N_3$	4.50	0.08	82.0406	$C_4H_9N_2$	5.32	0.11	85.0767
$C_4H_2O_2$	4.55	0.48	82.0054	C_5H_9O	5.73	0.33	85.0653
C_4H_4NO	4.91	0.29	82.0293	$C_5H_{11}N$	6.09	0.16	85.0892
$C_4H_6N_2$	5.27	0.11	82.0532	C_6H_{13}	6.86	0.20	85.1018
C_5H_6O	5.68	0.32	82.0419	C_7H	7.79	0.26	85.0078
C_5H_8N	6.04	0.14	82.0657				
C_6H_{10}	6.81	0.19	82.0783	**86**			
				CN_3O_2	2.30	0.41	85.9991
83				CH_2N_4O	2.66	0.21	86.0229
C_2HN_3O	3.39	0.24	83.0120	C_2NO_3	2.71	0.62	85.9878
$C_2H_3N_4$	3.75	0.06	83.0359	$C_2H_2N_2O_2$	3.07	0.43	86.0116
C_3HNO_2	3.80	0.45	83.0007	$C_2H_4N_3O$	3.43	0.24	86.0355
$C_3H_3N_2O$	4.16	0.27	83.0246	$C_2H_6N_4$	3.79	0.06	86.0594
$C_3H_5N_3$	4.52	0.08	83.0484	$C_3H_2O_3$	3.48	0.64	86.0003
$C_4H_3O_2$	4.57	0.48	83.0133	$C_3H_4NO_2$	3.84	0. 45	86.0242
C_4H_5NO	4.93	0.29	83.0371	$C_3H_6N_2O$	4.20	0.27	86.0480
$C_4H_7N_2$	5.29	0.11	83.0610	$C_3H_8N_3$	4.56	0.08	86.0719
C_5H_7O	5.70	0.33	83.0497	$C_4H_6O_2$	4.61	0.48	86.0368
C_5H_9N	6.06	0.15	83.0736	C_4H_8NO	4.97	0.30	86.0606
C_6H_{11}	6.83	0.19	83.0861	$C_4H_{10}N_2$	5.33	0.11	86.0845
				$C_5H_{10}O$	5.74	0.33	86.0732
84				$C_5H_{12}N$	6.10	0.16	86.0970
CN_4O	2.63	0.23	84.0073	C_6H_{14}	6.87	0.21	86.1096
$C_2H_2N_3O$	3.40	0.24	84.0198	C_6N	7.03	0.21	86.0031
$C_2N_2O_2$	3.04	0.43	83.9960	C_7H_2	7.80	0.26	86.0157
$C_2H_4N_4$	3.76	0.06	84.0437				
C_3O_3	3.45	0.64	83.9847	**87**			
$C_3H_2NO_2$	3.81	0.45	84.0085	CHN_3O_2	2.32	0.42	87.0069
$C_3H_4N_2O$	4.17	0.27	84.0324	CH_3N_4O	2.68	0.23	87.0308
$C_3H_6N_3$	4.53	0.08	84.0563	C_2HNO_3	2.73	0.62	86.9956
$C_4H_4O_2$	4.58	0.48	84.0211	$C_2H_3N_2O_2$	3.09	0.43	87.0195
C_4H_6NO	4.94	0.29	84.0449	$C_2H_5N_3O$	3.45	0.25	87.0433
$C_4H_8N_2$	5.30	0.11	84.0688	$C_2H_7N_4$	3.81	0.06	87.0672
C_5H_8O	5.71	0.33	84.0575	$C_3H_3O_3$	3.50	0.64	87.0082
$C_5H_{10}N$	6.07	0.15	84.0814	$C_3H_5NO_2$	3.86	0.45	87.0320
C_6H_{12}	6.84	0.19	84.0939	$C_3H_7N_2O$	4.22	0.27	87.0559
C_7	7.77	0.26	84.0000	$C_3H_9N_3$	4.58	0.08	87.0798
				$C_4H_7O_2$	4.63	0.48	87.0446
85				C_4H_9NO	4.99	0.30	87.0684
CHN_4O	2.65	0.23	85.0151				

	M + 1	M + 2	Mass		M + 1	M + 2	Mass
87 (*cont.*)				**90**			
$C_4H_{11}N_2$	5.35	0.11	87.0923	N_3O_3	1.23	0.60	89.9940
$C_5H_{11}O$	5.76	0.33	87.0810	$H_2N_4O_2$	1.59	0.40	90.0178
$C_5H_{13}N$	6.12	0.15	87.1049	CNO_4	1.64	0.80	89.9827
C_6HN	7.05	0.21	87.0109	$CH_2N_2O_3$	2.00	0.61	90.0065
C_7H_3	7.82	0.26	87.0235	$CH_4N_3O_2$	2.36	0.42	90.0304
				CH_6N_4O	2.72	0.23	90.0542
88				C_2H_2O4	2.41	0.82	89.9953
N_4O_2	1.56	0.41	88.0022	$C_2H_4NO_3$	2.77	0.63	90.0191
CN_2O_3	1.97	0.61	87.9909	$C_2H_6N_2O_2$	3.13	0.44	90.0429
$CH_2N_3O_2$	2.33	0.42	88.0147	$C_2H_8N_3O$	3.49	0.25	90.0668
CH_4N_4O	2.69	0.23	88.0386	$C_2H_{10}N_4$	3.85	0.06	90.0907
C_2O_4	2.38	0.82	87.9796	$C_3H_6O_3$	3.54	0.64	90.0317
$C_2H_2NO_3$	2.74	0.63	88.0034	$C_3H_8NO_2$	3.90	0.46	90.0555
$C_2H_4N_2O_2$	3.10	0.43	88.0273	$C_3H_{10}N_2O$	4.26	0.27	90.0794
$C_2H_6N_3O$	3.46	0.25	88.0511	$C_4H_{10}O_2$	4.67	0.48	90.0681
$C_2H_8N_4$	3.82	0.06	88.0750	C_4N_3	5.55	0.13	90.0093
$C_3H_4O_3$	3.51	0.64	88.0160	C_5NO	5.96	0.34	89.9980
$C_3H_6NO_2$	3.87	0.45	88.0399	$C_5H_2N_2$	6.32	0.17	90.0218
$C_3H_8N_2O$	4.23	0.27	88.0637	C_6H_2O	6.73	0.38	90.0106
$C_3H_{10}N_3$	4.59	0.08	88.0876	C_6H_4N	7.09	0.21	90.0344
$C_4H_8O_2$	4.64	0.48	88.0524	C_7H_6	7.86	0.26	90.0470
$C_4H_{10}NO$	5.00	0.30	88.0763				
$C_4H_{12}N_2$	5.36	0.11	88.1001	**91**			
$C_5H_{12}O$	5.77	0.33	88.0888	HN_3O_3	1.25	0.60	91.0018
C_5N_2	6.29	0.16	88.0062	$H_3N_4O_2$	1.61	0.41	91.0257
C_6O	6.70	0.38	87.9949	$CHNO_4$	1.66	0.81	90.9905
C_6H_2N	7.06	0.21	88.0187	$CH_3N_2O_3$	2.02	0.61	91.0144
C_7H_4	7.83	0.26	88.0313	$CH_5N_3O_2$	2.38	0.42	91.0382
				CH_7N_4O	2.74	0.23	91.0621
89				$C_2H_3O_4$	2.43	0.82	91.0031
HN_4O_2	1.58	0.41	89.0100	$C_2H_5NO_3$	2.79	0.63	91.0269
CHN_2O_3	1.99	0.61	88.9987	$C_2H_7N_2O_2$	3.15	0.44	91.0508
$CH_3N_3O_2$	2.35	0.42	89.0226	$C_2H_9N_3O$	3.51	0.25	91.0746
CH_5N_4O	2.71	0.23	89.0464	$C_3H_7O_3$	3.56	0.64	91.0395
C_2HO_4	2.40	0.82	88.9874	$C_3H_9NO_2$	3.92	0.46	91.0634
$C_2H_3NO_3$	2.76	0.63	89.0113	C_4HN_3	5.57	0.13	91.0171
$C_2H_5N_2O_2$	3.12	0.44	89.0351	C_5HNO	5.98	0.34	91.0058
$C_2H_7N_3O$	3.48	0.25	89.0590	C_5H_3N2	6.34	0.17	91.0297
$C_2H_9N_4$	3.84	0.06	89.0829	C_6H_3O	6.75	0.38	91.0184
$C_3H_5O_3$	3.53	0.64	89.0238	C_6H_5N	7.11	0.21	91.0422
$C_3H_7NO_2$	3.89	0.46	89.0477	C_7H_7	7.88	0.26	91.0548
$C_3H_9N_2O$	4.25	0.27	89.0715				
$C_3H_{11}N_3$	4.61	0.08	89.0954	**92**			
$C_4H_9O_2$	4.66	0.48	89.0603	N_2O_4	0.90	0.80	91.9858
$C_4H_{11}NO$	5.02	0.30	89.0841	$H_2N_3O_3$	1.26	0.60	92.0096
C_5HN_2	6.31	0.16	89.0140	$H_4N_4O_2$	1.62	0.41	92.0335
C_6HO	6.72	0.38	89.0027	CH_2NO_4	1.67	0.81	91.9983
C_6H_3N	7.08	0.21	89.0266	$CH_4N_2O_3$	2.03	0.61	92.0222
C_7H_5	7.85	0.26	89.0391	$CH_6N_3O_2$	2.39	0.42	92.0460

	M + 1	M + 2	Mass		M + 1	M + 2	Mass
92 (*cont.*)				**94** (*cont.*)			
CH$_8$N$_4$O	2.75	0.23	92.0699	C$_6$H$_6$O	6.79	0.38	94.0419
C$_2$H$_4$O$_4$	2.44	0.82	92.0109	C$_6$H$_8$N	7.15	0.22	94.0657
C$_2$H$_6$NO$_3$	2.80	0.63	92.0348	C$_7$H$_{10}$	7.92	0.27	94.0783
C$_2$H$_8$N$_2$O$_2$	3.16	0.44	92.0586				
C$_3$H$_8$O$_3$	3.57	0.64	92.0473	**95**			
C$_3$N$_4$	4.81	0.09	92.0124	H$_3$N$_2$O$_4$	0.95	0.80	95.0093
C$_4$N$_2$O	5.22	0.31	92.0011	H$_5$N$_3$O$_3$	1.31	0.60	95.0331
C$_4$H$_2$N$_3$	5.58	0.13	92.0249	CH$_5$NO$_4$	1.72	0.81	95.0218
C$_5$O$_2$	5.63	0.52	91.9898	C$_3$HN$_3$O	4.50	0.28	95.0120
C$_5$H$_2$NO	5.99	0.34	92.0136	C$_3$H$_3$N$_4$	4.86	0.10	95.0359
C$_5$H$_4$N$_2$	6.35	0.17	92.0375	C$_4$HNO$_2$	4.91	0.49	95.0007
C$_6$H$_4$O	5.76	0.38	92.0262	C$_4$H$_3$N$_2$O	5.27	0.31	95.0246
C$_6$H$_6$N	7.12	0.21	92.0501	C$_4$H$_5$N$_3$	5.63	0.13	95.0484
C$_7$H$_8$	7.89	0.27	92.0626	C$_5$H$_3$O$_2$	5.68	0.52	95.0133
				C$_5$H$_5$NO	6.04	0.35	95.0371
93				C$_5$H$_7$N$_2$	6.40	0.17	95.0610
HN$_2$O$_4$	0.92	0.80	92.9936	C$_6$H$_7$O	6.81	0.39	95.0497
H$_3$N$_3$O$_3$	1.28	0.60	93.0175	C$_6$H$_9$N	7.17	0.22	95.0736
H$_5$N$_4$O$_2$	1.64	0.41	93.0413	C$_7$H$_{11}$	7.94	0.27	95.0861
CH$_3$NO$_4$	1.69	0.81	93.0062				
CH$_5$N$_2$O$_3$	2.05	0.61	93.0300	**96**			
CH$_7$N$_3$O$_2$	2.41	0.42	93.0539	H$_4$N$_2$O$_4$	0.96	0.80	96.0171
C$_2$H$_5$O$_4$	2.46	0.82	93.0187	C$_2$N$_4$O	3.74	0.26	96.0073
C$_2$H$_7$NO$_3$	2.82	0.63	93.0426	C$_3$N$_2$O$_2$	4.15	0.47	95.9960
C$_3$HN$_4$	4.83	0.09	93.0202	C$_3$H$_2$N$_3$O	4.51	0.28	96.0198
C$_4$HN$_2$O	5.24	0.31	93.0089	C$_3$H$_4$N$_4$	4.87	0.10	96.0437
C$_4$H$_3$N$_3$	5.60	0.13	93.0328	C$_4$O$_3$	4.56	0.67	95.9847
C$_5$HO$_2$	5.65	0.52	92.9976	C$_4$H$_2$NO$_2$	4.92	0.49	96.0085
C$_5$H$_3$NO	6.01	0.35	93.0215	C$_4$H$_4$N$_2$O	5.28	0.31	96.0324
C$_5$H$_5$N$_2$	6.37	0.17	93.0453	C$_4$H$_6$N$_3$	5.64	0.13	96.0563
C$_6$H$_5$O	6.78	0.38	93.0340	C$_5$H$_4$O$_2$	5.69	0.53	96.0211
C$_6$H$_7$N	7.14	0.22	93.0579	C$_5$H$_6$NO	6.05	0.35	96.0449
C$_7$H$_9$	7.91	0.27	93.0705	C$_5$H$_8$N$_2$	6.41	0.17	96.0688
				C$_6$H$_8$O	6.82	0.39	96.0575
94				C$_6$H$_{10}$N	7.18	0.22	96.0814
H$_2$N$_2$O$_4$	0.93	0.80	94.0014	C$_7$H$_{12}$	7.95	0.27	96.0939
H$_4$N$_3$O$_3$	1.29	0.61	94.0253	C$_8$	8.88	0.34	96.0000
H$_6$N$_4$O$_2$	1.65	0.41	94.0491				
CH$_4$NO$_4$	1.70	0.81	94.0140	**97**			
CH$_6$N$_2$O$_3$	2.06	0.62	94.0379	C$_2$HN$_4$O	3.76	0.26	97.0151
C$_2$H$_6$O$_4$	2.47	0.82	94.0266	C$_3$HN$_2$O$_2$	4.17	0.47	97.0038
C$_3$N$_3$O	4.48	0.28	94.0042	C$_3$H$_3$N$_3$O	4.53	0.28	97.0277
C$_3$H$_2$N$_4$	4.84	0.09	94.0280	C$_3$H$_5$N$_4$	4.89	0.10	97.0515
C$_4$NO$_2$	4.89	0.49	93.9929	C$_4$HO$_3$	4.58	0.68	96.9925
C$_4$H$_2$N$_2$O	5.25	0.31	94.0167	C$_4$H$_3$NO$_2$	4.94	0.49	97.0164
C$_4$H$_4$N$_3$	5.61	0.13	94.0406	C$_4$H$_5$N$_2$O	5.30	0.31	97.0402
C$_5$H$_2$O$_2$	5.66	0.52	94.0054	C$_4$H$_7$N$_3$	5.66	0.13	97.0641
C$_5$H$_4$NO	6.02	0.35	94.0293	C$_5$H$_5$O$_2$	5.71	0.53	97.0289
C$_5$H$_6$N$_2$	6.38	0.17	94.0532	C$_5$H$_7$NO	6.07	0.35	97.0528

	M + 1	M + 2	Mass		M + 1	M + 2	Mass
97 (*cont.*)				**99** (*cont.*)			
$C_5H_9N_2$	6.43	0.17	97.0767	$C_4H_7N_2O$	5.33	0.31	99.0559
C_6H_9O	6.84	0.39	97.0653	$C_4H_9N_3$	5.69	0.13	99.0798
$C_6H_{11}N$	7.20	0.22	97.0892	$C_5H_7O_2$	5.74	0.53	99.0446
C_7H_{13}	7.97	0.27	97.1018	C_5H_9NO	6.11	0.35	99.0684
C_8H	8.90	0.34	97.0078	$C_5H_{11}N_2$	6.46	0.17	99.0923
				$C_6H_{11}O$	6.86	0.39	99.0810
98				$C_6H_{13}N$	7.23	0.22	99.1049
$C_2N_3O_2$	3.41	0.44	97.9991	C_7H_{15}	8.00	0.27	99.1174
$C_2H_2N_4O$	3.77	0.26	98.0229	C_7HN	8.16	0.29	99.0109
C_3NO_3	3.82	0.65	97.9878	C_8H_3	8.93	0.35	99.0235
$C_3H_2N_2O_2$	4.18	0.47	98.0116				
$C_3H_4N_3O$	4.54	0.28	98.0355	**100**			
$C_3H_6N_4$	4.90	0.10	98.0594	CN_4O_2	2.67	0.43	100.0022
$C_4H_2O_3$	4.59	0.68	98.0003	$C_2N_2O_3$	3.08	0.63	99.9909
$C_4H_4NO_2$	4.95	0.49	98.0242	$C_2H_2N_3O_2$	3.44	0.45	100.0147
$C_4H_6N_2O$	5.31	0.31	98.0480	$C_2H_4N_4O$	3.80	0.26	100.0386
$C_4H_8N_3$	5.67	0.13	98.0719	C_3O_4	3.45	0.84	99.9796
$C_5H_6O_2$	5.72	0.53	98.0368	$C_3H_2NO_3$	3.85	0.65	100.0034
C_5H_8NO	6.08	0.35	98.0606	$C_3H_4N_2O_2$	4.21	0.47	100.0273
$C_5H_{10}N_2$	6.44	0.17	98.0645	$C_3H_6N_3O$	4.57	0.28	100.0511
$C_6H_{10}O$	6.85	0.39	98.0732	$C_3H_8N_4$	4.94	0.10	100.0750
$C_6H_{12}N$	7.21	0.21	98.0970	$C_4H_4O_3$	4.62	0.68	100.0160
C_7H_{14}	7.98	0.26	98.1096	$C_4H_6NO_2$	4.98	0.49	100.0399
C_7N	8.14	0.27	98.0031	$C_4H_8N_2O$	5.34	0.31	100.0637
C_8H_2	8.91	0.33	98.0157	$C_4H_{10}N_3$	5.70	0.13	100.0876
				$C_5H_8O_2$	5.76	0.53	100.0524
99				$C_5H_{10}NO$	6.11	0.35	100.0763
$C_2HN_3O_2$	3.43	0.44	99.0069	$C_5H_{12}N_2$	6.47	0.18	100.1001
$C_2H_3N_4O$	3.79	0.25	99.0308	$C_6H_{12}O$	6.88	0.39	100.0888
C_3HNO_3	3.84	0.65	98.9956	$C_6H_{14}N$	7.24	0.22	100.1127
$C_3H_3N_2O_2$	4.20	0.47	99.0195	C_6N_2	7.40	0.23	100.0062
$C_3H_5N_3O$	4.56	0.28	99.0433	C_7H_{16}	8.01	0.28	100.1253
$C_3H_7N_4$	4.92	0.10	99.0672	C_7O	7.81	0.46	99.9949
$C_4H_3O_3$	4.61	0.68	99.0082	C_7H_2N	8.17	0.29	100.0187
$C_4H_5NO_2$	4.97	0.49	99.0320	C_8H_4	8.94	0.35	100.0313

6 Gas-Phase Ion Thermochemical Data of Molecules [9,10]

Molecules	IE $(eV)^a$	EA $(eV)^b$	PA $(kJ\,mol^{-1})^c$	GB $(kJ\,mol^{-1})^c$	ΔH°_{acid} $(kJ\,mol^{-1})^d$	ΔG°_{acid} $(kJ\,mol^{-1})^d$
Ar	15.6	—	369	346	—	—
Br_2	10.5	2.4	—	—	—	—
CO	14	1.4	426	402	—	—
CO_2	13.8	−0.6	540	515	—	—
CS_2	10	0.5	682	658	—	—
Cl_2	11.5	2.3	—	—	—	—
I_2	9.3	2.3	—	—	—	—
F_2	15.7	3.1	332	305	—	—
HBr	11.7	—	584	558	1353	1332
HCl	12.7	—	557	530	—	—
HF	16	—	484	457	1554	1530
HI	10.4	—	627	601	1315	1293
H_2	15.4	—	422	394	—	—
H_2O	12.6	—	691	660	1634	1607
H_2S	10.4	—	705	674	1469	1446
He	24.6	—	178	148	—	—
NH_3	10	—	854	819	1689	1660
N_2	15.6	—	494	464	—	—
O_2	12	0.4	421	396	—	—
PH_3	9.8	—	785	751	1551	1520
CH_4	12.6	—	543	521	1749	1715
C_2H_6	11.5	—	590	570	1758	1721
C_3H_8	10.9	—	625	607	1755	1721
$n\text{-}C_4H_{10}$	10.5	—	—	—	1739	1703
$i\text{-}C_4H_{10}$	10.7	—	677	671	1728	1697
$n\text{-}C_6H_{14}$	10.1	—	—	—	—	—
Cyclohexane	9.9	—	687	667	—	—
$CH_2{=}CH_2$	10.5	—	680	651	1713	1678
$CH_3CH{=}CH_2$	9.7	—	751	722	1635	1607
$CH_3CH{=}CHCH_3$	9.1	—	747	720	—	—
$CH_2{=}CHCH{=}CH_2CH_3$	8.6	—	834	804	1545	1525
$CH_2{=}CHCH{=}CH_2$	9	—	783	757	1672	1637
$HC{\equiv}CH$	12.4	—	641	616	1582	1547
$CH_3C{\equiv}CH$	10.4	—	748	723	1595	1562
$CH_3C{\equiv}CCH_3$	9.6	—	776	745	—	—
Benzene	9.2	—	750	725	1681	1644
$C_6H_5CH_3$	8.8	—	784	756	1593	1564
$C_6H_5CH{=}CH_2$	8.2	—	840	809	1636	1604
Biphenyl	8.2	—	814	783	—	—
Naphthalene	8.1	−0.2	803	779	—	—
CH_3F	12.5	—	599	571	1711	1676
CH_3Cl	11.3	—	647	621	1657	1628
CH_3Br	10.5	—	664	638	1660	1631
$CH_2{=}CHF$	10.4	—	729	700	1618	1586
$CH_2{=}CHCl$	10	—	—	—	—	—
$CH_2{=}CHBr$	9.8	—	—	—	—	—
CH_3OH	10.8	1.6	754	724	1592	1565

(*continued overleaf*)

Molecules	IE (eV)[a]	EA (eV)[b]	PA (kJ mol^{-1})[c]	GB (kJ mol^{-1})[c]	$\Delta H°_{acid}$ (kJ mol^{-1})[d]	$\Delta G°_{acid}$ (kJ mol^{-1})[d]
C_2H_5OH	10.5	1.7	776	746	1583	1555
n-C_3H_7OH	10.2	—	786	756	1572	1544
i-C_3H_7OH	10.2	—	793	763	1573	1545
C_6H_5OH	8.5	—	817	786	1456	1437
$C_6H_5CH_2OH$	8.3	2.1	778	748	1548	1520
CH_3SH	9.4	—	773	742	1654	1624
$(CH_3)_2O$	10.1	—	792	764	1703	1666
$(C_2H_5)_2O$	9.5	—	828	801	—	—
$C_6H_5OCH_3$	8.2	—	840	807	1679	1648
Furan	8.9	—	803	771	1624	1590
Tetrahydrofuran	9.4	—	822	795	—	—
$(CH_3)_2S$	8.7	—	831	801	1645	1615
CH_3NH_2	8.9	—	899	864	1687	1656
$C_2H_5NH_2$	8.9	0.7	912	878	1671	1639
n-$C_4H_9NH_2$	8.7	—	921	886	—	—
$C_6H_5NH_2$	7.7	—	882	851	1533	1502
$(C_2H_5)_2NH$	7.8	—	952	919	—	—
$(C_2H_5)_3N$	7.5	—	981	951	—	—
Pyrrole	8.2	2.3	875	843	1500	1468
Pyridine	9.3	—	930	898	1636	1601
H_2CHO	10.9	—	713	683	1646	1613
CH_3CHO	10.2	—	768	736	1531	1502
n-C_3H_7CHO	9.8	—	793	761	—	—
C_6H_5CHO	9.5	0.4	834	802	—	—
CH_2CO	9.6	—	825	794	1526	1497
CH_3COCH_3	9.7	—	812	782	1544	1514
$C_2H_5COCH_3$	9.5	—	827	795	1536	1508
$C_6H_5COCH_3$	9.3	0.3	861	829	1512	1483
HCOOH	11.3	—	742	710	1445	1415
CH_3COOH	10.6	3.1	784	753	1456	1427
n-C_3H_7COOH	10.2	3.2	—	—	1450	1420
C_6H_5COOH	9.3	3.5	821	790	1423	1393
CH_3COOCH_3	10.3	1.5	821	790	1556	1528
CH_3CONH_2	9.7	—	864	833	1515	1485
$CH_3CON(CH_3)_2$	9.2	—	908	877	1568	1540

[a] Ionization energy (IE) is defined as the 0 K enthalpy change required to remove an electron from a molecule. It is possible to have either adiabatic or vertical ionization energy, with the value of the vertical ionization energy being greater than or equal to the adiabatic ionization energy:

$$M \longrightarrow M^+ + e^- \quad \Delta H = IE$$

[b] Electron affinity (EA) is defined as the negative of the 0 K enthalpy change for the electron attachment reaction from a molecule. As with the ionization energy, adiabatic or vertical electron affinity can be possible:

$$M + e^- \longrightarrow M^- \quad \Delta H = -EA$$

[c] Proton affinity (PA) and gas-phase basicity (GB) are respectively the negative of the enthalpy change or the Gibbs energy change defined at 298 K for the protonation reaction:

$$M + H^+ \rightleftharpoons MH^+ \quad \Delta H = -PA \text{ and } \Delta G = -GB$$

[d] Gas-phase acidity. $\Delta G°_{acid}$ and $\Delta H°_{acid}$ are the Gibbs energy change or the enthalpy change defined at 298 K to remove a proton from a molecule. The $\Delta H°_{acid}$ and $\Delta G°_{acid}$ are, respectively, the proton affinity and the gas-phase basicity of the anion:

$$AH \rightleftharpoons A^- + H^+$$

7 Gas-Phase Ion Thermochemical Data of Radicals [9, 10]

Radicals	IE (eV)	EA (eV)	PA (kJ mol^{-1})	GB (kJ mol^{-1})
F$^\bullet$	17.4	3.4	340	315
Cl$^\bullet$	13	3.6	513	490
Br$^\bullet$	11.8	3.4	554	531
I$^\bullet$	10.4	3.1	608	583
H$^\bullet$	13.6	0.75	—	—
HO$^\bullet$	13	1.8	593	564
NC$^\bullet$	14.2	3.9	—	—
H$_2$N$^\bullet$	10.8	0.8	—	—
HS$^\bullet$	10.4	2.3	—	—
CH$_3$$^\bullet$	9.8	0.1	671	639
C$_2$H$_5$$^\bullet$	8.1	−0.3	616	583
i-C$_3$H$_7$$^\bullet$	7.4	−0.3	—	—
n-C$_3$H$_7$$^\bullet$	8.1	−0.1	—	—
$n-, i$-C$_4$H$_9$$^\bullet$	8	0.1	—	—
s-C$_4$H$_9$$^\bullet$	7.2	−0.1	—	—
t-C$_4$H$_9$$^\bullet$	6.7	−0.2	—	—
CH≡C$^\bullet$	11.6	2.9	753	721
CH≡CCH$_2$$^\bullet$	8.7	—	741	708
CH$_2$=CH$^\bullet$	8.3	0.7	755	720
CH$_2$=CHCH$_2$$^\bullet$	8.2	0.4	736	707
C$_6$H$_5$$^\bullet$	8.3	1.1	884	831
C$_6$H$_5$CH$_2$$^\bullet$	7.2	0.9	831	800
Tropyl$^\bullet$	6.3	0.5	832	800
BrCH$_2$$^\bullet$	8.6	0.8	—	—
ClCH$_2$$^\bullet$	8.7	0.8	—	—
FCH$_2$$^\bullet$	9	0.2	—	—
ICH$_2$$^\bullet$	8.4	—	—	—
F$_3$C$^\bullet$	8.7	—	—	—
Cl$_3$C$^\bullet$	8.1	—	—	—
Br$_3$C$^\bullet$	7.5	—	—	—
CH$_3$O$^\bullet$	10.7	—	—	—
HOCH$_2$$^\bullet$	7.6	—	695	662
CH$_3$S$^\bullet$	9.3	1.9	—	—
HSCH$_2$$^\bullet$	7.5	0.8	734	701
C$_6$H$_5$O$^\bullet$	8.6	2.2	858	827
CH$_3$OCH$_2$$^\bullet$	6.9	−0.1	756	724
CH$_3$SCH$_2$$^\bullet$	6.8	0.9	—	—
H$_2$NCH$_2$$^\bullet$	6.3	—	832	802
HCO$^\bullet$	8.1	0.3	636	601
CH$_3$CO$^\bullet$	7	0.4	653	620
HCOO$^\bullet$	8.2	3.2	623	590

8 Literature on Mass Spectrometry

The literature on mass spectrometry may be divided into three broad categories: journals, periodicals and books. Another important category is compilations of mass spectra, which have already been mentioned in Chapter 6.

Journals devoted only to mass spectrometry

1. *International Journal of Mass Spectrometry* (Elsevier), formerly *International Journal of Mass Spectrometry and Ion Processes* before 1998.

2. *Journal of Mass Spectrometry* (John Wiley & Sons), formerly *Organic Mass Spectrometry* incorporating *Biological Mass Spectrometry* before 1995. Each issue of this journal contains a Special Feature including 'Perspective' and 'Tutorial' articles that present authoritative materials on a featured topic in a succinct format.

3. *Journal of the American Society for Mass Spectrometry* (Elsevier).

4. *Rapid Communications in Mass Spectrometry* (John Wiley & Sons).

5. *European Mass Spectrometry* (IM Publications).

Journals containing substantial papers on mass spectrometry (not exhaustive)

1. *Analytical Chemistry* (American Chemical Society).

2. *Analytical Biochemistry* (Academic Press).

3. *Journal of Chromatography* (Elsevier).

Abstracting journals devoted only to mass spectrometry

1. *Mass Spectrometry Bulletin* (Royal Society of Chemistry) contains a comprehensive list of references with abstracts that are relevant to mass spectrometry and related ion processes. Published monthly. Available as a printed publication only.

2. *GC/MS Update and LC/MS Update* (HD Science) are more specialized abstracting services. They contain abstracts that emphasize the mass spectrometric aspects of the original publications. Also exist on CD-ROM with a search program.

3. *CA Selects: Mass Spectrometry* (American Chemical Society). Published bi-weekly. Also available for web subscription.

The most important periodicals on mass spectrometry

1. *Mass Spectrometry Reviews* (John Wiley & Sons)

2. *Specialist Periodical Reports on Mass Spectrometry* (Royal Society of Chemistry) reports on progress in mass spectrometry. Published every two years.

3. *Fundamental Reviews of Analytical Chemistry* (American Chemical Society) covers all mass spectrometry aspects in a condensed fashion. Published biennially in each even year.

4. *Proceedings of the nth ASMS Conference on Mass Spectrometry and Allied Topics* (American Society for Mass Spectrometry). These reports are published every year. Since the 1999 meeting, the proceedings are available only on CD-ROM.

5. *Advances in Mass Spectrometry* (Heyden, John Wiley & Sons and Elsevier). This series publishes the reports of the *x*th International Mass Spectrometry Conferences that occur every three years (16th in Edinburgh in 2003). The 2006 meeting is scheduled for Prague.

6. *The Encyclopedia of Mass Spectrometry* (Elsevier). This encyclopedia will be a 10-volume work published over a five-year period (2003–2006). Encyclopedia articles will provide unparalleled and comprehensive coverage of the full range of topics and techniques in the field of mass spectrometry.

Some useful mass spectrometry books

Adams, F., Gijbels, R. and Van Grieken, R. (eds) (1988) *Inorganic Mass Spectrometry*, John Wiley & Sons, Inc., New York.

Adams, N.G. and Babcock, L.M. (1992) *Advances in Gas Phase Ion Chemistry*, vol. 1, JAI Press, Greenwich, CT.

Adrey, B. (ed.) (1993) *Liquid Chromatography/Mass Spectrometry*, VCH, New York.

Ardey, R.E. (2003) *Liquid Chromatography—Mass Spectrometry: An Introduction*, John Wiley & Sons, Inc., New York.

Asamoto, B. and Dunbar, R.C. (1991) *Analytical Applications of Fourier Transform Ion Cyclotron Resonance Mass Spectrometry*, VCH, New York.

Barker, J. and Ando, D.J. (1999) *Mass Spectrometry: Analytical Chemistry by Open Learning*, John Wiley & Sons, Ltd, Chichester.

Benninghoven, A. (ed.) (1989) *Ion Formation from Organic Solids*, John Wiley & Sons, Inc., New York.

Beynon, J.H. and Gilbert, J.R. (1984) *Applications of Transition State Theory to Unimolecular Reactions*, John Wiley & Sons, Inc., New York.

Bowers, M.T. (ed.) (1979) *Gaz Phase Ion Chemistry*, vol. 3, Academic Press, New York.

Brown, M.A. (1998) *Liquid Chromatography/Mass Spectrometry Applications in Agricultural, Pharmaceutical and Environmental Chemistry*, Oxford University Press, Oxford.

Buchanan, M.B. (ed.) (1987) *Fourier Transform Mass Spectrometry*, American Chemical Society, Washington, DC.

Budde, W.L. (ed.) (2001) *Analytical Mass Spectrometry: Strategies for Environmental and Related Applications*, American Chemical Society, Washington, DC.

Burlingame, A.L. (2005) *Mass spectrometry: Modified Proteins and Glycoconjugates*, Academic Press, New York.

Burlingame, A.L. and Carr, S.A. (eds) (1996) *Mass Spectrometry in the Biological Sciences*, Humana Press, Totowa, NJ.

Burlingame, A.L. and McCloskey, J.A. (eds) (1990) *Biological Mass Spectrometry*, Elsevier, Amsterdam.

Burlingame, A.L., Carr, S. A. and Baldwin, M. A. (1999) *Mass Spectrometry in Biology and Medicine*, Humana Press, Totowa, NJ.

Bush, K.L. and Lehman, T.A. (1995) *Guide to Mass Spectrometry*, VCH, New York.

Bush, K.L., Glish, G.L. and McLuckey, S.A. (1988) *Mass Spectrometry/Mass Spectrometry*, VCH, New York.

Caprioli, R.M. (ed.) (1990) *Continuous-Flow Fast Atom Bombardment Mass Spectrometry*, John Wiley & Sons, Inc., New York.

Caprioli, R.M., Malorni, A. and Sindona, G. (eds) (1996) *Mass Spectrometry in the Biomolecular Sciences*, Kluwer Academic, Dordrecht.

Chapman, J.R. (1995) *Practical Organic Mass Spectrometry: A Guide for Chemical and Biological Analysis*, John Wiley & Sons, Ltd, Chichester.

Chapman, J.R. (ed.) (2000) *Methods in Molecular Biology, Mass Spectrometry of Proteins and Peptides*, vol. 146, Humana Press, Totowa, NJ.

Cole, R.B. (1997) *Electrospray Ionization Mass Spectrometry: Fundamentals, Instrumentation, and Applications*, John Wiley & Sons, Inc., New York.

Cooks, R.G., Beynon, J.H., Caprioli, R.M. and Lester, G.R. (1973) *Metastable Ions*, Elsevier, Amsterdam.

Cotter, R.J. (ed.) (1994) *Time-of-Flight Mass Spectrometry*, American Chemical Society, Washington, DC.

Cotter, R.J. (1997) *Time-of-Flight Mass Spectrometry: Instrumentation and Applications in Biological Research*, American Chemical Society, Washington, DC.

Dass, C. (2000) *Principles and Practice of Biological Mass Spectrometry*, John Wiley & Sons, Ltd, Chichester.

Dawson, P.H. (ed.) (1994) *Quadrupole Mass Spectrometry and its Applications*, American Institute of Physics, New York.

de Hoffmann, E. and Stroobant, V. (2001) *Mass Spectrometry: Principles and Applications*, 2nd edn, John Wiley & Sons, Ltd, Chichester.

de Laeter, J.R. (2001) *Applications of Inorganic Mass Chemistry*, John Wiley & Sons, Inc., New York.

Desiderio, D.M. (1991) *Mass Spectrometry of Peptides*, CRC Press, Boca Raton, FL.

Desiderio, D.M. (1993) *Mass Spectrometry: Clinical and Biomedical Applications (Modern Analytical Chemistry)*, vol. 2, Plenum Press, New York.

Farrar, J.M. and Saunders, W.H. (eds) (1988) *Techniques for the Study of Ion–Molecule Reactions*, John Wiley & Sons, Inc., New York.

Fenselau, C. (ed.) (1993) *Mass Spectrometry for the Characterization of Microorganisms*, American Chemical Society, Washington, DC.

Futrell, J.H. (ed.) (1986) *Gaseous Ion Chemistry and Mass Spectrometry*, John Wiley & Sons, Inc., New York.

Gaskell, S.J. (ed.) (1986) *Mass Spectrometry in Biomedical Research*, John Wiley & Sons, Ltd, Chichester.

Gerhatds, P., Bons, U., Sawazki, J., Szigan, J. and Wertman, A. (1999) *GC/MS in Clinical Chemistry*, John Wiley & Sons, Inc., New York.

Gilbert, J. (ed.) (1987) *Applications of Mass Spectrometry in Food Science*, Elsevier, London.

Gross, J.H. (2004) *Mass Spectrometry: A Textbook*, Springer-Verlag, Heidelberg.

Harrison, A.G. (1992) *Chemical Ionization Mass Spectrometry*, CRC Press, Boca Raton, FL.

Herbert, C.G. and Johnstone, R.A.W. (2003) *Mass Spectrometry Basics*, CRC Press, Boca Raton, FL.

Hites, R.A. (1992) *Handbook of Mass Spectra of Environmental Contaminants*, 2nd edn, Lewis, Boca Raton, FL.

Holland, G. (ed.) (1993) *Applications of Plasma Source Mass Spectrometry*, Royal Society of Chemistry, London.

James, P. (2000) *Proteome Research: Mass Spectrometry (Principles and Practice)*, Springer-Verlag, Heidelberg.

Jarvis, K.E. (ed.) (1990) *Plasma Source Mass Spectrometry*, Royal Society of Chemistry, London.

Jarvis, K.E., Grays, A.L. and Houk, R.S. (1991) *Handbook of Inductively Coupled Plasma Mass Spectrometry*, Chapman & Hall, London.

Johnstone, R.A. and Rose, M.E. (1996) *Mass Spectrometry for Chemists and Biochemists*, 2nd edn, Cambridge University Press, New York.

Kaltashov, I.A. and Eyles, S.J. (2005) *Mass Spectrometry in Biophysics: Conformation and Dynamics of Biomolecules*, John Wiley & Sons, Inc., New York.

Karasek, F.W. and Clement, R.E. (1988) *Basic Gas Chromatography–Mass Spectrometry Principles and Techniques*, Plenum Press, New York.

Karasek, F.W., Hutzinger, O. and Safe, S. (eds) (1985) *Mass Spectrometry in Environmental Sciences*, Plenum Press, New York.

Kienitz, H. (1968) *Massenspectrometrie*, Verlag Chemie, Weinheim.

Kienter, M. and Sherman, N.E. (2000) *Protein Sequencing and Identification Using Tandem Mass Spectrometry*, John Wiley & Sons, Ltd, Chichester.

Kitson, F.G., Larsen, B.S. and McEwen, C.N. (1996) *Gas Chromatography and Mass Spectrometry: A Practical Guide*, Academic Press, New York.

Korfmacher, W.A. (2005) *Using Mass Spectrometry for Drug Metabolism Studies*, CRC Press, Boca Raton, FL.

Larsen, B.S. and McEwen, C.N. (eds) (1990) *Mass Spectrometry of Biological Materials*, 2nd edn, Marcel Dekker, New York.

Lee, T.A. (1998) *A Beginner's Guide to Mass Spectral Interpretation*, John Wiley & Sons, Inc., New York.

Lias, S.G., Bartmess, J.E., Liebman, J.F., Holmes, J.L., Levin, R.D. and Mallar, G.M. (1988) *Gas-phase and neutral thermochemistry*, ACS, AIP and NBS Publication, *J. Phys. Chem. Ref. Data*, vol. 17.

Lindon, J.C., Tranter, G.E. and Holmes, J.L. (eds) (2000) *Encyclopedia of Spectroscopy and Spectrometry*, Academic Press, New York.

Longevialle, P. (1981) *Principes de la Spectrométrie de Masse Des Substances Organiques*, Masson, Paris.

Lubman, D.H. (ed.) (1990) *Lasers in Mass Spectrometry*, Oxford University Press, Oxford.

March, R.E. and Todd, J.F.J. (1995) *Practical Aspects of Ion Trap Mass Spectrometry (Modern Mass Spectrometry)*, vol. 3, CRC Press, Boca Raton, FL.

March, R.E. and Todd, J.F.J. (2005) *Quadrupole Ion Trap Mass Spectrometry*, 2nd edn, John Wiley & Sons, Inc., New York.

Marshall, A.G. and Verdun, F.T. (1990) *Fourier Transform in NMR, Optical and Mass Spectrometry*, Elsevier, Amsterdam.

Matsuo, T., Seyama, Y., Caprioli, R.M. and Gross, M.L. (eds) (1994) *Biological Mass Spectrometry: Present and Future*, John Wiley & Sons, Ltd, Chichester.

McCloskey, J.A. (ed.) (1990) *Methods in Enzymology, Mass Spectrometry*, vol. 193, John Wiley & Sons, Inc., New York.

McEwen, C.N. and Larsen, B.S. (eds) (1990) *Mass Spectrometry of Biological Materials*, Marcel Dekker, New York.

McLafferty, F.W. and Stauffer, D.B. (1989) *The Wiley/NBS Registry of Mass Spectral Data*, John Wiley & Sons, Inc., New York, Voir nouvelle édition en, 2005.

McLafferty, F.W. and Stauffer, D.B. (1991) *Important Peak Index of the Registry of Mass Spectral Data*, John Wiley & Sons, Inc., New York.

McLafferty, F.W. and Turecek, F. (1993) *Interpretation of Mass Spectra*, 4th edn, University Science Books, Mill Valley, CA.

McLafferty, F.W. and Venkataraghavan, R. (1982) *Advances in Chemistry Series, Mass Spectral Correlations*, vol. 40, 2nd edn, American Chemical Society, Washington, DC.

McMaster, M.C. (2005) *LC/MS: A Practical User's Guide*, John Wiley & Sons, Inc., New York.

McMaster, M.C. and McMaster, C. (1998) *GC/MS: A Practical User's Guide*, John Wiley & Sons, Inc., New York.

Mellon, F.A., Selh, R. and Startin, J.R. (2000) *Mass Spectrometry of Natural Substances*, Royal Society of Chemistry, London.

Meyers, R.A. (ed.) (2000) *Encyclopedia of Analytical Chemistry*, John Wiley & Sons, Ltd, Chichester.

Middletich, B.S., Missler, S.R. and Hines, H.B. (1981) *Mass Spectrometry of Priority Pollutants*, Plenum Press, New York.

Millard, B.J. (1978) *Quantitative Mass Spectrometry*, Heyden, London.

Montaudo, G. and Lattimer, R.P. (2001) *Mass Spectrometry of Polymers*, CRC Press, Boca Raton, FL.

Montaser, A. (1998) *Inductively Coupled Plasma Mass Spectrometry*, VCH, New York.

Murphy, R.C. (1993) *Mass Spectrometry of Lipids*, Plenum Press, New York.

Niessen, W.M.A. (1998) *Liquid Chromatography–Mass Spectrometry*, Marcel Dekker, New York.

Niessen, W.M.A. and Voyksner, R.D. (eds) (1998) *Current Practice of Liquid Chromatography–Mass Spectrometry*, Elsevier Science, Amsterdam.

Pasch, H. and Schrepp, W. (2004) *MALDI-TOF Mass Spectrometry of Synthetic Polymers*, Springer-Verlag, Heidelberg.

Pfleger, K., Maurer, H.H. and Weber, A. (1992) *Mass Spectral and GC Data of Drugs, Poisons, Pesticides, Pollutants and their Metabolites*, VCH, New York.

Platzner, I.T. (1997) *Modern Isotope Ratio Mass Spectrometry*, John Wiley & Sons, Ltd, Chichester.

Roboz, J. (2002) *Mass Spectrometry in Cancer Research*, CRC Press, Boca Raton, FL.

Russel, D.H. (ed.) (1994) *Experimental Mass Spectrometry*, Plenum Press, New York.

Schalley, C.A. (ed.) (2003) *Topics in Current Chemistry Series, Modern Mass Spectrometry*, vol. 225, Springer-Verlag, Heidelberg.

Schlag, E.W. (ed.) (1994) *Time-of-Flight Mass Spectrometry and Its Applications*, Elsevier, Amsterdam.

Silberring, J., Ekman, R., Desiderio, D.M. and Nibbering, N.M. (2002) *Mass Spectrometry and Hyphenated Techniques in Neuropeptide Research*, John Wiley & Sons, Inc., New York.

Silverstein, R.M., Bassler, G.C. and Morril, T.C. (1991) *Spectrometric Identification of Organic Compounds*, 5th edn, John Wiley & Sons, Inc., New York.

Suizdak, G. (1996) *Mass Spectrometry for Biotechnology*, Academic Press, San Diego, CA.

Suizdak, G. (2003) *The Expanding Role of Mass Spectrometry in Biotechnology*, Academic Press, San Diego, CA.

Smith, R.M. (2004) *Understanding Mass Spectra: A Basic Approach*, 2nd edn, John Wiley & Sons, Inc., New York.

Snyder, A.P. (ed.) (1998) *Biomedical and Biotechnological Applications of Electrospray Ionization Mass Spectrometry*, American Chemical Society, Washington, DC.

Snyder, A.P. (2000) *Interpreting Protein Mass Spectra: A Comprehensive Resource*, American Chemical Society, Washington, DC.

Sparkman, O.D. (2000) *Mass Spectrometry Desk Reference*, Global View Publishing, Pittsburgh.

Splitter, J.G. and Turecek, F. (eds) (1991) *Applications of Mass Spectrometry to Organic Stereochemistry*, VCH, New York.

Standing, K.G. and Ens, W. (eds) (1991) *Methods and Mechanisms for Producing Ions from Large Molecules*, Plenum Press, New York.

Stemmler, E.A. and Hites, R.A. (1988) *Electron Capture Negative Ion Mass Spectra of Environmental Contaminants and Related Compounds*, VCH, Weinheim.

Suelterm, C.H. and Watson, J.T. (eds) (1990) *Biomedical Applications of Mass Spectrometry*, John Wiley & Sons, Inc., New York.

Taylor, H.E. (2000) *Inductively Coupled Plasma–Mass Spectrometry*, Academic Press, New York.

Thomson, J.J. (1913) *Rays of Positive Electricity*, Longmans Green, London.

Tuniz, C., Tuniz, J.R. and Bird, D.F. (1998) *Accelerator Mass Spectrometry: Ultrasensitive Analysis for Global Science*, CRC Press, Boca Raton, FL.

Turecek, F. (ed.) (1993) *Applications of Mass Spectrometry to Organic Stereochemistry*, VCH, New York.

Vertes, A., Gijbels, R. and Adams, F. (1993) *Laser Ionisation Mass Analysis*, John Wiley & Sons, Inc., New York.

Watson, J.T. (1997) *Introduction to Mass Spectrometry: Biomedical, Environmental and Forensic Applications*, 3rd edn, Lippincot Raven Press, Philadelphia.

Williams, R. (1995) *Spectroscopy and the Fourier Transform: An Interactive Tutorial*, VCH, New York.

Willoughby, R., Sheehan, E. and Mitrovitch, S. (1998) *A Global View of LC/MS: How to Solve Your Most Challenging Analytical Problems*, Global View Publishing, Pittsburgh.

Yergey, A.L., Edmonds, C.G., Lewis, I.A.S. and Vestal, L.M. (1990) *Liquid Chromatography/Mass Spectrometry: Techniques and Applications*, Plenum Press, New York.

Yinon, J. (2004) *Advances in Forensic Applications of Mass Spectrometry*, CRC Press, Boca Raton, FL.

9 Mass Spectrometry on the Internet

The Base Peak site (http://www.spectroscopynow.com/coi/cda/home.cda?chId=4) is the most comprehensive web resource for mass spectrometrists. Developed by Kermit Murray and John Wiley & Sons, it contains a collection of links to mass spectrometry Internet sites.

Web sites for national and international mass spectrometry societies

Typical information that can be found on the society sites includes contact information, membership applications, information on meetings, and so on. Furthermore, member directories and employment information can be available at some of these sites:

American Society for Mass Spectrometry (ASMS):
www.asms.org/

Australian and New Zealand Society for Mass Spectrometry (ANZSMS):
http://www.latrobe.edu.au/anzsms/

British Mass Spectrometry Society (BMSS):
www.bmss.org.uk/

Canadian Society for Mass Spectrometry (CSMS):
www.csms.inter.ab.ca/

Dutch Society for Mass Spectrometry (NVMS):
www.denvms.nl/

European Society for Mass Spectrometry (EMS):
http://www.bmb.leeds.ac.uk/esms/

French Society for Mass Spectrometry (SFMS):
www.sfsm.info/

German Society for Mass Spectrometry (DGMS):
www.dgms.de/

International Mass Spectrometry Society (IMSS):
www.imss.nl/

Mass Spectrometry Society of Japan (MSSJ):
www.mssj.jp/

South African Association for Mass Spectrometry (SAAMS):
www.saams.up.ac.za/

Swiss Group for Mass Spectrometry (SGMS):
www.sgms.ch/

Web sites for mass spectrometry journals

Most of the mass spectrometry journal sites contain tables of contents and abstracts. Many of these sites offer on-line access to the full articles although it is typically restricted to subscribers only. General information such as aims and scope, editorial board, instructions to authors, contact and subscription information are also found on these sites:

European Mass Spectrometry (IM Publications):
http://www.impub.co.uk/ems.html

International Journal of Mass Spectrometry (Elsevier):
http://www.elsevier.com/homepage/saa/ijmsip/

Journal of the American Society for Mass Spectrometry (Elsevier):
http://www.elsevier.com/homepage/saa/webjam/

Journal of Mass Spectrometry (John Wiley & Sons):
http://www.interscience.wiley.com/jpages/1076-5174/

Mass Spectrometry Reviews (John Wiley & Sons):
http://www.interscience.wiley.com/jpages/0277-7037/

Rapid Communications in Mass Spectrometry (John Wiley & Sons):
http://www.interscience.wiley.com/jpages/0951-4198/

Web sites for useful mass spectrometry databases

The National Institute of Standards and Technology (NIST) provides thermochemical, thermophysical and ion energetics data for chemical species over the Internet. Indeed, the NIST Chemistry Web Book (http://www.webbook.nist.gov/chemistry/) contains, among other data, mass spectra for over 10 000 compounds and ion energetics data for over 14 000 compounds.

The WebElements site (www.webelements.com/) contains chemical and physical data of the elements. This site also has on-line isotope pattern calculators.

The IonSource site (www.ionsource.com) is a good on-line resource for the mass spectrometry and biotechnology community, including scientific tutorials, reference materials, conference schedules, and so on.

The PubMed site (http://www.ncbi.nlm.nih.gov/Entrez) is developed by the National Center for Biotechnology Information (NCBI) at the National Library of Medicine (NLM), located at the National Institutes of Health (NIH). It has been developed in conjunction with publishers of biomedical literature as a search tool for accessing literature citations and linking to full-text journals at web sites of participating publishers. This site allows on-line access to the Medline database containing bibliographic citations and abstracts from journals dating back to 1966.

References

1. Todd, J.F.J. (1991) Recommendation for nomenclature and symbolism for mass spectroscopy. *Pure Appl. Chem.*, **63**, 1541–66.
2. Price, P. (1991) Standard definitions of terms relating to mass spectrometry. *J. Am. Soc. Mass Spectrom.*, **2**, 336–48.
3. Sparkman, D., *Mass Spectrometry Desk Reference*, Global View Publishing, Pittsburgh.
4. Murray, K.K., Boyd, R.K., Eberlin, M.N., Langley, G.J., Li, L. and Naito, Y. IUPAC Standard Definitions of Terms Relating to Mass Spectrometry, IUPAC Project 2003-056-2-500.
5. http://mass-spec.lsu.edu/msterms/index.php/Main_Page (21 March 2007).
6. Busch, K.L. (2002) SAMS: speaking with acronyms in mass spectrometry. *Spectroscopy*, **17**, 55–62.
7. *Handbook of Chemistry and Physics*, 71st, edn, CRC Press, Boca Raton, FL, 1990.
8. NIST Reference on Constants, Units, and Uncertainty at http://www.physics.nist.gov/cuu/Constants/index.html (21 March 2007).
9. Thermodynamic data from Lias, S.G., Bartmess, J.E., Liebman, J.F., Holmes, J.L., Levin, R.D. and Mallard, W.G. (1988) Gas-phase ion and neutral thermochemistry. *J. Phys. Chem. Ref. Data*, **17**(Suppl. 1). Also available on CD-ROM from NIST, Washington, DC.
10. Mallard, W.G. and Linstrom, P.J. (eds) NIST Chemistry WebBook, NIST Standard Reference Database Number 69, November 1998, National Institute of Standards and Technology, Gaithersburg, MD, 20899 at www.webbook.nist.gov (21 March 2007).

Index